T0135282

Lecture Notes in Networks and Systems

Volume 90

The series “Lecture Notes in Networks and Systems” publishes the latest developments in Networks and Systems—quickly, informally and with high quality. Original research reported in proceedings and post-proceedings represents the core of LNNS.

Volumes published in LNNS embrace all aspects and subfields of, as well as new challenges in, Networks and Systems.

The series contains proceedings and edited volumes in systems and networks, spanning the areas of Cyber-Physical Systems, Autonomous Systems, Sensor Networks, Control Systems, Energy Systems, Automotive Systems, Biological Systems, Vehicular Networking and Connected Vehicles, Aerospace Systems, Automation, Manufacturing, Smart Grids, Nonlinear Systems, Power Systems, Robotics, Social Systems, Economic Systems and other. Of particular value to both the contributors and the readership are the short publication timeframe and the world-wide distribution and exposure which enable both a wide and rapid dissemination of research output.

The series covers the theory, applications, and perspectives on the state of the art and future developments relevant to systems and networks, decision making, control, complex processes and related areas, as embedded in the fields of interdisciplinary and applied sciences, engineering, computer science, physics, economics, social, and life sciences, as well as the paradigms and methodologies behind them.

**** Indexing: The books of this series are submitted to ISI Proceedings, SCOPUS, Google Scholar and Springerlink ****

More information about this series at http://www.springer.com/series/15179

Nenad Mitrovic · Milos Milosevic ·
Goran Mladenovic
Editors

Computational and Experimental Approaches in Materials Science and Engineering

Proceedings of the International Conference of Experimental and Numerical Investigations and New Technologies, CNNTech 2019

Editors
Nenad Mitrovic
Faculty of Mechanical Engineering,
Department for Process Engineering
and Environmental Protection
University of Belgrade
Belgrade, Serbia

Milos Milosevic
Innovation Center of Faculty
of Mechanical Engineering
Belgrade, Serbia

Goran Mladenovic
Faculty of Mechanical Engineering,
Department for Production Engineering
University of Belgrade
Belgrade, Serbia

ISSN 2367-3370 ISSN 2367-3389 (electronic)
Lecture Notes in Networks and Systems
ISBN 978-3-030-30852-0 ISBN 978-3-030-30853-7 (eBook)
https://doi.org/10.1007/978-3-030-30853-7

This Springer imprint is published by the registered company Springer Nature Switzerland AG
The registered company address is: Gewerbestrasse 11, 6330 Cham, Switzerland

Preface

This book is a collection of high-quality peer-reviewed research papers presented at the International Conference of Experimental and Numerical Investigations and New Technologies (CNNTech 2019) held at Zlatibor, Serbia, from 3 to 5 July 2019. The conference is organized by the Innovation Center of the Faculty of Mechanical Engineering, Faculty of Mechanical Engineering at the University of Belgrade and Center for Business Trainings. Over 60 delegates were attending the CNNTech 2019—academicians, practitioners and scientists from 15 countries—presenting and authoring more than 60 papers. The conference program included seven keynote lectures with seven invited lectures, two mini-symposia, four sessions (oral and poster) and one workshop. Twenty-eight selected full papers went through the double-blind reviewing process.

The main goal of the conference is to make positive atmosphere for the discussion on a wide variety of industrial, engineering and scientific applications of the engineering techniques. Participation of a number of domestic and international authors, as well as the diversity of topics, has justified our efforts to organize this conference and contribute to exchange of knowledge, research results and experience of industry experts, research institutions and faculties which all share a common interest in the field in experimental and numerical investigations.

The CNNTech 2019 focused on the following topics:

- Mechanical Engineering,
- Materials Science,
- Chemical and Process Engineering,
- Experimental Techniques,
- Numerical Methods,
- New Technologies,
- Industry and Sustainable Development: Contemporary Management Perspectives.

We express our gratitude to all people involved in conference planning, preparation and realization, especially to:

- All the authors, specially keynote speakers and invited speakers, who have contributed to the high scientific and professional level of the conference,
- All members of the Organizing Committee,
- All members of the International Scientific Committee for reviewing the papers and chairing the conference sessions,
- Ministry of Education, Science and Technological Development of Republic of Serbia for supporting the conference.

We wish to express our special gratitude to Ms. Dragana Perovic for her effort in preparing and managing the conference in the best way.

Organization

Scientific Committee

Miloš Milošević (Chairman)	University of Belgrade, Faculty of Mechanical Engineering, Serbia
Nenad Mitrović (Co-chairman)	University of Belgrade, Faculty of Mechanical Engineering, Serbia
Aleksandar Sedmak	University of Belgrade, Faculty of Mechanical Engineering, Serbia
Hloch Sergej	Technical University of Košice, Faculty of Manufacturing Technologies, Slovakia
Dražan Kozak	University of Osijek, Faculty of Mechanical Engineering, Slavonski Brod, Croatia
Nenad Gubeljak	University of Maribor, Faculty of Mechanical Engineering, Slovenia
Monka Peter	Technical University of Kosice, Faculty of Manufacturing Technologies, Slovakia
Snežana Kirin	University of Belgrade, Innovation Center of faculty of Mechanical Engineering, Serbia
Ivan Samardžić	University of Osijek, Faculty of Mechanical Engineering, Slavonski Brod, Croatia
Martina Balać	University of Belgrade, Faculty of Mechanical Engineering, Serbia
Ludmila Mládková	University of Economics, Prague, Czech Republic
Johanyák Zsolt Csaba	Pallasz Athéné University, Faculty of Engineering and Computer Science, Hungary
Igor Svetel	University of Belgrade, Innovation Center of Faculty of Mechanical Engineering, Serbia

Contents

List of Contributors

Sinisa M. Arsic Faculty of Organizational Sciences, University of Belgrade, Belgrade, Republic of Serbia

Petar Avdalovic Mixed Holding Power Utility of the Republic of Srpska, Parent Joint - Stock Company, Trebinje, RS, Bosnia and Herzegovina

Sebastian Balos Faculty of Technical Sciences, Department of Production Engineering, University of Novi Sad, Novi Sad, Serbia

Marija Baltic Faculty of Mechanical Engineering, University of Belgrade, Belgrade, Serbia

Jozef Bucha Faculty of Mechanical Engineering, Institute of Transport Technology and Engineering Design, Slovak University of Technology in Bratislava, Bratislava, Slovakia

Ivan M. Buzurovic Harvard Medical School, Harvard University, Boston, MA, USA

Andrej Chríbik Faculty of Mechanical Engineering, Institute of Transport Technology and Engineering Design, Slovak University of Technology in Bratislava, Bratislava, Slovakia

Robert A. Cormack Harvard Medical School, Harvard University, Boston, MA, USA

Ján Danko Faculty of Mechanical Engineering, Institute of Transport Technology and Engineering Design, Slovak University of Technology in Bratislava, Bratislava, Slovakia

Dragutin Lj. Debeljkovic Faculty of Civil Aviation, Megatrend University, Belgrade, Serbia

Tsanka Dikova Faculty of Dental Medicine, Medical University of Varna, Varna, Bulgaria

Jasmina Dlacic Faculty of Economic and Business, University of Rijeka, Rijeka, Croatia

Vladimir Dodevski Laboratory for Materials Sciences, Institute of Nuclear Sciences "Vinča", University of Belgrade, Belgrade, Serbia

Aleksandra Dragicevic Faculty of Mechanical Engineering, University of Belgrade, Belgrade, Serbia

Olivera Eric Cekic Faculty of Mechanical Engineering, Innovation Centre, University of Belgrade, Belgrade, Serbia

Aleksandar Grbović Faculty of Mechanical Engineering, University of Belgrade, Belgrade, Serbia

Toni Ivanov Faculty of Mechanical Engineering, University of Belgrade, Belgrade, Serbia

Elisaveta Ivanova Nikola Vaptsarov Naval Academy, Varna, Bulgaria

Petar Janjatovic Faculty of Technical Sciences, Department of Production Engineering, University of Novi Sad, Novi Sad, Serbia

Bojan Jankovic Department of Physical Chemistry, Institute of Nuclear Sciences "Vinča", University of Belgrade, Belgrade, Serbia

Zorana Jeli Faculty of Mechanical Engineering, University of Belgrade, Belgrade, Serbia

Đorđe Jovanović Faculty of Technical Sciences, University of Novi Sad, Novi Sad, Serbia

Rastko Jovanovic Vinca Institute of Nuclear Sciences, Laboratory for Thermal Engineering and Energy, University of Belgrade, Belgrade, Serbia

Vladimir Jovanovic Faculty of Mechanical Engineering, Fuel and Combustion Laboratory, University of Belgrade, Belgrade, Serbia

Gordana Kastratović Faculty of Transport and Traffic Engineering, University of Belgrade, Belgrade, Serbia

Boris Kosić Faculty of Mechanical Engineering, University of Belgrade, Belgrade, Serbia

Tatjana Kosic Innovation Center of Faculty of Mechanical Engineering, University of Belgrade, Belgrade, Serbia

Zdravko Krivokapic Faculty of Mechanical Engineering Podgorica, University of Montenegro, Podgorica, Montenegro

Vladislav Krstic "Ljubex International" d.o.o, Belgrade, Serbia

Katarina Maksimovic Secretariat for Utilities and Housing Services Water Management, Belgrade, Serbia

Mirko Maksimovic Belgrade Waterworks and Sewerage, Belgrade, Serbia

Nebojša Manic Faculty of Mechanical Engineering, Fuel and Combustion Laboratory, University of Belgrade, Belgrade, Serbia

Kocareva Marina The College of Textile - Design, Technology and Management, Belgrade, Serbia

Gabriel–Catalin Marinescu Faculty of Mechanics, University of Craiova, Craiova, Romania

Zlatko Marković Faculty of Civil Engineering, University of Belgrade, Belgrade, Serbia

Marko M. Mihic Faculty of Organizational Sciences, University of Belgrade, Belgrade, Republic of Serbia

Dragan Milcic Faculty of Mechanical Engineering in Niš, University of Niš, Niš, Serbia

Miodrag Milcic Faculty of Mechanical Engineering in Niš, University of Niš, Niš, Serbia

Tomás Milesich Faculty of Mechanical Engineering, Institute of Transport Technology and Engineering Design, Slovak University of Technology in Bratislava, Bratislava, Slovakia

Vesna Miletic University of Belgrade School of Dental Medicine, Belgrade, Serbia

Borut Milfelner Faculty of Economics and Business, University of Maribor, Maribor, Slovenia

Aleksandra Milić Lemić University of Belgrade, Faculty of Dental Medicine, Belgrade, Serbia

Marko Milovanovic Deutsches Elektronen-Synchrotron (DESY), Zeuthen, Germany

Matej Minárik Faculty of Mechanical Engineering, Institute of Transport Technology and Engineering Design, Slovak University of Technology in Bratislava, Bratislava, Slovakia

Nikola Mirkov Institute of Nuclear Sciences - Vinča, Laboratory for Thermal Engineering and Energy, University of Belgrade, Belgrade, Serbia

Zarko Miskovic Faculty of Mechanical Engineering, University of Belgrade, Belgrade 35, Serbia

Aleksandra Mitrovic University Union "Nikola Tesla", Faculty of Information Technology and Engineering, Belgrade, Serbia;
The College of Textile - Design, Technology and Management, Belgrade, Serbia;
Faculty of Information Technology and Engineering, University Union "Nikola Tesla", Belgrade, Serbia

Nenad Mitrović Faculty of Mechanical Engineering, University of Belgrade, Belgrade, Serbia

Radivoje Mitrovic Faculty of Mechanical Engineering, University of Belgrade, Belgrade 35, Serbia

Goran Mladenovic Faculty of Mechanical Engineering, Department of Production Engineering, University of Belgrade, Belgrade, Serbia

Mauro Overend Department of Engineering, Cambridge University, Cambridge, UK

Ognjen Pekovic Faculty of Mechanical Engineering, University of Belgrade, Belgrade, Serbia

Sanja Pekovic Faculty of Tourism and Hospitality, University of Montenegro, Kotor, Montenegro

Jasmina Perisic Union University Belgrade, Belgrade, Serbia

Danilo Petrasinovic Faculty of Mechanical Engineering, University of Belgrade, Belgrade, Serbia

Milos Petrasinovic Faculty of Mechanical Engineering, University of Belgrade, Belgrade, Serbia

Toni Petrinic Domeni Ltd., Matulji, Croatia

Milena M. Pijovic Department of Physical Chemistry, Institute of Nuclear Sciences "Vinča", University of Belgrade, Belgrade, Serbia

Milos Pjevic Faculty of Mechanical Engineering, Department of Production Engineering, University of Belgrade, Belgrade, Serbia

Marián Polóni Faculty of Mechanical Engineering, Institute of Transport Technology and Engineering Design, Slovak University of Technology in Bratislava, Bratislava, Slovakia

Dejana Popovic University of Belgrade, Vinča Institute of Nuclear Sciences, Belgrade, Serbia

Mihajlo Popovic Faculty of Mechanical Engineering, Department of Production Engineering, University of Belgrade, Belgrade, Serbia

Radovan Puzovic Faculty of Mechanical Engineering, Department of Production Engineering, University of Belgrade, Belgrade, Serbia

Milos B. Radojevic Faculty of Mechanical Engineering, Fuel and Combustion Laboratory, University of Belgrade, Belgrade, Serbia

Dragan Rajnovic Faculty of Technical Sciences, Department of Production Engineering, University of Novi Sad, Novi Sad, Serbia

Boško Rašuo Faculty of Mechanical Engineering, University of Belgrade, Belgrade, Serbia

Mirjana Reljic CIS Institute, Belgrade, Serbia;
The College of Textile - Design, Technology and Management, Belgrade, Serbia

Marko Ristic University of Belgrade, Institute Mihajlo Pupin, Belgrade, Serbia

Jelena Sakovic Jovanovic Faculty of Mechanical Engineering Podgorica, University of Montenegro, Podgorica, Montenegro

Slavisa Salinic Faculty of Mechanical and Civil Engineering, University of Kragujevac, Kraljevo, Serbia

Leposava Sidjanin Faculty of Technical Sciences, Department of Production Engineering, University of Novi Sad, Novi Sad, Serbia

Ljiljana Tihaček Šojić University of Belgrade, Faculty of Dental Medicine, Belgrade, Serbia

Predrag Šojić University of Belgrade, Faculty of Dental Medicine, Belgrade, Serbia

Dragomir Stamenkovic University of Belgrade, Faculty for Special Education and Rehabilitation, Belgrade, Serbia

Jovana N. Stasic University of Belgrade School of Dental Medicine, Belgrade, Serbia

Slavenko Stojadinovic Faculty of Mechanical Engineering, Department of Production Engineering, University of Belgrade, Belgrade, Serbia

Dragoslava Stojiljkovic Faculty of Mechanical Engineering, Fuel and Combustion Laboratory, University of Belgrade, Belgrade, Serbia

Milena Stojiljkovic University of Nis, Faculty of Technology, Leskovac, Serbia

Stanisa Stojiljkovic Faculty of Technology, University of Nis, Leskovac, Serbia

Igor Svetel Innovation Center of Faculty of Mechanical Engineering, University of Belgrade, Belgrade, Serbia

Jelena Svorcan Faculty of Mechanical Engineering, University of Belgrade, Belgrade, Serbia

Ivan Tanasić Medical College of Applied Studies Belgrade, Belgrade, Serbia

Ljubodrag Tanovic Faculty of Mechanical Engineering, Department of Production Engineering, University of Belgrade, Belgrade, Serbia

Tihomir Vasilev Nikola Vaptsarov Naval Academy, Varna, Bulgaria

Ivana Vasovic Lola Institute, Belgrade, Serbia

Nenad Vidanović Faculty of Transport and Traffic Engineering, University of Belgrade, Belgrade, Serbia

Dragiša Vilotić Faculty of Technical Sciences, University of Novi Sad, Novi Sad, Serbia

Aleksandar Vujovic Faculty of Mechanical Engineering Podgorica, University of Montenegro, Podgorica, Montenegro

Marija Zivkovic Vinca Institute of Nuclear Sciences, Laboratory for Thermal Engineering and Energy, University of Belgrade, Belgrade, Serbia

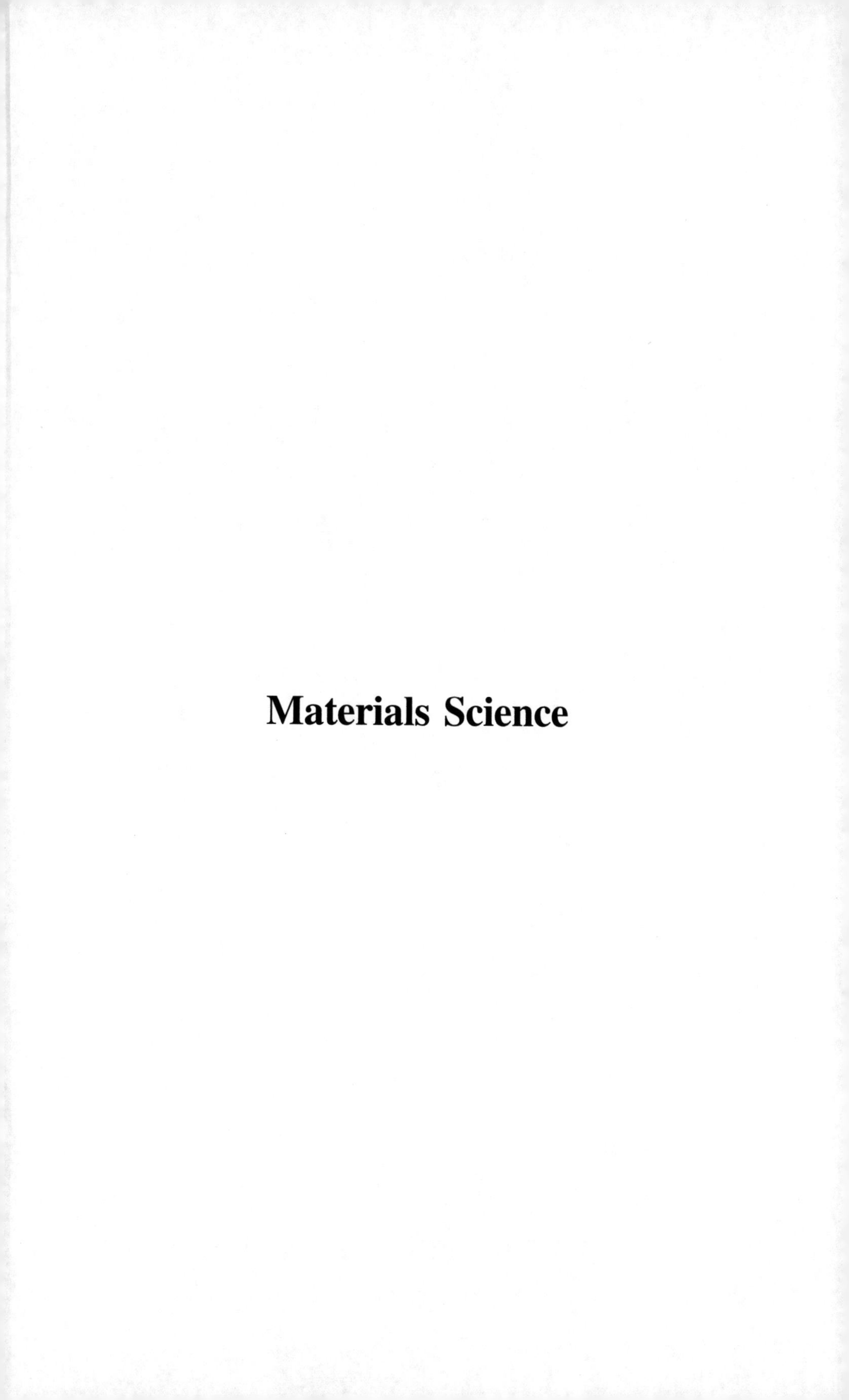

Materials Science

Residual Fatigue Life Estimation of Structural Components Under Mode-I and Mixed Mode Crack Problems

Ivana Vasovic[1(✉)], Mirko Maksimovic[2], and Katarina Maksimovic[3]

[1] Lola Institute, Kneza Viseslava 70a, Belgrade, Serbia
ivanavvasovic@gmail.com
[2] Belgrade Waterworks and Sewerage, Kneza Milosa 27, Belgrade, Serbia
[3] Secretariat for Utilities and Housing Services Water Management, Kraljice Marije 1, 11120 Belgrade, Serbia

Abstract. The work presents a residual fatigue life prediction methodology of cracked structural components under interspersed mode-I and mixed-modes (I and II). In this paper the numerical computation methods and procedures for predicting the fatigue crack growth trajectories and residual fatigue lives of notched structural components are analyzed. Special attention in this work is focused on notched structural components such as aircraft wing skin type structural components under mixed modes and cracked aircraft lugs under mode-I. Stress intensity factor (SIF) solutions are required for assessment fracture strength and residual fatigue life for defects in structures or for damage tolerance analysis recommend to be performed at the stage of aerospace structure design. A variety of methods have been used to estimate the SIF values, such as approximate analytical methods, finite element (FE), finite element alternating, weight function, photo elasticity and fatigue tests. In this work the analytic/numerical methods and procedures were used to determine SIF and predicting the fatigue crack growth life of damaged structural components with notched cracks. For this purpose, finite element method (FEM) is used to derive analytic expressions for SIF of cracked structural components. To obtain the stress intensity factors of cracked structural components special singular finite elements are used. The strain energy density method (SED) and MTS criterions are used for determination of the crack growth trajectories in thin-walled structures. Computation results are compared with experiments.

Keywords: Fracture mechanics · Residual fatigue life · FEM · Mixed modes · Crack growth trajectory

1 Introduction

Surface and through-thickness cracks frequently initiate and grow at notches, holes in structural components. Such cracks are present during a large percentage of the useful life of these components. Hence, understanding the severity of cracks is important in the development of life prediction methodologies. Current methodologies use the stress

N. Mitrovic et al. (Eds.): CNNTech 2019, LNNS 90, pp. 3–21, 2020.
https://doi.org/10.1007/978-3-030-30853-7_1

intensity factor (SIF) to quantify the severity of cracks and the development SIF solutions for notched structural components using analytical, numerical and semi-analytical methods. These methodologies are in use for the last three decades.

Methods for design against fatigue failure are under constant improvement. Fracture mechanics has developed into a useful discipline for predicting strength and life of cracked structures. Surface and through-thickness cracks frequently initiate and grow at notches, holes in structural components. Such cracks are present during a large percentage of the useful life of these components. Hence, understanding the severity of cracks is important in the development of life prediction methodologies [9].

The adoption of the damage tolerance design concept [1, 2] along with an increased demand for accurate residual structure and notched component life predictions have provided growing demand for the study of fatigue crack growth in aircraft mechanical components. The damage tolerance approach assumes that the structure contains an initial crack or defect that will grow under service usage. The crack propagation is investigated to ensure that the time for crack growth to a critical size takes much longer than the required service life of notched structural components. For a damage tolerance program to be effective it is essential that fracture data can be evaluated in a quantitative manner. Since the establishment of this requirement not only the understanding of fracture mechanics has greatly improved, but also a variety of numerical tools have become available to the analyst. These tools include Computer Aided Design (CAD), Finite Element Modeling (FEM) and Computation Fluid Dynamics (CFD) [18, 19]. Fracture mechanics software provides the engineering community with this capability. Computer codes can be used to predict fatigue crack growth and residual strength in aircraft structures. They can also be useful to determine in-service inspection intervals, time-to-onset of widespread fatigue damage and to design and certify structural repairs. Used in conjunction with damage tolerance programs fracture analysis codes can play an important role in extending the life of "high-time" aircraft. Traditional applications of fracture mechanics have been concerned on cracks growing under an opening or mode I mechanism. However, many service failures occur from cracks subjected to mixed mode loadings. A characteristic of mixed mode fatigue cracks is that they usually propagate in a non-self similar manner. Therefore, under mixed mode loading conditions, not only the fatigue crack growth rate is of importance, but also the crack growth direction. Several criteria have been proposed regarding the crack growth direction under mixed mode loadings. In this work the maximum strain energy density criterion [5, 6] and the maximum tangential stress criterion [14] are considered. This S-criterion allows stable and unstable crack growth in a mixed mode. The application of this criterion can be found in the works by several authors [7, 8]. The aim of this work is to investigate the strength behavior of important aircraft notched structural elements such as cracked lugs and riveted skin. The attention is focused on crack growth behavior of cracked structural components under mode I and the mixed mode.

2 Numerical Simulation of Crack Growth

Numerical simulation of crack growth provides a powerful predictive tool to use during the design phase as well as for evaluating the behavior of existing cracks. These simulations can be used to complement experimental results and allow engineers to economically evaluate a large number of damage scenarios. Numerical methods are the most efficient way to simulate fatigue crack growth because crack growth is an incremental process where stress intensity factor (SIF) values are needed at each increment as an input to crack growth equations.

In order to simulate mixed-mode crack growth an incremental type analysis is used where knowledge of both the direction and the size of the crack increment extension are necessary. For each increment of the crack extension, a stress analysis is performed using the quarter-point singular finite elements (Q-E) [4, 16, 17] and SIFs are evaluated. The incremental direction and size along the crack front for the next extension are determined by fracture mechanics criteria involving SIFs as the prime parameters. The crack front is re-meshed and the next stress analysis is carried out for a new configuration.

2.1 Stress Intensity Factor Solution of Cracked Lugs

In general, geometry of notched structural components and loading is too complex for the stress intensity factor (SIF) to be solved analytically. The SIF calculation is further complicated because it is a function of the position along the crack front, crack size and shape, type loading and geometry of the structure. In this work analytic and FEM were used to perform a linear fracture mechanics analysis of the pin-lug assembly. The analytic results are obtained using the relations derived in this paper. Good agreement between finite element and analytic results is obtained. It is very important because we can use analytic derived expressions in crack growth analyses. Lugs are essential components of an aircraft for which a proof of damage tolerance has to be undertaken. Since the literature does not contain the stress intensity solution for lugs which are required for proofs of damage tolerance, the problem posed in the following investigation are: selection of a suitable method of determining SIF, determination of SIF as a function of crack length for various forms of lugs and setting up a complete formula for calculating the SIF for lug, allowing essential parameters. The stress intensity factors are the key parameters to estimate the characteristics of the cracked structure. Based on the stress intensity factors, fatigue crack growth and structural life predictions have been investigated. The lug dimensions are defined in Fig. 1.

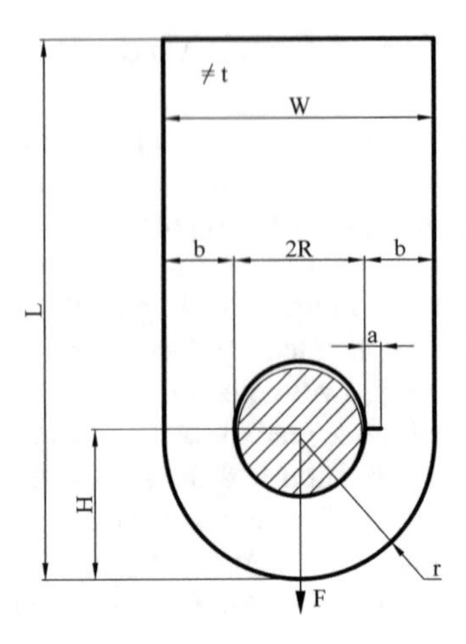

Fig. 1. Geometry and loading of lugs

To obtain the stress intensity factor for the lugs it is possible to start with the general expression for the SIF, K, in the next form

$$K = Y_{SUM}\,\sigma\,\sqrt{\pi\,a} \tag{1}$$

where: Y_{SUM} – the correction function, a- the crack length. This function is essential in determining the stress intensity factor. Primary, this function depends on the stress concentration factor, k_t and the geometric ratio a/b. The correction function is defined using experimental and numerical investigations. This function can be defined in the next form [11, 12]:

$$Y_{SUM} = \frac{1.12 \cdot k_t \cdot A}{A + \frac{a}{b}} \cdot k \cdot Q \tag{2}$$

$$k = e^r\,\sqrt{a/b} \tag{3}$$

$$b = \frac{w - 2 \cdot R}{2} \tag{4}$$

$$r = -3.22 + 10.39 \cdot \left[\frac{2 \cdot R}{w}\right] - 7.67 \cdot \left[\frac{2 \cdot R}{w}\right]^2 \tag{5}$$

$$Q = \frac{U \cdot \frac{a}{b} + 10^{-3}}{\frac{a}{b} + 10^{-3}} \tag{6}$$

$$U = 0.72 + 0.52 \cdot \left[\frac{2 \cdot R}{H}\right] - 0.23 \cdot \left[\frac{2 \cdot R}{H}\right]^2 \tag{7}$$

$$A = 0.026 \cdot e^{1.895 \cdot \left(1 + \frac{a}{b}\right)} \tag{8}$$

The stress concentration factor k_t is very important in the calculation of the correction function, Eq. 2. In this investigation a contact finite element stress analysis was used to analyze the load transfers between the pin and the lug.

2.2 Incremental Direction Growth Criteria

In mixed mode fatigue crack propagation, an important part of investigation is not only the description of the fatigue crack growth rate but also the analysis of the crack paths – the angular direction (crack branching) of fatigue crack growth (θ).

Several criteria have been considered to describe the direction of crack propagation for mixed mode crack growth. The most widely used criteria for mixed mode fatigue crack propagation are: maximum energy release rate (MERR), [9, 14] the maximum tangential stress (MTS), [10, 14] and strain energy density (SED), [5, 6, 11]. The MTS criterion assumes that the direction of crack growth is connected with the direction, where the tangential stress reaches its maximum value. For the unification of the results comparison in terms of mixed mode (I + II) loading, the elastic-mixity parameter Me has been defined as $M_e = \frac{2}{\pi} a \tan \frac{K_I}{K_{II}}$, The value Me = 0 corresponds with pure shear state (mode II) and the value Me = 1 means that the specimen is loaded in pure mode I. K_I and K_{II} are stress intensity factors for mode I and mode II.

2.3 Strain Energy Density Criterion

The minimum strain energy density criterion [5, 6] is discussed in this work. The strain energy density criterion is based on the postulate that the direction of crack propagation at any point along the crack front toward the region where in the direction of minimum strain energy density factor. The strain energy density factor, S, is given as

$$S(\theta) = a_{11} K_I^2 + a_{12} K_I K_{II} + a_{33} K_{III}^2 \tag{9}$$

where the factors a_{ij} are the functions of the angle θ, and are defined as

$$a_{11} = \frac{1}{16G\pi} [(1 + \cos\theta)(k - \cos\theta)]$$

$$a_{12} = \frac{1}{16G\pi} \sin\theta[2\cos\theta - (k - 1)] \tag{10}$$

$$a_{22} = \frac{1}{16G\pi} [(k+1)(1 - \cos\theta) + (1 + \cos\theta)(3\cos\theta - 1)]$$

where G is the shear modulus and k is a constant depending upon stress state, and is defined as: k = (3 – ν)/(1 + ν) for plane stress. The direction of crack growth is determined by minimizing this equation with respect to the angle theta (θ). In the mathematical form, the strain energy density criterion can be stated as

$$
\begin{aligned}
&[2(1+k)\mu]\tan^4\frac{\theta}{2}+\left[2k(1-\mu^2)-2\mu^2+10\right]\tan^3\frac{\theta}{2}\\
&-24\mu\tan^2\frac{\theta}{2}+\left[2k(1-\mu^2)+6\mu^2-14\right]\tan\frac{\theta}{2}\\
&+2(3-k)\mu=0
\end{aligned}
\tag{11}
$$

$$
\begin{aligned}
&[2(k-1)\mu]\sin\theta-8\mu\sin 2\theta+\left[(k-1)(1-\mu^2)\right]\cos\theta\\
&+\left[2(\mu^2-3)\right]\cos 2\theta\rangle 0
\end{aligned}
\tag{12}
$$

$$
\mu=K_I/K_{II} \tag{13}
$$

Once S is established, crack initiation will take place in a radial direction θ, from the crack tip, along which the strain energy density is minimum. The main advantage of this criterion is simplicity and its ability to handle various combined loading situations.

2.4 MTS Criterion

This criterion [14] states that the direction of crack initiation coincides with the direction of the maximum tangential stress (MTS) along a constant radius around the crack tip. It can be stated mathematically as

$$
\frac{\partial\sigma_\theta}{\partial\theta}=0,\frac{\partial^2\sigma}{\partial\theta^2}\leq 0 \tag{14}
$$

Using the stress field in the polar co-ordinates and applying the MTS –criterion the following equation is obtained

$$
\tan^2\frac{\theta}{2}-\frac{\mu}{2}\tan\frac{\theta}{2}-\frac{1}{2}=0 \tag{15}
$$

$$
\begin{aligned}
&\frac{3}{2}\left[\left(\frac{1}{2}\cos^3\frac{\theta}{2}-\cos\frac{\vartheta}{2}\sin^2\frac{\theta}{2}\right)+\right.\\
&\left.+\frac{1}{\mu}\left(\sin^3\frac{\theta}{2}-\frac{7}{2}\sin\frac{\theta}{2}\cos^3\frac{\theta}{2}\right)\right]\langle 0
\end{aligned}
\tag{16}
$$

where μ is defined in Eq. (13). This criterion is the simplest of all but very effective. The MTS criterion has been found to be good for brittle fracture.

The crack growth direction angle in the local coordinate plane perpendicular to the crack front can then be determined for each point along the crack front. In this work, the crack inclination angle is taken into account in the calculations by means of the values of the SIF K_I and K_{II}, because their values are the function of the orientation of the crack plane.

This criterion states that a crack propagates in a direction corresponding to the direction of maximum tangential stress along a constant radius around the crack-tip. Using the Westergaard stress field in the polar coordinates and applying the (MTS) criterion, the Eq. (17) is obtained to predict the crack propagation direction in each incremental step [20].

$$\Theta_o = 2\tan^{-1}\left(\frac{K_I}{4K_{II}} - \frac{1}{4}\sqrt{\left(\frac{K_I}{K_{II}}\right)^2 + 8}\right) \quad \text{for } K_{II}\rangle 0$$

$$\Theta_o = 2\tan^{-1}\left(\frac{K_I}{4K_{II}} + \frac{1}{4}\sqrt{\left(\frac{K_I}{K_{II}}\right)^2 + 8}\right) \quad \text{for } K_{II}\langle 0 \tag{17}$$

To initiate crack propagation, the maximum circumferential tensile stress σ needs to reach a critical value. This results in an expression for the equivalent stress intensity factor SIF in mixed mode condition as:

$$K_{eq} = K_I\cos^3\frac{\Theta_0}{2} - \frac{3}{2}K_{II}\cos\frac{\Theta_0}{2}\sin\Theta_0 \tag{18}$$

However, when the plastic zone size cannot be ignored, it is necessary to use the stress state at a material dependent finite distance from the crack tip.

3 Numerical Examples

In order to demonstrate the accuracy and efficiency of the methodology discussed in the preceding sections two crack growth applications are described. The first application describes crack growth in an aircraft wing lug and the second illustrates the use of the finite element methodology to simulate crack trajectory under mixed-mode.

Example 1. Fatigue Crack Growth in an Aircraft Wing Lug
This example describes the analytical and numerical methods for obtaining the stress intensity factors and for predicting the fatigue crack growth life for cracks at attachment lugs. Straight-shank male lug is considered in the analysis, Fig. 2. Three different head heights of lugs are considered. The straight attachment lugs are subjected to axial pin loading only. The material properties of lugs are (7075 T7351) [12]: $\sigma_m = 432$ N/mm^2 ⇔ Ultimate tensile strength, $\sigma_{02} = 334$ N/mm^2, $C_F = 3 \times 10^{-}7$, $n_F = 2.39$, $K_{IC} = 2225$ [N/mm$^{3/2}$].

The stress intensity factors of cracked lugs are calculated under the stress level: $\sigma g = \sigma_{max} = 98.1$ N/mm^2, or the corresponding axial force, $F_{max} = \sigma g$ (w-2R) t = 63716 N. The presented finite element analysis of cracked lug is modeled with special singular quarter-point six-node finite elements around the crack tip, Fig. 3. The load of the model, a concentrated force, Fmax, was applied at the centre of the pin and reacted at the other end of the lug. Spring elements were used to connect the pin and the lug at each pairs of nodes having identical nodal coordinates all around the periphery.

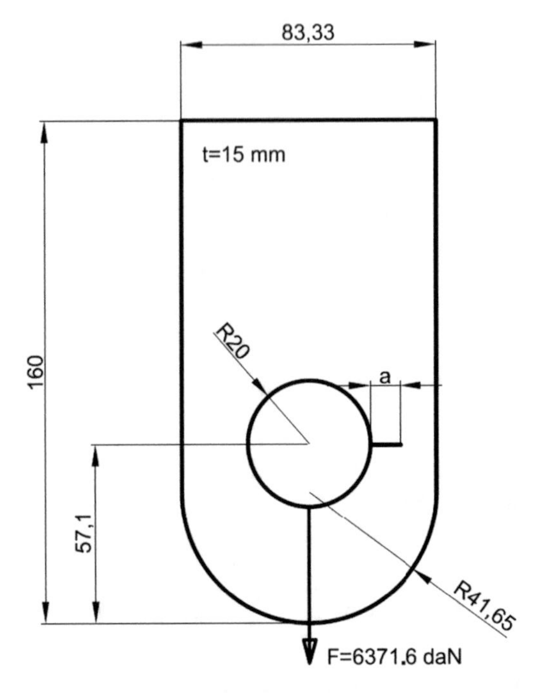

Fig. 2. Geometry of cracked lug No. 6 (H = 57.1 mm)

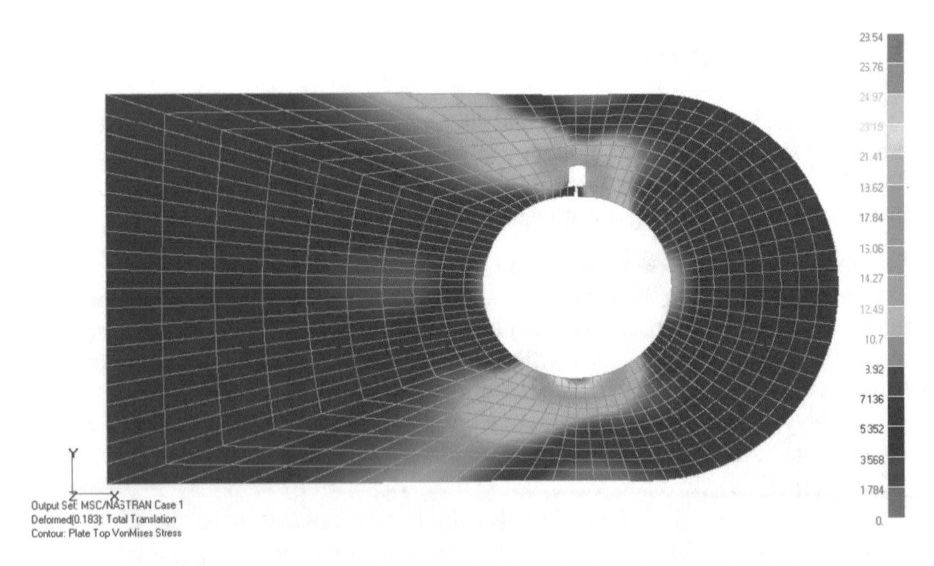

Fig. 3. Finite element model of a cracked lug No. 6 with stress distribution

Table 1. Geometric parameters of lugs [12]

Lug no.	Dimensions [mm]				
	2R	W	H	L	t
2	40	83.3	44.4	160	15
6	40	83.3	57.1	160	15
7	40	83.3	33.3	160	15

The area of contact was determined iteratively by assigning a very high stiffness level to spring elements, which were in compression and a very low stiffness level (essentially zero) to spring elements, which were in tension. The stress intensity factors for through the thickness cracks of lugs, analytic and finite elements, are shown in Table 2. Analytic results are obtained using the relations from the previous sections, Eq. (1).

Table 2. Comparisons of analytic and FE results for SIF, K_I

Lug no.	a [mm]	$K_{I_{max}}^{FEM}$	$K_{I\,\max}^{ANAL.}$
2	5.00	68.784	65.621
6	5.33	68.124	70.246
7	4.16	94.72	93.64

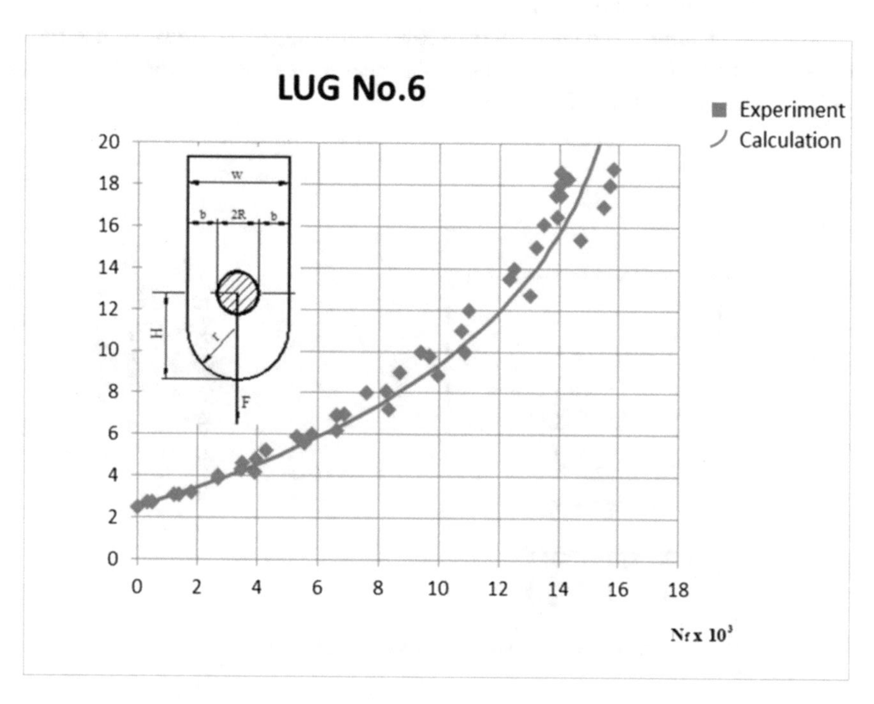

Fig. 4. Crack propagation at the lug No. 6 – comparisons analytic results and experimental results (H = 57.1 mm); k_t = 2.6

Figure 4 shows a comparison between the experimentally determined crack propagation curves and the load cycles calculates to Walker low [3] for several crack lengths. A relatively close agreement between the test and the presented computation results are obtained. The analytic computation methods presented in this work can satisfy requirements for damage tolerance analyses of notched structural components such as lugs-type joints.

Example 2. Crack Growth from the Riveted Holes

In this section, we consider the modeling of crack propagation in a plate with cracks emanating from one hole subjected to a far-field tension, σ, Fig. 5. In the initial configuration the left crack has 0.1 in and is oriented at angle θ = 33.6o to the left hole. The change in the crack length for each iteration is taken to be a constant, $\Delta a = 0.1\ 1$ in, and the cracks grow for eight steps. In this analysis the maximum tangential stress (MTS-criterion) is used to determine the crack trajectory or the angle of crack propagation.

In this work, the crack inclination angle is taken into account in the calculations by means of the values of the SIF K_I and K_{II}, because their values are a function of the orientation of the crack plane. These parameters were calculated numerically with the finite element method.

Figure 7 shows the stress contour and the crack trajectory for the last configuration. In this crack growth analysis quarter-point (Q-P) singular finite elements are used with the MTS-criterion. These results are compared with an extended finite element method (X-FEM) [13], Table 3 and Fig. 8.

The extended finite element method allows modeling of arbitrary geometric features independently of the finite element mesh, because, it allows the modeling of crack growth without remeshing [13, 15] (Fig. 6).

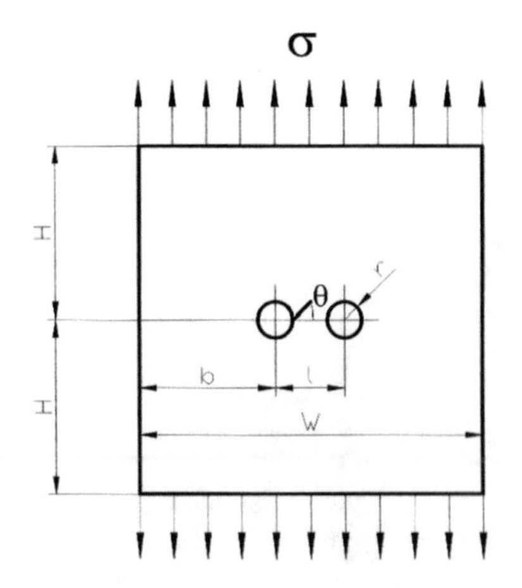

W = 5 in
b = 2 in
l = 1 in
H = 2.5 in
r = 5/64 in
θ = 33.6°
E = 3 .107 psi
ν = 0.3
σ = 2 psi

Fig. 5. Geometry and load of the riveted crack problem

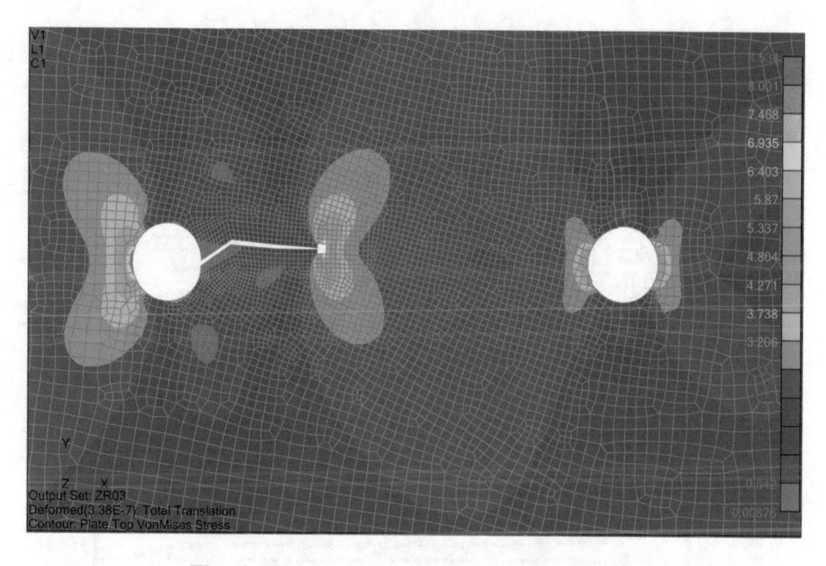

Fig. 6. The crack trajectory after the third step

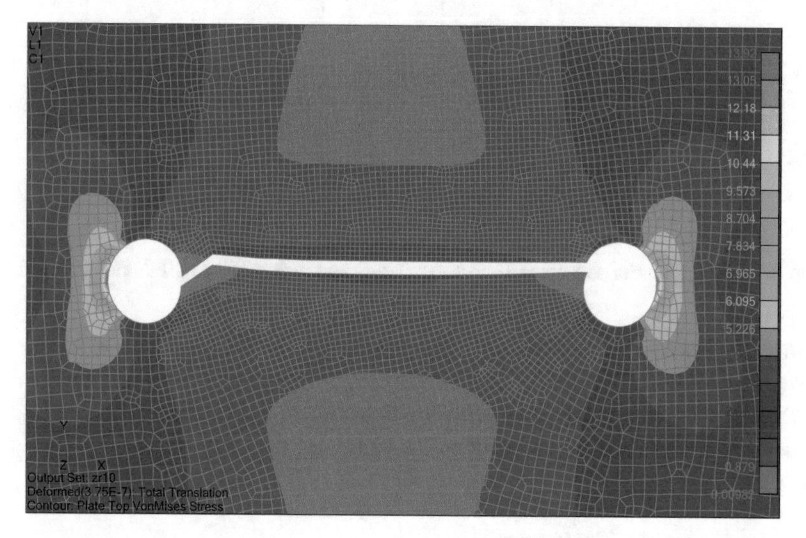

Fig. 7. The crack trajectory using the Q-P elements and the MTS-criterion

In this example the MTS criterion is used. The predicted crack trajectories using the Q-P singular finite elements and the X-FEM method are nearly identical. These computation results for the crack growth trajectory under mixed modes are compared with experiments. Good agreements are obtained.

Table 3. Position of the left crack tip during crack growth

X-FEM [13]		Presented Q-P singular FE solutions	
Xc [in]	Yc [in]	Xc [in]	Yc [in]
2.144	2.544	2.144	2.544
2.260	2.538	2.2436	2.535
2.376	2.531	2.3435	2.531
2.493	2.531	2.4435	2.5299
2.610	2.534	2.5435	2.5294
2.727	2.533	2.6435	2.5294
2.840	2.530	2.7435	2.5303
2.92	2.51	2.8436	2.5321

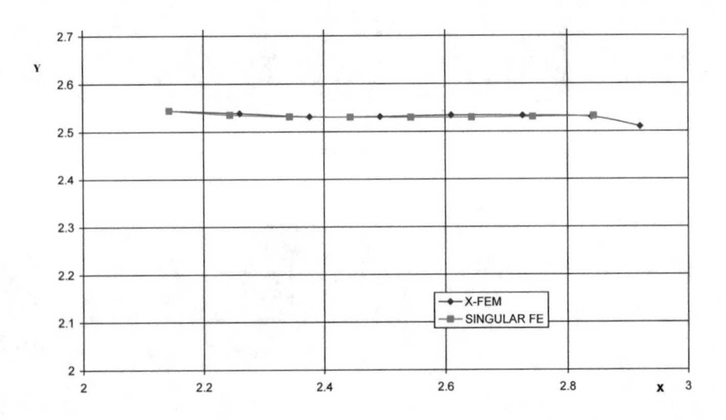

Fig. 8. Comparison of the crack trajectory using the present QP singular FE with X-FEM

Example 3. Validation of Computation Procedure in Domain Mixed Modes. To validate computation procedures in domain crack growth behavior under mixed modes in this work two type problems are considered: I) Determination crack growth trajectory and II) Residual life estimation along „curve" mixed mode crack growth trajectory.

3.1 Crack Growth Trajectory

To illustrate the determination crack growth trajectory under mixed modes I/II, duraluminum plate with two holes and initial crack under tension load Fy as shown in Fig. 9 is considered. To determine stress intensity factors KI and KII, Msc/Nastran software code [21] is used. In Figs. 10 and 11 finite element models with stress distributions of cracked specimen are shown.

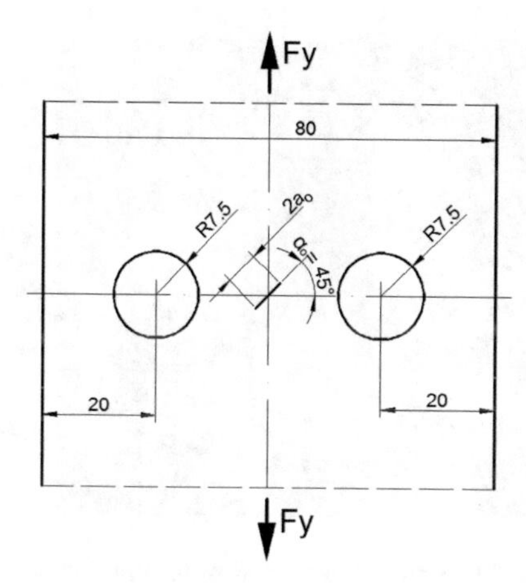

Fig. 9. Geometry of specimen for modeling of crack growth trajectory

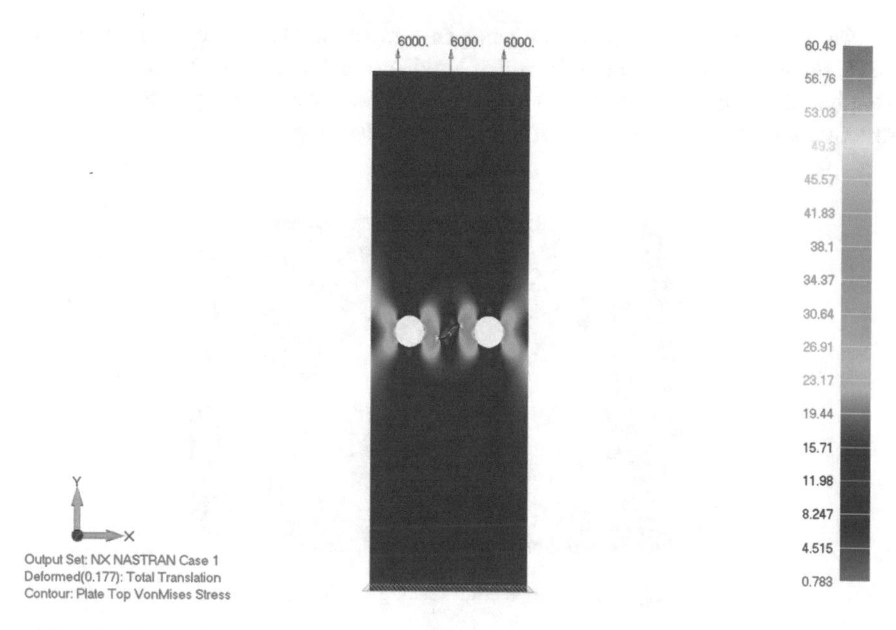

Fig. 10. Stress distributions of cracked specimen using finite elements (F_y = 60000 N)

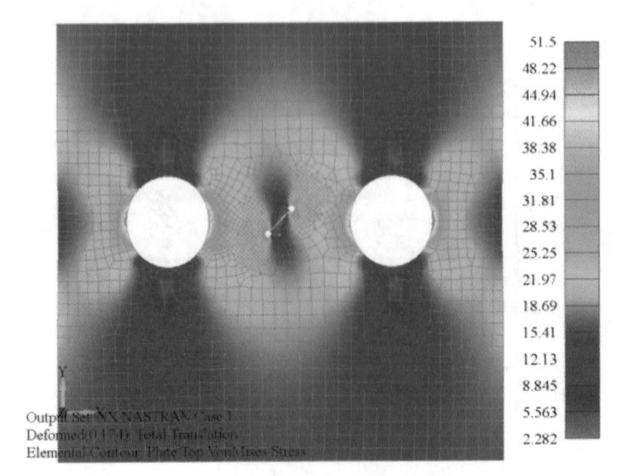

Fig. 11. Stress distributions of cracked specimen using finite elements (F_y = 60000 N), detail view

To predict the crack growth direction in a mixed-mode problem, in this analysis MTS criterion is used. This criterion is used in combining Msc/Nastran code [21]. Combining finite elements for determination of the stress intensity factors and MTS criterion the computation crack trajectory is obtained, Fig. 12.

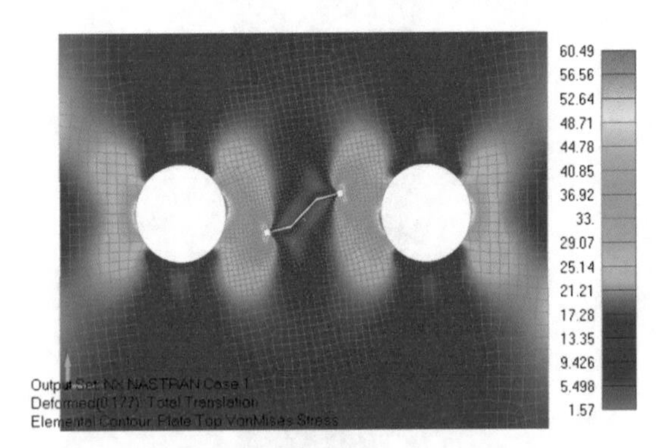

Fig. 12. Computation crack growth trajectory

To validate computation procedure for determination of the crack growth trajectory experimental test is included. Experimental test is carried out using servohydraulic MTS system, Figs. 13 and 14.

Figure 15 illustrates good agreement between computation crack growth trajectory with experiments.

Fig. 13. Specimen in servo hydraulic MTS system

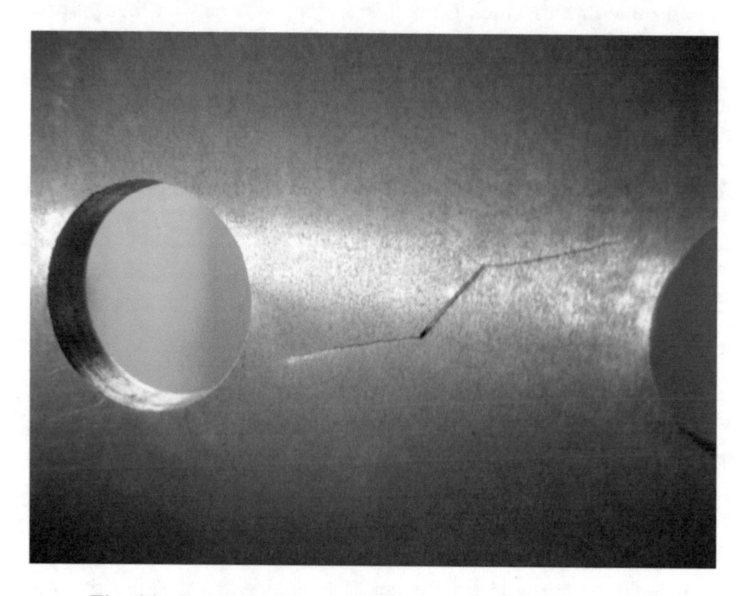

Fig. 14. Experimentally determined crack growth trajectory

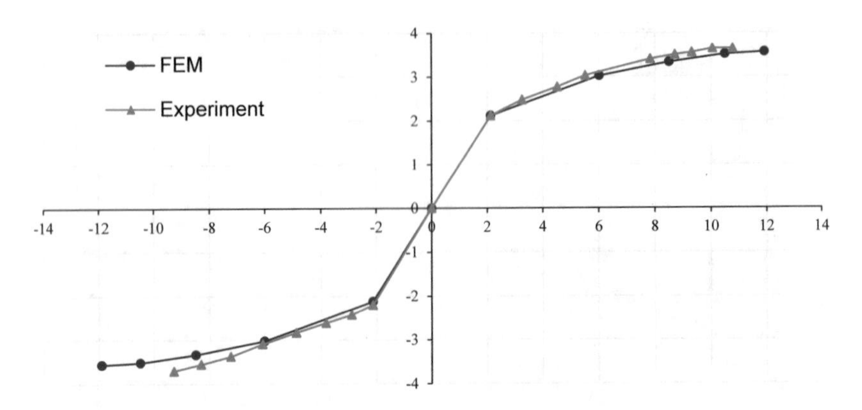

Fig. 15. Comparisons of computation based on FEM with experimental crack growth trajectory

3.2 Residual Life Estimation

In this paper, residual life of cracked structural element, is considered: numerical and experimental. Figure 14 shows experimentally determined crack growth trajectory. For computation determination of residual life for this structural element Paris crack growth low [22] along mixed mode crack growth trajectory is used. For that purpose, analytic formula for equivalent stress intensity factor K_{eq} is necessary. For determination analytic formula of K_{eq} discrete values of the stress intensity factors along crack trajectory are used. Discrete values of SIF`s are given in Table 4:

Table 4. Discrete values of SIF's along mixed mode crack growth trajectory

a [mm]	$a_o = 3$	$a_o + a_1 = 7$	$a_o + .. + a_2 = 9.5$	$a_o + .. + a_3 = 11.5$	$a_o + .. + a_4 = 12.9$
K_I[daN/mm$^{3/2}$]	37.6	90.8	123.5	162.5	177
K_{II}[daN/mm$^{3/2}$]	21	9.5	2.2	−1	2.5
α_i [°]	45	32	5.9	0.7	3.2

Using discrete values of SIF's from Table 4 and relation for equivalent SIF in the next form [21]:

$$K_{eq} = \left[K_I^4 + 8K_{II}^4\right]^{1/4} \tag{19}$$

we can obtain analytic formula for the stress intensity factor along the crack growth trajectory, in accordance to Fig. 15, in the next form:

$$K_{eq} = -2E + 07a^3 + 820616a^2 - 5217.1a + 20.311 \tag{20}$$

in which a is the crack length along crack trajectory.

To determine residual life of cracked structural component, in this paper, analytic formula Eq. (20) has been used in Paris's low. Paris's constants for considered steel

(1.7225) are: C = 0.00000000058, n = 2.57. Specimens are tested under cyclic load of constant amplitude in which σmax = 250 MPa and σmin = 25 MPa. The crack length versus number of loading cycles *min* and *max* is shown in Fig. 16.

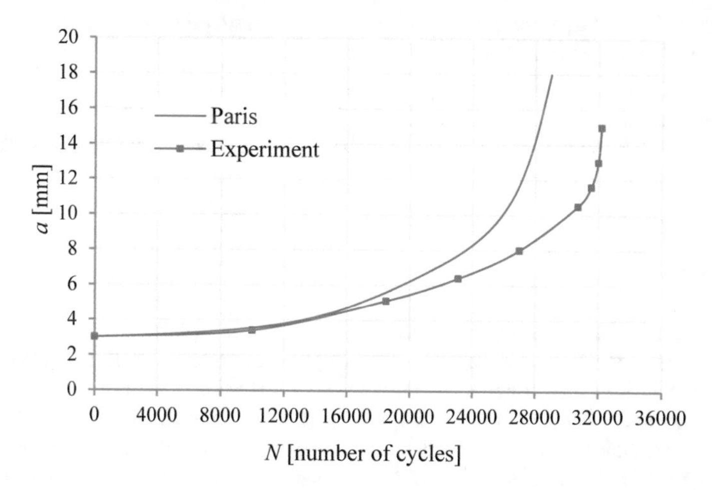

Fig. 16. Comparisons of computation based on Paris method with experimental crack growth trajectory

Experimental determined number of cycle before failure is N_{exp} = 32200 cycles as shown in Fig. 16. Residual life estimation under mixed modes crack growth is computed from point a_0 to point a_4 as shown in Fig. 15.

4 Conclusions

This work considers the crack growth analyses of damaged structural components under fracture mechanics for mode I and the mixed modes. The finite element method is a robust and efficient method that can be used to investigate the impact crack on the performance of notched structural components.

The aim of this work is to investigate the strength behavior of the notched structural elements such as the cracked lugs. In the fatigue crack growth and the fracture analysis of lugs, an accurate calculation of SIFs is essential. An analytic expression for the stress intensity factor of the cracked lug is derived using the correction function and FEM. The contact finite element analysis for the true distribution of the pin contact pressure is used for the determination of stress concentration factors that are used in the correction function. Good agreement between the derived analytic SIFs of the cracked lug with finite elements is obtained.

Two applications were considered in this work to demonstrate the effectiveness of finite element based computer codes in evaluating the impact of fatigue crack growth on structural components. Applications described fatigue crack growths analysis of lugs with complex geometry and loading. In this paper the predicted crack trajectory using quarter-point (Q-P) singular finite elements together with the MTS criteria were nearly identical to the trajectories predicted with X-FEM. This approach can be effective use in practical analysis crack trajectory of structural component under mixed modes. The computation results of damaged lug type structural components are compared with the experiments. Good correlations between the computation and the experiments [2] are obtained as well.

Acknowledgement. The authors would like to thank the Ministry of Education, Science and Technological Development of Serbia for financial support under the projects TR 35024 and TR 35045.

References

1. Jankovic, D., Maksimovic, S., Kozic, M., Stupar, S., Maksimovic, K., Vasovic, I., Maksimovic, M.: CFD Calculation of helicopter tail rotor airloads for fatigue strength experiments. J. Aerosp. Eng. **5**(30), 04017032-1–04017032-11 (2017)
2. Maksimović, K., Stamenković, D., Boljanović, S., Maksimović, M., Vasović, I., Modeling fracture mechanics parameters of cracked structural elements under thermomechanical loads. In: Gvero, P. (ed.) 13th International Conference on Accomplishments in Mechanical and Industrial Engineering, University of Banja Luka, Faculty of Mechanical Engineering, Banja Luka, Republic of Srpska, Bosnia and Herzegovina, May 2017
3. Walker, K.: The effect of stress ratio during crack propagation and fatigue for 2024 T3 Aluminum, Effects of environment and complex loading history on fatigue life, pp. 1–15. ASTM STP 462, Philadelphia (1970)
4. Barsoum, R.S.: Triangular quarter-point elements as elastic and perfectly plastic crack tip elements. Int. J. Numer. Meth. Eng. **11**, 85–98 (1977)
5. Sih, G.C.: Mechanics of Fracture Initiation and Propagation. Kluwer, Springer, Netherlands (1991)
6. Sih, G.C.: Strain-energy-density factor applied to mixed mode crack problems. Eng. Fract. Mech. **5**, 365–377 (1974)
7. Gdoutos, E.E.: Fracture Mechanics Criteria and Applications. Kluwer, Springer, Netherlands (1990)
8. Jeong, D.Z.: Mixed mode fatigue crack growth in test coupons made from 2024-T3 aluminum. Theor. Appl. Frac. Mech. **42**, 35–42 (2004)
9. Maksimović, S.: Fatigue life analysis of aircraft structural components. Sci. Tech. Rev. **5**(1), 15–22 (2005)
10. Maksimović, K., Nikolić-Stanojević, V., Maksimović, S.: Efficient Computation Method in Fatigue Life Estimation of Damaged Structural Components, FACTA UNIVERSITATIS Niš, Series Mechanics, Automatic control and Robotics, vol. 4, no. 16, pp. 101–114 (2004)
11. Maksimović, K.: Estimation of residual strength for aircraft structural elements. J. Techn. Diagn. **3**, 54–57 (2002)
12. Geier, W.: Strength behavior of fatigue cracked lugs, Royal Aircraft Establishment, LT 20057, Farnborough, Hants, UK (1980)
13. Maksimović, S., Maksimović, K., Vasović, I., Đurić, M., Maksimović, M.: Residual life estimation of aircraft structural components under load spectrum. In: 8th International Scientific Conference on Defensive Technologies, OTEH 2018, Military Technical Institute, Belgrade, Serbia (2018). www.vti.mod.gov.rs/oteh

14. Erdogen, F., Sigh, G.C.: On the crack extension in plates under plane loading and transverse shear. J. Basic Eng. **85**, 519–527 (1963)
15. Jovičić, G., Živković, M., Maksimović, K., Đorđević, N.: The crack growth analysis on the real structure using the X-FEM and EFG methods. Sci. Tech. Rev. **8**(2), 21–26 (2008)
16. Maksimović, M., Vasović, I., Maksimović, K., Maksimović, S., Stamenković, D.: Crack growth analysis and residual life estimation of structural components under mixed modes, ECF22 - Loading and Environmental effects on Structural Integrity, Procedia Structural Integrity 13, pp. 1888–1894. Elsevier (2018)
17. Sherry, A.H., Wilkes, M.A., Ainsworth, R.A.: Numerical modeling of mixed-mode ductile fracture. In: ASME Pressure Vessels and Piping Conference ASME PVP-Vol. 412, pp. 15–24 (2000)
18. Maksimovic, S., Kozic, M., Stetic-Kozic, S., Maksimovic, K., Vasovic, I., Maksimovic, M.: Determination of load distributions on main helicopter rotor blades and strength analysis of its structural components. J. Aerosp.Eng. **6**(27), 04014032-1–04014032-8 (2014)
19. Maksimović, M., Vasović, I., Maksimović, K., Trišović, N., Maksimović, S.: Residual life estimation of cracked structural components. FME Trans. **46**(1), 124–128 (2018)
20. Shafique, S.M.A., Marwan, K.K.: Analysis of mixed mode crack initiation angles under various loading conditions. Engineering Fracture Mechanics **5**(67), 397–419 (2000)
21. Msc/NASTRAN Software code, Theoretical Manuals
22. Paris, P.C., Erdogan, F.: A critical analysis of crack propagation laws. Trans. ASME J. Basic Eng. **4**(85), 528–533 (1963). https://doi.org/10.1115/1.3656900

Dual Phase Austempered Ductile Iron - The Material Revolution and Its Engineering Applications

Olivera Eric Cekic[1(✉)], Dragan Rajnovic[2], Leposava Sidjanin[2], Petar Janjatovic[2], and Sebastian Balos[2]

[1] Faculty of Mechanical Engineering, Innovation Centre, University of Belgrade, Kraljice Marije 16, 11000 Belgrade, Serbia
olivera66eric@gmail.com
[2] Faculty of Technical Sciences, Department of Production Engineering, University of Novi Sad, Trg Dositeja Obradovica 6, 21000 Novi Sad, Serbia

Abstract. During the last few years, there has been an increased interest in improving the strength of ductile iron by means of heat-treating to obtain dual phase microstructures that enhance the properties found in the Austempered Ductile Irons (ADI). Therefore, additional research int this field is well motivated. An example of this is the innovative dual phase austempered irons - DP-ADI, that has a greater ductility than the conventionally heat treated ADI, due to its microstructure composed of ausferrite (regular ADI microstructure) and free (proeutectoid) allotriomorphic ferrite. Additionally, dual phase ADI could provide a wide range of mechanical properties as a function of the relative proportion of proeutectoid ferrite and ausferrite constituents, thereby replacing conventional ductile iron. Hence, dual phase ADI will be appropriate for new applications in the critical parts, where a combination of high strength and ductility is a pressing requirement. Furthermore, with the introduction of free ferrite into the matrix increased machinability of ADI could be achieved. The aim of this paper was to review works that have been published over the past years on the effects of process variables, mechanical properties, microstructure and applications of Dual phase Austempered Ductile Iron and to present some own results. This review is an attempt to assemble the results of the worldwide sudden increase in research and development that followed the announcement of the first production of the DP-ADI material.

Keywords: Ductile iron · Dual phase Austempered Ductile Iron · Microstructure · Mechanical properties

1 Introduction

In the world, there has been great interest in the processing and developing of austempered ductile irons (ADI). The ADI materials have a unique microstructure of ausferrite, produced by the heat treatment (austempering) of ductile irons. The ausferrite is a mixture of ausferritic ferrite and carbon enriched retained austenite [1].

N. Mitrovic et al. (Eds.): CNNTech 2019, LNNS 90, pp. 22–38, 2020.
https://doi.org/10.1007/978-3-030-30853-7_2

Due to this unique microstructure, the ADI materials have a remarkable combination of high strength, ductility and toughness together with good wear, fatigue resistance and machinability [1–4].

The combination of strength, ductility, fatigue resistance, machinability and wear resistance makes ADI a unique engineering material that may substitute steel (cast, forged and/or heat treated) or aluminum in applications where the strength/weight ratio is important [5–7].

During the last few years, there has been an increased interest in improving the strength of ductile iron by means of heat-treating to obtain dual phase microstructures that enhance the properties found in conventional Austempered Ductile Irons (ADI). Nowadays, a new type of ductile iron, called Dual Phase ADI (DP-ADI), has been developed to improve mechanical properties.

An important characteristic of a new ADI is the development of dual-phase microstructures by the appropriate heat treatment. This dual phase microstructure is obtained by intercritical annealing (partial austenitization) in the ($\alpha + \gamma +$ graphite) region whereby colonies of free (proeutectoid) ferrite are introduced. Partial austenitization is finally followed by austempering in the conventional temperature range 250–400 °C. As a result, depending on the temperature of austenitization, the matrix of the produced Dual Phase ADI contains different amounts of free ferrite and ausferrite. Compared to the conventional ductile irons, ADI with dual phase microstructure achieve a higher ratio of strength to ductility [8]. Therefore, additional research in this field is well motivated.

The emphasize of this paper was to review works that have been published over the past years on the effects of process variables, mechanical properties, microstructure and applications of Dual phase Austempered Ductile Iron.

Also, this review describes the own results realized through the research in the Department of Production Engineering, Faculty of Technical Sciences, University of Novi Sad.

2 Dual Phase Austempered Ductile Iron

2.1 Microstructure

The Dual Phase Austempered Ductile Iron (DP-ADI) is commonly referred to as a special ADI material due to unconventional austempering process. The curiosity of this ductile iron is its mixed microstructure, which is consisted of different amounts and morphologies of ausferrite (AF) and free (proeutectoid) allotriomorphic ferrite (FF). The microstructure of dual phase austempered ductile iron is shown in Fig. 1

The dual phase microstructure obtained during some special and unconventional austempering processes is shown in Fig. 1. It can be described as ausferrite matrix containing networks of allotropic free ferrite.

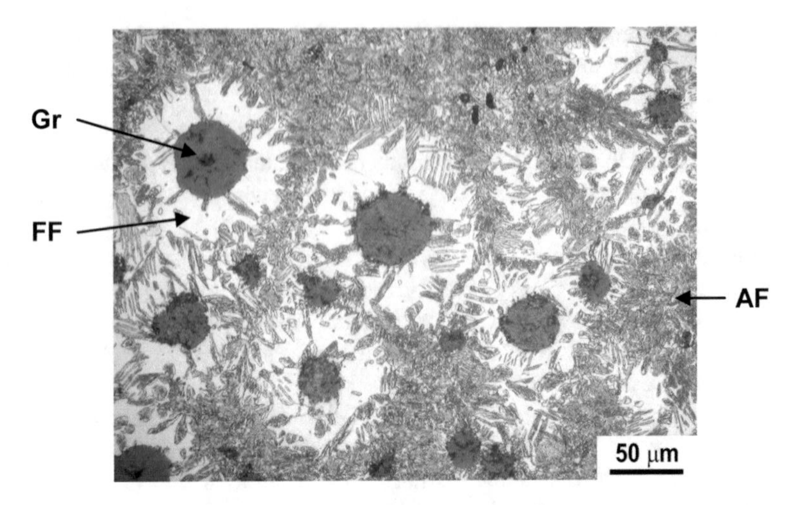

Fig. 1. Microstructure (light microscope) of a DP-ADI obtained after austenitization at 820 °C within intercritical interval; and austempering at 400 °C (AF – ausferrite, FF – free ferrite, Gr – graphite nodules)

The different thermal cycles are used to obtain dual phase ADI with different percentages of free ferrite and ausferrite in its microstructure. The different amounts of free ferrite and ausferrite have been confirmed to have improved fracture toughness compared with conventional ADI where "only" ausferrite phase is present [9].

2.2 Intercritical Heat Treatment in Dual Phase Ductile Iron

The simplest method to obtain dual phase microstructure in ADI seems to be that described by Basso et al. [10, 11] and Kilicli et al. [12, 13] which consists of an incomplete austenization stage at temperatures within the intercritical interval (apx. 750–860 °C), where austenite (γ), ferrite (α) and graphite (Gr) exist (Fig. 2).

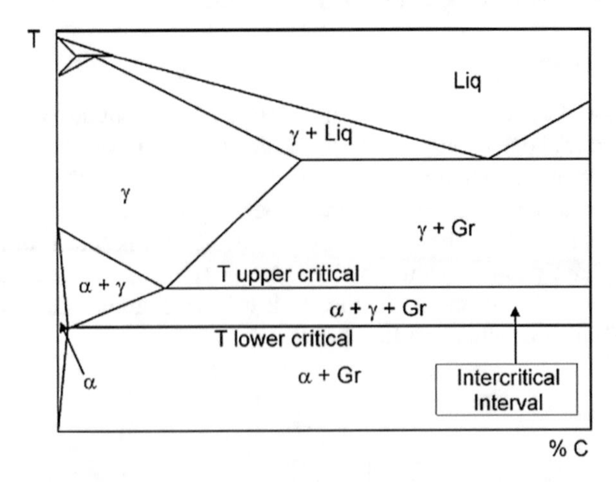

Fig. 2. Schematic pseudo binary Fe-C phase diagram at a constant amount of Si [10]

This region is called intercritical interval, and it is limited by the upper and lower critical temperatures. Such temperatures define the starting point at which ferrite transforms into austenite and austenite into ferrite in heating and cooling processes, respectively. The intercritical austenitization is followed by an austempering stage (apx 250–400 °C) in a salt bath, which is carried out to transform the austenite into ausferrite. This heat treatment yields microstructures with different morphologies and amounts of free ferrite and ausferrite, depending on the intercritical austenitizing temperature.

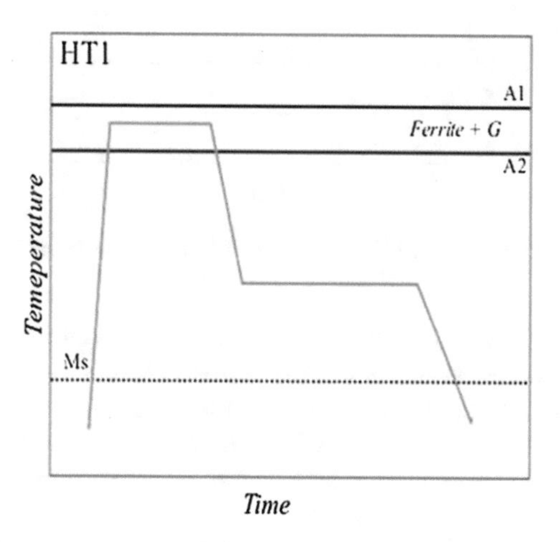

Fig. 3. Heat treatment HT1: partial austenitization [14]

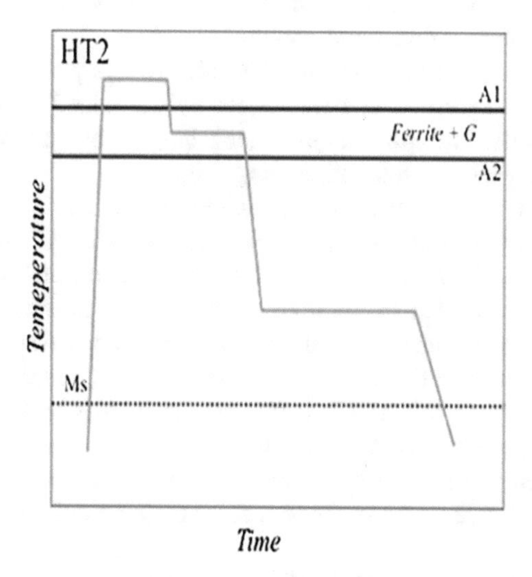

Fig. 4. Heat treatment HT2: complete austenitization, second partial austenitization [14]

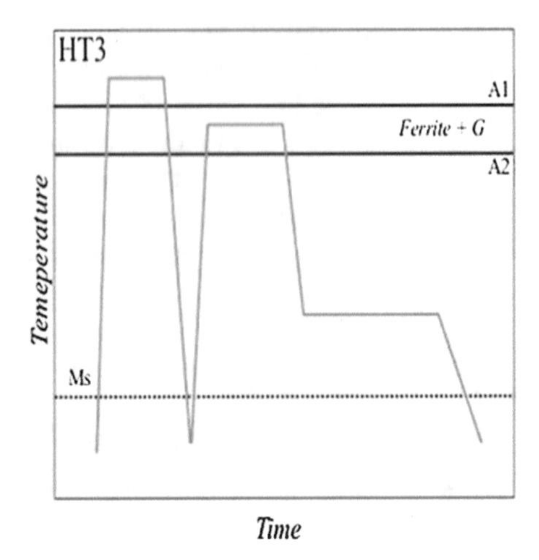

Fig. 5. Heat treatment HT3: a typical austenitization-quench cycle [14]

The 3 different types of processes of producing DP-ADI are shown in Figs. 3, 4 and 5, and further explained in the following text [14]:

HT1: is a heat treatment based on a partial austenization in the intercritical interval, defined by A1 and A2 temperatures, where the ferrite formation starts. Thereafter a quick quenching to the austempering temperature precedes the austempering dwell and cooling, Fig. 3.

HT2: is a two steps austenitization process where a complete austenization is performed above the upper critical temperature before a second austenization in the intercritical interval. Thereafter a quick quenching to the austempering temperature precedes the austempering dwell and cooling, Fig. 4.

HT3: In HT3 a typical austenization-quench cycle is performed before a partial austenization in the intercritical interval, defined by A1 and A2 temperatures, where the ferrite formation starts. Thereafter a quick quenching to the austempering temperature precedes the austempering dwell and cooling, Fig. 5.

The intercritical temperatures austenitization step is carried out to cause ferrite formation. The interval temperature is apx. 750 and 860 °C, for lower and upper temperature, respectively, however this directly depends on the ductile iron composition. It should be noted, that in the conventional ADI heat treatments the complete austenitization process is carried out above the upper intercritical temperature.

As discussed for conventional ADI, ausferrite phase optimization is a key issue on fracture toughness [7]. In DP-ADI it is also a considerable factor that must be optimized along with the ferrite volume amount.

In HT1, allotropic ferrite formed shows a dispersed formation within the matrix, whereas in HT2 the ferrite formation occurs at the grain boundaries of the recrystallized austenite.

The HT2 treatment was proved [15] to result in an interesting combination of strength and ductility, better than in the HT1. Depending on process variables, combinations of elongations over 24% and yield strength around 600 MPa could be obtained.

In HT3 the properties improvement are not so related with the allotropic ferrite site formation. The martensitic microstructure introduced by quenching [16] provides a large number of formation sites for the acicular ferrite to form, and thus a uniform fine ausferrite microstructure is obtained. Moreover, the stability of retained austenite in the final microstructure is increased due to the fact that alloying elements concentrate in the austenite phase during the holding in the ($\alpha + \gamma$) temperature range, showing lower subsequent carbide precipitation which also depends on the increment on the holding time within this range.

2.3 Mechanical Properties

The selection and application of the dual phase austempered ductile irons significantly depend on their properties. Taking into account this reason, the car suspension parts (high strength and elongation until failure and toughness), are made of dual phase austempered ductile iron [17].

However, in all the literature reviewed in relation to the dual phase ADI, it is indicated that the dual phase ADI can offer a wide range of the mechanical properties (tensile strength, yield stress, elongation until failure and hardness) depending on the relative percentage of ausferrite and free ferrite present in the microstructure [12, 18, 19]. Particularly, they found that the presence of 20% ausferrite in the microstructure increases yield and tensile strength (apx. 30%) as compared to fully ferritic ductile iron.

Valdes et al. [20] studied the mechanical properties of Dual phase ADI austempered at 375 °C with different percentages of ferrite and ausferrite in its microstructure. Especially, samples with 45% ausferrite and 65% ferrite yielded the best combination of strength and ductility.

The effect of several variables, such as the number and morphology of phases, austempered temperature and cast section size (or solidification rate), among others, on the final microstructure and mechanical properties was given by Basso et al. [10, 11]. The results show that as the amount of ausferrite increases, the yield and tensile strength increases also, while the elongation decreases for all the austempered temperatures used. The best combination of strength and elongation (UTS = 690 MPa, YS = 550 MPa, E = 22%) was obtained for samples austenitizated at 820 °C/1 h and subsequently austempered at 350 °C/1.5 h [10].

Fracture toughness tests have also revealed encouraging results when compared to those of fully ferritic or fully ausferritic matrices [21]. An increase of ausferrite in the microstructure promotes higher values of K_{IC}, up to the value found for fully ausferritic matrix. The relationship of $(K_{IC}/YS)^2$ yielded its highest values for dual phase ADI microstructure composed of free ferrite and less than 25% ausferrite. This was attributed to the encirclement effect of ausferrite located around last-to-freeze zones where small defects are present. The ausferrite acts as a reinforcing phase of the weakest zones, increasing fracture toughness.

Regarding the influence of section size on mechanical properties, strength and elongation decrease as the size of the piece increases, while yield strength remains unchanged [22].

Kilicli et al. [23] and Sahin et al. [18] reported that the mechanical properties of dual phase ADI on a wide range of microstructures composed of different ausferrite volume fractions and morphologies austempered at (365 °C). These microstructures showed, once again, that the yield stress and tensile strength increased when the quantity of ausferrite was incremented and exhibited higher ductility than the fully ausferritic ones.

Druschitz et al. [24] reported that dual phase ADI ("called machinable austempered cast iron or MADI") improved strength and ductility when compared to conventional ductile iron. The improved strength and ductility are attributable to the microstructure composed of a continuous matrix of equiaxed ferrite with islands of ausferrite.

Avishan et al. [27] studied the influence of depth of cut on machinability of an alloyed austemepred ductile iron. This study indicated that contrary to the behavior of many other metallic alloys, in astempered ductile irons reducing the depth of cut will not improve the machinability, or, vice-versa, the increase in the depth of cut will not affect the machinability.

However, in literature there is not currently any study on the machinability of ADI with dual phase microstructure. The proeutectoid fraction in the dual phase microstructure has very important effects on machinability. The machinability of ADI can be enhanced by controlling of proeutectoid ferrite and ausferrite volume fraction.

Ovalia and Mavib [28] investigated the influence of continuity of ausferrite microstructure along intercellural boundaries on machinability of austempered ductile iron. Their study also showed that dual phase austempered ductile iron can provide a wide range of machinabilities based on the relative percentage of ausferrite and proeutectoid ferrite present in the matrix. They found that austenitising temperature is an important factor in controlling the phase volume fractions and the surface roughness and primary cutting force.

Verdu et al. [22] indicated that tensile strength and yield stress increased when the amount of ausferrite was increased, while elongation and toughness exhibited the best values when the intercritical heating was carried out between (800–830 °C). The variation of tensile strength, yield stress, elongation as a function of V_{AU} for 300 C, 330 and 3507 °C austempering temperatures employed are plotted in Fig. 6. The increase in tensile strength and yield stress indicate that V_{AU} (the amount of ausferrite) plays an important role in the strength. In the same time there is a decrease of elongation for all the austempering temperatures when the amount of ausferrite increases. The optimal mechanical properties of dual phase ADI materials (tensile strength and yield stress and elongation) may be achieved if the process austempering is performed at 350 °C.

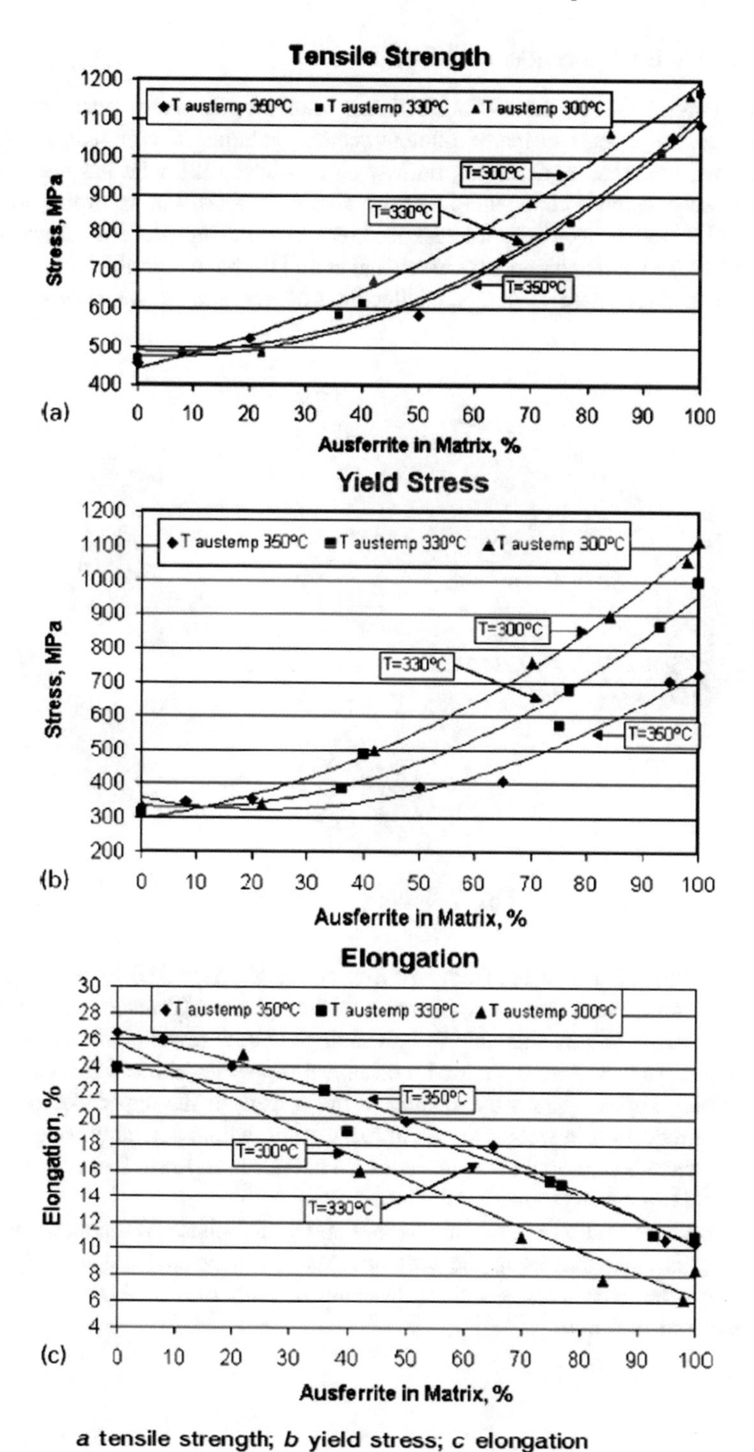

Fig. 6. Relationship between mechanical properties and amounts of ausferrite in the matrix [10]

2.4 Possible Engineering Application

Since its introduction in the 1970s, ADI has been used in many wear resistant and engineering components in many different sectors including automotive, trucks, construction, earthmoving, agricultural, railway and military [25]. ADI has been used for numerous suspension components in heavy goods vehicles, but according to Seaton and Li [26] there has been little interest until recently in using ADI for automotive and light truck suspension components. Aranzabal et al. [19] have described recent research to develop a dual phase austempered ductile iron for automotive suspension parts (Fig. 7).

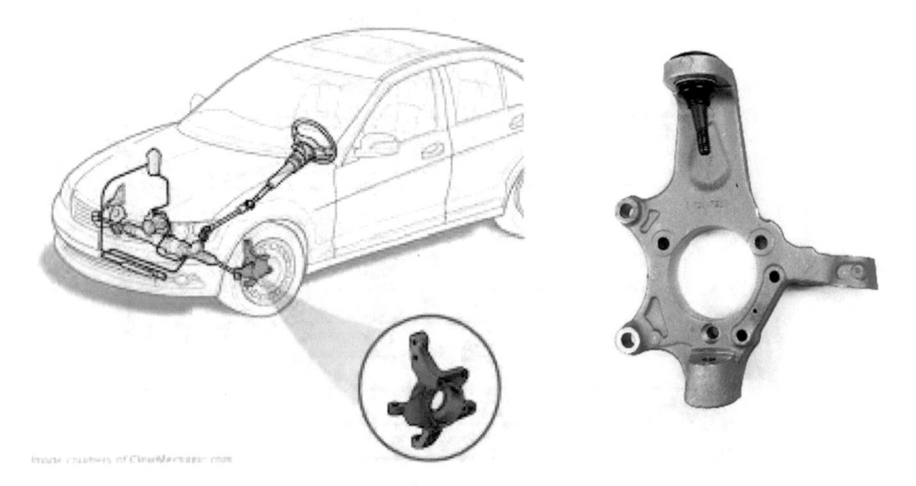

Fig. 7. Steering knuckle [25]

Steering knuckle fabricated from cast iron and dual-phase ADI materials is shown in Fig. 8. It can be seen that the application of dual-phase ADI results in dramatically increased strength and considerable weight savings in both raw material and machined weight of the product. Also, increased mechanical properties allowed for the alteration of the steering knuckle geometry, mainly in the respect of the upper element configuration. The reduction in weight has an influence on the final assembly weight savings and also on the whole product weight savings. That means, a lower fuel consumption is expected and a greener product is obtained.

Suspension control arm made of cast iron and a dual-phase ADI material is shown in Fig. 9. It is clear that as in the previous case (steering knuckle, Fig. 8), weight reduction can be expected,, as well as mechanical properties of ADI that are nearly double those of cast iron.

GH 60-38-10 Pearlitic – ferritic cast iron	
Rp02 (MPa)	370
Rm (MPa)	590
A (%)	10

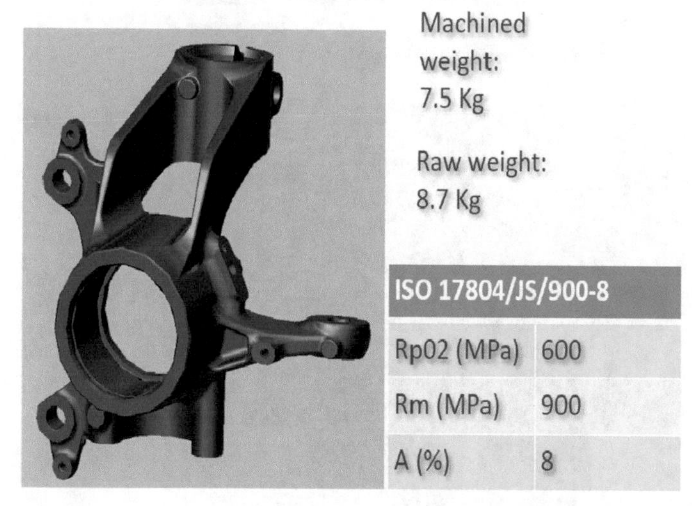

ISO 17804/JS/900-8	
Rp02 (MPa)	600
Rm (MPa)	900
A (%)	8

Fig. 8. Cast iron and dual-phase ADI knuckle weights and mechanical properties [25]

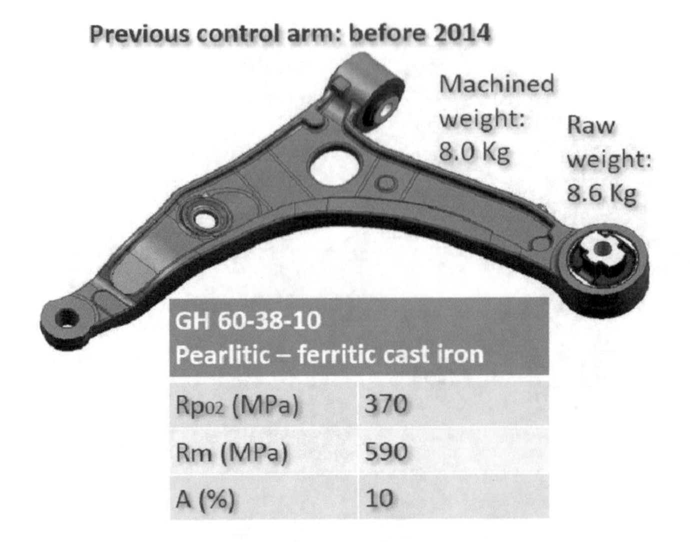

GH 60-38-10 Pearlitic – ferritic cast iron	
Rp02 (MPa)	370
Rm (MPa)	590
A (%)	10

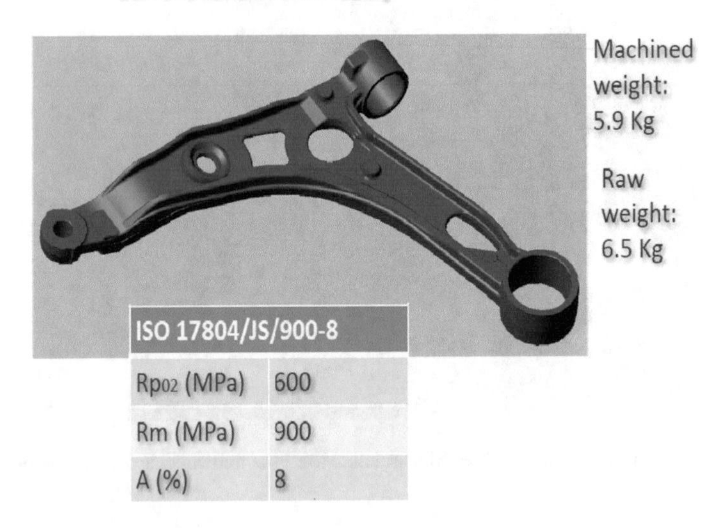

ISO 17804/JS/900-8	
Rp02 (MPa)	600
Rm (MPa)	900
A (%)	8

Fig. 9. Cast iron and dual-phase ADI material suspension control arm weights and mechanical properties [25]

3 Experimental Work

3.1 Melt

Ductile iron keel blocks with a chemical composition in wt%: 3.53C; 2.53Si; 0.347Mn; 0.045Cu; 0.069Ni; 0.055Mg; 0.031P; 0.015S; were produced in a commercial electro induction foundry furnace. The melt was poured from about 1420 °C into a standard 25 mm-thick Y-block-shaped sand moulds (ASTM A897M-06), which ensured sound castings. Round and prismatic samples were cut from the Y-blocks and used to prepare test specimens.

3.2 Heat Treatments

Four different Dual Phase ADI microstructures, containing different relative quantities of free ferrite and ausferrite were obtained by heat treatment, consisting of a partial austenitization of the test specimens, by holding them into the furnace within the intercritical interval at temperatures of 840 °C, 820 °C, 800 °C and 780 °C for two hour, followed by an austempering step in a salt bath at 400 °C for 1 h, in every case. For comparison, conventional ADI samples were obtained by means of a heat treatment comprising a complete austenitization stage in a protective argon atmosphere at 900 °C for 2 h, austempering at 400 °C for 1 h.

The mechanical properties (ISO 6892) of as-cast ductile iron was YS = 326 MPa, UTS = 473 MPa, E = 22.2%, HV10 = 164, while conventional, fully ausferritic ADI (900 °C/2 h + 400 °C/1 h) has YS = 757 MPa, UTS = 1042 MPa, E = 14.2%, HV10 = 296.

The Vickers hardness HV10 (ISO 6507) was determined for dual phase ADI materials on n a machine HPO 250, WEB Leipzig, with a test load of 98.07 N (10 kg) and a dwell time of 15 s. The average of the five hardness measurements was reported.

Standard metallographic preparation techniques (mechanical grinding and polishing followed by etching in Nital) were applied prior to light microscopy (LM) examinations. A "Leitz-Orthoplan" metallographic microscope was used for microstructure characterization.

The relationship between the amount (% in volume) of free ferrite, ausferrite, and graphite in Dual Phase ADI microstructures was quantified by JMicroVision software. The reported values are the average of at least five fields of view on each sample.

4 Results and Discussions

4.1 Material Characteristics

The morphology of graphite nodules in the microstructure of the as-cast ductile iron (Fig. 10a and b) is fully spherical, with the nodule count from 150 to 200 nodules/mm^2. Spheroidization is evident (>90%) with an average nodule size of 25 μm to 30 μm (Fig. 10a). The microstructure of the as-cast material is mostly ferritic, over 90% (Fig. 10b).

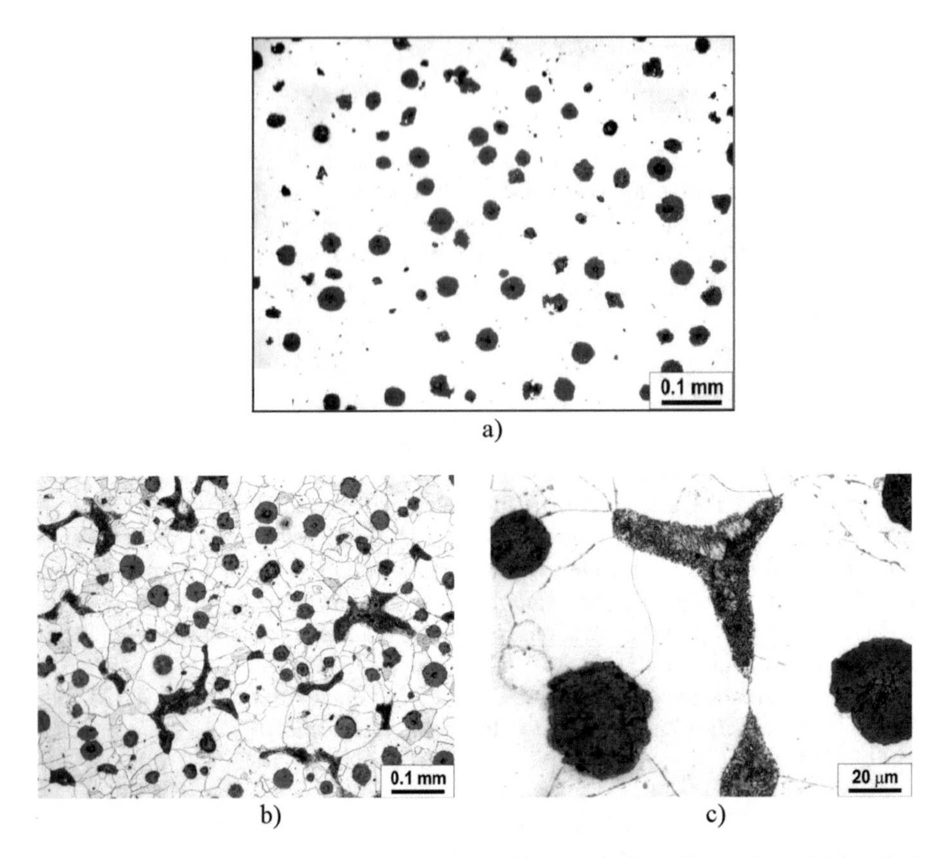

Fig. 10. LM Microstructure of as-cast specimen: (a) un-etched specimen; (b) and (c) etched specimen

Figure 11 illustrates the microstructures obtained for all samples austenitized at different temperatures within the intercritical interval and austempered at 400 °C/1 h.

It is obvious that the heat treatment cycle allowed the attainment of microstructures composed of different percentages of ausferrite and free ferrite depending on the austenitizing temperature.

The relative amount of free ferrite increase with decrease of austenitization temperature from 18.4 to 78.8%, while amount of ausferrite decrease from 71.2 to 9.2%.

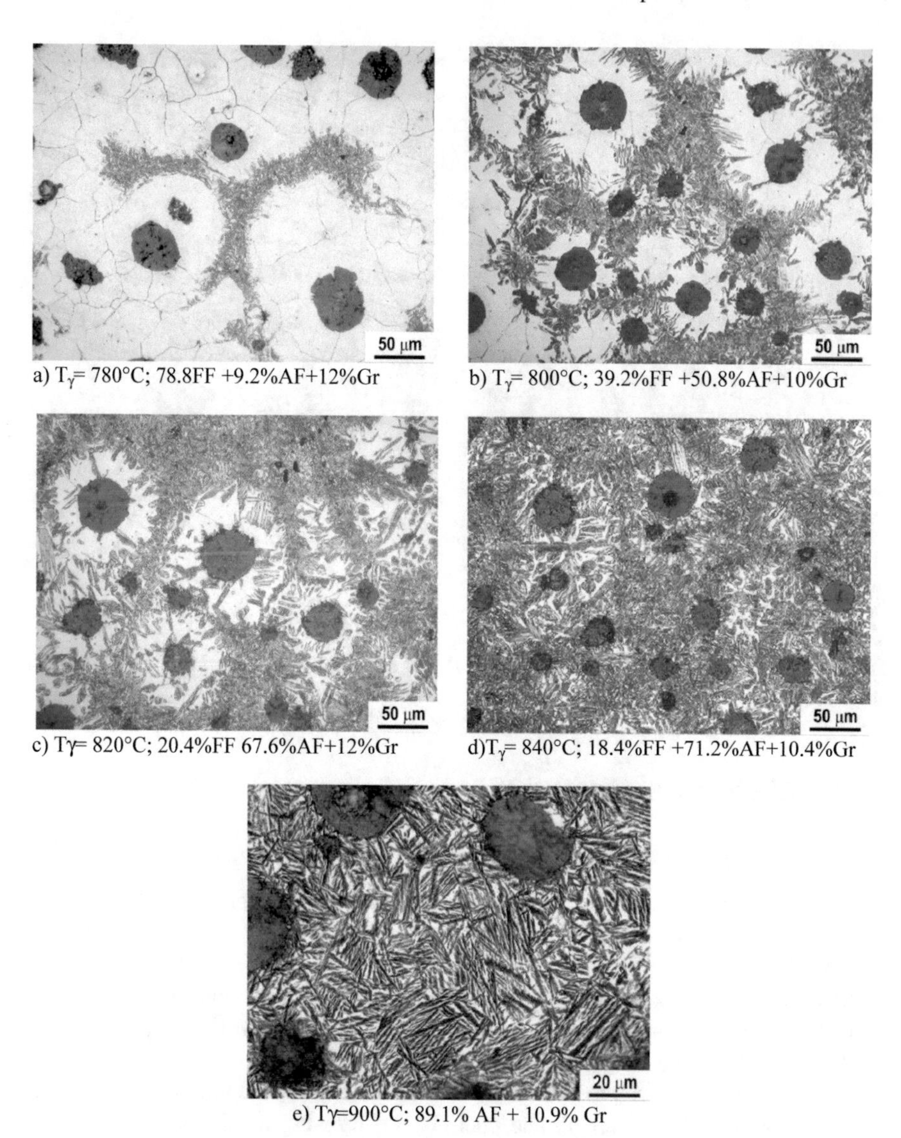

a) T_{γ}= 780°C; 78.8FF +9.2%AF+12%Gr

b) T_{γ}= 800°C; 39.2%FF +50.8%AF+10%Gr

c) Tγ= 820°C; 20.4%FF 67.6%AF+12%Gr

d)T_{γ}= 840°C; 18.4%FF +71.2%AF+10.4%Gr

e) Tγ=900°C; 89.1% AF + 10.9% Gr

Fig. 11. Microstructure obtained for samples austenitized at different austenitization temperatures Tγ within intercritical interval; austempering temperature 400 °C/1 h in all cases

The variation of the hardness HV10 values for samples austenitized at 840 °C, 820 °C, 800 °C and 780 °C as a function of final percentages of the matrix micro-constituents is plotted in Fig. 12. The hardness values decreases from 300 to 178 HV10 with an increase of free ferrite in the matrix indicating the increase in ductility but the decline of the tensile strength.

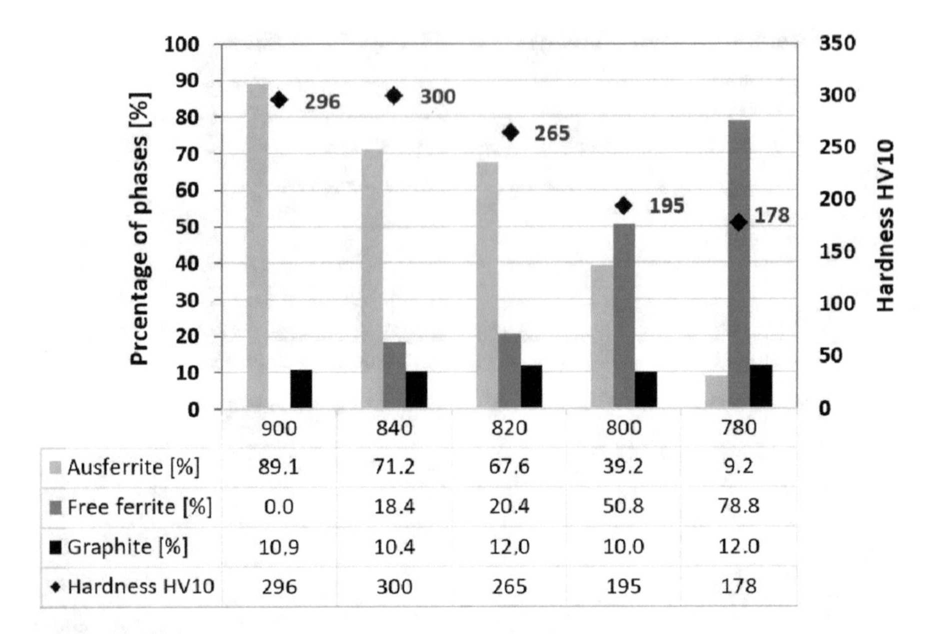

	900	840	820	800	780
Ausferrite [%]	89.1	71.2	67.6	39.2	9.2
Free ferrite [%]	0.0	18.4	20.4	50.8	78.8
Graphite [%]	10.9	10.4	12.0	10.0	12.0
Hardness HV10	296	300	265	195	178

Fig. 12. Effect of microconstituents percentages on the values hardness of dual phase austempered ductile iron

5 Concluding Remarks

On the basis of previous considerations, the following concluding remarks can be done:

- It is shown that dual phase ADI could provide a wide range of mechanical properties as a function of the relative proportion of free ferrite and ausferrite constituents,
- Dual-phase ADI materials have a considerable potential for replacing conventional ductile iron and conventional ADI materials in vehicles suspension parts, some of them being steering knuckles and control arms.
- These materials are appropriate for new applications in the critical parts, where a combination of high strength and ductility is a pressing requirement.

Acknowledgment. The authors gratefully acknowledge research funding from the Ministry of Education, Science and Technological Development of The Republic of Serbia under grant number TR34015.

References

1. Sidjanin, L., Smallman, R.E., Young, J.M.: Electron microstructure and mechanical properties of silicon and aluminium ductile irons. Acta Metallurgica Materialia **42**(9), 3149 (1994)

2. Eric, O., Rajnovic, D., Sidjanin, L., Żec, S., Jovanovic, T.: An austempering study of ductile iron alloyed with copper. J. Serb. Chem. Soc. **70**(7), 1015–1022 (2005)
3. Sidjanin, L., Rajnovic, D., Eric, O., Smallman, R.E.: Austempering study of unalloyed and alloyed ductile irons. Mater. Sci. Technol. **26**(5), 567 (2010)
4. Eric, O., Rajnovic, D., Zec, S., Sidjanin, L., Jovanovic, T.: Microstructure and fracture of alloyed austempered ductile iron. Mater. Charact. **57**, 211 (2006)
5. Rajnovic, D., Eric, O., Sidjanin, L.: Transition temperature and fracture mode of as-cast and austempered ductile iron. J. Microsc. **232**(3), 605 (2008)
6. Sidjanin, L., Smallman, R.E.: Metallography of bainitic transformation in austempered ductile iron. Mater. Sci. Technol. **8**(12), 1095 (1992)
7. Eric, O., Sidjanin, L., Miskovic, Z., Zec, S., Jovanovic, T.: Microstructure and toughness of CuNiMo austempered ductile iron materials. Mater. Lett. **58**, 2707–2711 (2004)
8. ASM Handbook: Properties and Selections: Irons and Steel, vol. 1, 9th edn., p. 33. ASM, Ohio (1992)
9. Basso, A., Caldera, M., Massone, J.: Development of high silicon dual phase austempered ductile iron. ISIJ Int. **55**(5), 1106–1113 (2015)
10. Basso, A.D., Martinez, R.A., Sikora, J.: Influence of austenitising and austempering temperatures on microstructure and properties of dual phase ADI. Mater. Sci. Technol. **23** (11), 1321–1326 (2007)
11. Basso, A., Martinez, R., Sikora, J.: Influence of section size on dual phase ADI microstructure and properties: comparison with fully ferritic and fully ausferritic matrices. Mater. Sci. Technol. **25**(10), 1271 (2009)
12. Kilicli, V., Erdogan, M.: Tensile properties of partially austenitised and austempered ductile irons with dual matrix structures. Mater. Sci. Technol. **22**(8), 919 (2006)
13. Erdogan, M., Kilicli, V., Demir, B.: The influence of the austenite dispersion on phase transformation during the austempering of ductile cast iron having a dual matrix structure. Int. J. Mater. Res. **99**(7), 751–760 (2008)
14. Hidalgo, G.J.: Improving the fracture toughness of dual phase austempered ductile iron. Universidad Carlos III De Madrid, 2008(5) Master thesis: 27/2008, Department of Materials and Manufacturing Technology Chalmers University of Technology, SE-412 96, Göteborg, Sweden (2009)
15. Basso, A., Martínez, R., Sikora, J.: Analysis of mechanical properties and its associated fracture surfaces in dual-phase austempered ductile iron. Fatigue Fract. Eng. Mater. Struct. **36**(7), 650–659 (2013)
16. Kobayashi, T., Yamada, S.: Effect of Holding Time in the (alfa+gamma) Temperature Range on Toughness of Specially Austempered Ductile Irons. Metall. Mater. Trans. A **27**(7), 1961–1971 (1996)
17. Basso, A., Caldera, M., Chapetti, M., Sikora, J.: Mechanical characterization of dual phase austempered ductile iron. ISIJ Int. **50**(2), 302–306 (2010)
18. Sahin, Y., Erdogan, M., Kilicli, V.: Wear behaviour of austempered ductile irons with dual matrix structures. Mater. Sci. Eng., A **444**, 31–38 (2007)
19. Aranzabal, J., Serramoglia, G., Rousiere, D.: Development of a new mixed (ferritic-ausferritic) ductile iron for automotive suspension parts. Int. J. Cast Metals Res. **16**(1–3), 185–190 (2003)
20. Valdes, E.- C., Perez Lopez, M.J., Figueroa, M., Ramirez, L.E.: Austempered ductile iron with dual matrix structures. Revista Mexicana de Fisica **S55**(1), 48–51 (2009)
21. Basso, A., Sikora, J.: Review on production processes and mechanical properties of dual phase austempered ductile iron. Int. J. Met. Cast. **6**, 7–14 (2012)
22. Verdu, C., Adrien, J., Reynaud, A.: Contributions of dual phase heat treatments to fatigue properties of SG cast irons. Int. J. Cast Met. Res. **18**(6), 346–354 (2005)

23. Kilicli, V., Erdogan, M.: Effect of ausferrite volume fraction and morphology on tensile properties of partially austenitised and austempered ductile irons with dual matrix structures. Int. J. Cast Metals Res. **20**(4), 202–214 (2007)
24. Druschitz, A.P., Fitzgerald, D.C.: Machinable austempered cast iron article having improved machinability, fatigue performance and resistance to environmental cracking and a method of making the same. U.S. Patent No. 7,070,666 (2006)
25. Harding, R.A.: The production, properties and automotive applications of austempered ductile iron. Kov. Mater. **45**, 1–16 (2007)
26. Seaton, P.B., Li, X.M.: In: Proceedings 2002 World Conference on ADI. Louisville, KY, Ductile Iron Society and the American Foundry Society 2002, FC02CD on CD ROM, p. 129 (2002)
27. Avishan, B., Yazdani, S., Vahid, D.J.: The influence of depth of cut on the machinability of an alloyed austempered ductile iron. Mater. Sci. Eng., A **523**, 93–98 (2009)
28. Ovalia, I., Mavib, A.: Investigating the machinability of austempered ductile irons with dual matrix structures. Int. J. Mater. Res. **103**, 1–7 (2012)

The Pyrolysis of Waste Biomass Investigated by Simultaneous TGA-DTA-MS Measurements and Kinetic Modeling with Deconvolution Functions

Nebojša Manic[1(✉)], Bojan Jankovic[2], Vladimir Dodevski[3], Dragoslava Stojiljkovic[1], and Vladimir Jovanovic[1]

[1] Faculty of Mechanical Engineering, Fuel and Combustion Laboratory, University of Belgrade, Kraljice Marije 16, P.O. Box 35, 11120 Belgrade, Serbia
nmanic@mas.bg.ac.rs

[2] Department of Physical Chemistry, Institute of Nuclear Sciences “Vinča”, University of Belgrade, Mike Petrovića Alasa 12-14, P.O. Box 522, 11001 Belgrade, Serbia

[3] Laboratory for Materials Sciences, Institute of Nuclear Sciences “Vinča”, University of Belgrade, Mike Petrovića Alasa 12-14, P.O. Box 522, 11001 Belgrade, Serbia

Abstract. As waste biomass from fruit processing industry, apricot kernel shells have a potential for conversion to renewable energy through a thermochemical process such as pyrolysis. However, due to major differences of biomass characteristics as the well-known issue, it is extremely important to perform detailed analysis of biomass samples from the same type (or same species) but from different geographical regions. Regarding full characterization of considered biomass material and to facilitate further process development, in this paper, the advanced mathematical model for kinetic analysis was used. All performed kinetic modeling represents the process kinetics developed and validated on thermal decomposition studies using simultaneous thermogravimetric analysis (TGA) – differential thermal analysis (DTA) – mass spectrometry (MS) scanning, at four heating rates of 5, 10, 15 and 20 °C min^{-1}, over temperature range 30–900 °C and under an argon (Ar) atmosphere. Model-free analysis for base prediction of decomposition process and deconvolution approach by Fraser-Suzuki functions were utilized for determination of effective activation energies (E), pre-exponential factors (A) and fractional contributions (φ), as well as for separation of overlapping reactions. Comparative study of kinetic results with emission analysis of evolved gas species was also implemented in order to determine the more comprehensive pyrolysis kinetics model. Obtained results strongly indicated that the Fraser-Suzuki deconvolution provides excellent quality of fits with experimental ones, and could be employed to predict devolatilization rates with a high probability. From energy compensation effect properties, it was revealed the existence of unconventional thermal lag due to heat demand by chemical reaction.

Keywords: Waste lignocellulosic biomass · Pseudo-components · Kinetics · Pyrolysis · Fraser-Suzuki deconvolution · Unconventional thermal lag

N. Mitrovic et al. (Eds.): CNNTech 2019, LNNS 90, pp. 39–60, 2020.
https://doi.org/10.1007/978-3-030-30853-7_3

1 Introduction

In order to meet energy demands and environmental principles, especially in developing countries, it is necessary to use agricultural residues (as a waste lignocellulosic biomass) available in large quantities for the production of value-added products [1]. The major bottlenecks of using this type of biomass as a renewable energy source are the imparity of the characteristics, forms and availability. Taking into mind that there are major differences in the characteristics of biomass (from different geographical regions), it is necessary to consider all technical and technological possibilities of application as well as environmental impact for each concrete case. Therefore, if agricultural residues are used as a raw material, it is necessary to consider all inequalities of characteristics and shapes and to perform detailed analysis of the samples belonging to the same biomass group but from different areas [2–4]. Also, it is well known that chemical energy from biomass can be transformed into heat and/or electrical energy by application of various processes of thermal conversion of biomass such as combustion, gasification and pyrolysis [5]. The advantage of pyrolysis in relation to the combustion process is that this process represents a more flexible and efficient process of thermal conversion of biomass chemical energy, the production of biofuels and numerous chemical compounds [6, 7]. According to the larger number of possible products obtained by the pyrolysis process, it is also extremely important to analyze in detail the raw material in order to efficiently perceive all the aspects and products of this thermochemical conversion. Additionally, it is possible to define important process parameters in order to use the specific raw material in the most efficient way.

This paper focuses on the thermal conversion study related to the slow pyrolysis devolatilization kinetics of the waste lignocellulosic biomass that reflects its energy characteristics. For this purpose, the waste lignocellulosic biomass, the apricot (Prunus armeniaca) kernel shells was used. In contrast, the pyrolysis of apricot kernel shells (with heating rates of 10–50 °C min^{-1} and in static atmosphere incorporating sweep gas flow rates of 50–200 cm^3 min^{-1}) (Turkey apricot variety) was performed in the literature, but in order to determine the main characteristics and quantities of liquid and solid products [8]. The work was focused on the effects of pyrolysis temperature, heating rate, and sweeping gas flow rate on the product yields, as well as to the characterization of the obtained bio-oil, and the characterization of the chars as solid products for possible use of solid fuels or activated carbon. Namely, most of the studies related to stone-fruiting waste lignocellulosic biomass are focused on the char, the solid, residual product of pyrolysis, which could be harnessed for energy production like coal or for production of low-cost adsorbents [9–11].

The main goal of this research represents showing the most probable devolatilization kinetics during slow pyrolysis process of the raw apricot kernel shell samples, which is monitored by simultaneous TGA (Thermogravimetric analysis) – DTA (Differential thermal analysis) techniques, coupled with MS (Mass spectrometry) technique, for analyzing the gas products at the various heating rates, in a dynamic, the linear regime of heating. The description of devolatilization kinetics for apricot kernel shells during slow pyrolysis was made possible by application and comparison of multicomponent modeling using combined "model-free" kinetics, multi-dimensional nonlinear regression analysis and Fraser-Suzuki deconvolution procedure performed in MATLAB software program. The "model-free" methods [12–17] were used in order to check the degree of complexity of the process (one-step or multi-step kinetics — assuming that the pyrolysis process of the waste lignocellulosic biomass is complex caused by the interaction of devolatilization, diffusion effect, catalyst, and secondary reactions [18]), while implemented the objective function (OBF) minimization procedure in multicomponent nonlinear regression and Fraser-Suzuki deconvolution provides reliable determination of kinetic parameters, forming the entire pyrolysis model. Presented approach is capable - of working on multi-step reactions, involved in thermal decomposition of assorted biomass where overlapping mechanisms occur.

2 Materials and Methods

The apricot kernel shells as waste lignocellulosic biomass residues originating from a local fruit processing plant were used. The sample preparation was done according to a standard procedure defined by ISO 14780:2017. Prepared sample was used for ultimate (carbon, hydrogen and nitrogen content according to ISO 16948:2015 and proximate analyses (total moisture, ash, volatile matter and char content according to ISO 17225-1:2004 and relevant standards) and for the simultaneous thermal analysis (STA) coupled with mass spectrometry (MS) measurements.

The STA experimental tests were done by NETZSCH STA 445 F5 Jupiter system under the following conditions:

- sample mass: 5.0 ± 0.3 mg
- temperature range: from room temperature up to 900 °C
- the heating rate: 5, 10, 15 and 20 °C min^{-1}
- carrier gas: argon (the high purity – Class 5.0)
- the total carrier gas flow rate: 30 mL min^{-1}

The evolved gases from STA was identified simultaneously using the quadrupole mass spectrometer NETZSCH QMS 403 D Aëolos (QMS). Evolved gas composition was monitored whereby atomic mass units (amu) were scanned in the range from 1 to 80. Accordingly, further analysis and discussion cover the molecules with amu of 2, 16, 18, 28, 30, 44, 58 and 72 which correspond to H_2, CH_4, H_2O, CO, C_2H_6, C_3H_8 (CO_2), C_4H_{10} and C_5H_{12} respectively, which are commonly considered as evolving gasses from biomass raw material.

3 Theoretical Background

Typical kinetic analysis of the thermal decomposition of lignocellulosic materials under the non-isothermal conditions is usually written in a form of following rate-law equation:

$$\frac{d\alpha}{dt} \equiv \beta \cdot \left(\frac{d\alpha}{dT}\right) = A \cdot \exp\left(-\frac{E}{RT}\right) \cdot f(\alpha) \tag{1}$$

where α is the conversion degree which is dimensionless quantity (the conversion degree increases from 0 to 1 during the process and reflects the overall progress of the reactant transition into the products), dα/dt is the rate of the process, T is the absolute temperature, β = dT/dt is the heating rate, A is the pre-exponential factor, E is the apparent (effective) activation energy, R is the gas constant (8.314 J mol^{-1} K^{-1}), and f (α) is the conversion (reaction model) function (a solid state reaction model that depends on the controlling mechanism and conversion degree). In the literature, a large number of reaction models can be found, however, they can be resolved by isoconversional ("model free") method approach which could provide the initial decomposition pattern of the selected material [19]. This isoconversional analysis allows determination of Ea in a function of conversion degrees (α), without considering the reaction model, f(α). This group of methods most often involves performing the experiments at the different heating rates, whereby the kinetic parameters can be calculated for each selected conversion degree value, assuming that the reaction rate at a constant α's is only a function of the reaction temperature. In this research performed "model-free" methods were based on Friedman (FR) [12, 13], Kissinger-Akahira-Sunose (KAS) [14, 15] and Ozawa-Flynn-Wall (OFW) methods [16, 17].

The more comprehensive approach for kinetic analysis of thermal decomposition of biomass presented in the literature is related to multicomponent kinetic modeling, which compromise the identification of each pseudo-component or each stage in biomass material decomposition process, by application of conversion rate curve fitting together with statistical analysis [20]. Overall sample conversion rate with n-th reaction order model as conversion function can be expressed as:

$$\frac{d\alpha}{dt} = \sum_{i=1}^{4} \frac{d\alpha_i}{dt} = \sum_{i=1}^{4} \varphi_i \cdot A_i \cdot \exp(-\frac{E_i}{RT}) \cdot (1 - \alpha_i)^{n_i} \tag{2}$$

where dα/dt is the overall conversion rate, T is the temperature, φ_i is contribution of i-th reaction component, A_i is the pre-exponential factor related to considered i-th reaction component, E_i is the apparent activation energy related to i-th reaction component, while n_i represents the reaction order attached to i-th reaction component; αi is the partial conversion fraction.

In this study, the deconvolution approach for describing the complex conversion rate curves was applied. The peak deconvolution method was performed using Fraser-Suzuki function, as an asymmetrically skewed Gaussian with 4-parameters for each heating rate, and this function is capable for efficient separation of overlapping reactions in a complex pyrolytic scheme. Overall relative conversion rate with application of Fraser-Suzuki deconvolution function can be defined as:

$$\left(\frac{d\alpha}{dT}\right)_{overall} = \sum_{i=1}^{4} \frac{d\alpha_i}{dT} = \sum_{i=1}^{4} \left(h_i \cdot exp[-\frac{\ln 2}{s_i^2} \cdot [\ln(1 + 2 \cdot s_i \cdot \frac{(T - p_i)}{w_i}]^2]\right), \tag{3}$$

where $d\alpha/dT$ is the relative conversion rate, T is the temperature and hi, s_i, pi and w_i are the function parameters [21–23]. Overview of the function parameters for Fraser-Suzuki deconvolution procedure used in this study for each considered heating rate was presented in Table 1.

Table 1. The function parameters for Fraser-Suzuki deconvolution procedure

Heating rate	5 °C min^{-1}			
Parameter	1	2	3	4
h [°C $^{-1}$]	0.028	0.0395	0.0037	0.0019
p [°C]	275	335	415	470
w [°C]	70	80	68	500
s [-]	−0.20	−0.27	−0.45	−0.25
Heating rate	10 °C min^{-1}			
Parameter	1	2	3	4
h [°C $^{-1}$]	0.058	0.0772	0.0063	0.0045
p [°C]	284	345	429	470
w [°C]	77	80	65	500
s [-]	−0.29	−0.27	−0.39	−0.25
Heating rate	15 °C min^{-1}			
Parameter	1	2	3	4
h [°C^{-1}]	0.090	0.1145	0.0078	0.0070
p [°C]	288	348	430	473
w [°C]	70	80	68	500
s [-]	−0.20	−0.27	−0.45	−0.25
Heating rate	20 °C min^{-1}			
Parameter	1	2	3	4
h [°C $^{-1}$]	0.131	0.159	0.011	0.010
p [°C]	293	354	434	457
w [°C]	70	80	68	500
s [-]	−0.20	−0.27	−0.45	−0.25

Based on these function parameters, the position of functions are all positive ($p > 0$), where at the various heating rates, the function widths (w) are different suggesting on the different reactivities of separated reactions. The heights of the functions (h) were also differing at all heating rates where intensities of reaction evaluation are differing from one to another. In addition, all functions are left-skewed (where $s < 0$ at all considered heating rates). The presented values suggest the existence of four separated reactions involved in the pyrolysis process of apricot kernel shells.

However, kinetics cannot reveal true chemical components in a reactive system, so we can operate with "pseudo-components" that we can interpret afterward based on knowledge from other sources (e.g. "mainly lignin decomposition"). Pseudo-component is the sum of those decomposing species that can be described by the same reaction kinetic parameters in a given model.

4 Results and Discussion

Table 2 shows the results of the ultimate and proximate analysis of the raw apricot kernel shell sample studied in this work.

Table 2. Results of ultimate and proximate analysis of raw apricot kernel shell samples.

Proximate analysis (wt%)		Ultimate analysis[b] (wt%)	
Moisture	9.71	C	46.88
Volatile matter	73.84	H	6.38
Fixed carbon	15.51	O[c]	45.45
Ash	0.94	N	0.25
HHV (MJ kg^{-1})	20.26	S	0.00
LHV[a] (MJ kg^{-1})	18.72		

[a]Calculated according to Ref. [23].
[b]On a dry basis.
[c]By the difference.

From obtained results (Table 2), the volatile matter content of apricot kernel shell studied shows a high value (73.84%) which indicated that tested material produces more tar (the tars include any of several high molecular weight products that are volatile at pyrolyzing temperatures, but condense onto any surface near room temperature), having an impact on a lower percentage of fixed carbon. In this content, the fixed carbon content can be related to calorific value, since it has a positive effect on the energy potential of biomass. However, it should be noted that high volatile matter content does not guarantee high calorific value since some of the ingredients of volatile matter are formed from non-combustible gases such as CO_2 and H_2O. For considered sample, the content of volatile matter (Table 2) is close to volatile matters contents of red lentil hull (74.73%), broad been husk (74.88%) and the pea stem (74.67%) [24]; the obtained value of volatile matter content is typical for herbaceous plant biomasses. As a result of fairly high volatile matter content may have consequences in the large quantities of gaseous products released during the process. In addition, the fixed carbon content (Table 2) is very close to those related to coconut shell (15.97%), almond shell (14.82%) and apple pulp (15.13%) [24]. However, the obtained content of fixed carbon of 15.51% suggests that apricot kernel shell can be used to obtain the coke residue (the solid bio-char), which would also indicate on the material with good energy characteristics. On the other hand, the ash content of 0.94% (Table 2) is almost between the ash contents of elaeagnus (0.90%) and peach stone (0.97%) [24].

The obtained result is in very good agreement with ash contents related to the flowering plant biomass, just like apricot. The HHV for apricot kernel shell (Table 2) almost coincides with HHV value related to coconut shell (20.24%) and peach (Prunus persica L.) stones (20.50%) [24, 25], so the estimated value belongs to those HHV values that have heat values greater than some known agricultural wastes, and falls within the limit for production of steam in electricity generation [26].

Also, the moisture content can have different effects on the pyrolysis product yields depending on conditions. The obtained value of apricot kernel shell moisture content (9.71%, Table 2) is lower than those typical for the wood moisture content (which appears in the range of 15–20%) [6]. Therefore, the moisture in the reaction affects the char properties and this has been used to produce activated carbons, through the pyrolysis process of biomass. So, based on the estimated moisture content value (Table 2), it was obtained a fairly dry biomass feedstock (around 10%), and this is an acceptable value considering that slow pyrolysis is more tolerant for the moisture, since the main issue is related on its effect on the process energy requirements.

From the ultimate analysis, it can be noticed a dominant contribution of carbon (46.88%) and oxygen (45.45%) (Table 2 and this is in good agreement with the typical ranges of these elements in most biomass samples (40–60% for C, and around 45% for O) [27].

Simultaneous TGA (thermogravimetric analysis) – DTG (derivative thermogravimetry) curves in an argon (Ar) atmosphere at the heating rates of β = 5, 10, 15 and 20 °C min^{-1} are shown on the Fig. 1a–d.

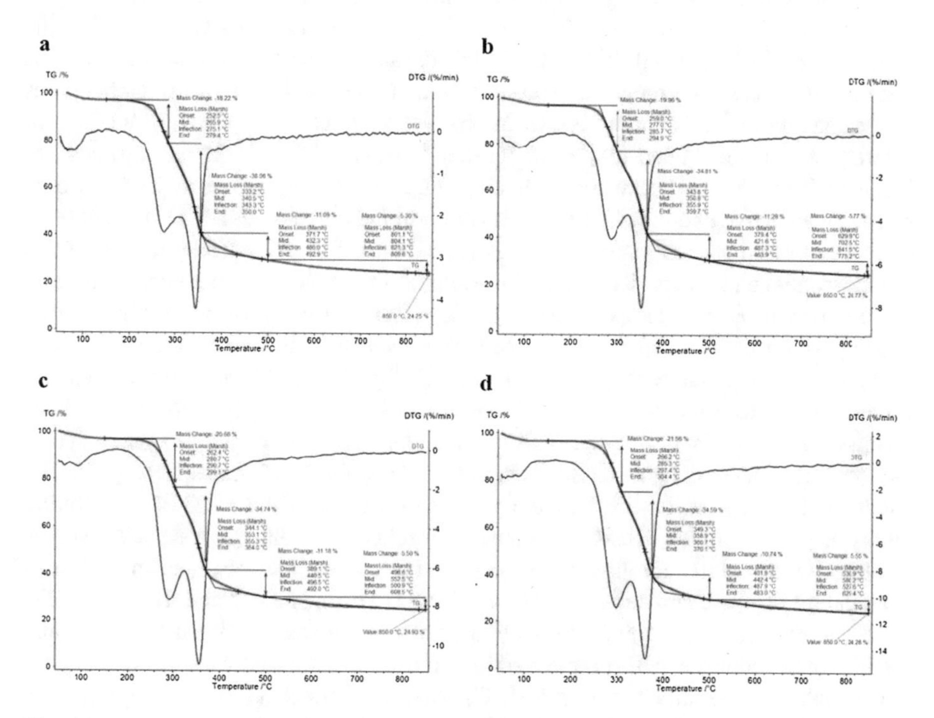

Fig. 1. (a–d) Simultaneous TGA-DTG curves in an argon (Ar) atmosphere at β = 5, 10, 15 and 20 °C min^{-1} for apricot (*Prunus armeniaca*) kernel shell pyrolysis

At all observed heating rates, after the mass loss (TG-%) up to the temperature of approximately 120 °C which can be attributed to removing of moisture (in the range of

3.00–3.41% and this amount of mass loss was recorded due to loss of water present in the sample and external water bound by surface tension), the non-isothermal decomposition of apricot kernel shell occurs through four successive processes accompanied by designated mass losses in Fig. 1a–d.

Above temperature 150 °C, the thermal stability of apricot kernel shells gradually decreases, and decomposition takes place (Fig. 1a–d). The decomposition begins with extracting the volatile products that respectively accompanied by mass reducing. The total mass loss completes at about 850 °C, with residual mass loss in the range of 24.25–24.93% considering all heating rates. Since that apricot kernel shells represent the lignocellulosic material, the devolatilization process includes the decomposition reactions of the basic structural components of the biomass. The first two stages include decomposition of structural polymers, mainly hemicelluloses and cellulose [8] in the temperature range of 250–360 °C with about 56.28% as the largest mass loss of the sample (Fig. 1a–d). The third stage can be attributed to the slow decomposition of aromatic polymer (lignin), which decomposes in the widest temperature range [28]. In a general manner, in accordance with results reported by Gasparovič et al. [29], the thermal decomposition of lignocellulosic material takes place in three main stages: evaporation of moisture and light hydrocarbons, active and passive pyrolysis. The decomposition of hemicelluloses and cellulose takes place through active pyrolysis zone in the temperature ranges of 160–305 °C and 305 °C–370 °C, with temperatures at the maximum rate of mass losses as Tp1 = 285.7 °C and Tp2 = 355.9 °C (considering the values related to $\beta = 10$ °C min^{-1}, where “1” and “2” are attributed to hemicelluloses and cellulose, respectively) (Fig. 1). The lignin decomposes in both stages, in an active and passive pyrolysis zones, in the range of 160–900 °C, without a clearly expressed characteristic DTG peak (Fig. 1a–d). The pyrolyzable fraction of this pseudo-component starts its devolatilization at the lowest temperature, but it is decomposed in a wide temperature range. The lignin decomposition is hindered by decomposition of hemicelluloses and cellulose, excepting at high temperature (especially in the temperature range of 380–480 °C, Fig. 1a–d). These issues are clearly pieces of evidence for existence of multi-step process, which is typical for behaviour of complex reactions, involving the multiple, parallel, and consecutive processes during the decomposition of lignocellulosic biomass under the inert atmosphere [30]. The effect of the increase of heating rate can be observed in the DTG curves during pyrolysis process. In actual case, the increase resulted in an enlargement of the curves resulting in the shifts through the size of DTG curves, and changes in the main temperature profile characteristics such as: Tonset, Tp (inflection) and Tend (Fig. 1).

It can be seen that pyrolyzable fraction of hemicelluloses has much lower thermal stability than celluloses pyrolyzable fraction. In addition, the char fractions decompose at the higher temperatures (above 500 °C), with the hemicelluloses non-pyrolyzable fraction being the one that is burnt at the lower temperatures, followed by the cellulose char. Lignin char was decomposed at the highest temperature and its yield may be very high. However, in this case, the analysis is especially difficult due to the formation of complex phenolic species during lignin decomposition.

In addition, the heating rate affects the rate of volatile evolution from biomass sample. It can be observed from DTG curves (Fig. 1) that higher heating rate above 300 °C promotes the rapid volatile evolution. The molecular disruption is extremely fast and volatile fragments are released so rapidly that successive adjustments and equilibrium leading to further primary reactions that yield char have less opportunity to take place. As the heating rate increases, the amount of volatiles released also increases. The DTA (differential thermal analysis) curves of the apricot kernel shell pyrolysis at the various heating rates (as βj, j = 1, 2, 3, 4) and at the one (selected) specific heating rate value (β = 10 °C min^{-1}) are shown in Fig. 2a–b.

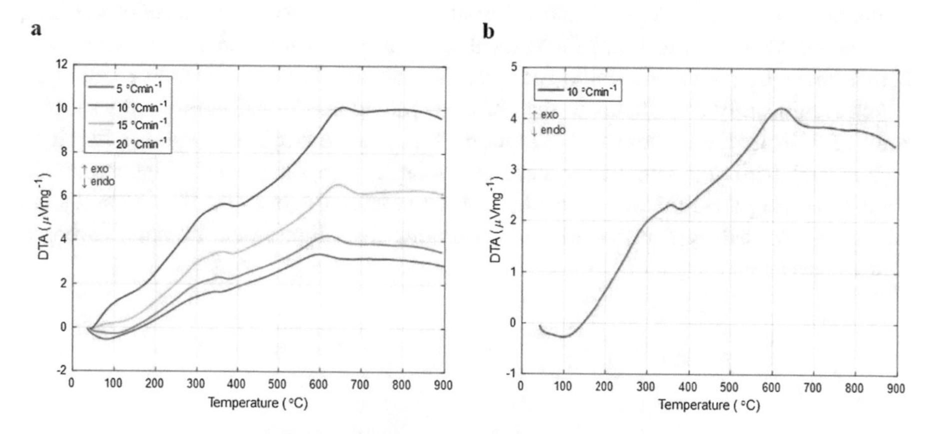

Fig. 2. (a–b) The DTA curves of apricot kernel shell pyrolysis at the various heating rates and the DTA curve at single heating rate value (10 °C min^{-1})

According to Fig. 2a–b, the initial curve deep from 4 °C to 12 °C (Fig. 2b) is due to moisture evaporation, which is the endothermic reaction. The depth of this peak drastically decreases with the increase in the heating rate (Fig. 2a), which means that the higher heating rate is very stimulating for evaporation. At much higher temperatures, the heat released during pyrolysis can be explained by decomposition of lignocellulosic structure of the sample.

In the second phase, in the temperature range of 250–390/400 °C (Fig. 2a–b), the broader DTA peak can be observed at all heating rates (where the largest mass loss of the sample occurs) and this can be attributed to the thermal decomposition of hemicelluloses and cellulose. The next (-exo) peak can be observed in the temperature range of 580–680 °C, where its strict position depends on the applied heating rate (Fig. 2a–b). This exo-peak may correspond to the combustion of complex, the aromatic structure of lignin polymer. At temperature of approximately 850 °C, the char was produced from each pyrolysis process monitoring, and produced char generally contains two entities. These entities are fixed carbon and ash. The summation of the fixed carbon and ash contents obtained for the apricot kernel shells (Table 1) is in very good agreement with the char yield that was obtained from every TGA measurement (Fig. 1). The values of char yield obtained at the various operating temperatures and residence times were calculated from following relation: Char Yield (%) = [(mf*)/(mi)] × 100, where mf* is the mass at desired temperature and time, while mi is the mass at time t = 0 (min). Figure 3 shows the char yield of apricot kernel shell pyrolysis obtained at the different operating temperatures and different residence times (t1, t2, …, t8) at the heating rate of 10 °C min^{-1}. From obtained results, a very high yield of the char was obtained at lowest operating temperature and at shortest residence time (250 °C and t1 = 21 min. ~95.55%), where with a rise in operating temperature and the prolongation of residence time, the char yields exponentially decrease and reaches a minimum value at 600 °C for t8 = 56 min (~27.83%). It can be observed from Fig. 3 that the high char yields were obtained in the decomposition stage which occurs between 200 and 350 °C and can be related to depolymerization reactions during cellulose decomposition. From these results, it can be mostly concluded that apricot kernel shells as waste lignocellulosic material contains higher amounts of cellulose, which significantly contributes to the char accumulation. Furthermore, above temperature of 350 °C, the random fragmentation of glucosidic bonds in cellulose, hemicelluloses and their oligomers may generate volatile compounds with higher oxygen content, leaving a carbonaceous residue. At temperatures above 400 °C, CO and CO2 gases are released by depolymerization reactions from lignin-rich aromatic carbonaceous matrix.

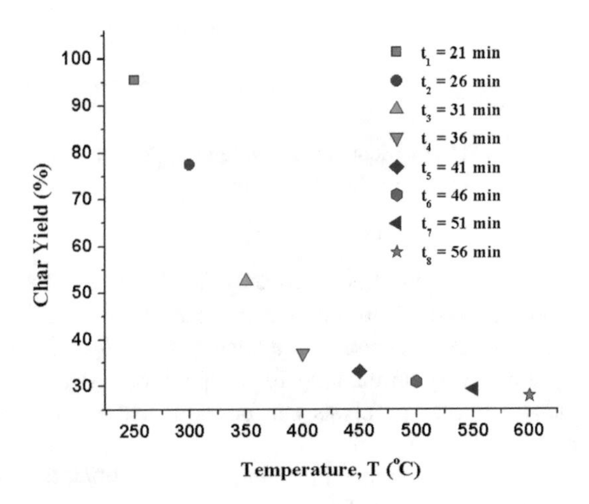

Fig. 3. The char yield of apricot kernel shell pyrolysis obtained at the different operating temperatures and different residence times at the heating rate of 10 °C min^{-1}

In addition, the operating temperature represents the most important factor controlling the properties of resulting char. So, the yield of product decreases as operating temperature increases, as a result of greater thermal decomposition of organic constituents (it seems that chain-scission reaction requires a higher temperature and the content of lignin in the sample can also be one of the majors), while high yields obtained at lower operating temperatures are due to partially pyrolyzed biomass (Fig. 3).

According to Novak et al. [31], the decrease in the char yield with an increasing of the operating temperature is due to dehydration of the hydroxyl groups and the thermal decomposition of cellulose and lignin structures with a great yield reduction which occurs above 400 °C (Fig. 3). Therefore, the char obtained from lower operating temperatures, are characterized by the higher volatile matter contents with an easily decomposable material constituent [32] and, therefore, the amount of volatile matter in the char is supposed to be the highest with lowest operating temperature used in the experiments.

The pyrolysis gas mainly contains CO2, CO, CH4, H2, C2H6, C4H10, C5H12, the minor amounts of higher gaseous organics and water vapor. In addition, the primary decomposition of biomass occurs mainly below 400 °C including various degradation reactions, whereas the secondary pyrolysis (above 400 °C) involves an aromatization process.

Figure 4 shows the simultaneous analysis of TGA and MS measurements of the apricot kernel shell pyrolysis process monitored at every considered heating rate: Fig. 4a–c (5 °C min^{-1}), Fig. 4d–f (10 °C min^{-1}), Fig. 4g–i (15 °C min^{-1}), and Fig. 4j–l (20 °C min^{-1}), respectively.

In general, for water and CO_2, the shape of the mass spectrometric curves resembles the DTG curves. Shape similarities of the mass spectra of these compounds with shapes of DTG curves at all heating rates are obvious (Figs. 1 and 4). On the other hand, however, hydrogen signal has a different behavior (Fig. 4). Considering the fixed heating rate value, the ion current of hydrogen increases from the lower temperature (around 250 °C) to approximately 350 °C, and then decreases to the higher temperature zones, continuously to the very end of the process.

For lower heating rates (5 and 10 °C min^{-1}), a slight transition at about 300 °C can be observed, while at higher heating rates (>10 °C min^{-1}), this transition disappears. Such behavior, however, indicates the existence of certain differences in mass spectrometric signals with DTG ones. From these results (Fig. 4a, d, g and j), it can be concluded that the main source for the hydrogen production arises from the thermal volatilization of biomass pseudo-components.

The methane (CH_4) mass spectrometric signal at all heating rates is similar to the shape of DTG curves. The CH_4 signal shows the main peak in the temperature range of 300–400 °C and one additional "shoulder" peak signal, which is positioned at about 420/430 °C, respectively. This means that the most pronounced sources for CH_4 production come from two origins: volatilization at the lower temperatures (dominantly from hemicelluloses and cellulose decomposition), and the charring process at high temperature. At high temperatures, the CH_4 production is attributed to charring processes which involve aliphatic and aromatic chars. The broad CH_4 evolution beyond 500 °C can be explained on the basis on the fact that methane could be obtained from char within the high temperature ranges. Mass spectrometric signals for ethane (C_2H_6) at all observed heating rates coincide with the shapes of DTG curves (Figs. 1 and 4).

At the lower temperatures, the mass signals of H_2O, CO and CO_2 (C_3H_8) evolutions (Fig. 4b, e, h, and k) are due to the volatilization of hemicelluloses and cellulose (CO was most likely generated from cellulose decomposition). Mass spectrometric signals for CO and CO_2 are very similar, and in both cases, the considered gases liberate from cellulose thermal volatilization. The peak at about 350 °C can be attributed to the decomposition of aromatic and aliphatic carboxyl groups in cellulose. It can be assumed that below 500 °C, the abundant presence of $C = O$ groups in hemicelluloses can be favorable for CO_2 emission. As temperature increased, the more stable ether structures and oxygen-bearing hetero-cycles in lignin are also decomposed into CO_2 [33].

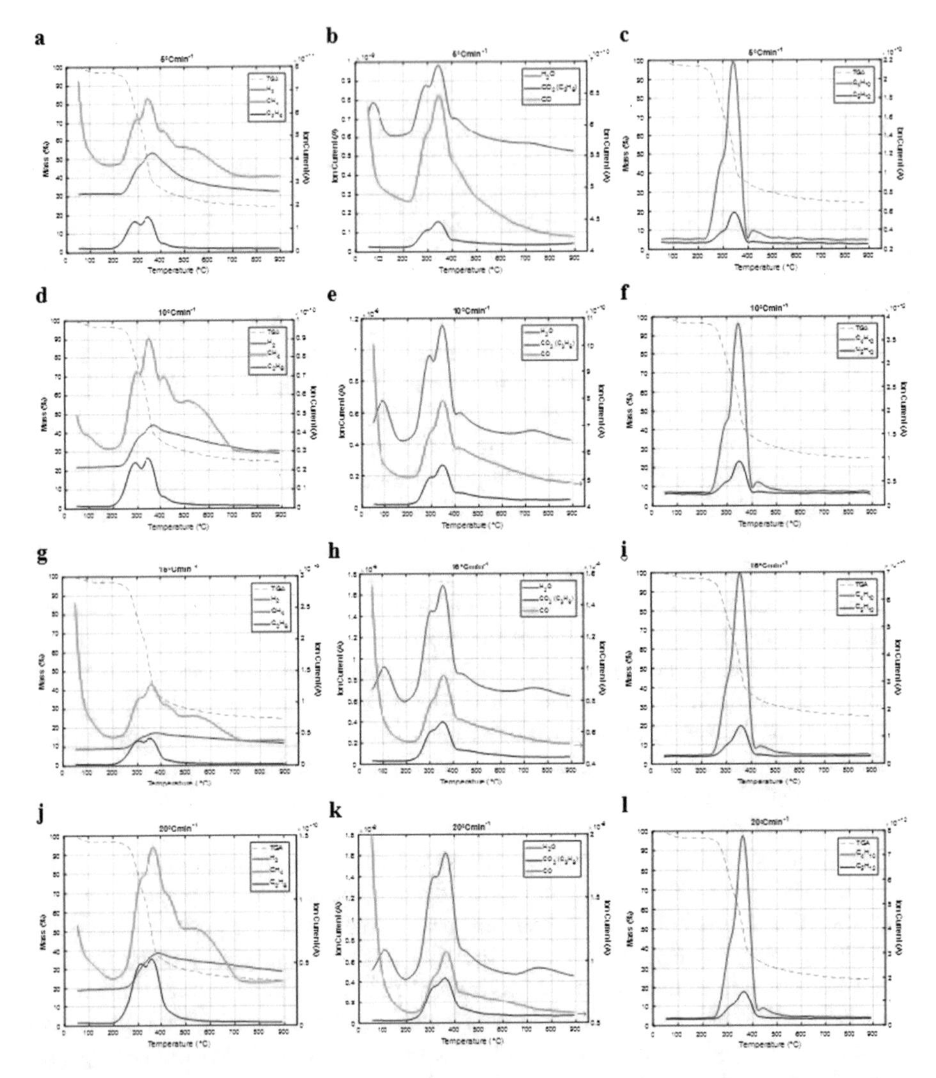

Fig. 4. Simultaneous analysis of TGA and MS measurements of apricot kernel shell pyrolysis process monitored at every considered heating rate

It can be observed that mass spectrometric signals of lighter hydrocarbons (C_4–C_5) (Fig. 4c, f, i, and l) fall within temperature range for hemicelluloses and cellulose volatilizations. Consistent with the distribution of mass signals related to H_2O, CO and CO_2 evolutions, we can assume that the most important volatiles sources are cellulose and hemicelluloses, followed by the lignin. Compared to hemicelluloses and lignin, cellulose released most of the gaseous products in a narrow temperature range (300–400 °C), since on its higher abundance in the sample.

Also, it can be noted that the highest concentration of CH_4 emission occurs about 350 °C. The CH_4 emission at relatively lower temperatures was most likely released from C—C bond cleavage in aliphatic chains whereas the CH_4 emission at higher temperatures was mainly produced from cracking of a weakly bonded methoxyl-O-CH_3 group as well as the break of having higher bond energy of the methylene group –CH_2. Further, in the lower temperature zones, the H_2 emission was mainly generated from thermal decomposition of cellulose and hemicelluloses. At $T > 450$ °C, decreasing trends in the H_2 mass signal may reveal that lignin pyrolysis could also generate the hydrogen. Generally, in the third process stage (beyond 450 °C, that represents the slow decomposition stage), the relative concentration of all volatile components gradually decreased.

Distribution of activation energy (E) values in a function of conversion degree (α), estimated by FR, KAS and OFW methods for apricot kernel shell pyrolysis was presented in Fig. 5a–c. There are some differences in obtained E values calculated from the KAS and OFW methods, with those calculated from FR method. This can be clearly seen from the mean values of the effective activation energy, which are as follows: E_{mean}^{FR} (kJ mol^{-1}) = 229.68, E_{mean}^{KAS} (kJ mol^{-1}) = 221.42, and E_{mean}^{OFW} (kJ mol^{-1}) = 220.02, respectively. It can be seen that $E_{mean}^{FR} \neq E_{mean}^{KAS} \approx E_{mean}^{OFW}$, so that actual difference in mean values between differential and integral isoconversional methods suggests on the presence of reaction complexity of the process, which can also be seen from the shape of $E - \alpha$ dependency (Fig. 5).

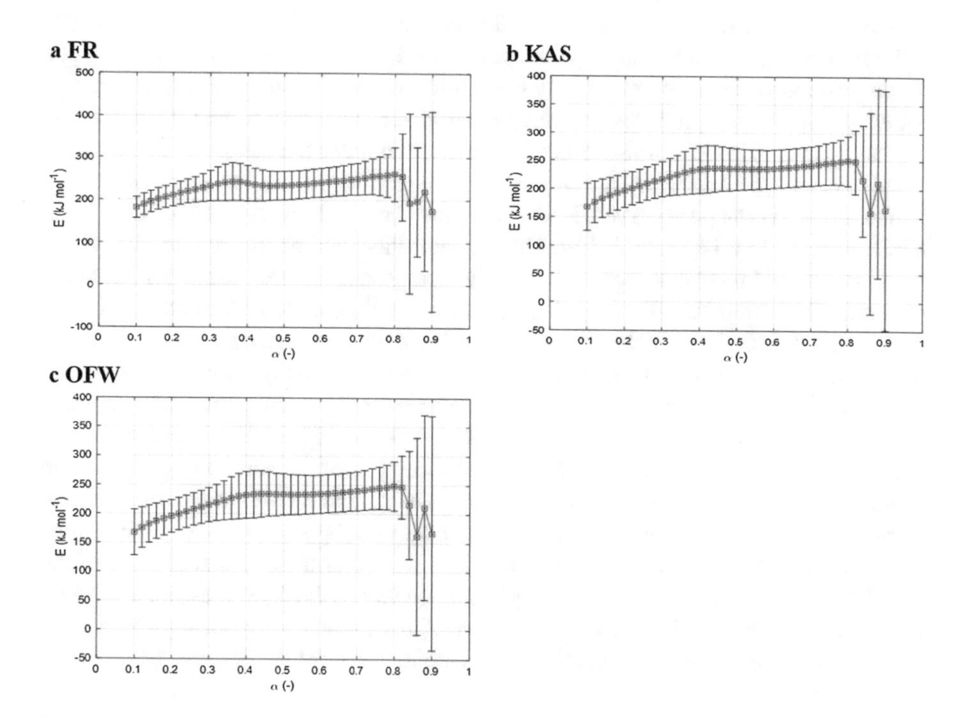

Fig. 5. (a–c) Variation of the effective activation energy (E) with respect to conversion degree (α) estimated by the FR, KAS and OFW methods, for the studied pyrolysis process

The first process area, from $\alpha = 0.10$ to $\alpha = 0.40$ (Fig. 5), the E value increases from 168.0 to 235.0 kJ mol^{-1} for the KAS method, and from 168.0 to 232.0 kJ mol^{-1} for the OFW method, as conversion degree increased, and this could be mainly attributed to hemicelluloses decomposition. In the initial stage, the decomposition started easily on weakly linked sites inherent to the polymeric lineal chain of the hemicelluloses, which led to the lower effective activation energy. After the weaker bonds broke, the random scission on the lineal chain can be expected, which may cause that E increases. Meanwhile, it can be expected that there is an interaction between hemicelluloses and lignin, where the obtained value of E in the actual stage, is much higher than the value for the single component of xylan (represents the hemicelluloses) usually analyzed in pyrolysis process (87.65 kJ mol^{-1} and 69.39 kJ mol^{-1}) [34].

The second process area, which encompass the range of $0.40 < \alpha < 0.80$, shows that there are some stabilized E values, observing all three isoconversional methods (Fig. 5a–c), with average values of $E = 240.0$ kJ mol^{-1} for KAS method, and $E = 237.0$ kJ mol^{-1} for OFW method, respectively. The indicated conversion range falls in the experimental temperature range of $\Delta T = 310$–360 °C, and this area was just located between two mass loss rates peaks in DTG curves. For cellulose decomposition in this reaction stage, the cellulose initially pyrolyzed to active cellulose according to the Briodo-Shafizadeh extended model, and this can lead to reducing of the degree of polymerization and the length of molecule chain.

The active cellulose represents the intermediate product before further pyrolysis. This fact was validated from the certain increasing of E value during transition from the first to the second reaction stage. In considered α range ($0.40 < \alpha < 0.80$), the active cellulose continues to decompose with approximately constant E value. The particular decomposition process may eventually splits into two parallel and competitive reaction pathways, one which produces char favored at lower temperatures, and the other may produces tar and gas that *prior* to occur at higher temperatures [35].

In the final process area, for $\alpha > 0.80$, the significant variation in E value with decreased trends can be observed. In this area, the E value drops at 158.0 kJ mol^{-1} (for KAS) and at 161.8 kJ mol^{-1} (for OFW), and this can be related to the lignin decomposition. Since that lignin is mainly composed of three kinds of benzene-propane, which were heavily cross-linked and make it an extremely thermal stable.

Using the algorithm which was previously described, the simulation between 3- and 5- pseudo-components was performed, but the best fit quality results show - data related to 4- pseudo-component model. Simulations for 4-pseudo-components (marked by PSC-1, PSC-2, PSC-3 and PSC-4) with corresponding model fit parameters at different heating rates, together with overall kinetic model values are presented in Table 3 and Fig. 6a–d. Among all pseudo-component models (with three, four and five - components), the 4-pseudo-component model showed the lowest values of statistical parameters and modeling quality parameters, in the case of the best assessments of the model accuracy. Therefore, the lowest QOF (the quality of the fit) values and extremely low values of AIC for the maximum number of the pseudo-components which were used without disturbing the fit quality at all heating rates (Table 3).

Table 3. Kinetic parameters of the multi component kinetic model and quality of fit for the different heating rates (5, 10, 15 and 20 °C min^{-1})

Pseudo-component	5 °C min^{-1}			
	φ	A (min^{-1})	E_a (kJ mol^{-1})	n
PSC-1	0.2983	9.995×10^9	128.5	1.51
PSC-2	0.2984	9.996×10^9	118.3	1.27
PSC-3	0.2984	1.876×10^2	48.6	1.19
PSC-4	0.1049	1.169×10^2	57.5	1.38
QOF	1.7359			
AIC	−2385.47			
Pseudo-component	10 °C min^{-1}			
	φ	A (min^{-1})	E_a (kJ mol^{-1})	n
PSC-1	0.2992	9.998×10^9	116.8	1.29
PSC-2	0.2991	9.998×10^9	126.8	1.55
PSC-3	0.2990	1.266×10^2	61.2	1.28
PSC-4	0.1027	2.475×10^2	51.5	1.18
QOF	1.4458			
AIC	−2190.48			
Pseudo-component	15 °C min^{-1}			
	φ	A (min^{-1})	E_a (kJ mol^{-1})	n
PSC-1	0.2311	9.259×10^9	123.6	1.75
PSC-2	0.2990	1.134×10^{10}	116.4	1.30
PSC-3	0.2993	3.117×10^2	53.8	1.15
PSC-4	0.1706	1.285×10^2	64.8	1.14
QOF	1.3697			
AIC	−2065.94			
Pseudo-component	20 °C min^{-1}			
	φ	A (min^{-1})	E_a (kJ mol^{-1})	n
PSC-1	0.2704	8.435×10^9	123.5	1.65
PSC-2	0.2957	1.765×10^{10}	118.0	1.31
PSC-3	0.2983	6.003×10^2	57.0	1.14
PSC-4	0.1356	1.327×10^2	64.5	1.16
QOF	1.2675			
AIC	−1969.24			

In accordance with necessary and sufficient conditions for minimum number of processes taking place during pyrolysis, the AIC values (Table 3) clearly shows that (excluding dehydration process) there are four unique reactions which occur in the pyrolysis, and these reactions correspond to cellulose (PSC-1), hemicelluloses (PSC-2), extractives (PSC-3) and lignin (PSC-4) decompositions, respectively.

Within the active pyrolysis zone, the decomposition of PSC-1 and PSC-2 occurs (cellulose and hemicelluloses) in the temperature range of approximately 150–400 °C (Fig. 6a–d). On the other hand, the decomposition of extractives PSC-3 represents a smaller part of non-structural component of biomass which - occurs in a very narrow temperature range (Fig. 6a–d). Also this decomposition is characterized with a very small deconvoluted peak, which intensity decreased with increasing of the heating rate. In addition, PSC-4 (lignin) processing occurs through a very wide temperature range including both, the active and passive pyrolysis zones (Fig. 6a–d).

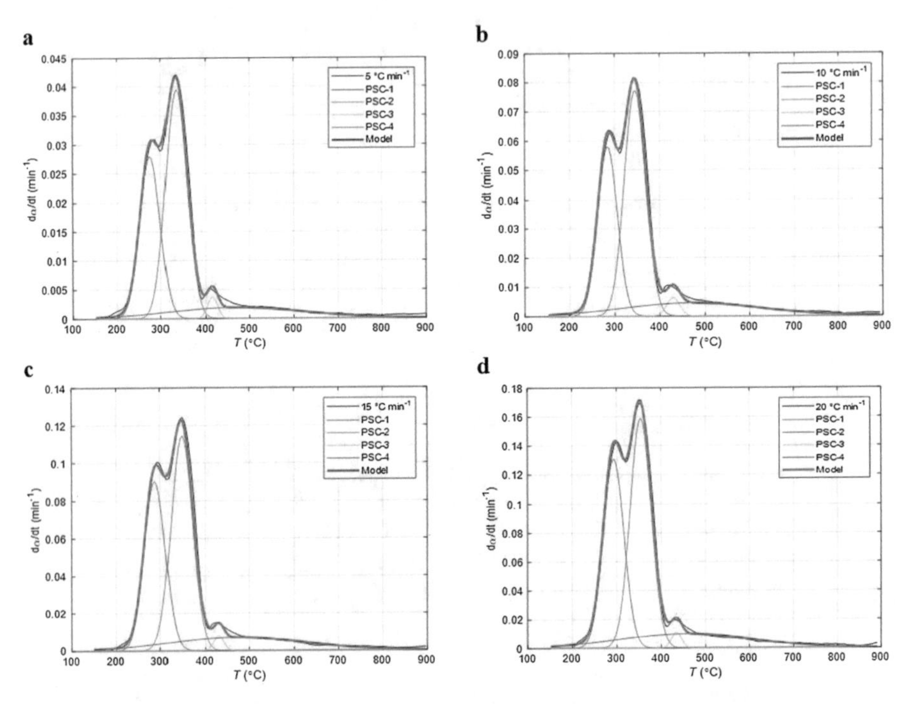

Fig. 6. (a–d) The experimental and kinetic model pyrolysis rate curves for 4-pseudo-components model (PSC-1, PSC-2, PSC-3 and PSC-4) at the different heating rates (5, 10, 15, 20 °C min^{-1}). The full red line represents the sum of all 4-pseudo-component curves obtained according to deconvolution by Fraser-Suzuki function

The PSC-3 overlapping the PSC-4 peak decomposition region, so this reaction probably can be accelerated by the catalytic effect of the present extractives, making the lignin decomposed through two reaction pathways, one of which occurs with the lower E (liberating H_2O, CO_2, CO and char producing), while the other occurs with the higher E (producing monomers). The cleavage of the aryl-ether linkages may result in the formation of highly reactive and unstable free radicals that may further react through re-arrangement, electron abstraction or radical-radical interactions, to form products with increased stability. This can especially be expressed for the sample that contains a high oxygen concentration in composition, as for apricot kernel shells (Table 2). Furthermore, the char oxidation can occur, followed by exothermic effect (usually for an oxidative surface reactions) where for apricot kernel shell pyrolysis it can be expected about/or above 600 °C (Fig. 2a).

For PSC-1 and PSC-2 decompositions, the obtained apparent activation energies (Table 3) are lower than those for pure cellulose (between 140.00 and 240.23 kJ mol^{-1}) and pure hemicelluloses (xylan) (179.84 kJ mol^{-1}) [23, 36]. This difference may be due to the temperature difference of various reaction stages, used in the pyrolysis experiments.

From results presented in Table 3, it can be seen that for PSC-1 and PSC-2 decompositions, the non-integer reaction orders were obtained. These values are characteristic for the description of random scission mechanisms, where the non-integer reaction orders indicate that the processes are complex and comprise several elementary reaction stages. So, this behaviour is not unusual considering that both, cellulose and hemicelluloses decomposition includes a series of reaction pathways which can not be described by the simple integer reaction-order kinetics, because the complicated reactions related to entire apricot kernel shell pyrolysis. Also, from actual results, it can be seen that the contribution of cellulose decomposition in overall kinetics dominates, which is confirmed by the results established through the MS analysis (see above).

As can be seen from Fig. 6a–d, the fits of single Fraser-Suzuki function (Table 1) to PSC's in the rate curve profiles are almost perfect. Regarding the width (w) of Fraser-Suzuki function, it is a characteristic parameter for a width of reaction temperature range. The increase of w_i values attributes to the residence time of the sample, since higher heating rate means the shorter exposure to the certain temperature or the temperature domain. Therefore, the sample actually needs more time to reach higher temperatures for completion of overall decomposition, so we have a following order in this sense, such as PSC-3 < PSC-1 < PSC-2 < PSC-4.

The advantage of this methodology in comparison with previous investigations on the same topic [8, 37], consists in the fact, that based on combined use of isoconversional and multicomponent approaches using deconvolution procedure, it is possible to accurately segregate the process phases in the temperature areas through the existence of energy compensation effect. This meets the relationship criteria $\ln A_\alpha = a \cdot E_{a,\alpha} + b$, where a and b are unknown compensation parameters. Linear regression of $\ln A_\alpha$ against $E_{a,\alpha}$ gives a straight line and parameters a and b can be obtained from the slope and intercept, respectively.

Figure 7 shows the energy compensation effects (also known as kinetic compensation effect) of entire pyrolysis process that obeys to four components of the decomposition process.

However, according to physical explanations, the compensation effect is caused by a correlation between the change of enthalpy and the underway of entropy change from reagents to transition state of the reaction. Parmon [38] studied the kinetic compensation effect in complex stepwise reactions and found that the existence of the "isokinetic" temperature in homogeneous systems caused the compensation effect.

It can be seen from Fig. 7 that the energy compensation effects of entire pyrolysis process are in very good linear correlations. This demonstrates that multi-step reaction model is suitable for studied pyrolysis process. It can be observed that all reaction stages are interconnected and monitor the respective regions of temperature change.

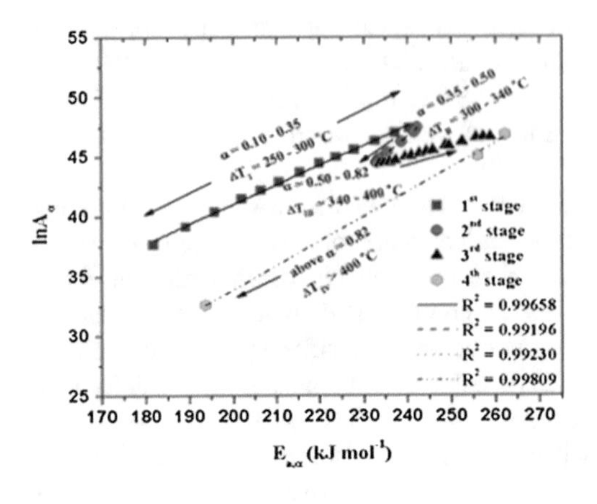

	1st stage
	$\Delta\alpha = 0.10 - 0.35$
a [mol (kJ)$^{-1}$]	0.16416 ± 0.00277
b	8.22219 ± 0.59566
	2nd stage
	$\Delta\alpha = 0.35 - 0.50$
a [mol (kJ)$^{-1}$]	0.30394 ± 0.01116
b	-26.20635 ± 2.64216
	3rd stage
	$\Delta\alpha = 0.50 - 0.82$
a [mol (kJ)$^{-1}$]	0.09638 ± 0.00227
b	21.89729 ± 0.55672
	4th stage
	$\Delta\alpha > 0.82$
a [mol (kJ)$^{-1}$]	0.20487 ± 0.00634
b	-7.06901 ± 1.51599

Fig. 7. Energy compensation effects of the entire apricot kernel shell pyrolysis process, with indicated process stages, as well as the temperature and conversion degree range changes

The position of all stages is typical for compensation effect plots which arise from lignocellulosic content distinctions. This behaviour identified for the apricot kernel shell pyrolysis is characteristic for the emergence of the “pseudo kinetic compensation effect” [39].

If we consider only pseudo-components of the studied sample, it seems that the largest contribution belongs to cellulose, which agrees with conclusion that in indicated temperature range corresponding to this fraction in Fig. 7 and largest yield of gaseous products originates from this zone (see MS spectra data).

According to abovementioned it could be observed that when different kinetic parameters sets (E and A) are found to satisfy a specific kinetic compensation effect relation, it is still possible that they support different kinetic descriptions, which can also be seen by the values of parameters a, and b, shown also on Fig. 7.

Also, it could be noticed that the distribution of the effective activation energy (E) and the pre-exponential (A) values exists, and could be obtained by the means of the combination of the proposed optimization method and the energy compensation effect.

The compensation parameter a can be related to a strength of the bond being broken when gaseous products are formed, and a small value of a, is associated with rupturing of a strong bond. Considering the results presented in Fig. 7, the primary event represents the breaking a weaker bond-links in the cellulose molecule (the high a value), and then breaking the stronger bond-links in a lignin molecule (the low a value). Hemicelluloses bond rupturing for apricot kernel shell pyrolysis stands between cellulose and lignin bonds rupturing. For the last stage, a value is quite high, and is accompanied by a high variation in E with α, and this may indicate on the existence of the complex mechanisms with the various E values. This transformation requires additional energy input for rupturing of a large number of weakly bonded molecular units. It seems that the reaction order (n) is a parameter that compensates for inaccuracy in the pyrolysis modeling of included chemical reactions. In other words, it is a parameter that lumps several reactions together and defines the sharpness of TGA curve. This indicates that parameter n is very important in the case of consideration of pyrolysis process monitored on the lab-scale level.

Furthermore, the very interesting phenomena was noticed, and it represents the behaviour of parameter b (Fig. 7) (that can be correlated to the "isokinetic" rate constant, $k_{iso} = \exp(b)$). The k_{iso} values are changed for second and third stages (Fig. 7) from 4.157×10^{-12} min^{-1} (2nd stage) up to 3.235×10^{9} min^{-1} (3rd stage), and represents an extremely great variation in magnitudes by a factor of 10^{21}. This tremendous change has been probably propagated by thermal lag which is underlying cause of compensation effect in kinetic values. From compensation effect properties, the considered system was characterized by unconventional thermal lag due to the heat demand by the chemical reaction (which is governed by the Arrhenius kinetics). The un-identified thermal lag in thermo-analytical (TA) experiments lowers the values of effective activation energy and pre-exponential factor associated with experimental data, but approximately retains the true value of their ratio. Variation of kinetic parameters for cellulose pyrolysis over applied heating rate range may be the reason for appearance of thermal lag characteristics in TA experiments.

5 Conclusions

The multicomponent kinetic properties of apricot kernel shells during pyrolysis process were examined. Kinetic parameters of pyrolysis process were estimated using 4-pseudo-component model, together with applied isoconversional approaches (FR, KAS and OFW methods). The multicomponent kinetic model has successfully executed the pyrolysis rate curves at various heating rates, where 4-pseudo-components PSC-1, PSC-2, PSC-3 and PSC-4 were recognizable for cellulose, hemicelluloses, extractives and lignin decomposition reactions, respectively. The Fraser-Suzuki deconvolution procedure showed successfully separation of reactions in multi-stage process, where the fitting of reaction rate profiles is almost perfect compared to the experimental ones. The pseudo kinetic compensation effect was identified for studied pyrolysis process. From this phenomenon, it was found that hemicelluloses bond rupturing is located between cellulose and lignin bonds rupturing. The great change in the values of isokinetic rate constants for second and third decomposition stages is a consequence of the presence of thermal lag, which is underlying cause of the compensation effect in kinetic values.

Proposed algorithm for kinetic modeling was strongly supported by the mass spectrometry (MS) analysis of gaseous products and *vice versa*. Conclusions drawn from this study can be important guidelines for future investigations of pyrolysis properties of the waste lignocellulosic materials. The actual investigation shows that apricot kernel shells as waste fruit residues can be recovered as the potential fuel or chemical feedstock (which is characterized by dominant content of volatile matter) through the pyrolysis. Also, the content of fixed carbon is not low, so there is potential to be pyrolyzed to produce the char for the use as activated carbon or catalyst support. Results showed that apricot kernel shells can be pyrolyzed at temperatures higher than

350 °C to convert the majority of its content into volatiles for recovery as useful biogas, and the remaining solid mass can be recovered as *bio-char*.

Acknowledgments. Authors would like to acknowledge financial support of Ministry of Education, Science and Technological Development of the Republic of Serbia under the Projects III42010, 172015 and III45005.

References

1. Dodić, S.N., Popov, S.D., Dodić, J.M., Ranković, J.A., Zavargo, Z.Z., Golušin, M.T.: An overview of biomass energy utilization in Vojvodina. Renew. Sustain. Energy Rev. **14**(1), 550–553 (2010)
2. Antal, M.J.: Biomass pyrolysis: a review of the literature Part 1—carbohydrate pyrolysis. In: Böer, K.W., Duffie, J.A. (eds.) Advances in Solar Energy, pp. 61–111. Springer, New York (1983)
3. Al Arni, S.: Comparison of slow and fast pyrolysis for converting biomass into fuel. Renew. Energy **124**, 197–201 (2018)
4. Waheed, Q.M.K., Nahil, M.A., Williams, P.T.: Pyrolysis of waste biomass: investigation of fast pyrolysis and slow pyrolysis process conditions on product yield and gas composition. J. Energy Inst. **86**(4), 233–241 (2013)
5. Janković, B., Dodevski, V.: The combustion performances and thermo-oxidative degradation kinetics of plane tree seeds (PTS) (Platanus orientalis L.). Energy **154**, 308–318 (2018)
6. Diebold, J., Power, A.: Engineering aspects of the vortex pyrolysis reactor to produce primary pyrolysis oil vapors for use in resins and adhesives. In: Bridgwater, A.V., Kuester, J. L. (eds.) Research in Thermochemical Biomass Conversion, pp. 609–628. Springer, Dordrecht (1988)
7. Mohan, D., Pittman, C.U., Steele, P.H.: Pyrolysis of wood/biomass for bio-oil: a critical review. Energy Fuels **20**(3), 848–889 (2006)
8. Demiral, İ., Kul, Ş.Ç.: Pyrolysis of apricot kernel shell in a fixed-bed reactor: characterization of bio-oil and char. J. Anal. Appl. Pyrolysis **107**, 17–24 (2014)
9. Abbas, M., Aksil, T.: Adsorption of malachite green (MG) onto apricot stone activated Carbon (ASAC)-Equilibrium, kinetic and thermodynamic studies (2017)
10. Taghizadeh-Alisaraei, A., Assar, H.A., Ghobadian, B., Motevali, A.: Potential of biofuel production from pistachio waste in Iran. Renew. Sustain. Energy Rev. **72**, 510–522 (2017)
11. Şentorun-Shalaby, Çd., Uçak-Astarlıogˇlu, M.G., Artok, L., Sarıcı, Ç.: Preparation and characterization of activated carbons by one-step steam pyrolysis/activation from apricot stones. Microporous Mesoporous Mater. **88**(1–3), 126–134 (2006)
12. Friedman, H.L.: Kinetics of thermal degradation of char-forming plastics from thermogravimetry: Application to a phenolic plastic. J. Polym. Sci. Part C Polym. Symp. **6**, 183–195 (1964)
13. Berčič, G.: The universality of Friedman's isoconversional analysis results in a model-less prediction of thermodegradation profiles. Thermochim. Acta **650**, 1–7 (2017)
14. Kissinger, H.E.: Reaction kinetics in differential thermal analysis. Anal. Chem. **29**(11), 1702–1706 (1957)
15. Akahira, T., Sunose, T.: Method of determining activation deterioration constant of electrical insulating materials. Res. Rep. Chiba Inst. Technol. (Sci Technol) **16**, 22–31 (1971)
16. Ozawa, T.: A new method of analyzing thermogravimetric data. Bull. Chem. Soc. Jpn **38** (11), 1881–1886 (1965)

17. Janković, B., Manić, N., Stojiljković, D., Jovanović, V.: TSA-MS characterization and kinetic study of the pyrolysis process of various types of biomass based on the Gaussian multi-peak fitting and peak-to-peak approaches. Fuel **234**, 447–463 (2018)
18. Wang, X., Hu, M., Hu, W., Chen, Z., Liu, S., Hu, Z., et al.: Thermogravimetric kinetic study of agricultural residue biomass pyrolysis based on combined kinetics. Bioresour. Technol. **219**, 510–520 (2016)
19. Li, J., Qiao, Y., Zong, P., Wang, C., Tian, Y., Qin, S.: Thermogravimetric analysis and isoconversional kinetic study of biomass pyrolysis derived from land, coastal zone and marine. Energy Fuels **33**, 3299–3310 (2019)
20. Oladokun, O., Ahmad, A., Abdullah, T.A.T., Nyakuma, B.B., Bello, A.A.-H., Al-Shatri, A. H.: Multicomponent devolatilization kinetics and thermal conversion of Imperata cylindrica. Appl. Therm. Eng. **105**, 931–940 (2016)
21. Cheng, Z., Wu, W., Ji, P., Zhou, X., Liu, R., Cai, J.: Applicability of Fraser-Suzuki function in kinetic analysis of DAEM processes and lignocellulosic biomass pyrolysis processes. J. Therm. Anal. Calorim. **119**(2), 1429–1438 (2015)
22. Perejón, A., Sánchez-Jiménez, P.E., Criado, J.M., Pérez-Maqueda, L.A.: Kinetic analysis of complex solid-state reactions. A new deconvolution procedure. J. Phys. Chem. B **115**(8), 1780–1791 (2011)
23. Hu, M., Chen, Z., Wang, S., Guo, D., Ma, C., Zhou, Y., et al.: Thermogravimetric kinetics of lignocellulosic biomass slow pyrolysis using distributed activation energy model, Fraser-Suzuki deconvolution, and iso-conversional method. Energy Convers. Manage. **118**, 1–11 (2016)
24. Özyuğuran, A., Yaman, S.: Prediction of calorific value of biomass from proximate analysis. Energy Procedia **107**, 130–136 (2017)
25. Arvelakis, S., Gehrmann, H., Beckmann, M., Koukios, E.: Preliminary results on the ash behavior of peach stones during fluidized bed gasification: evaluation of fractionation and leaching as pre-treatments. Biomass Bioenergy **28**(3), 331–338 (2005)
26. Manić, N., Janković, B., Stojiljković, D., Jovanović, V.: TGA-DSC-MS analysis of pyrolysis process of various biomasses with isoconversional (model-free) kinetics. In: Mitrovic, N., Milosevic, M., Mladenovic, G. (eds.) Experimental and Numerical Investigations in Materials Science and Engineering, pp. 16–33. Springer, Cham (2019)
27. Trninić, M., Todorović, D., Jovović, A., Stojiljković, D., Skreiberg, Ø., Wang, L., Manić, N.: Mathematical modelling and performance analysis of a small-scale combined heat and power system based on biomass waste downdraft gasification. In: Mitrovic, N., Milosevic, M., Mladenovic, G. (eds.) Experimental and Numerical Investigations in Materials Science and Engineering, pp. 159–173. Springer, Cham (2019)
28. Yang, H., Yan, R., Chen, H., Lee, D.H., Zheng, C.: Characteristics of hemicellulose, cellulose and lignin pyrolysis. Fuel **86**(12–13), 1781–1788 (2007)
29. Gašparovič, L., Koreňová, Z., Jelemenský, Ľ.: Kinetic study of wood chips decomposition by TGA. Chem. Pap. **64**(2), 174–181 (2010)
30. Zapata, B., Balmaseda, J., Fregoso-Israel, E., Torres-García, E.: Thermo-kinetics study of orange peel in air. J. Therm. Anal. Calorim. **98**(1), 309–315 (2009)
31. Novak, J.M., Busscher, W.J., Laird, D.L., Ahmedna, M., Watts, D.W., Niandou, M.A.S.: Impact of biochar amendment on fertility of a southeastern coastal plain soil. Soil Sci. **174** (2), 105–112 (2009)
32. Jindo, K., Mizumoto, H., Sawada, Y., Sanchez-Monedero, M.A., Sonoki, T.: Physical and chemical characterization of biochars derived from different agricultural residues. Biogeosciences **11**(23), 6613–6621 (2014)
33. Wang, T., Yin, J., Liu, Y., Lu, Q., Zheng, Z.: Effects of chemical inhomogeneity on pyrolysis behaviors of corn stalk fractions. Fuel **129**, 111–115 (2014)

34. Janković, B., Manić, N., Dodevski, V., Popović, J., Rusmirović, J.D., Tošić, M.: Characterization analysis of Poplar fluff pyrolysis products. Multi-component kinetic study. Fuel **238**, 111–128 (2019)
35. Broido, A., Nelson, M.A.: Char yield on pyrolysis of cellulose. Combust. Flame **24**, 263–268 (1975)
36. Radojević, M., Janković, B., Jovanović, V., Stojiljković, D., Manić, N.: Comparative pyrolysis kinetics of various biomasses based on model-free and DAEM approaches improved with numerical optimization procedure. PLoS ONE **13**(10), e0206657 (2018)
37. Aljoumaa, K., Tabeikh, H., Abboudi, M.: Characterization of apricot kernel shells (Prunus armeniaca) by FTIR spectroscopy, DSC and TGA. J. Indian Acad. Wood Sci. **14**(2), 127–132 (2017)
38. Parmon, V.N.: Kinetic compensation effects: a long term mystery and the reality. A simple kinetic consideration. React. Kinet. Mech. Catal. **118**(1), 165–178 (2016)
39. Budrugeac, P.: On the pseudo compensation effect due to the complexity of the mechanism of thermal degradation of polymeric materials. Polym. Degrad. Stab. **58**(1–2), 69–76 (1997)

Comparative Numerical and Experimental Modal Analysis of Aluminum and Composite Plates

Marija Baltic, Jelena Svorcan(✉), Ognjen Pekovic, and Toni Ivanov

Faculty of Mechanical Engineering, University of Belgrade, 11120 Belgrade, Serbia
jsvorcan@mas.bg.ac.rs

Abstract. The paper presents the comparative analysis of the dynamic behavior of two rectangular plates of different material, aluminum and composite. While their global geometric dimensions (length, width and thickness) are similar, their inner structures are quite different. Whereas aluminum plate can be considered isotropic, the composite plate is a unidirectional carbon-epoxy laminate. Modal characteristics of the two plates were determined both numerically and experimentally and a comparative analysis of the obtained results was performed. Responses of the plates were documented by an optical, contactless 3D digital image correlation (DIC) system that contains a set of high-speed cameras capable of recording the movement of the white-and-black stochastic pattern applied to the upper surfaces of the plates. Numerical simulations were performed by the finite element method (FEM) in the commercial software package ANSYS. The plates were excited by a modal hammer and allowed to freely oscillate. In order to determine the natural frequencies of the plates the recorded time-domain responses were post-processed, i.e. converted to the frequency domain by fast Fourier transform (FFT). The first three natural modes were successfully experimentally established and compared to the corresponding numerical values. Since the differences between the two sets of results are less than 5%, the applied experimental technique can be considered valid and suitable for a wide range of engineering problems involving vibrations.

Keywords: Aluminum plate · Composite plate · DIC · FEM · FFT

1 Introduction

Favorable characteristics and numerous advantages of both aluminum and composite based structures have led to their wide application in many branches of industry, particularly in aeronautics. Throughout their multidecade working life, these elements are supposed to endure various operating conditions. It is, therefore, necessary to perform regular checks and inspections of their condition. Preventive maintenance usually employs non-invasive and non-destructive techniques, e.g. visual, ultra-sound, etc. [1–5]. One of the possible methods for determination of mechanical properties (such as elasticity modules), displacement field under loading and dynamic characteristics of aluminum and composite parts is digital image correlation (DIC) [2, 4–10].

N. Mitrovic et al. (Eds.): CNNTech 2019, LNNS 90, pp. 61–75, 2020.
https://doi.org/10.1007/978-3-030-30853-7_4

The paper presents a comparative experimental and numerical study of the behavior of two distinct plates exposed to both static and dynamic loading. While the dimensions of the plates are similar, they are made of different materials: aluminum and composite (carbon-epoxy laminate). In order to make a correct estimation of the structural response under various load cases (tension, pressure, bending, torsion or their combination) it is necessary to have a full understanding of its mechanical properties (in all directions for heterogeneous materials). However, for composite materials (both laminates and sandwich structures that offer wide freedom of choice during design and manufacturing phases) this procedure is not straightforward, and a lot of experimental and computational work is required. Some examples of analytical and computational approaches to composite plates' behavior can be found in [11–14]. This study, therefore, describes only a portion of the complete investigation that deals with pure static bending and free vibrations. The tests are performed over two geometries: isotropic aluminum and 8-layer unidirectional laminate (that behaves similarly to isotropic structures along the main, longitudinal direction).

A great number of measurements was performed – two distinct load cases over two geometries. The DIC measuring method, equipment, and procedure were first verified through comparison with numerical values obtained by Finite Element Method (FEM) in the case of static displacements induced by the force acting on the free end of the composite plate. Later, the observed modal shapes and values of natural frequencies obtained by post-processing the recorded spatial and temporal displacement fields by fast Fourier transform (FFT) of both plates in different conditions were compared to the corresponding numerical results.

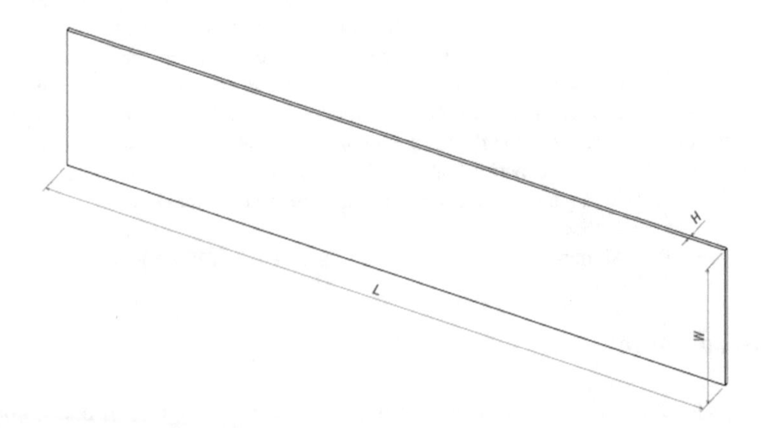

Fig. 1. Geometric model of the tested plates.

2 Geometric Models

Both specimens are rectangular in shape with nearly the same length L, width W and height H (here, adjective nearly is used since aluminum plate is cut out from a standard thickness sheet, while the composite plate is manufactured by the 8 lamina lay-up).

Global dimensions of the aluminum plate are (350 × 79 × 2) mm, and for the composite plate (380 × 80 × 2.4) mm. A sketch of the tested geometric models is illustrated in Fig. 1.

Lay-up sequence of the composite plate is [0°/0°/0°/0°/0°/0°/0°/0°]. Laminas CU 300-0/SO with an areal weight of 300 g/m^2 and approximate layer width is 0.3 mm, are formed from unidirectional carbon fibers with the fiber marking HTS40-12k 800 tex. Matrix is “Hexion Specialty Chemicals Stuttgart” epoxy marked by L285 and hardener H287.

3 Experimental Testing

Experimental measurement was performed by a relatively novel and quite attractive DIC method [2, 4–10]. This recent and increasingly widespread contactless system, based on the principles of photogrammetry and modern computer technology, is used for points displacement measurement. DIC systems use a series of sequentially recorded digital images to determine (calculate) the displacement and the corresponding stress fields. The surface of the tested object is colored (speckled) by a stochastic pattern whose points are transformed to corresponding pixels in the images. A great advantage of the DIC method is the visualization of the complete displacement field, even in the proximity of the structural discontinuities [15]. The measuring equipment contains a set of high-speed cameras capable of capturing high-frequency modes that are quickly damped. Experimental procedure involves several activities: preparation and positioning of the specimen and the measuring equipment, optics set-up and recording parameter definition, excitation, data acquisition and results analysis [4, 6].

Fig. 2. Experimental equipment and set-up.

The DIC system used in this investigation, presented in Fig. 2, contains: two high-tech, ultra-fast cameras (FASTCAM SA6 75K-M3 with 32 GB memory, max. frame-rate 75000 fps, max. resolution 1920 × 1440 pixel) with the frame-rate of 3000 fps and image resolution of 1024 × 512 pixels capable of very short recording intervals and high levels of accuracy that enable the measurement of 3D displacement field, a synchronizer, an additional light unit/source and the data acquisition, processing and analysis system. The aluminum sample was prepared for measurement by first applying a layer of white color, followed by a fine dispersion of black spots that enable the process of correlation. Similarly, the carbon composite plate, due to its natural black color, was only speckled by white paint, Fig. 3.

Fig. 3. Tested plates – (a) aluminum, (b) composite.

Image recording is performed by the PHOTRON software package [8], while the image correlation process and subsequent analysis are executed in the ARAMIS software package [9]. A complete, unsteady, 3D displacement field is obtained by correlating the series of images from the left and right camera.

Prior to the measurement, it is necessary to perform the hardware calibration (prepare the sensors – cameras), i.e. define the frame rate, adjust the surrounding lighting with the shutter speed, choose an adequate image resolution and focus the lenses. This is followed by the positioning of the specimen and software calibration that is necessary for measurement accuracy and precision. It implies the conformity of the camera's field of vision with the specimen's position as well as the definition of the control volume (a spatial region in which the displacements can be measured).

Software calibration is performed by means of a calibration plate – a black board with regularly distributed white, reference spots, Fig. 4. The distance between the reference points determines the dimensions of the control volume as well as the scale in the recorded images. The projected 2D positions of the reference points in the images from both cameras are combined to form 3D coordinate space. The calibration procedure is strictly defined [8]: the plate is repositioned and reoriented 13 times with a new set of images being recorded every time. The images are then processed in the ARAMIS software [9] and the result is a calibration file. To be considered successful, the calibration deviation should be below 0.04 pixels.

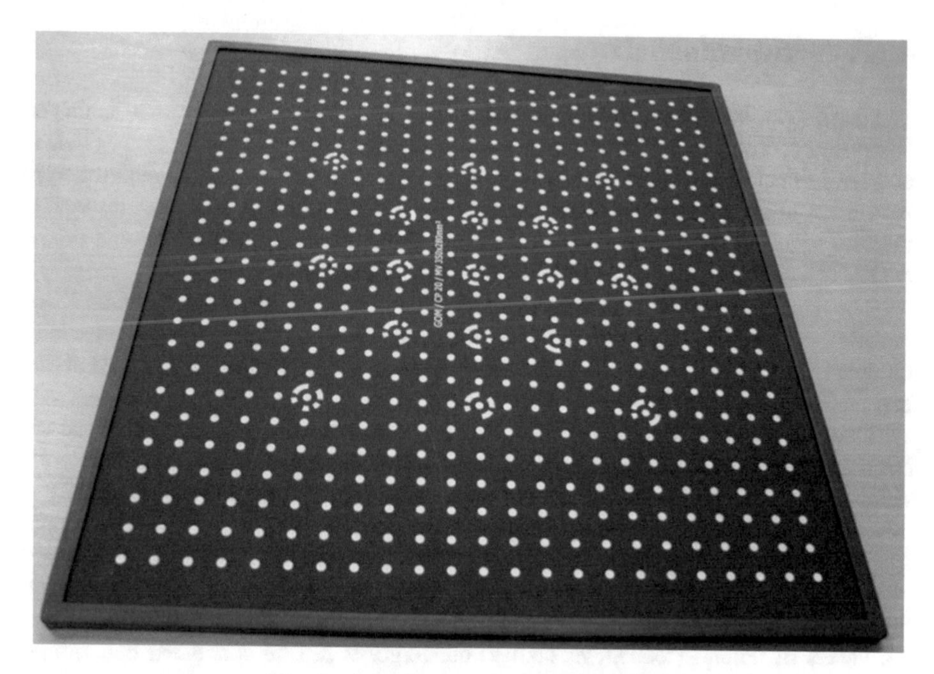

Fig. 4. An example of a calibration plate.

In all performed experiments, the plates were clamped at one edge and their movement could be regarded as 1 DOF. In the case of static loading, the force was applied to the other end. In the case of dynamic analysis, the free edge was excited and the plates allowed to oscillate freely. In both cases, the plate movement was recorded by DIC system since modal analysis, in its simplest form, requires only the measurement of the response of the structure [16].

For static displacement measurements it is sufficient to record only two images, of unloaded and deformed structure. On the other hand, the measurements of the structure's response lasted approximately T = 1.2 s with the frame-rate FR = 3000 fps. The recording started right after the excitation in order to capture higher-frequency modes that get damped quickly. The point of excitation was varied for the same reason.

Since the gathered data are numerous, usually, only a representative portion of the surface was processed. This was achieved by defining a mask that defines the borders of the domain of interest. That domain is then split into quadratic facets of particular dimensions, here (16 $\times$ 12) pixels. The points belonging to each facet are correlated (by statistical methods) in the "left" and "right" image to form a 3D displacement field [8].

The resulting displacement field can be either analyzed in whole in each stage, or information on the movement of a single point can be gathered throughout the stages to form a time-dependent displacement curve. Such a function can then be Fourier transformed to the frequency domain. High accuracy of the DIC system enables a precise determination of the natural frequencies and oscillation modes of the tested plate.

4 Numerical Simulations

Although very interesting numerical analysis of free vibration of various laminated composite geometries performed by shear deformation theory of higher order (TSDT) and iso-geometric analysis can be found in [11–13], here numerical simulations were performed in the commercial software package ANSYS. Its separate modules enable the user to create the geometry, computational grid and perform structural and modal analysis by FEM.

The virtual geometry corresponds in full to the tested geometry. The assigned material in the first case is aluminum with the following characteristics: elasticity modulus $E = 71000$ MPa and Poisson coefficient $v = 0.33$. The plate is clamped at one end and free-modal analyses are performed.

Due to the simple, regular shape of the specimen, the generated meshes are structured and consist of straight prismatic elements. Mesh density is controlled by the number of elements defined along the three outer edges. A grid convergence study is conducted to ensure the independence of numerical results on the computational meshes, Table 1. The number of elements column provides information on the number of elements created along the plate width, length and height respectively. By doubling the overall element number in every step, the accuracy of computation increases and the values of natural frequencies start to converge. It can be concluded that further division/refinement of the meshes is not necessary, since after the 3rd iteration the values are insignificantly affected by the computational meshes.

Table 1. The first 4 natural frequencies of the aluminum plate computed on different meshes.

No of elements	v_1 [Hz]	v_2 [Hz]	v_3 [Hz]	v_4 [Hz]
4 × 18 × 1	13.69	85.62	119.20	240.34
8 × 36 × 2	13.60	85.04	118.44	238.71
16 × 72 × 4	13.58	84.95	118.33	238.46
32 × 144 × 8	13.58	84.94	118.31	238.42

In the case of the composite plate, the geometry was modeled as planar and plate-type finite elements with assigned layering sequences were used. Characteristics of the defined material – Epoxy Carbon UD (230 GPa) Prepreg are listed in Table 2. Again, structured meshes were generated and a mesh convergence study was performed to assure the quality of numerical results, Table 3. Final meshes number over three thousand elements.

Table 2. Mechanical characteristics of Epoxy Carbon UD (230 GPa) Prepreg.

Quantity	Unit	Value
Density ρ	[g/cm^3]	1.49
Elasticity modulus E_x	[GPa]	121
Elasticity modulus E_y	[GPa]	8.6
Elasticity modulus E_z	[GPa]	8.6
Poisson coefficient ν_{xy}		0.27
Poisson coefficient ν_{yz}		0.4
Poisson coefficient ν_{zx}		0.27
Shear modulus G_{xy}	[GPa]	4.7
Shear modulus G_{yz}	[GPa]	3.1
Shear modulus G_{zx}	[GPa]	4.7
Layer thickness	[mm]	0.3

Figure 5 illustrates the global coordinate system (where *x*-axis corresponds to the main longitudinal direction of the fibers), as well as the defined element size and density of the finally adopted computational grid.

Table 3. The first 4 natural frequencies of the composite plate computed on different meshes.

Edge size [mm]	No of elements	ν_1 [Hz]	ν_2 [Hz]	ν_3 [Hz]	ν_4 [Hz]
10	304	24.216	83.805	151.75	282.84
5	1216	24.214	83.643	151.55	282.02
3	3429	24.213	83.509	151.51	281.56
2	7520	24.213	83.421	151.49	281.29

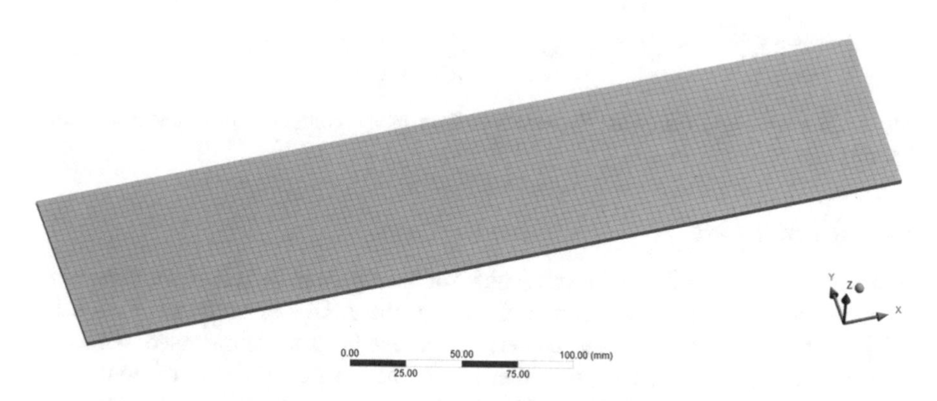

Fig. 5. An example of the generated computational mesh.

5 Results and Discussion

5.1 Static Case

As previously mentioned, the described testing procedure was first tried in a case of static loading, in particular, pure bending introduced in a form of a concentrated force $F = 3.914$ N (corresponding to a weight of 400 g) acting perpendicularly to the free edge of the composite plate. In the case of isotropic structure, the displacement field would follow a polynomial function of the 3rd degree whose shape is determined by the intensity of the acting force, geometrical properties and mechanical characteristics of the plate. Although, in this case, a composite plate was investigated, due to its ply lay-up sequence with all orientations at 0° it behaves in a similar manner. Figure 6 demonstrates the consistency between the measured and computed values by FEM. The measured equivalent bending elasticity modulus is $E = 115.61$ GPa, in comparison to the nominal value for a single lamina of $E_x = 121$ GPa.

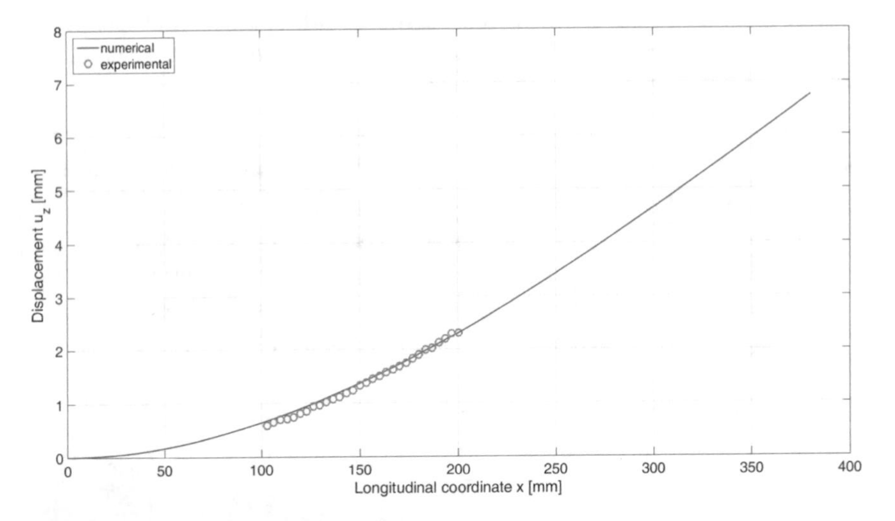

Fig. 6. Measured and computed displacement field of the composite plate under pure static bending.

5.2 Dynamic Case

In the case of the experimental investigations of the plate's modal characteristics, Fig. 7 illustrates the processing of recorded images in the ARAMIS software. Each set of images (from the left and right camera) corresponds to an enumerated stage. The resulting displacement in the *z*-direction (along the smallest plate dimension) in one moment in time measured in the global coordinate system of the control volume on a chosen portion of the plate surface is colored. Blue corresponds to the smallest, and red to the highest measured values. The graphs in the lower left parts of the figure represent the generated exemplary response curves.

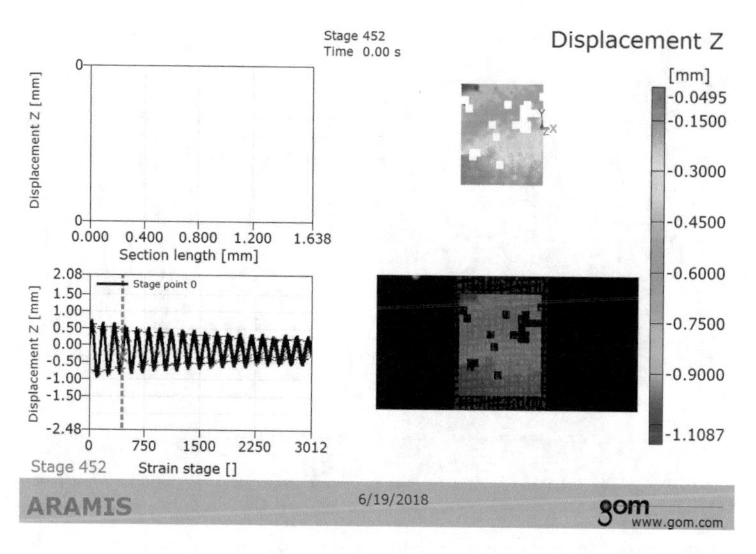

Fig. 7. Initial analysis in ARAMIS software.

The most representative response curves were then additionally analyzed. The complex vibration shape appears as a result of the momentary excitation of the structure. The superposition of several first natural modes of aluminum plate is clearly depicted in Fig. 8a. With the adopted experimental set-up and after the necessary Fourier transformations, it was possible to establish the values of the first three natural frequencies of the aluminum plate, which are clearly illustrated in Fig. 8b by the sudden peaks in the oscillation amplitudes. Since higher modes require high levels of energy for excitation and get damped very quickly, only first three natural frequencies could be registered. Experience gained during the experiment suggests that the point of excitation also significantly affects the number of modes that can be captured.

The composite plate was tested in a very similar manner to the previously described. Obtained and computed responses in time- and frequency-domain are presented in Fig. 9. It can be observed that the tested structure is quite rigid, i.e. damping coefficients are much higher for composite than for aluminum plate.

Tables 4 and 5 represent all analyzed data, both experimentally and numerically obtained, referring to both aluminum and composite plate. Overall, the obtained results can be considered quite satisfactory. The discrepancies that appear between the experimental and numerical values can, at least partially, be explained by the certain irregularities in the real structure that are not simulated by the idealized geometry (e.g. perfectly straight edges, material uniformity, etc.). Nonetheless, the trend of the results is accurately captured.

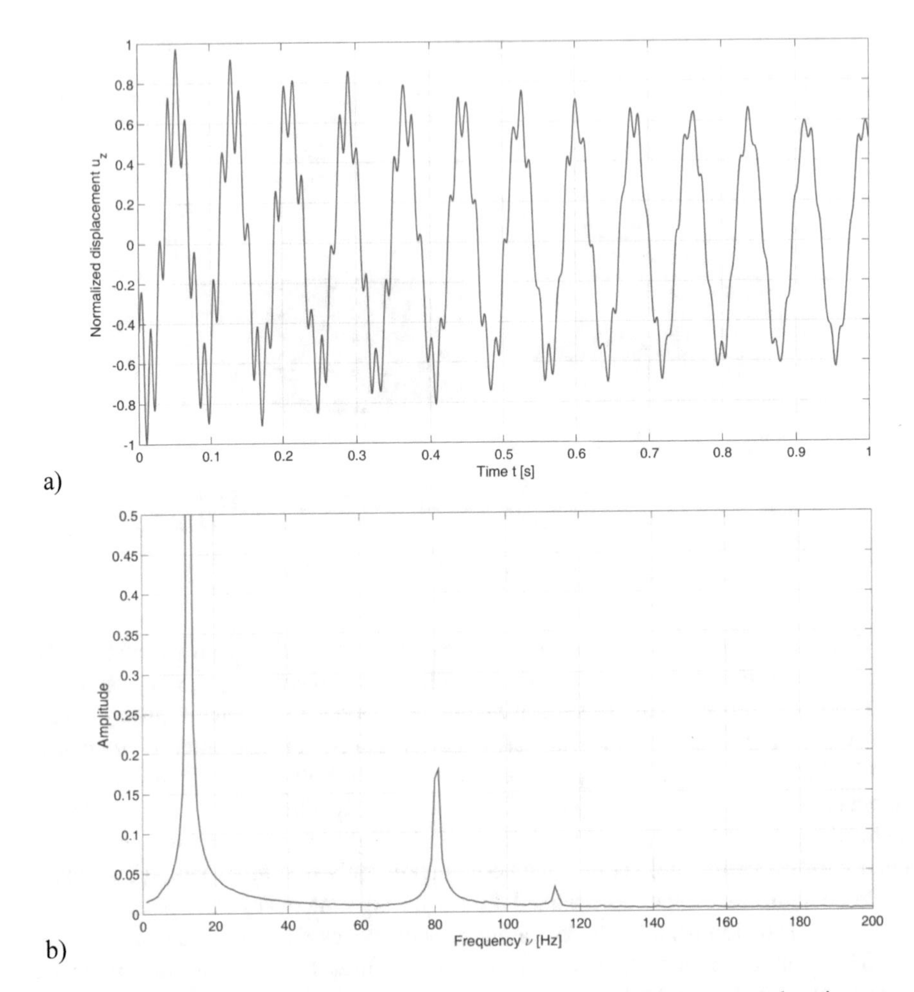

Fig. 8. Response of the aluminum plate in – (a) time and (b) frequency domain.

Table 4. Comparison between the first 3 natural frequencies of the aluminum plate obtained by experimental and numerical method.

Method	Experimental	Numerical
ν_1 [Hz]	13.0 ± 0.5	13.58
ν_2 [Hz]	81.0 ± 0.5	84.94
ν_3 [Hz]	113.0 ± 0.5	118.31

Table 5. Comparison between the first 3 natural frequencies of the composite plate obtained by experimental and numerical method.

Method	Experimental	Numerical
ν_1 [Hz]	24.0 ± 0.5	24.21
ν_2 [Hz]	80.0 ± 0.5	83.50
ν_3 [Hz]	150.0 ± 0.5	151.51

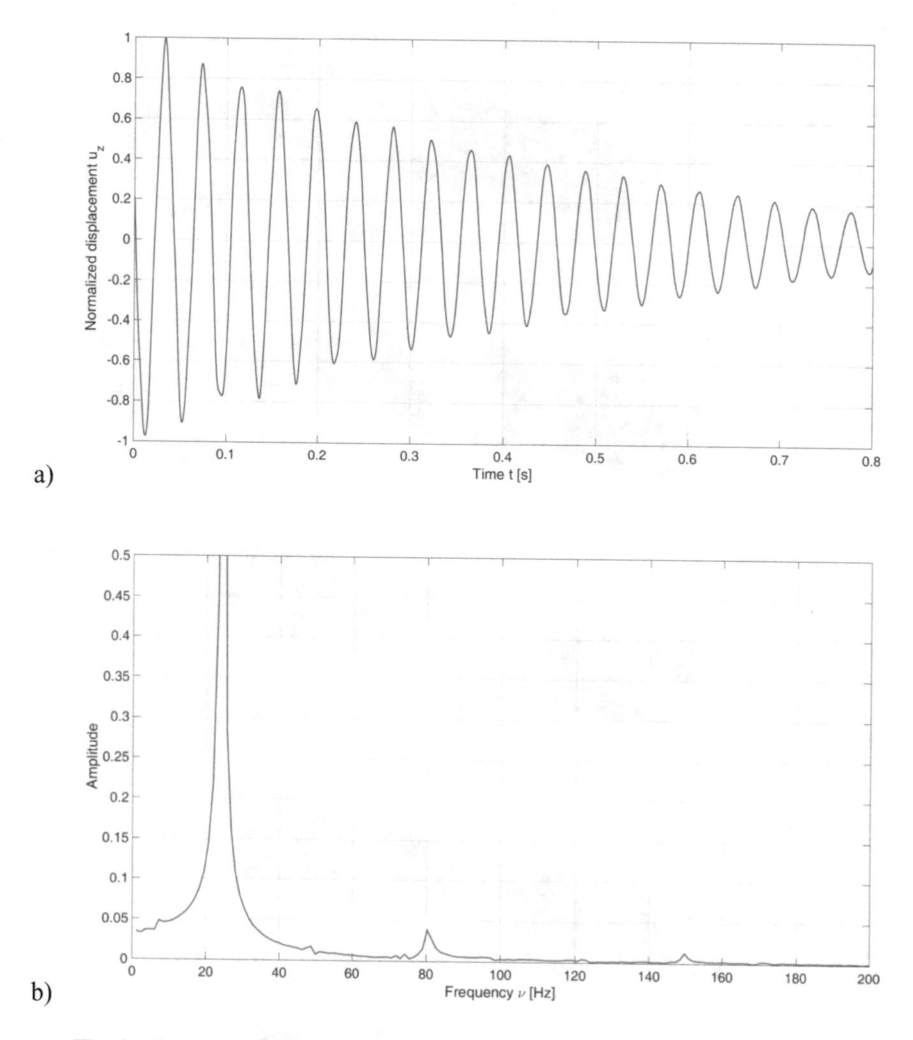

Fig. 9. Response of the composite plate in – (a) time and (b) frequency domain.

Another observation that can be made is that, for aluminum plate, there is a tendency of growing deviation between the two sets of results at higher frequencies (i.e. $\Delta\nu_1 \approx 0$ Hz, $\Delta\nu_2 \approx 3$ Hz, $\Delta\nu_3 \approx 5$ Hz, etc.). Nonetheless, the relative difference between the experimental and numerical results is below 5% for all modes which is acceptable and falls within the safety margin of engineering applications.

The situation is different for the considered composite plate. The first bending mode is absolutely dominant, regardless of the point of impact/excitation, while higher tones get damped very quickly and even some difficulties with their post-processing may appear. This statement can be confirmed by Fig. 9b that shows the significant contrast between the first compared to the other two registered amplitudes. Overall, bending modes can be captured with higher accuracy ($\Delta\nu_1 \approx 0$ Hz, $\Delta\nu_3 \approx 1$ Hz) while the difference between the computed and measured value of the second natural, torsional

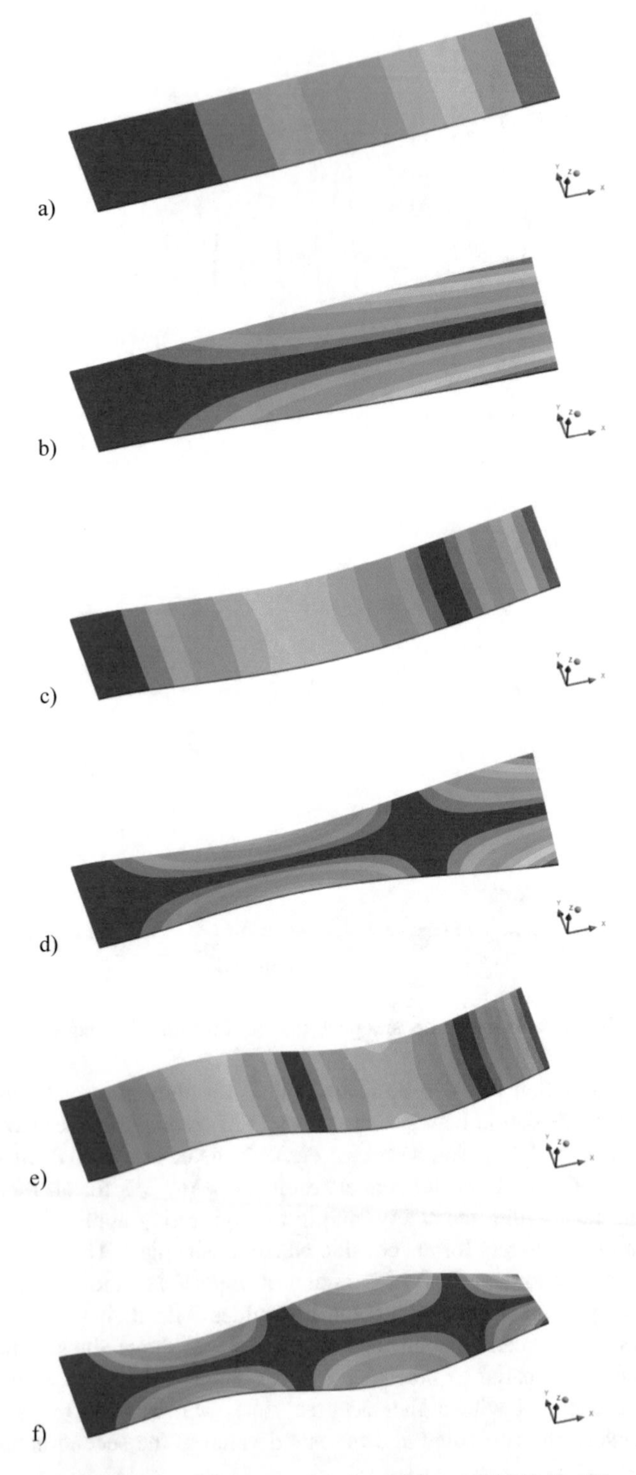

Fig. 10. Computed mode shapes of the composite plate – (a) ν_1 = 24.213 Hz, (b) ν_2 = 83.509 Hz, (c) ν_3 = 151.51 Hz, (d) ν_4 = 281.56 Hz, (e) ν_5 = 423.36 Hz, (f) ν_6 = 568.68 Hz.

frequency is $\Delta\nu_2 \approx 3$ Hz. Still, the discrepancy between the two sets of data in percents remains below 4%.

Another type of results that can be extracted from numerical analyses is modal shapes. They are not limited by real-time responses and can produce an arbitrary number of natural modes. Here, the first six modes for the composite plate are illustrated in Fig. 10. Depicted natural modes of the composite plate correspond to the first three flexural (modes 1, 3 and 5) and torsional (modes 2, 4 and 6).

Apart from the values of the natural frequencies, an additional check of the experimental results that should be performed is mode shape inspection (i.e. whether the captured values correspond to the actual mode shapes or are a mere consequence of a noisy signal). One of the possible quantifiers that can be used is Modal Assurance Criteria (MAC) [17].

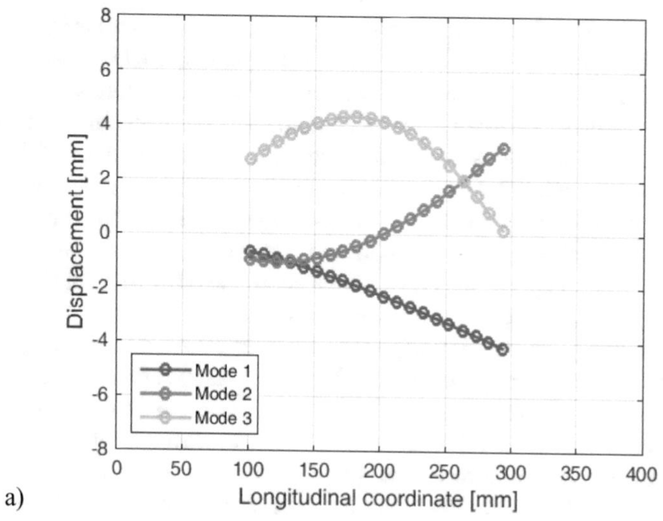

a)

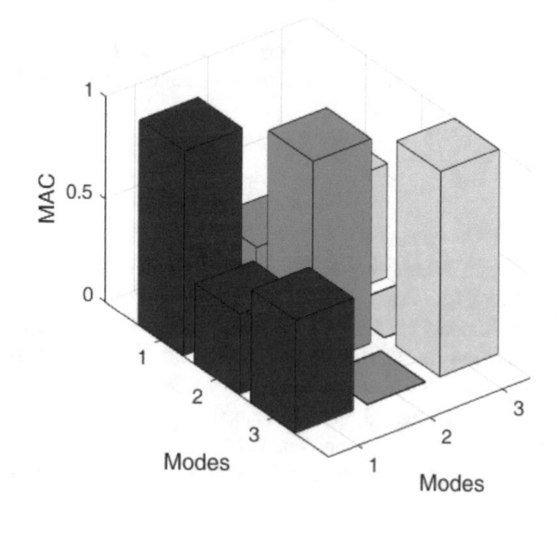

b)

Fig. 11. (a) Filtered plate response at particular modes, (b) Computed MAC.

MAC represents a scalar value that equals 1 for the two identical mode shapes and 0 for two completely dissimilar mode shapes. Since measurements are always performed in a finite number of points, a part of the original information gets lost in the process. Therefore, this criterion proves quite useful for comparison of registered/measured displacements to the corresponding numerically computed ones. In this case, it was estimated from the values of displacement along an arbitrary longitudinal cross-section, Fig. 11. The values of MAC very close to 1 along the mode matrix diagonal prove that the captured signals do actually correspond to the first three natural mode shapes of the composite plate.

6 Conclusion

Modal analysis of a given structure is very important for the determination of the structure's functionality and behavior under different expected and unpredicted static and dynamic load cases. For accurate determination of structure's mechanical characteristics, it is often necessary to combine numerical and experimental techniques. In this work, the DIC method was used for recording and measurement of the structure's response to excitation. After the transformation to the frequency domain, several first natural frequencies and oscillation modes were detected and quantified. In order to validate the applied technique, the obtained experimental results were compared to numerical values computed by FEM. Verification was also performed in the case of static loading, i.e. pure bending of the composite plate.

Although the presented investigation was performed on simple geometry, the obtained results confirm its procedures and methods can be applied to a wide spectrum of structures for its simplicity, efficiency and short performance time. Neither structure geometry nor size presents a limitation to the tested measurement technique. Another demonstrated application is to anisotropic composite materials and structures whose mechanical properties differ along with the referent directions and whose response to applied loading may be difficult to predict.

Acknowledgement. The research work is funded by the Ministry of Education, Science, and Technological Development of Republic of Serbia through Technological Development Project no. 35035.

References

1. Adams, D.E.: Health Monitoring of Structural Materials and Components: Methods with Applications. Wiley, Chichester (2007)
2. Goidescu, C., Welemane, H., Garnier, C., Fazzini, M., Brault, R., Péronnet, E., Mistou, S.: Damage investigation in CFRP composites using full-field measurement techniques: combination of digital image stereo-correlation, infrared thermography and X-ray tomography. Compos. Part B Eng. **48**, 95–105 (2013)
3. Spasic, D.M., Stupar, S.N., Simonovic, A.M., Trifkovic, D., Ivanov, T.D.: The failure analysis of the star-separator of an aircraft cannon. Eng. Fail. Anal. **42**, 74–86 (2014)

4. Reu, P.: Introduction to digital image correlation: best practices and applications. Exp. Tech. **36**(1), 3–4 (2012)
5. Hild, F., Roux, S.: Comparison of local and global approaches to digital image correlation. Exp. Mech. **52**(9), 1503–1519 (2012)
6. Caminero, M.A., Lopez-Pedrosa, M., Pinna, C., Soutis, C.: Damage assessment of composite structures using digital image correlation. Appl. Compos. Mater. **21**(1), 91–106 (2014)
7. Gerbig, D., Bower, A., Savic, V., Hector, L.G.-Jr.: Coupling digital image correlation and finite element analysis to determine constitutive parameters in necking tensile specimens. Int. J. Solids Struct. **97–98**, 496–509 (2016)
8. PONTOS/ARAMIS HighSpeed PHOTRON SA1 User Information. GOM mbH, Braunschweig, Germany (2008)
9. ARAMIS v6.3 and higher User Manual-Software. GOM mbH, Braunschweig, Germany (2013)
10. Baltić, M., Svorcan, J., Perić, B., Vorkapić, M., Ivanov, T., Peković, O.: Comparative numerical and experimental investigation of static and dynamic characteristics of composite plates. J. Mech. Sci. Technol. **33**(6), 2597–2603 (2019)
11. Chaubey, A.K., Kumar, A., Chakrabarti, A.: Vibration of laminated composite shells with cutouts and concentrated mass. AIAA J. **56**(4), 1662–1678 (2018)
12. Peković, O., Stupar, S., Simonović, A., Svorcan, J., Komarov, D.: Isogeometric bending analysis of composite plates based on a higher-order shear deformation theory. J. Mech. Sci. Technol. **28**(8), 3153–3162 (2014)
13. Peković, O., Stupar, S., Simonović, A., Svorcan, J., Trivković, S.: Free vibration and buckling analysis of higher order laminated composite plates using the isogeometric approach. J. Theor. Appl. Mech. **53**(2), 453–466 (2015)
14. Posteljnik, Z., Stupar, S., Svorcan, J., Peković, O., Ivanov, T.: Multi-objective design optimization strategies for small-scale vertical-axis wind turbines. Struct. Multidiscip. Optim. **53**(2), 277–290 (2016)
15. Tarigopula, V., Hopperstad, O.S., Langseth, M., Clausen, A.H., Hild, F.: A study of localisation in dual-phase high-strength steels under dynamic loading using digital image correlation and FE analysis. Int. J. Solids Struct. **45**(2), 601–619 (2008)
16. Geradin, M., Rixen, D.J.: Mechanical Vibrations: Theory and Application to Structural Dynamics, 3rd edn. Wiley, Chichester (2015)
17. Ewins, D.J.: Modal Testing: Theory, Practice and Application. Research Studies Press Ltd., Letchworth (2000)

Numerical Simulation of Crack Propagation in Seven-Wire Strand

Gordana Kastratović[1(✉)], Nenad Vidanović[1], Aleksandar Grbović[2], Nikola Mirkov[3], and Boško Rašuo[2]

[1] Faculty of Transport and Traffic Engineering, University of Belgrade, Vojvode Stepe 305, 11010 Belgrade, Serbia
g.kastratovic@sf.bg.ac.rs

[2] Faculty of Mechanical Engineering, University of Belgrade, Kraljice Marije 16, 11120 Belgrade, Serbia

[3] Institute of Nuclear Sciences - Vinča, Laboratory for Thermal Engineering and Energy, University of Belgrade, Mike Petrovića Alasa 12-14, 11351 Belgrade, Serbia

Abstract. This paper discusses certain aspects of numerical simulation of crack propagation in wire ropes subjected to axial loading, with the aim to explore and to demonstrate the capacity, performances and difficulties of crack propagation modeling by usage of numerical computational methods in such complex structures. For this purpose, the finite element method (FEM) was used, and 3D numerical analyses were performed in Ansys Workbench software. In order to validate and verify performed numerical modeling, crack growth rate based on calculated stress intensity factors (SIFs) along the crack fronts was obtained for the model for which experimental results could be found in the available literature. Finally, using the advanced modeling techniques, the parametric 3D model of seven-wire strand was analyzed. Conducted analysis showed that FEM could be a powerful tool for fatigue life predictions in order to reduce the need for experiments, which are still the only successful method for fatigue life estimation of wire ropes.

Keywords: Crack propagation · Finite element method · Stress intensity factors · Wire rope strand

1 Introduction

Because of their flexibility and high strength, wire ropes are very important structural members, which are used for transmitting tensile forces. They are in widespread uses in various industries. Their long service life make them highly susceptible to fatigue, which increases the possibility of diminishing, or even the total loss of their structural integrity due to fatigue crack occurrence. This requires a good estimate of fatigue life after crack initiation. However, experimental work is still the only successful method for fatigue life estimation of complex damaged structures like wire ropes. Since this kind of experimental work requires specific, large and expensive testing devices,

N. Mitrovic et al. (Eds.): CNNTech 2019, LNNS 90, pp. 76–91, 2020.
https://doi.org/10.1007/978-3-030-30853-7_5

numerical analysis such as finite element analysis (FEA), as non-destructive method, has to be employed in the wire rope behavior analysis.

Over the last few decades, many numerical models of wire ropes were developed to analyze mechanical behavior of wire ropes under various load conditions.

One of the first finite element model of simple straight strand has been presented in [1, 2] where contact and friction were taken into consideration. Three-dimensional solid elements were used for structural discretization and helical symmetry of the strand was used to establish accurate boundary conditions. In papers [3, 4] Páczelt and Beleznai presented the application of the p - version of beam finite elements, together with special contact elements in the case of one - and two - layered wire rope strands. They applied Hertz theory and Coulomb dry friction law in the newly developed computer code, and considered relative displacement, wear and Poisson-effect, contact and friction between the wires.

In [5] authors presented 3D solid model and numerical analysis of independent wire rope core subjected to axial loading. They have also been taking in consideration frictional effects between the wires. Stanova et al. [6, 7] studied the elastic behaviors of a multi-layered strand with a construction of 1 + 6 + 12 + 18 wires under tensile loads by using Abaqus/Explicit software. They made the mathematical representation of single and double helixes and used parametric equations with varied input parameters to determine the centerline of an arbitrary circular wire. Kastratović et al. [8] explored some aspects of 3D modeling of a sling wire rope using the FEM based computer program with special emphasis on different types of contact and different types of axial loading. Same authors carried out the load distributions analysis of wire strands and ropes in case of two different types of axial loading (evenly distributed axial force and axial strain) and two different types of contacts (linear bonded and nonlinear frictional contact) [9]. Their work shows that the obtained numerical results were affected by the way the load is applied and by the selected contact model. Authors in [10] proposed a new parametric geometric model of a spiral triangular strand by using parametric equations for the wire center lines implemented in Pro/Engineer software. Based on the proposed geometric model, 3D finite element models of the spiral triangular strand were established in which some important influence factors, such as the elasto-plastic property of wire, inter-wire friction and contact, were considered.

Mouradi et al. in [11] did an extensive survey of literature regarding investigation on the main degradation mechanisms of steel wire rope. They have been determined that, when fatigue is concerned, all conducted studies were based on experimental analyses. This trend is still prevailing in more recent studies, like in [12], where conducted research was also based on experimental work. It dealt with the experimental investigation of mechanical response and fracture failure behavior of wire rope with different given surface wear. But, in [13] FEA and simulations were used to determine internal stress fields and locations of maximal contact pressure in order to define where fretting fatigue damage will occur. Also, Chang et al. in [14] used FEA together with experimental analysis to study the wire rope strand with different wear scars and to analyze its fracture failure behavior. However, none of the mentioned studies used FEA to investigate fatigue crack growth in wire ropes strands.

It is well known, that the stress intensity factor (SIF) is one of the most important fracture mechanics parameter, since it provides data on the crack initiation and propagation, i.e. for the fatigue life estimation. In complex geometries, it is almost impossible to find an exact solution for SIFs; therefore, the numerical methods, such FEM are needed for their estimation. The FEM has been used for decades for calculating SIFs. It has been constantly improving and upgraded by the researchers [15, 16] with the intent to simplify and quicken its implementation in SIFs computations. This has led to successful application of improved FEM, in number of problems regarding crack growth simulation in complex 3D structures [17–20].

On the other hand, it is clear that the researchers are still struggling to apply FEA to simulate fatigue crack growth in wire ropes in order to estimate their fatigue life after crack initiation. The main reason for this problem is the complex geometry of wire ropes and difficulties in creating theirs suitable 3D models. The complexity of this issue is additionally increased by the crack growth modeling, as well. Therefore, the aim of this study is to explore and to demonstrate the capacity, performances and difficulties of crack propagation modeling in wire ropes strands by the usage of FEM.

2 Finite Element Models and Analyses

2.1 Validation Analysis

In order to validate proposed numerical modeling, FEA was carried out on a model for which experimental results could be found in the available literature. The FEM was applied thru ANSYS software. It has to be noted that, when wire ropes are concerned, experimental testing is usually conducted on a single wire (round bar) [13, 14, 21–24]. However, according to Zheng et al. [23] a unified test standard with highly efficient operations is still absent for this type of research. Thus, they proposed a single edge cracked sheet test method for fast determination of the crack propagation characteristics of steel wires. Those experimental results were used here for validation purposes.

The model used in this analysis is based on the specimen that represents a modified round bar with the diameter of 7 mm, total length of 150 mm and thickness of the central section of 2 mm. A 5 mm diameter arc chamfered the two sides of the central section. A notch is positioned on one side of the sheet, with the depth of 0.9 ± 0.1 mm, and the root radius of 0.1 mm.

All other data, regarding the test specimen geometry, manufacturing, material data and loading conditions applied in the experiment can be found in detail in [23]. Those data were used for the creation of the 3D geometry model (Fig. 1) and finite element model (Fig. 2) of the specimen.

The parametric geometrical model was created by using CAD commercial software and imported to Ansys Workbench. This program provides generation of finite element mesh, application of loads, and contact definition and it was used for solving and obtaining necessary output data. The patch conforming mesh method with tetrahedral elements was applied for resolving high stress concentrations occurring in the crack region.

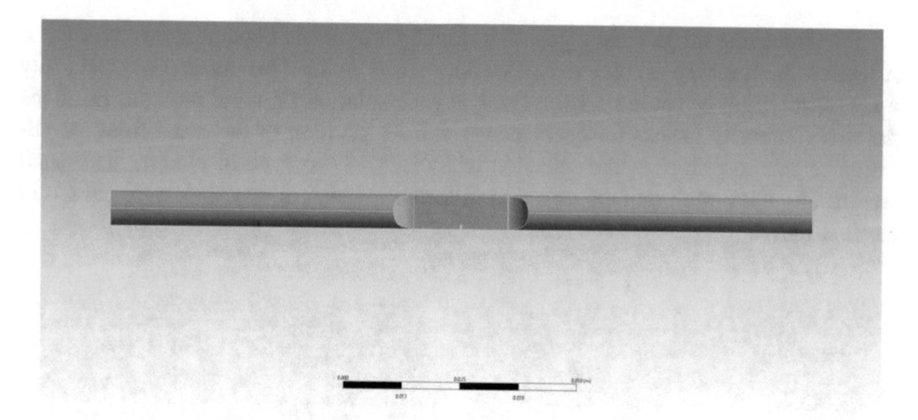

Fig. 1. 3D geometry model of analyzed specimen.

The mesh was created with higher order 3D10-nodes solid element SOLID 187, that exhibit quadratic displacement behavior. The element is defined by 10-nodes having three degrees of freedom per node: translations in the nodal *X*, *Y*, and *Z* directions. The element supports plasticity, hyper-elasticity, creep, stress stiffening, large deflection, and large strain capabilities [25]. Since very important issue in any FEA is, beside element selection, the size of the mesh, a special attention was paid to this and 3D model were made with optimum numbers of elements, which in this case was 332549 (480728 nodes). In addition, a very refined mesh was used in the region around the crack, in order to capture rapidly varying stress and deformation fields around the crack front (Fig. 2).

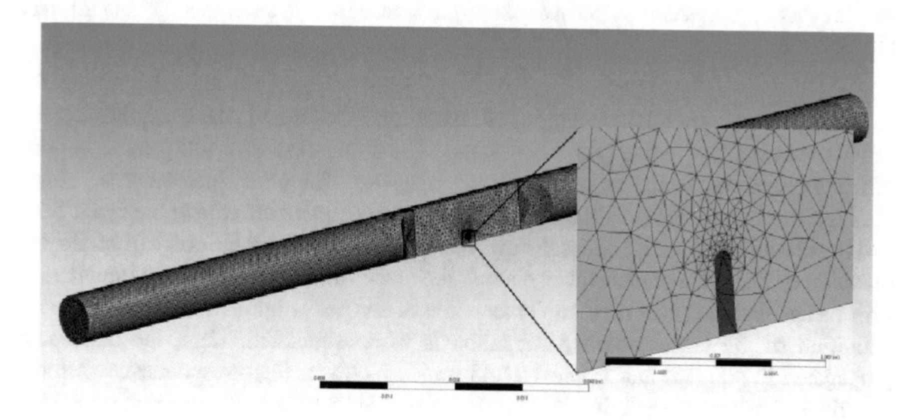

Fig. 2. FEM model of specimen.

Applied boundary conditions and load were in accordance with testing conditions, where the crack propagation characteristics under maximal stress of 500 MPa and stress R ratio of 0.7 were analyzed.

For simulating fatigue crack growth a recently introduced **S**eparating **M**orphing and **A**daptive **R**e-meshing **T**echnology (SMART) was used. This feature automatically updates the mesh but only near the crack at each solution step and provides solutions for stress intensity factors for every newly defined position of the crack front. Application of this tool implies the selection of SOLID 187 as type of element and patch conforming tetra mesh method as meshing method, which is why they were chosen in the first place.

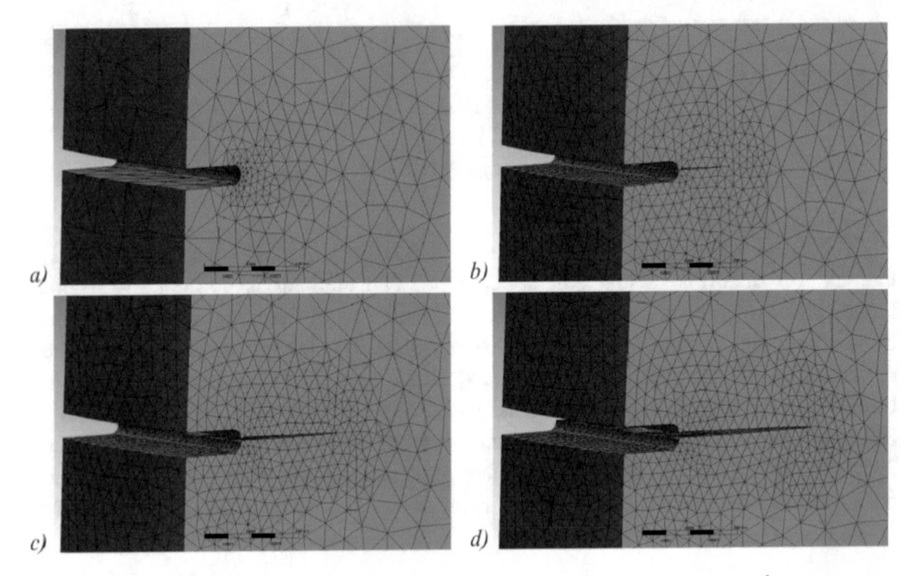

Fig. 3. Crack propagation: (a) initial crack; (b) crack after 15th; (c) crack after 30th and (d) crack after 40th step.

Fatigue crack growth calculations are based on Paris' Law, which requires special fatigue material properties (Paris constants). The simulation of crack growth is done thru the calculation steps in which crack propagates. The crack increment is defined between minimum, which is 0.25 times the average element size along the crack front and maximum which is 1.5 times the average element size along the crack front. So, the selection of the average element size along the crack front is of paramount importance. This means that the successful implementation of SMART highly depends on the mesh definition of the crack which propagation is to be simulated. Here, the crack was propagated in 40 steps and reached 2.743 mm, which was in correspondence with the experiments where the maximum cracks size reached was 3 mm. Its shape after 15th, 30th and 40th steps is shown in Fig. 3.

As already explained, the finite element analysis was validated by comparing obtained results with the results available in the literature. The results for the stress intensity factor range ΔK as a function of crack size a are shown in Fig. 4. The stress intensity factor range can be defined as:

$$\Delta K = (1 - R) \cdot K_{\max} \tag{1}$$

where R is stress ratio and K_{max} is SIF for maximum load stress.

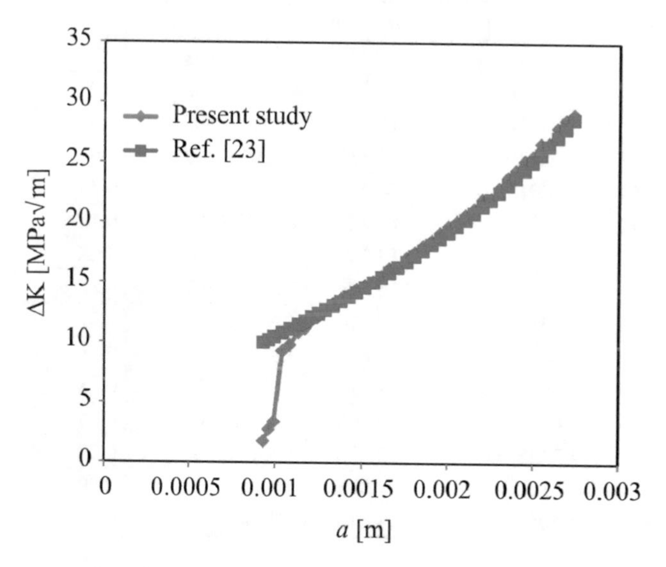

Fig. 4. Stress intensity factors of analyzed model.

As shown in Fig. 4, after first five steps, when the stable crack growth was established, an excellent agreement was obtained for stress intensity factors. The differences are well under 5%.

Crack growth rates were also determined and compared. Results for crack length up to 2.743 mm in FEM analysis and in experiment are shown in detail in Table 1 and in Fig. 5. It can be seen the crack growth rate determined by improved finite element method is in well coherence with the results obtained in the experiment. Here the differences are under 10%. From engineering point of view, it is also important to notice that crack growth rate obtained numerically is lower than in the experiment, i.e. numerical simulation provides conservative estimation. All mentioned leads to the conclusion that numerical model is very well established.

Table 1. FEM and experimental results for analyzed model.

FEM analysis			Experiment Ref. [22]		ΔK diff.	Crack growth rate diff.
No. of step	ΔK [MPa$\sqrt{m}$]	Crack growth rate [m/cycles]	ΔK [MPa$\sqrt{m}$]	Crack growth rate [m/cycles]	[%]	[%]
4	9.312	3.4758E−08	9.529	3.8467E−08	2.28	9.64
6	10.872	4.5772E−08	10.767	5.7992E−08	0.97	21.07
8	11.821	7.0591E−08	11.902	7.8168E−08	0.68	9.69
11	13.247	1.0883E−07	13.323	1.0937E−07	0.57	0.49
14	14.443	1.5001E−07	14.318	1.5884E−07	0.87	5.56
15	14.794	1.6226E−07	14.820	1.6799E−07	0.18	3.41
17	15.506	1.9078E−07	15.340	1.8442E−07	1.08	3.45
18	16.208	2.1111E−07	16.179	2.2225E−07	0.18	5.01
20	17.075	2.5596E−07	17.010	2.6784E−07	0.38	4.44
21	17.524	2.7667E−07	17.662	3.0521E−07	0.78	9.35
22	18.083	3.1631E−07	18.110	3.4137E−07	0.15	7.34
24	19.168	3.6785E−07	19.160	4.0379E−07	0.04	8.90
26	20.134	4.3328E−07	20.208	4.5163E−07	0.37	4.06
29	21.971	5.8121E−07	21.991	6.3190E−07	0.09	8.02
39	28.614	1.4087E−06	28.610	1.3540E−06	0.02	4.04

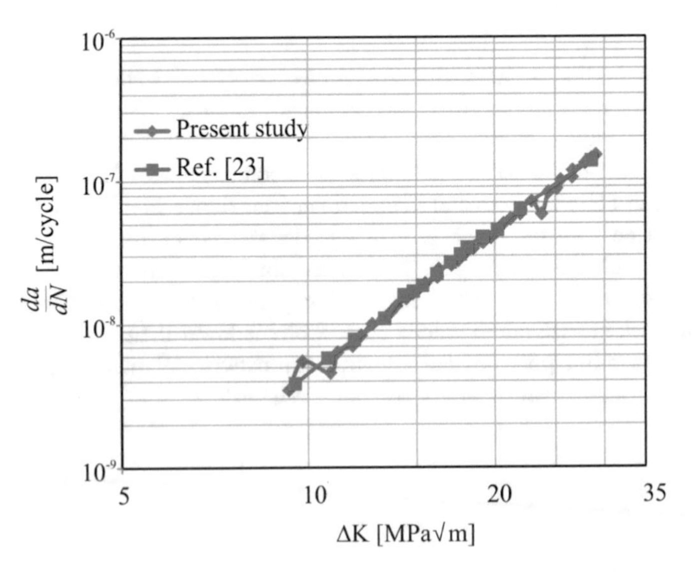

Fig. 5. Crack growth rate vs. stress intensity factor (log scale).

2.2 Crack Propagation in Seven-Wire Strand

Further, the seven-wire strand has been analyzed. The radius of center wire is $r_1 = 1.97$ mm, and outer wire $r_2 = 1.865$ mm, and pitch length was 70 mm, which corresponds to helix angle $\alpha = 71.005^\circ$. The overall length of the strand model is L = 70 mm (Fig. 6). The parametric geometrical model of seven-wire strand was created by using CAD commercial software and imported to Ansys Workbench.

Creating of the CAD model required a special effort. The wires had to be created as solids which do not penetrate each other, with continuous line contact between center wire and outer wires. This was very important since inadequate CAD geometry extremely affects FEA. In this case, according to [26]:

$$r_2\sqrt{1+\frac{tg^2\left(\frac{\pi}{2}-\frac{\pi}{m_2}\right)}{\sin^2\alpha}}<r_1+r_2 \tag{2}$$

where m_2 is the number of the outer wires, there was no contact between the outer wires. Initially, wires were created as swept surfaces.

The guideline for center wire was a straight line (axis of the center wire). Then, first auxiliary surface was created with profile length $r_1 + r_2$ (Fig. 6a). Second auxiliary surface was created with the same profile length, but with the first auxiliary surface as a reference and with the reference angle which was equal to helix angle α (Fig. 6b). The outer edge of this surface represents the helix guideline for creating the surface of the first outer wire (red line in Fig. 6c). The procedure is then repeated for all remaining outer wires, having in mind that reference angle is increased for every next wire by 60° (since there are 6 outer wires around the center one) (Fig. 6d and e). In this manner it was certain that line contact between the wires was achieved. At the end, solids were created from obtained surfaces (Fig. 6f).

All settings regarding finite element modeling, like mesh generation, boundary conditions and crack propagation settings were adjusted as in previous analysis.

Here, another important issue are the contacts between the wires, which must be taken into consideration. They determine how the wires move relative to one another and the distribution of load between them. As already mentioned, the only existing contacts were the ones between the center wire and outside wires. The model was meshed with 178242 elements (262670 nodes) and the prescribed boundary conditions were used for solving. At one end of the model the degree of freedom in all three directions were constrained, and on the other end constant 500 MPa stress was applied (Fig. 7).

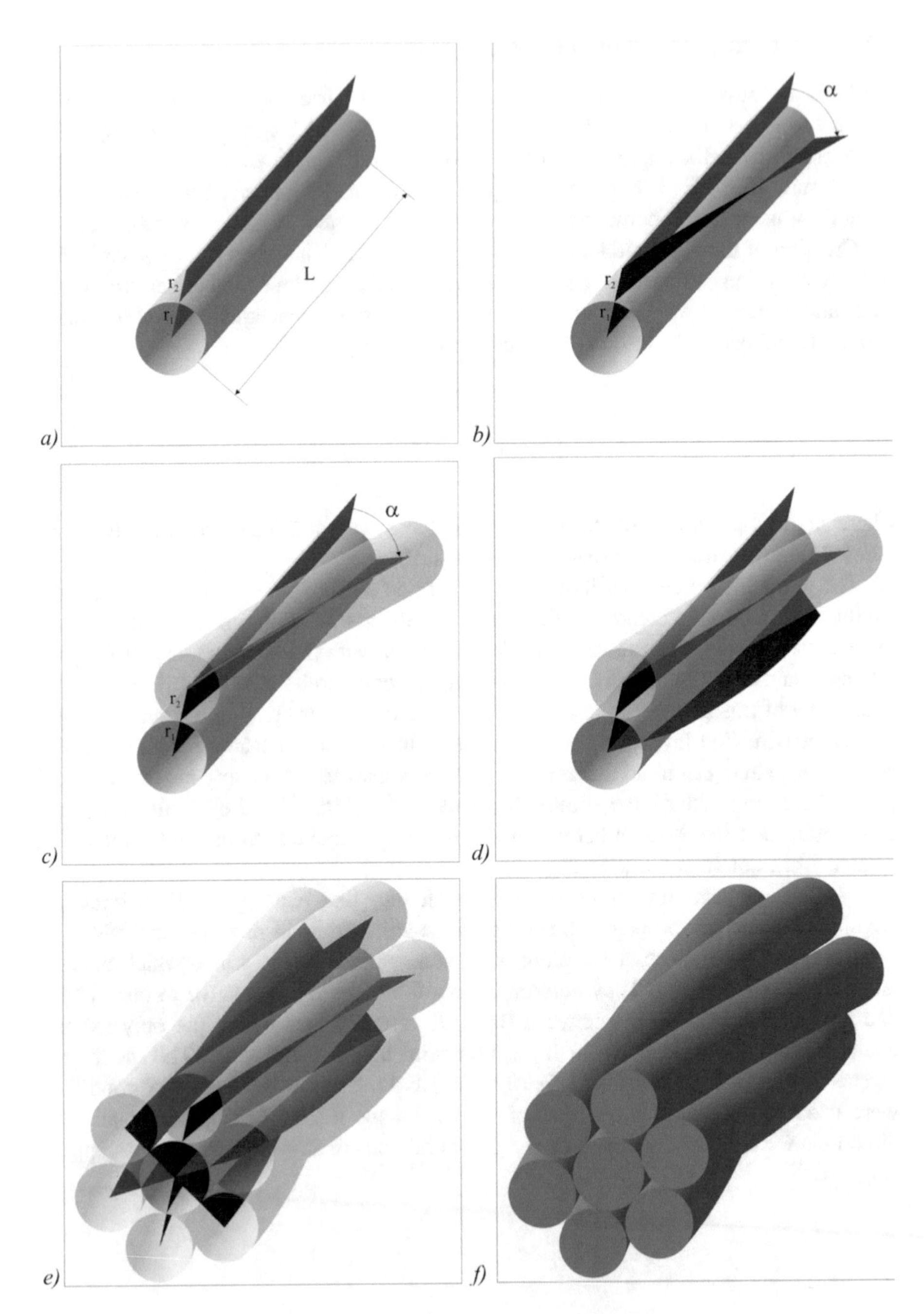

Fig. 6. CAD model of seven-wire rope.

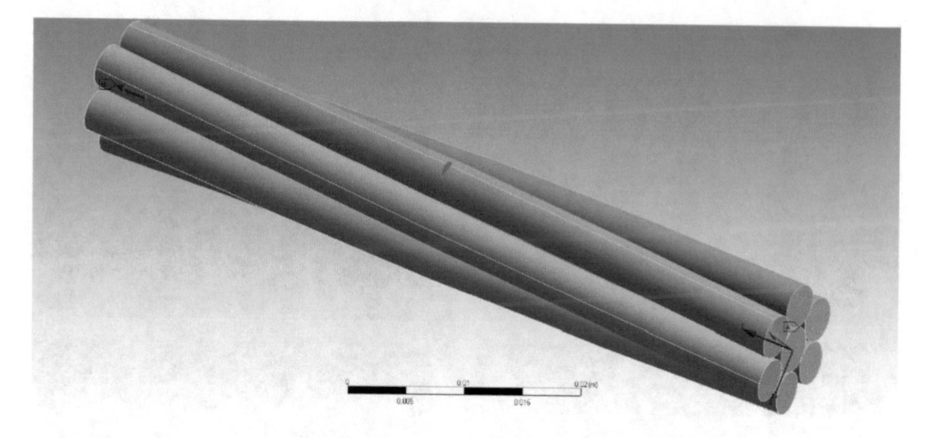

Fig. 7. 3D model and boundary conditions

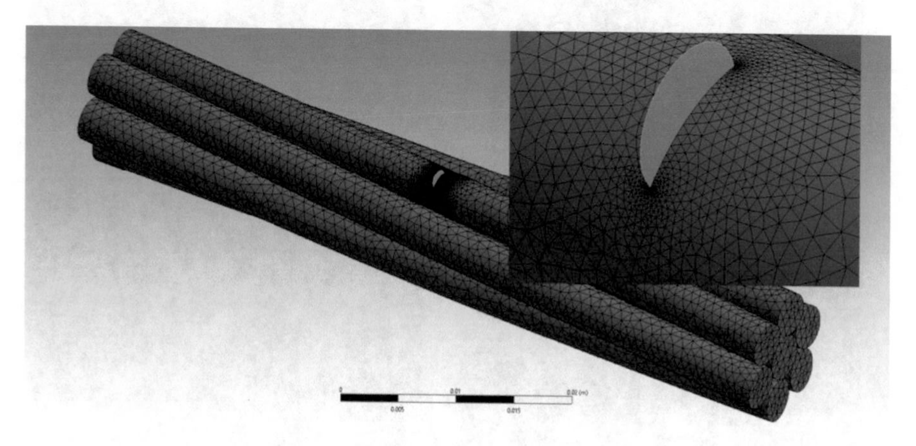

Fig. 8. FEM model of seven-wire strand

The crack propagation under the axial loading behavior was analyzed. The material characteristics were as in previous analysis, i.e. they were taken from [23]. The initial crack was a 0.8 mm in radius penny shaped crack. A very refined mesh was used in this case as well, in the region around the crack (Fig. 8).

As before, for simulating fatigue crack growth SMART feature was used. It has to be noted that contact between the wires were modelled as bonded. Bonded contact is a linear type of contact. It is the default configuration of contact and applies to all contact regions (surfaces, solids, lines, faces, edges). If contact regions are bonded, then no sliding or separation between faces or edges is allowed. For this type of contact a multipoint constraint (MPC) formulation is available. The program builds MPC equations internally based on the contact kinematics. This method enforces contact compatibility by using internally generated constraint equations to establish a relationship between the two surfaces. No normal or tangential stiffness is required. Since there is no penetration or contact sliding within a tolerance, MPC represents "true linear contact"

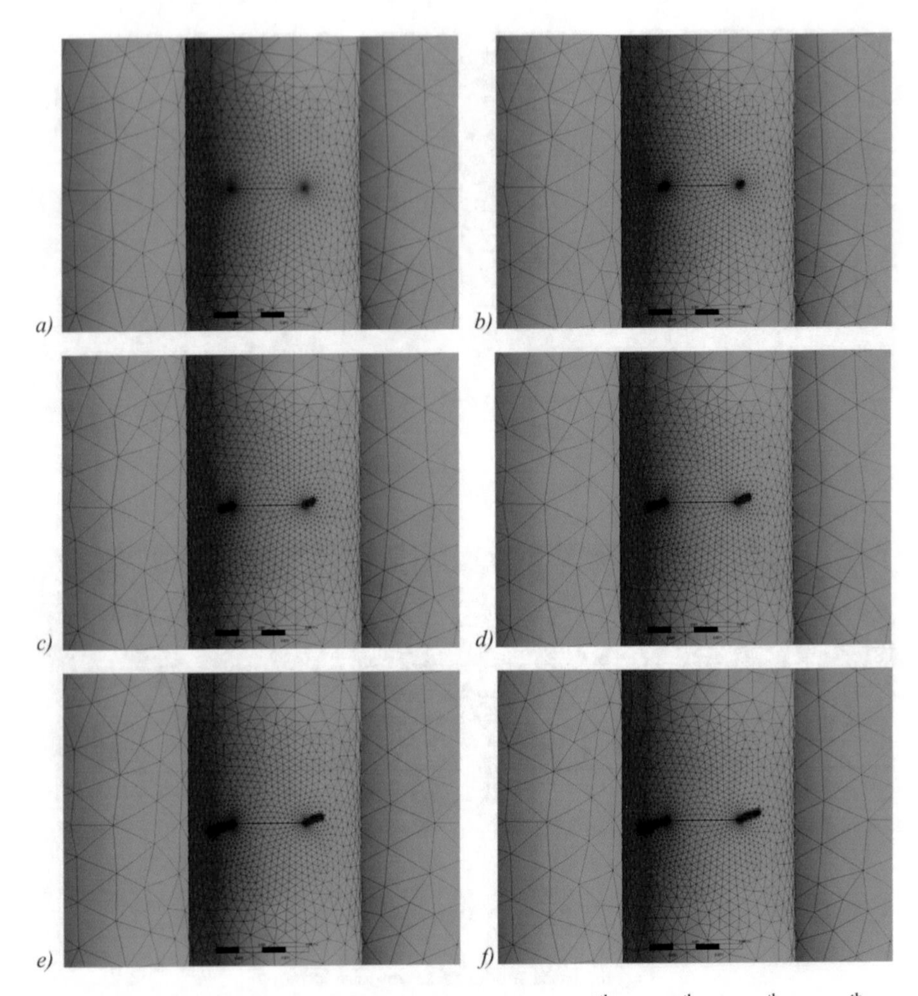

Fig. 9. Crack propagation: (a) initial crack; (b) crack after 8th, (c) 15th, (d) 24th, (e) 33th and (d) 40th step.

behavior. Although fictional contact was also available, the friction between the wires was not taken into consideration, since this type of contact is not supported by SMART.

Again, the crack was propagated in 40 steps. It was limited by the computer resources. In this analysis crack reached the size of 1.6 mm. Its shape after 8th, 16th, 24th 33th and 40th steps is shown in Figs. 9 and 10. It should be noted that during the simulation for every propagation step new elements were created in the vicinity of the crack, but as it can be observed in Figs. 9 and 10 there was no re-meshing of the other wires in the strand. At the end of the simulation, total number of element was 526419 (726916 nodes).

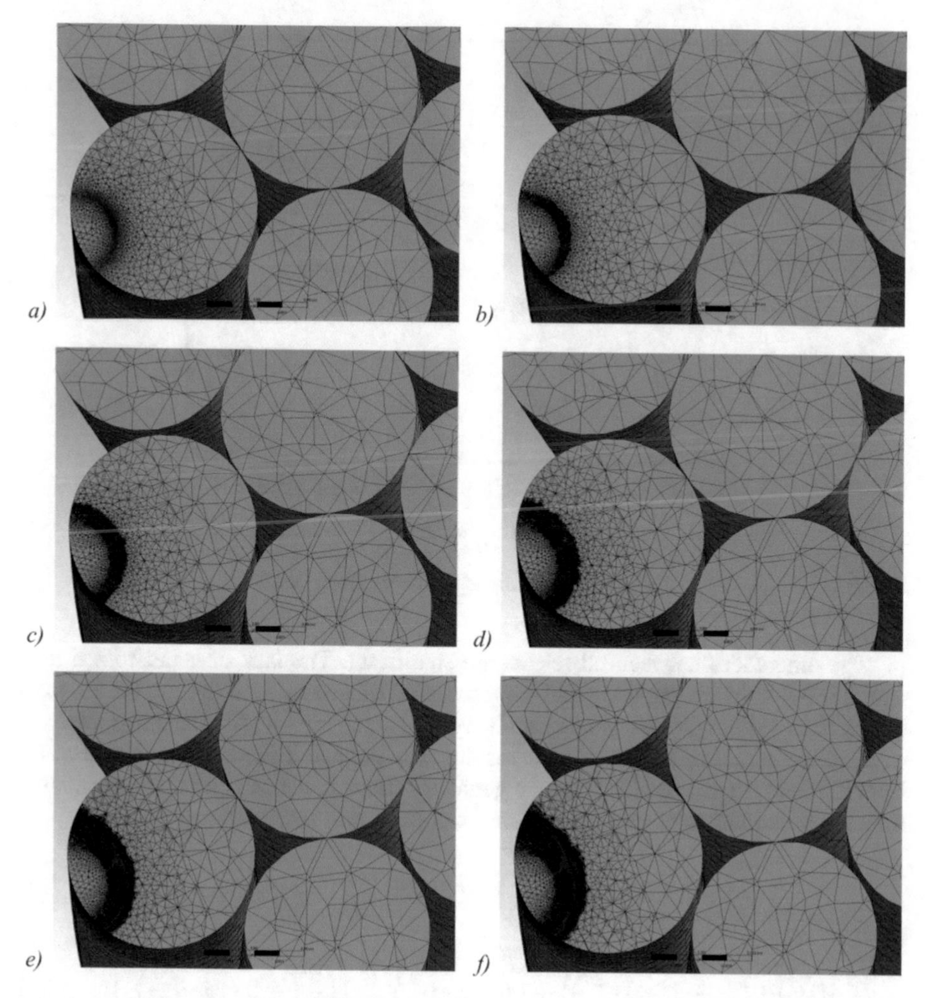

Fig. 10. Crack front: (a) initial crack; (b) crack after 8th, (c) 15th, (d) 24th, (e) 33th and (d) 40th step.

3 Analysis of the Results

As in previous case, the solutions are obtained for stress intensity factors (Fig. 11). They are presented along with the results obtained from the studies by Zheng et al. [23] and Llorka et al. [24].

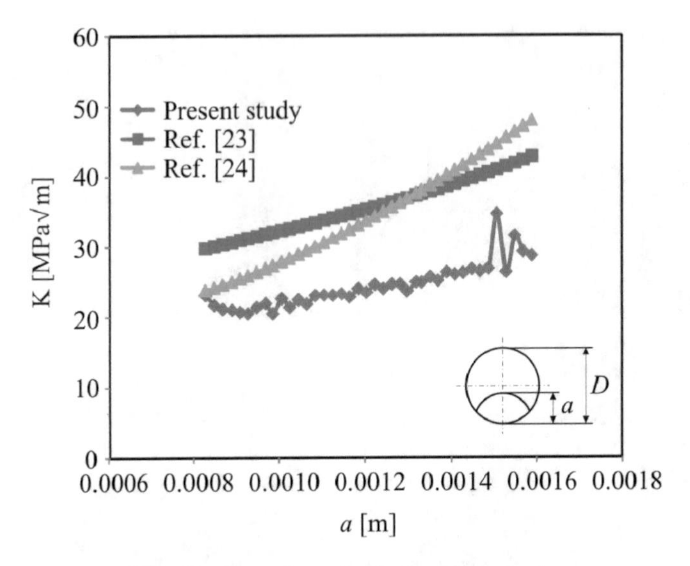

Fig. 11. Stress intensity factors along crack.

The differences between all results are significant. The lack of standardized tests and processing of the experimental results caused difference between the results from the literature, although both studies analyzed the round-bar with the same diameter and crack shape and size. Based on their results, they defined different equations (Eq. (2) from [23] and Eq. 3 from [24]) for the geometry factor β, which is used for calculating SIFs for round bar subjected to uniform axial loading.

$$\beta = 1.4835 - 2.9219 \cdot \left(\frac{a}{D}\right) + 5.9602 \cdot \left(\frac{a}{D}\right)^2 + 1.6304 \cdot \left(\frac{a}{D}\right)^3 \tag{3}$$

$$\beta = \left[0.473 - 3.286 \cdot \left(\frac{a}{D}\right) + 14.797 \cdot \left(\frac{a}{D}\right)^2\right]^{\frac{1}{2}} \cdot \left[\left(\frac{a}{D}\right) - 0.621 \cdot \left(\frac{a}{D}\right)^2\right]^{-\frac{1}{4}} \tag{4}$$

where geometry factor β, i.e. normalized stress intensity factor can be expressed as:

$$\beta = \frac{K}{\sigma\sqrt{\pi \cdot a}} \tag{5}$$

The SIFs were calculated with previous equations for D = 4.975 mm which corresponds to the round bar with the same cross-section area as the seven-wire strand diameter, and with the crack size as in conducted analysis.

Taking all mentioned into account, it can be assumed that the SIFs calculated by FEA in this study are obtained with acceptable accuracy.

If SIFs calculated in this analysis are expressed thru previously mentioned geometry factor β, that factor can be then presented as a function of dimensionless relation a/D (Fig. 12).

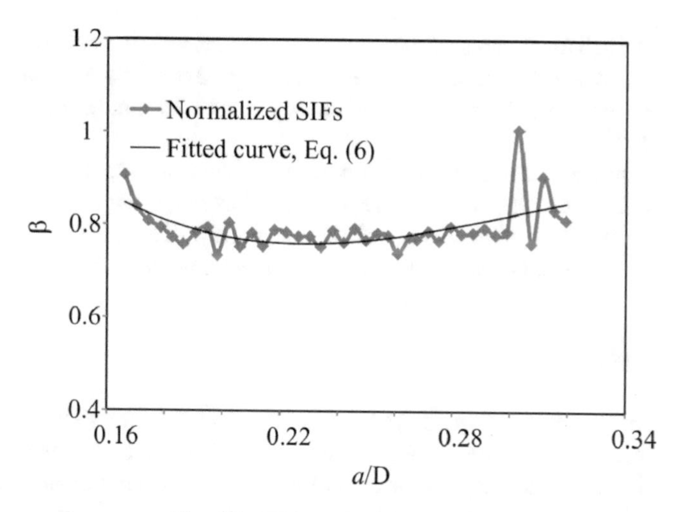

Fig. 12. Fitting of normalized SIFs

Based on the fitting of those results, the geometry factor equation for calculating SIFs for seven-wire strand subjected to uniform axial loading, have been proposed:

$$\beta = 2.6085 - 20.251 \cdot \left(\frac{a}{D}\right)^2 + 70.989 \cdot \left(\frac{a}{D}\right)^2 - 77.757\left(\frac{a}{D}\right)^3 \tag{6}$$

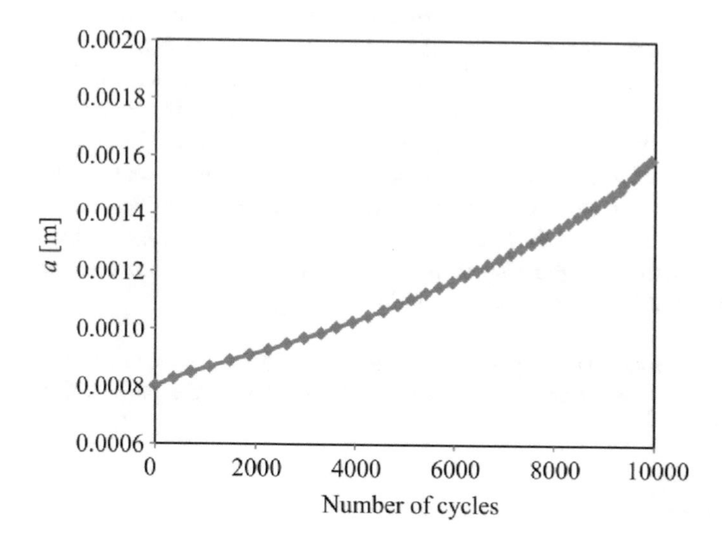

Fig. 13. Crack vs. number of cycles.

Finally, the crack growth as a function of number of cycles was obtained in the simulation and it is shown in Fig. 13. The estimated number of cycles for crack size of

1.6 mm was 9929. This number of cycles is in agreement with the experimental results reported in [23], where for the similar load conditions ($\Delta\sigma$ = 456 MPa and R = 0.29) and initial crack size of 1.09 mm and final crack size of 2.85 mm estimated number of cycles was 12000.

It has to be noted that total deformation of the seven-wire strand was 0.31054 mm, i.e. that the total strain was 0.00444. This means that seven-wire strand exhibits the elastic behavior, since the value of the strain is well under 0.008 [5, 8, 9], and that bonded type of contact, as linear contact, was actually quite suitable for this type of analysis.

4 Conclusion

Improved FEM, implemented in Ansys Workbench was used for conducting numerical simulation of crack propagation in seven-wire strand. First, the FEA was carried out on a model for which experimental results could be found in the literature and the obtained solutions for SIFs and crack growth rate were in very good agreement with those results. The settings, which were approved and acknowledged thru this analysis, were used for numerical modeling of the crack propagation in seven-wire strand subjected to axial loading. In order to avoid problems in finite element modeling and analysis and to achieve the best possible result special attention was paid to the creation of parametric 3D CAD model of the seven-wire strand. The SIFs and crack growth as a function of number of cycles were obtained. Also, based on the SIFs results equation for geometry factor determination is proposed.

A specialized feature that was used for adaptive re-meshing, has proven to be very useful, since it enabled the fatigue crack growth simulation in complex structure such as wire rope. However, it can be used only with specific type of element, with tetrahedral meshes, and only in the presence of bonded type of contact, which can be a large limiting factor when the mechanical behaviour of wire rope is concern. Also, it has to be pointed out that creating adequate geometrical model can still be an issue, especially in the case of more complex wire ropes, with multiple wire layers and strands.

Nevertheless, conducted analysis showed that FEM, besides its shortcomings, could be a powerful tool for fatigue life predictions of wire ropes in order to reduce the need for experiments, but also to improve them, since they are not standardized in procedures, nor in data acquisitions and processing.

Acknowledgements. This work is partially supported by the Ministry of Education, Science and Technological Development of the Republic of Serbia under the projects ON 174004 and TR-33036.

References

1. Jiang, W.G., Henshall, J.L.: The analysis of termination effects in wire strand using the finite element method. J. Strain Anal. Eng. Des. **34**(1), 31–38 (1999)
2. Jiang, W.G., Henshall, J.L., Walton, J.M.: A concise finite element model for three-layered straight wire rope strand. Int. J. Mech. Sci. **42**(1), 63–86 (2000)

3. Páczelt, I., Beleznai, R.: Nonlinear contact-theory for analysis of wire rope strand using high-order approximation in the FEM. Comput. Struct. **89**(11–12), 1004–1025 (2011)
4. Beleznai, R., Páczelt, I.: Design curve determination for two-layered wire rope strand using p-version finite element code. Eng. Comput. **29**(3), 273–285 (2012)
5. Imrak, C.E., Erdönmez, C.: On the problem of wire rope model generation with axial loading. Math. Comput. Appl. **15**(2), 259–268 (2010)
6. Stanova, E., Fedorko, G., Fabian, M., Kmet, S.: Computer modeling of wire strands and ropes Part I: theory and computer implementation. Adv. Eng. Softw. **42**(6), 305–315 (2011)
7. Stanova, E., Fedorko, G., Fabian, M., Kmet, S.: Computer modeling of wire strands and ropes Part II: finite element-based application. Adv. Eng. Softw. **42**(6), 322–331 (2011)
8. Kastratović, G., Vidanović, N.: 3D finite element modeling of sling wire rope in lifting and transport processes. Transport **30**(2), 129–134 (2015)
9. Kastratović, G., Vidanović, N., Bakić, V., Rašuo, B.: On finite element analysis of sling wire rope subjected to axial loading. Ocean Eng. **88**, 480–487 (2014)
10. Chen, Y., Meng, F., Gong, X.: Parametric modeling and comparative finite element analysis of spiral triangular strand and simple straight strand. Adv. Eng. Softw. **90**, 63–75 (2015)
11. Mouradi, H., Barkany, A.E., Biyaali, A.E.: Investigation on the main degradation mechanisms of steel wire ropes: a literature review. J. Eng. Appl. Sci. **11**, 1206–1217 (2016)
12. Chang, X.D., Peng, Y.X., Zhu, Z.C., Gong, X.S., Yu, Z.F., Mi, Z.T., Xu, C.M.: Experimental investigation of mechanical response and fracture failure behavior of wire rope with different given surface wear. Tribol. Int. **119**, 208–221 (2018)
13. Salleh, S., Abdullah, M.A., Abdulhamid, M.F., Tamin, M.N.: Methodology for reliability assessment of steel wire ropes under fretting fatigue conditions. J. Mech. Eng. Sci. **11**(1), 2488–2502 (2017)
14. Chang, X.D., Peng, Y.X., Zhu, Z.C., et al.: Breaking failure analysis and finite element simulation of wear-out winding hoist wire rope. Eng. Fail. Anal. **95**, 1–17 (2019)
15. Belytschko, T., Black, T.: Elastic crack growth in finite elements with minimal remeshing. Int. J. Methods Eng. **45**, 601–620 (1999)
16. Moës, N., Dolbow, J., Belytschko, T.: A finite element method for crack growth without remeshing. Int. J. Methods Eng. **46**, 131–150 (1999)
17. Aldarwish, M., Grbović, A., Kastratović, G., et al.: Stress intensity factors evaluation at tips of multi-site cracks in unstiffened 2024-T3 aluminium panel using XFEM. Tech. Gaz. **25**, 1616–1622 (2018)
18. Grbović, A., Kastratović, G., Sedmak, A., et al.: Fatigue crack paths in light aircraft wing spars. Int. J. Fatigue **123**, 96–104 (2019)
19. Grbović, A., Sedmak, A., Kastratović, G., et al.: Effect of laser beam welded reinforcement on integral skin panel fatigue life. Eng. Fail. Anal. **101**, 383–393 (2019)
20. Balac, M., Grbovic, A., Petrovic, A., Popovic, V.: Fem analysis of pressure vessel with an investigation of crack growth on cylindrical surface. Eksploatacja i Niezawodnosc – Maint. Reliab. **20**(3), 378–386 (2018)
21. Beretta, S., Matteazzi, S.: Short crack propagation in eutectoid steel wires. Int. J. Fatigue **18**, 451–456 (1996)
22. Mahmoud, K.M.: Fracture strength for a high strength steel bridge cable wire with a surface crack. Theoret. Appl. Fract. Mech. **48**, 152–160 (2007)
23. Zheng, X.L., Xie, X., Li, X.Z., Tang, Z.Z.: Fatigue crack propagation characteristics of high-tensile steel wires for bridge cables. Fatigue Fract. Eng. Mater. Struct. **42**, 256–266 (2019)
24. Llorca, J., Sanchez-Galvez, V.: Fatigue threshold determination in high strength cold drawn eutectoid steel wires. Eng. Fract. Mech. **26**(6), 869–882 (1987)
25. ANSYS Workbench, Release 19, ANSYS, Inc.
26. Costello, G.A.: Theory of Wire Rope, 2nd edn. Springer, Berlin (1990)

Specifics in Production of Fixed Partial Dentures Using 3D Printed Cast Patterns

Tsanka Dikova(✉)

Faculty of Dental Medicine, Medical University of Varna,
84 Tsar Osvoboditel Blvd, 9000 Varna, Bulgaria
Tsanka.Dikova@mu-varna.bg

Abstract. Present paper deals with the specifics in production of fixed partial dentures (FPD) using 3D printed cast patterns. The cast patterns of four-part dental bridges were manufactured of polymer *NextDent Cast* using *RapidShape D30* printer. Two cases of application of cast patterns were discussed – for production of press-ceramic and metallic constructions. The metallic samples were cast by centrifugal casting of Co-Cr and Ni-Cr dental alloys using different investment materials and heating regimes of the casting mold. The dimensions of polymeric cast patterns and cast bridges were measured. It was established that for production of FPD with high accuracy and high adhesion of porcelain coating, precise cast patterns should be manufactured by 3D printing. The dimensions of virtual model should be corrected with coefficients, specific for each axis. The increased roughness of 3D printed cast patterns is disadvantage in dental constructions with high smoothness requirements and advantage for metal-ceramic FPD. Therefore, the position of patterns with respect to the building direction should be different for FPD of press-ceramics and cast infrastructures for metal-ceramics. In the first, vertical axes of teeth must be parallel to the print direction Z-axis, and in the second, they have to be at an angle between 45°–70° to the base. For ensuring high adhesion strength of porcelain coating in metal-ceramic restorations, surface smoothing operations should not be applied to 3D printed cast patterns. The revealed specifics would be very useful in dental practice for manufacturing of accurate FPD using 3D printed cast patterns.

Keywords: Fixed partial dentures · 3D printing · Cast patterns · Casting · Press-ceramic

1 Introduction

The lost-wax casting process is still the mostly used technique for production of metal frameworks for metal-ceramic fixed partial dentures (FPD). Usually the wax models are produced manually, which is precondition for generating errors and decreasing the quality of the cast metallic constructions. The newly developed Additive Technologies (AT), which are part of the CAD-CAM systems, offer a number of opportunities for effective manufacturing of complex dental constructions with high precision and high quality. AT can be used in two ways for production of metallic dental restorations – for manufacturing of cast patterns by stereolithography, multi-jet modelling, etc. or direct

N. Mitrovic et al. (Eds.): CNNTech 2019, LNNS 90, pp. 92–102, 2020.
https://doi.org/10.1007/978-3-030-30853-7_6

fabrication from powder of dental alloys by selective laser melting/sintering. Three groups of factors influence on the quality of dental alloys, cast with 3D printed models: the properties of the materials for manufacturing of cast patterns, the peculiarities of the 3D printing and casting processes [1].

In order to obtain a quality cast in the lost-wax casting process, the first requirement to the material for patterns fabrications is to burn without residue [2, 3]. For this reason, the special polymers are designed, intended especially for manufacturing of cast models by different 3D printing processes [4]. These polymers can be used in production of patterns for casting of dental alloys or manufacturing dental constructions by pressing of ceramics [1, 5, 6]. The other main requirement to materials for cast patterns is to have minimal or zero expansion during heating of the casting mold. Otherwise, it is possible to get the cracked mold resulting in a defective casting [1]. In order to avoid deformation of the model in the process of the mold manufacturing, it is necessary that the used material possesses sufficiently high mechanical properties. A liquid of light-curing methacrylate monomer is used as a starting material in the stereolithographic processes. In order to guarantee the necessary mechanical properties of the cast pattern, a final photo polymerization is applied after 3D printing [4–7].

The 3D printing process is characterized by building of the objects layer by layer using polymerization, melting or sintering [8–12]. The quality of the 3D printed details depends on the process type, the layer thickness and the position of the specimen relative to the build direction [1, 5, 7, 10, 12–15]. Decreasing the thickness of the 3D printing layer results in reduced surface roughness. However, even after sand blasting, the traces of the layers of the 3D printed pattern can be observed on the surface of the casting [1, 15–17]. Therefore, for production of precise castings with high smoothness, it is necessary to use cast patterns, manufactured by stereolithography or multi-jet modelling processes with the possible minimum layer thickness.

For manufacturing of high quality castings from dental alloys, it is important:

1. To select the material for the 3D printed cast pattern and the investment material for the casting mold, matching the available printer and the dental alloy;
2. All operations throughout the casting process to be strictly performed in accordance with the materials used and the manufacturer's instructions [1, 15, 16].

As each type of 3D printer uses a specific polymer type, it is necessary that the investment material is selected depending on the polymer type. The heating regime of the casting mold should be determined by the type of investment material and the alloy to be cast.

Up to now, there is a great variety of 3D printing processes, working with many polymers types. However, there is still a lack of information about fabricating of dental restorations with sufficient accuracy by casting with additively manufactured patterns. The aim of the present paper is to reveal the specifics in production of precise fixed partial dentures using 3D printed cast patterns.

2 Materials and Methods

2.1 Materials and Samples Manufacturing

Two groups of samples are manufactured by lost-wax casting process from dental alloys - four-part dental bridges from 1-st premolar to 2-nd molar. The specimens of the first group are cast from Co-Cr *i-Alloy* of *i-Dental* (Co-64, Cr-30, Mo-5, C-0.5 mass%), while that of the second – from Ni-Cr alloy *Wiron light* of *Bego* (Ni-64.6, Cr-22, Mo-10, Si-2.1, Mn < 1, B < 1, Nb < 1 mass%). The casting molds are fabricated from two types of refractory materials - *Sherafina-Rapid* in casting of *Wiron light* alloy and *Sheravest RP* for *i-Alloy*. The casting molds are heated up to maximal temperatures using four different thermal modes (Fig. 1). The cast patterns are manufactured by 3D printing of *NextDent Cast* polymer using *Rapidshape D30* device, working on the stereolithography with digital light projection principle. The models are printed with two layer thicknesses - 35 μm and 50 μm recommended by the producer. No additional treatment for surface smoothing of the polymeric cast patterns is applied. The cast patterns are produced using 3D virtual model, generated by scanning of a conventionally cast base four-unit bridge. After casting, the base bridge-model and all samples are sandblasted for 8 s with 250 μm corundum (Al2O3) particles.

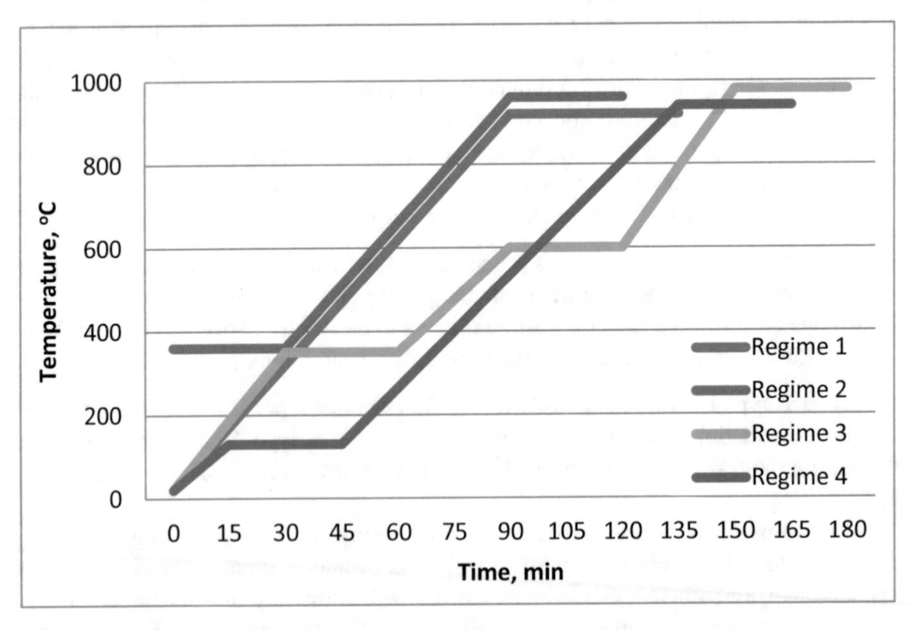

Fig. 1. Heating regimes of the casting mold: Regime 1 – investment material *Sherafina-Rapid*, conventional casting with wax pattern; Regime 2 – investment material *Sherafina-Rapid*, casting with polymer patterns; Regime 3 – investment material *Sherafina-Rapid*, Ni-Cr alloy Wiron-light, casting with 3D printed patterns of the polymer *Nextdent Cast*; Regime 4 – investment material *Sheravest RP*, Co-Cr alloiy *i-Alloy*, casting with 3D printed patterns of the polymer *Nextdent Cast*.

2.2 Dimensional Accuracy Measurements

The dimensional accuracy of the cast patterns and bridges, cast of dental alloys, is studied by measurements of the connectors between the bridge retainers and the pontics (a1, a2, a3), the pontics' width (b1, b2) and the bridges' length – L [13–15]. Three samples of each group are measured. The relative difference in % of the dimensions of 3D printed cast patterns and cast dental bridges with that of the bridge-base model is calculated via Excel software and evaluated.

3 Results and Discussion

3.1 Specifics of the 3D Printing Process

It is known that the quality of the additively manufactured details depends on the process type, the layer thickness and the position of the object towards the print direction. It was established in our previous research that the surface roughness of the cast patterns, fabricated by stereolithography, is about 2–3 times higher than that of the conventionally cast base model [15]. Decreasing the polymerization layer's thickness leads to decrease of the surface roughness. Consequently, precise castings with high smoothness can be obtained by casting with patterns made using the stereolithography method with the possible minimum thickness of the layer [1, 15]. However, their production requires almost twice more time - 122 min for dental bridges, printed with lower layer's thickness of 35 μm, comparing to the printing with 50 μm layer (recommended by the producer) - 68 min.

When working with the thickness of the 3D printing layer, recommended by the manufacturer, relatively good smoothness can be achieved by:

1. Proper positioning of the model's walls relative to the printing direction - parallel to the build direction - *Z* axis or the printer's base depending on the construction shape;
2. Additional treatment of the 3D printed cast pattern to increase the smoothness of the surfaces by coating with a thin wax layer, using a chemical solvent or mechanical grinding.

On the other hand, our previous study showed that the adhesion strength of the porcelain coating to dental alloys, cast with 3D printed patterns, was higher than that of the conventionally cast using wax models: 63.4 MPa–84.6 MPa and 67.5 MPa respectively [16, 17]. Optimal regimes for 3D printing of cast patterns were obtained when using the layer thickness of 50 μm, recommended by the manufacturer. According to the results, the position of the models, inclined at 56.3° and 67.5° towards the printer's base, ensure adhesion strength of the porcelain coating of 77.9 MPa and 79.9 MPa, respectively.

It should be noted that the 3D printed patterns were not subjected to further treatments for increasing the surface smoothness. Consequently, in manufacturing of metal-ceramic FPD, the increased roughness of the casting, respectively the cast pattern has a favorable effect on the adhesion strength of the ceramic coating. In this case, the cast models should be positioned at an angle between 45°–70° to the base.

However, the increased roughness of the 3D printed cast patterns is a disadvantage in dental constructions with high smoothness requirements, such as in pressing ceramics. Therefore,

1. It is necessary that the walls of the patterns are parallel to the printer base or the building direction - Z axis during 3D printing process and
2. The above-mentioned operations about increasing the surface smoothness have to be applied.

3.2 Specifics of the Casting Process

For the high quality of dental constructions, made of Co-Cr and Ni-Cr alloys, it is important that all operations throughout the casting process are strictly performed, depending on the materials used and the manufacturer's instructions.

During casting process, the casting mold is heated up to temperatures of about 800–1050 °C, which are selected depending on the alloy used. The conventional procedure when using *Sherafina-Rapid* investment material consists of heating up to maximal temperature with rate of 9–20 °C/min and holding for about 45 min (regime 1, Fig. 1) [18]. However, when using polymeric cast patterns, it is recommended the casting mold to be placed in a preheated oven at 360 °C and held for about 30 min to burn the plastic (regime 2, Fig. 1). The heating is then continued until the maximal temperature is reached.

The *Sherafina-Rapid* investment material, used for casting of constructions from Ni-Cr alloy Wiron-light with cast patterns, 3D printed of *NextDent Cast* polymer, did not give very good results. When heating the casting molds under conventional procedures, some of them were cracked [1]. Therefore, a new heating regime was proposed and used, which is a combination of the two above. It involves slow heating (9 °C/min) of the casting mold from room temperature to 350 °C with holding at this temperature for 30 min to burn the plastic. To minimize the risk of subsequent deformations, the heating continues with the same rate up to 600 °C, followed by holding for 30 min, heating up to a maximum temperature of 980 °C and final holding for 45 min (regime 3, Fig. 1).

The application of the new proposed mode has delivered much better results while keeping all other requirements in manufacturing the casting mold. However, in order to perform a normal casting process and to obtain high-quality castings using 3D printed cast patterns of the *Nextdent Cast* polymer, it is necessary to use the recommended *Sheravest RP* investment material and the heating of the casting mold to be performed using the corresponding thermal mode (regime 4, Fig. 1). This procedure has ensured casting of FPD frameworks of Co-Cr *i-Alloy* of the required quality.

3.3 Specifics in Manufacturing of FPDs by Casting with 3D Printed Patterns

According to the majority of authors, the high roughness of the cast patterns, made by 3D printing, is their disadvantage. Our studies, in which 3D printed cast models without additional processing for increasing their smoothness were used, showed that the adhesion strength of the porcelain coating to dental alloys, cast with them, is higher than that of the conventionally cast [16, 17].

Table 1. Relative difference (%) of the dimensions of dental bridges, cast of *i-Alloy* using 3D printed patterns, with that of the bridge-base model.

Position of dimension during 3D printing	XZ	XZ	XZ	Average value	XZ	XZ	Average value	Average value in XZ	Y
Dimension	a1	a2	a3	a	b1	b2	b	(a + b)/2	L
Layer thickness of cast pattern 50 μm	3.30	4.50	9.14	5.65	5.90	3.94	4.92	5.29	−0.17
35 μm	0.44	1.96	6.38	2.97	2.75	3.01	2.88	2.93	−0.38
Dimension type	Connector's width 4–6 mm				Bridge bodies' width 7–9 mm				L = 34 mm

In fact, it was proven that the higher roughness of dental alloys, cast with 3D printed patterns, is their advantage and not a drawback in production of FPD of metal-ceramics.

Therefore, for dental alloys, cast with additively manufactured patterns, it is not recommended to apply additional treatments for increasing the smoothness of their surfaces. Thus, on the one hand, time and materials are saved, and on the other - higher adhesion strength of the porcelain coating is guaranteed.

To obtain a casting with precise dimensions, the shrinkage of dental alloys during cooling after casting process is compensated by the expansion of the refractory material of casting mold [19]. Our research has shown that the dimensions of the dental bridges, cast with patterns, 3D printed with 50 μm layer thicknesses, are larger than those of the basic bridge model with 3.30%–9.14% (Table 1). The average value of the relative difference of the dimensions in the XZ plane is 5.29%. Decreasing the layer thickness of the 3D printed cast patterns to 35 μm results in lower average relative difference – 2.93%, i.e. higher accuracy. However, it needs twice more time for 3D printing of the cast models. The larger dimensions of the cast bridges are most likely due to the 1.10%–5.17% larger sizes of the 3D printed patterns (Table 2) from one hand and from another - the expansion of the investment material. Therefore, in order to obtain a precise casting, it is necessary the dimensions of the cast patterns to be with high accuracy. When working with 50 μm layer recommended by the manufacturer, it is necessary to make adjustments of the sizes along the three axes. Based on the results, shown in Tables 1 and 2, the correction coefficients, specific for each axis, are calculated and proposed.

Table 2. Relative difference (%) of the dimensions of 3D printed cast patterns, made of *Nextdent Cast* polymer, with that of the bridge-base model.

Position of dimension along axes or plane	XZ	XZ	XZ	Average value	XZ	XZ	Average value	Average value in XZ	Y
Dimension	a1	a2	a3	a	b1	b2	b	(a + b)/2	L
Layer thickness 50 μm	1.10	1.96	5.17	2.74	2.34	−0.93	0.71	1.73	−0.53
35 μm	−1.76	−0,59	3.97	0.54	0.00	1.62	0.81	0.65	−0.47
Dimension type	Connector's width 4–6 mm				Bridge bodies' width 7–9 mm				L = 34 mm

The peculiarities of the technological process for production of FPD of dental alloys by casting or pressing of ceramics with 3D printed models are presented in Table 3, Figs. 2 and 3. The possibility for manufacturing casting mold with the widely used *Sherafina-Rapid* refractory material and heating it under a developed thermal mode for fabricating quality casting with 3D printed patterns is also shown. For the best of our knowledge, it is proposed for the first time not to apply operations for smoothing of the surface of 3D printed cast patterns in casting of frameworks for metal-ceramic FPDs. The presented process ensures high construction accuracy and a 25% higher adhesion strength of porcelain coating compared to conventional casting with hand-made wax models.

Table 3. Technological processes for manufacturing of fixed partial dentures by metal casting or ceramic pressing with 3D printed patterns.

Technological feature		Crown	Bridge
3D printing process			
Printer		*RapidShape D30*	
Polymer		*NextDent Cast*	
Layer thickness		50 μm (recommended by the producer)	
Design of the virtual model			
Position		Casting of dental alloys	
		Vertical axes of the teeth must be inclined at 45°–70° towards the base	
		Ceramic pressing	
		Vertical axes of the teeth must be parallel to the printing direction Z-axes	
Supports number		≥4	≥4 of each tooth
Corrections of the dimensions	Along *X* or *Y* axes	0%	An increase with 0.50% on the axis with the largest size; 0% on the other axis
	Along *Z* axis	Decrease with 2%	
Additional treatment			
Final photopolymerization		The construction must be placed on a working model	
Increasing surface smoothness		Casting of metal framework for metalceramic FPDs: NOT APPLICABLE	
		Ceramic pressing: Cover with thin wax layer or treatment with solvent	
Manufacturing of casting mold (casting of dental alloys)			
Investment material		*Sherafina-Rapid*	*Sheravest RP*
Heating regime of the casting mold		Newly proposed regime: Start: 22 °C – room temperature ↑9 °C/min Hold 30 min at 350 °C ↑9 °C/min Hold 20 min at 600 °C ↑9 °C/min Hold 30 min at 980 °C	From the manufacturer: Start 22 °C – room temperature ↑9 °C/min Hold 30 min at 130 °C ↑9 °C/min Hold 30 min at 950 °C
Casting			
Cleaning of the casting			
Sandblasting		Alumina oxide (Al_2O_3, 250 μm) under 6 atm pressure for 8 s	

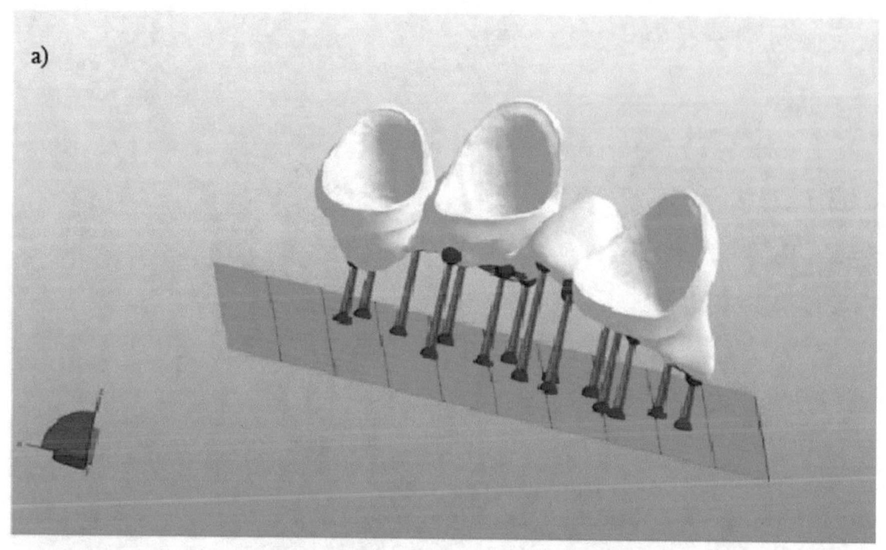

Fig. 2. Stages of generating and design of virtual model as well as preparation for 3D printing of cast pattern: supports of the virtual model (a) and position of virtual model on the printer's base (b).

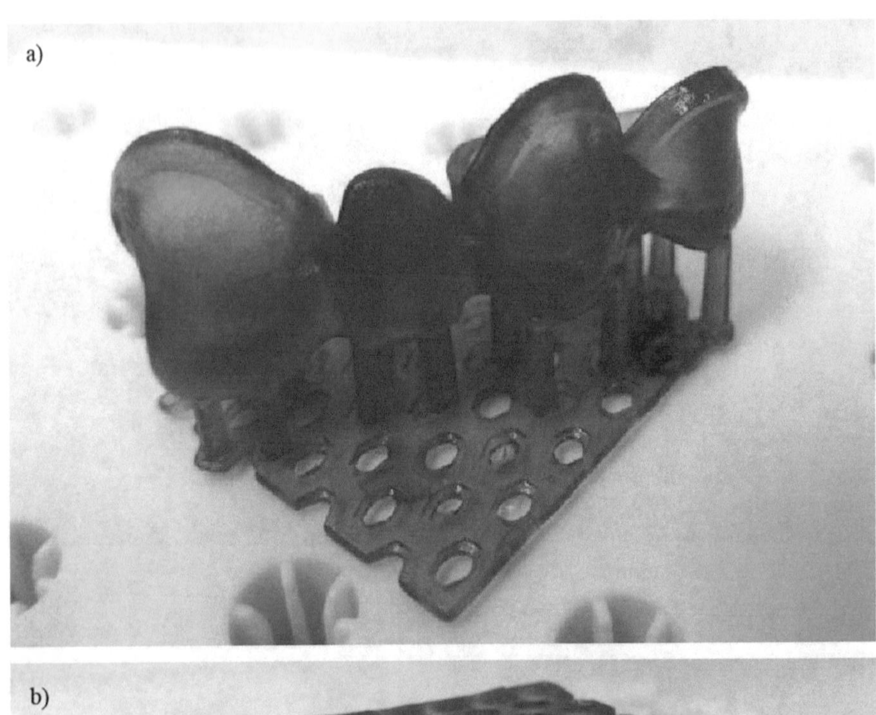

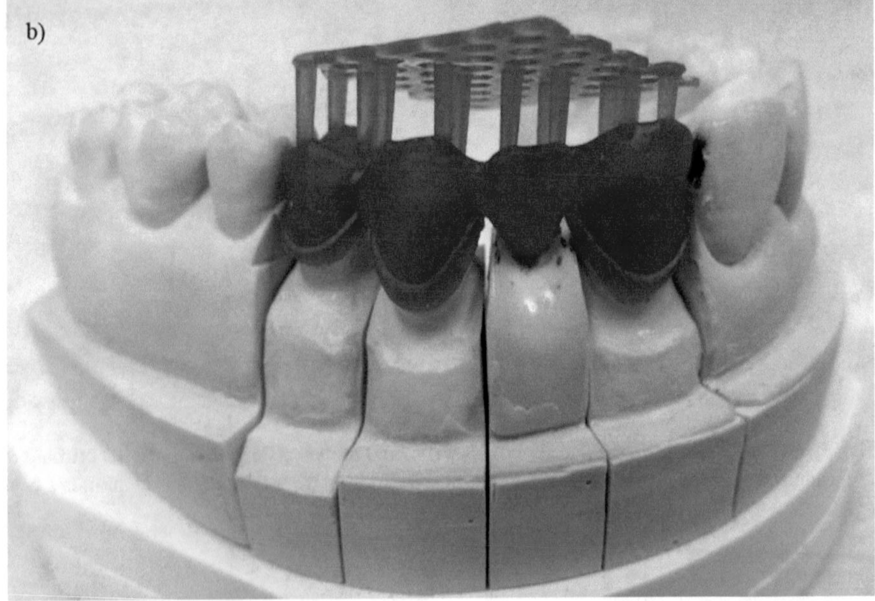

Fig. 3. Printed cast pattern (a) and construction on a working gypsum model after final photo-polymerization (b).

4 Conclusion

The present study has shown that in order to obtain ceramic or metal-ceramic FPDs with high accuracy and adhesion strength of the porcelain coating using AT in their production, it is necessary to produce precise cast patterns by 3D printing, as the dimensions being corrected with the proposed coefficients on the virtual model stage.

The increased roughness of the 3D printed cast patterns is a disadvantage in dental constructions with high smoothness requirements and an advantage in metal-ceramic dental prostheses. Therefore, the position of the patterns relative to the printing direction should be different for FPDs of pressed ceramics and in casting frameworks of dental alloys for metal-ceramics. For the first, the vertical axes of the teeth must be parallel to the built direction of the Z-axis, and for the second, they should be inclined at an angle of 45° to 70° to the base.

It is proposed for the first time not to apply operations for increasing the surface smoothness of the 3D printed cast patterns in casting of frameworks for metal-ceramic FPDs from dental alloys. The peculiarities of the technological process for production of fixed partial dentures from dental alloys by casting or pressing ceramics with 3D printed patterns are summarized. The presented technological process provides high precision of the constructions and a 25% higher adhesion strength of porcelain coating compared to conventional casting with handmade wax models. The revealed specifics would be very useful in dental practice for manufacturing of accurate FPD using 3D printed cast patterns.

References

1. Dikova, T.: Factors, influencing on the quality of Co-Cr dental alloys, cast using 3D printed patterns. Foundry **1**(1), 58–62 (2017). (in Bulgarian)
2. Minev, E., Yankov, E., Minev, R.: The RepRap 3D printers for metal casting pattern making – capabilities and application. In: On Innovative Trends in Engineering and Science, SFITES 2015, Kavala, Greece, pp. 122–127. Parnas Publishing House (2015). ISBN 978-954-8483-35-6
3. Technical Application Guide: Investment Casting with FDM Patterns, Doc. No. TAG 06-02, 19 April 2011, Stratasys, 11 p. www.stratasys.com. Accessed 20 Mar 2019
4. NextDent Cast, Instruction for Use, NextDent, 4 p. http://nextdent.com/wp-content/uploads/2016/05/IFU-NextDent-Cast-INC201601UK.pdf. Accessed 20 Mar 2019
5. Shamseddine, L., Mortada, R., Rifai, K., Chidiac, J.J.: Fit of pressed crowns fabricated from two CAD-CAM wax pattern process plans: a comparative in vitro study. J. Prosthet. Dent. **118**(1), 49–54 (2017)
6. Katreva, I., Dikova, T., Tonchev, T.: 3D printing – an alternative of conventional crown fabrication: a case report. J. IMAB **24**(2), 2048–2054 (2018)
7. Karalekas, D., Aggelopoulos, A.: Study of shrinkage strains in a stereolithography cured acrylic photopolymer resin. J. Mater. Process. Technol. **136**, 146–150 (2003)
8. Dobrzański, L.A.: Overview and general ideas of the development of constructions, materials, technologies and clinical applications of scaffolds engineering for regenerative medicine. Arch. Mater. Sci. Eng. **69**(2), 53–80 (2014)

9. Dobrzański, L.A.: The concept of biologically active microporous engineering materials and composite biological-engineering materials for regenerative medicine and dentistry. Arch. Mater. Sci. Eng. **80**(2), 64–85 (2016)
10. Milovanović, A., Milošević, M., Mladenović, G., Likozar, B., Čolić, K., Mitrović, N.: Experimental dimensional accuracy analysis of reformer prototype model produced by FDM and SLA 3D printing technology. In: Mitrovic, N., Milosevic, M., Mladenovic, G. (eds.) Experimental and Numerical Investigations in Materials Science and Engineering, CNNTech 2018. LNNS, vol. 54, pp. 84–95. Springer, Cham (2019)
11. Milovanović, J., Trajanović, M., Vitković, N., Stojković, M.: Rapid Prototyping Tehnologije i Materijali za Izradu Implantata. IMK-14-Istraživanje i razvoj **15**(1-2), 23–30 (2009)
12. Minev, R., Minev, E.: Technologies for rapid prototyping (RP) - basic concepts, quality issues and modern trends. Scripta Scientifica Medicinae Dentalis **2**(1), 29–39 (2016)
13. Dikova, T., Dzhendov, D., Katreva, I., Pavlova, D.: Accuracy of polymeric dental bridges manufactured by stereolithography. Arch. Mater. Sci. Eng. **78**(1), 29–36 (2016)
14. Dikova, T., Dzhendov, D., Ivanov, D., Bliznakova, K.: Dimensional accuracy and surface roughness of polymeric dental bridges produced by different 3D printing processes. Arch. Mater. Sci. Eng. **94**(2), 65–75 (2018)
15. Dikova, T., Dzhendov, D., Katreva, I., Tonchev, T.: Study the precision of fixed partial dentures of Co-Cr alloys cast over 3D printed prototypes. Arch. Mater. Sci. Eng. **90**(1), 25–32 (2018)
16. Dikova, T., Dolgov, N., Dzhendov, D., Simov, M.: Adhesion strength evaluation of ceramic coatings on cast and selective laser melted Co-Cr dental alloys using tensile specimens. Int. J. Mater. Sci. Non-Equilib. Phase Transform. **3**(2), 49–52 (2017)
17. Dikova, T., Dolgov, N., Vasilev, T., Katreva, I.: Adhesion strength of ceramic coatings to dental Ni-Cr alloy fabricated by casting with 3D printed patterns. Russ. Metall. (Metally) **4**, 385–391 (2019). © Pleiades Publishing, Ltd., ISSN 0036-0295; Russian Text © The Author (s) published in Deformatsiya i Razrushenie Materialov 9, 33–39 (2018)
18. Sherafina-Rapid, Instruction for Use, SHERA Werkstoff-Technologie GmbH & Co. KG, 3 p. https://www.shera.de/var/StorageShera/Pdf_neu/GA%20SHERAFINARAPID%20(englisch).pdf. Accessed 20 Mar 2019
19. Anusavice, K.J.: Philips' Science of Dental Materials, 11th edn. Saunders, Elsevier Sciense, St. Louis (USA) (2003)

Mechanical Properties of Direct and Indirect Composite Materials Used in Prosthodontics

Ivan Tanasić[1(✉)], Ljiljana Tihaček Šojić[2], Aleksandra Milić Lemić[2], and Predrag Šojić[2]

[1] Medical College of Applied Studies Belgrade, 11000 Belgrade, Serbia
ivan.tanasic@vzsbeograd.edu.rs
[2] University of Belgrade, Faculty of Dental Medicine, 11000 Belgrade, Serbia

Abstract. This study was conducted to investigate composite materials using mechanical equipment corresponded to masticatory system composed of two antagonistic parts that simulates bimaxilar contact in occlusion and articulation. The aim of this study was to investigate restorative materials with different chemical properties and to find out which material showed the best mechanical properties. The experiment was conducted using 4 types × 20 specimens of direct resin composite materials and equal number of indirect resin composites (IRCs). The following composites were tested: Tetric (Ivoclar Vivadent), Kerr Herculite XRV (Kerr), Charisma (Heraeus Kulzer), Heliomolar (Ivoclar Vivadent), Artglass (Heraeus Kulzer), Targis/Vectris (Ivoclar Vivadent), Vita Zeta LC (Vita) and Kerr Herculite Lab (Kerr). The universal dimensions of all specimens were 5 mm × 5 mm with 2 mm thickness. This was achieved using silicone molds. A total of 160 specimens (80 direct composite specimens and 80 indirect composite specimens) were tested using tensile testing machine. Artglass showed minimal change in thickness (30.20 ± 23.0 μm) compared to others. The highest change in thickness was found in Heliomolar (86.2 ± 32.6 μm). In the group of direct composite materials, Tetric experienced the lowest thickness change (31.0 ± 8.8 μm), while Charisma showed thickness variation of 65.2 ± 27.3 μm. Among indirect composites, Vita Zeta LC showed the highest value of thickness variation (76.2 ± 31.3 μm). In this experimental study, indirect composite specimens showed superior mechanical performance compared to direct composite specimens highlighted Artglass as the material of choice viewed from mechanical aspect.

Keywords: Resin based composites · Occlusion · Prosthodontics · Mechanical properties

1 Introduction

Last few decades, dental industry experienced wide expansion and released a large number of different dental materials to satisfy prosthetic demands. Beside esthetics, the mechanical properties of dental materials are highlighted as important factors for long term survival rate of restorative materials with application in oral rehabilitation and contemporary prosthodontics. Different restorative materials for anterior and posterior dental regions were proposed and composite resins are proved to be an alternative to

N. Mitrovic et al. (Eds.): CNNTech 2019, LNNS 90, pp. 103–118, 2020.
https://doi.org/10.1007/978-3-030-30853-7_7

amalgam, cast restorations or ceramics. Resin based composites (RBCs) can replace amalgam as a posterior restoration due to better esthetics and satisfied mechanical properties. Unless there is an absence of teeth in the intercanine region, anterior teeth can be efficiently restored using various composite materials [1]. Beside ceramics, posterior teeth could also be restored employing direct or indirect composite restorations. Direct composites are indicated for restoration of relatively small cavities due to polymerization stress. Indirect resin composites (IRCs) i.e. prosthetic composites or laboratory composites are also called composite inlays due to their main role in fabrication inlays. IRCs are indicated for large posterior restorations [2]. Their introduction in the field of restorative dentistry was due to poor wear resistance of direct composite resins [3]. IRCs resolved problem with marginal micro leakage i.e. marginal gap [4] associated with polymerization shrinkage. These laboratory resins and fiber-reinforced composites are treated as indirectly cured resin composite materials. Clinical advantages of using indirect instead direct technique indicated improvement in contour and proximal contacts, absence of postoperative sensibility due to fact that dental restoration was not made directly in the cavity preparation [5, 6].

Structurally, dental composites in their composition have three chemically different materials: the organic matrix or organic phase; the inorganic matrix, filler or disperse phase; and an organo-silane or bonding agent which connects the filler and the organic resin matrix [7]. The organic matrix of composite resins is a system of mono-, di- or tri-functional monomers. Frequently used monomer is bisphenol A-glycidyl methacrylate (Bis-GMA) and its' combinations and substitutions with triethylene glycol dimethacrylate (TEGDMA) and urethane dimethacrylate (UEDMA), all responsible for cross-linking in polymer [8]. During photo-free radical (an alpha diketone-camphoroquinone applied together with a tertiary aliphatic amine reducing agent 4-n, n-dimethylamino-phenyl-ethanol, DMAPE) or chemical polymerization (benzoil peroxide is combined with an aromatic tertiary amine (n,n-dihydroxyethyl-p-toluidine) [9], shrinkage is developing in certain degree which depends on the organic matrix [10].

In order to suppress negative effect of polymerization shrinkage, different monomers were proposed: spiroortho-carbonates (SOCs), expand composites [11], epoxy-polyol system, decrease shrinkage for 40%–50%, the siloxane-oxirane resins (3M-Espe) [12]. In addition, molecules with high molecular weight e.g. multiethylene glycol dimethacrylate and copolymers can reduce the C = C bonds to achieve conversion to polymer of 90%–100%. Introducing 3D cross-linked copolymers-ormocers (organic modified ceramics) as modified direct composite resins with organic and inorganic fillers reduced curing shrinkage, in certain extent [13].

The first-generation of IRCs and direct resin composites had identical composition with organic resin matrix, inorganic filler, and coupling agent. Through years, indirect composites changed. For IRCs, there is an additional, extra-oral cure. This fact decreases the degree of conversion (DC) and mitigates the negative effects of inevitable polymerization shrinkage. Considering clinical side, there is only the shrinkage of the luting cement [14].

Another factor that contributes to material properties and mechanical behavior of resin composites corresponds to the filler content. The filler particles indicate different dimensions, chemical composition and morphology. The most common filler is silicon dioxide (quartz), though boron silicates and lithium aluminum silicates have also been commonly employed. Barium, strontium, zinc, aluminum or zirconium, are sometimes used to replace certain amount of quartz due to phenomenon of radio-opacity. Additionally, it should be considered that less hard filler-materials, like calcium metaphosphate compared to glass indicate less wear on the opposing tooth/teeth [15].

In order to improve physical properties, the composition of the restorative materials was changing over the time. Basically, some components of these materials were eliminated and some new were added to improve material properties as much as possible. Actually, confirmation on material structural reinforcing was done by employing various tests. Thus, restorative materials have been usually tested using rigorous strategies for biomedical materials testing [16–18]. Regarding mechanical properties, materials were subjected to high limited values of masticatory force. Restorative materials were actually exposed to different forces due to dynamic nature of masticatory system [16]. Thus, beside chemical and biological/biocompatibility requests upon these materials, mechanical resistance has found to be very important for long term clinical prognosis. In dentistry research field, the mechanical testing is subjected to determination of mechanical properties of whether tested materials, dental tissues or both, materials and tissues.

Many factors in restorative dentistry influence the selection of adequate restorative material in conservative rehabilitation and conventional prosthetics: an individual patient choice for non-metal, natural-looking restorations, the less invasive procedures, the significant improvement in composite resin material leading to increased durability and longevity of such restorations. Structural integrity and composition of composites are the crucial factors when interpreting physical, mechanical and aesthetic properties and the clinical behavior of composites. The choice which composite should be picked for tooth-prosthetics is often associated with the amount of remaining tooth substance. Actually, direct composite resins are inadvisable for restoration of the extensive cavities i.e. huge lesions. Though, in such case, bulk-fill materials are appropriate solution, but it is recommended to use IRC whenever possible.

In prosthetic dentistry, a significant difference of direct resin composites to IRCs is the chair-side modeling of the composite resin fillings compared to laboratory fabrication (lab-side) of composite inlays. In this case, different conditions are established regarding work in the oral cavity (humidity and temperature) compared to dry laboratory fabrication of properly designed inlays to be placed in the appropriate occlusal rest-seat preparation. This is also known as preparation of abutment teeth and adjacent supporting dental tissues to receive retentive and supportive elements of different removable partial denture (RPD)-modalities. The mentioned preparation usually requests IRC due to high precision technique of fabrication and lower polymerization shrinkage. After fabrication, the indirect resin composite restoration is bonded (fixed, cemented) using adequate resin composite cement. The supporting dental tissues and composite resin inlays or fillings are then modified to receive the elements of RPD. Thus another application of composite resin lay in the fact that resin composite materials can be utilized in remodeling of the former tooth contours. Composite resin may also be used in daily clinical practice for the restoration of abutment teeth, vertical support, and re-establishment of a patient's vertical dimension.

Clinical disadvantages of direct composite restoration are addressed to polymerization shrinkage and inadequate contour of tooth to be restored [4]. However, economic, comfort, esthetic, efficiency and concerns of teeth conservation should also be considered when making choice about the best therapy solution for patient.

This research was conducted to investigate composite materials using mechanical equipment corresponded to masticatory system composed of two antagonistic parts that simulates bimaxilar contact in occlusion and articulation. In real, physiological conditions where masticatory muscles are involved in the action of bite, occlusion of antagonistic teeth is happened thanks to the contraction of the masticatory muscles and elevation of the mandible towards the maxilla. In this study, tensile testing machine was an alternative to masticatory muscles due to tensile performances of this equipment. Additionally, occlusal contact was achieved between either direct resin composite materials or indirect resin composite materials and Co-Cr-Mo alloy on the opposite sides to simulate occlusal surface of tooth restored using resin composite materials and rest seats made of Co-Cr-Mo alloy. In normal central occlusion, maximal contact between antagonistic teeth is performed. The occlusal rests and composite restorative materials experience the strongest interface contact which is followed by possibility for thickness and volume loss of material with increased elastic properties. In this study, changes of thickness and volume in resin composite specimens were caused by Co-Cr-Mo alloy, which is proved to be a harder material than resin composites.

Considering the fact that antagonistic teeth can contact each other over 2000 times per day then it means that about 1 095 000 contacts can reach over one-year period [17, 18]. Regarding the aforementioned, teeth prepared using different composite materials for accepting the dental elements of RPDs, should be able to withstand the forces produced by clasps or active segments of RPDs. This refers to mechanical properties of applied composite materials.

Different adhesive materials were introduced for preparing abutment teeth to accept the elements of RPDs. Every attempt to simulate real conditions and testing material properties is of high significance for dental practitioners and dental industry, to improve techniques and materials composition to withstand occlusal forces in dynamic system of the oral cavity.

The aim of this study was to investigate composite restorative materials using loading (compression) test for determination thickness and volume changes.

2 Materials and Methods

The experiment was conducted using 4 types × 20 specimens of direct resin composites and equal number of IRCs. Four types of direct composite materials were 4 tested: Tetric (Ivoclar Vivadent), Kerr Herculite XRV (Kerr), Charisma (Heraeus Kulzer), Heliomolar (Ivoclar Vivadent). Following types of IRCs were employed in this study: Artglass (Heraeus Kulzer), Targis/Vectris (Ivoclar Vivadent), Vita Zeta LC (Vita), Kerr Herculite Lab (Kerr). Their compositions [1, 2, 7–9, 14] were presented in Table 1.

The specific mechanical properties of direct and IRCs (Table 2) were obtained using the literature data from numerous sources [1–6, 12, 14].

The universal dimensions of all specimens were 5 mm × 5 mm with 2 mm thickness. This was achieved using silicone molds. A total of 160 specimens (80 direct

Table 1. The composition of tested composite resin materials.

Material designation	Composition
Tetric (Ivoclar Vivadent)	ytterbium trifluoride ≥ 10–<20% w/w, Bis-GMA ≥ 5–<10% w/w, urethane dimethacrylate ≥ 5–<10% w/w, triethylene glycol dimethacrylate 1–5% w/w. Irregular-shaped particles
Kerr Herculite XRV (Kerr)	Bis-GMA, TEGDMA, Uncured Methacrylate Ester Monomers. Three fillers—prepolymerized filler (PPF), silica nanofillers (20–50 nm), and Barium glass (0.4 micron). Irregular-shaped particles. Total filler content 75%
Charisma (Heraeus Kulzer)	Based on a BIS-GMA matrix and contains 64% filler by volume: Bariumaluminum fluoride glass (0.02–2 microns) Highly dispersive Siliciumdioxyde (0.02–0.07 microns) 64% in volume (Ba and Al glass – 0.02–2 μ; silica dioxide - 0.02–0.07 μ) Irregular-shaped particles
Heliomolar (Ivoclar Vivadent)	Bis-GMA 10–30% w/w (Bis-EMA, UDMA, TEGDMA) urethane dimethacrylate 1–5% w/w 1,10-decandiol dimethacrylate 1–5% w/w ytterbium trifluoride ≥ 10–<13% w/w 46vol% silicon dioxide, copolymer Microfilled composite. Pre-polymerized particles <. 1 μm Total filler content in Heliomolar is 46%
Artglass (Heraeus Kulzer)	Filler - 70wt% filler of bariumsilicate glass of 0.7 μ. Matrix - 30wt% organic resin. Additional to conventional bifunctional molecules, Artglass contains four to six functional groups which provide the opportunity for more double-bond conversions. Monomer: Multifunctional methacrylic acid ester. Filler: Barium alumina silica glass and silicone dioxide. Filler particle size: 0.7–2.0 μm. Percentage of filler: 68 wt. % and 54 vol. %. Irregular-shaped particles. Total filler content in Artglass is 87%
Kerr Herculite Lab (Kerr)	Contents app. 79vol.% inorganic filler (by weight) Ba-Al-F-B, silicate glass. with an average particle size of 0.6 μm. Irregular-shaped particles. Total filler content in Kerr Lab is 87%
Targis/Vectris (Ivoclar Vivadent)	Monomer matrix: BisGMA and TEGDMA (24–39Wt%), decandioldimethacrylate UDMA - 0.3&0.1wt%. preimpregnated E&R glass - 60Wt% for pontic and around 45–50% for the other materials, glass fibers. Irregular-shaped particles. Total filler content in Targis is 55%.
Vita Zeta LC (Vita)	Monomer: BisGMA, TEGDMA and UDMA; Filler: Multiphase feldspar frits and silicone dioxide; Filler particle size: 0.04–1.5 μm; Percentage of filler: 44.3 wt. % - 27 vol. %. Round particles.

composite specimens and 80 indirect composite specimens) were tested using tensile testing machine. The polymerization of specimens was done using UniXS (Heraeus, Kulzer, Germany) for 60 s with a polymerization rate between 450–500 nm and strobe frequency of 20 Hz in each 10 ms according to manufacturer. Further, IRCs were treated using water vapor under pressure to eliminate superficial monomer layer on the surface of IRCs. After polymerization, fabrication of specimens was done. Afterwards, specimens were submitted to a three-point bend test with a universal testing machine (Instron 825 University Ave, Norwood, MA, Ilinois, USA; Fig. 1) with a crosshead speed of 2 mm/min.

Table 2. Mechanical properties of direct composite resins tested in this study [2, 5, 6, 12, 17].

Material designation	Compressive strength (MPa)	Tensile strength (MPa)	Flexural strength (MPa)	Elastic modulus/Flexural modulus (GPa)	Vickers hardness (HV)
Tetric (Ivoclar Vivadent)	122–300	/	140 105.0 ± 5.6	11.5	63.74–80
Kerr Herculite XRV (Kerr)	225	/	105-110	24.6	47.166
Charisma (Heraeus Kulzer)	101.7–130.6	33	127.39 (±11.77)	5.28 (±0.73)	55.71 ± 0.03
Heliomolar (Ivoclar Vivadent)	340	114	100	6	350
Artglass (Heraeus Kulzer)	224	/	94.76	11–14.03	55
Kerr Herculite Lab (Kerr)	163.02	49.166	112.0 ± 4.2	21.1	68
Targis/Vectris (Ivoclar Vivadent)	163.39	32.862	111.23 ± 17.02	19.48–20	72
Vita Zeta LC (Vita)	110	/	120	6	28

The universal electro tensile testing machine was used for simulation of 5 years wearing of removable partial denture with 3000 antagonistic contacts per day. All measurements were done under the load of 50 N with speed interval of 200 mm/min considering the movable bridge of Instron. On the stationary bridge of the Instron head, investigated samples of composite materials were placed. The lower surface of the Instron head was coated using dentin-like-layer composed of residual dentin debris specially produced for the experiment purposes and bonded using OptiBond® Solo Plus™ (Kerr, Bolzano, Italy). Each tested specimen was then bonded and fixed to dentin-like-layer, in the same position using whether OptiBond® Solo Plus™ (Kerr, Bolzano, Italy) in the case of direct resin composite specimens or RelyX U200 (3M ESPE, St. Paul, MN, USA) during cementation of IRC specimens. The abovementioned materials were used to simulate real conditions as much as possible. In that way,

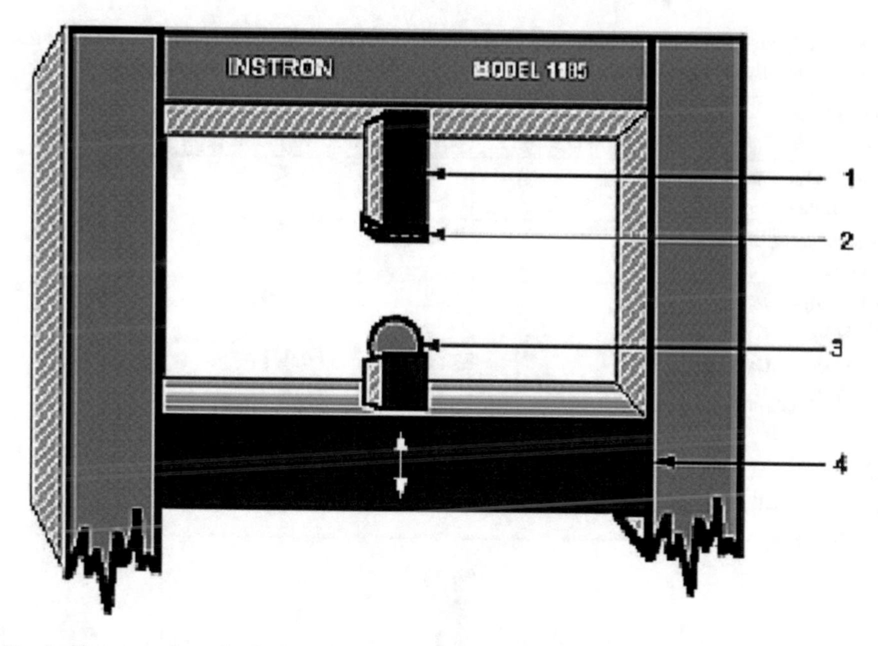

Fig. 1. Schema of mechanical testing of composite specimen: stationary bridge of Instron (1); specimen (2); CoCrMo-globe (ball) (3); movable bridge of Instron (4).

tested resin composite specimens were positioned on the lower surface of Instron head and directed towards movable bridge of Instron. A bowl with 6 mm diameter made of Co-Cr-Mo dental alloy which usually served for RPDs fabrication was fixed on the movable bridge of Instron. This ball simulated the dental element of the clasp removable partial denture directed upon antagonistic jaw.

After testing in the Instron, the experimental analysis of 3D volume and thickness changes was done using 3D scanning profilometer (Perthometer S3P, Mahr GmbH, Göttingen, Germany) which detects and tracks changes at 1.25 μm. The thickness and volume were measured to obtain the real changes immediately after load application.

3 Results

The highest change in thickness was found in Heliomolar (86.2 ± 32.6 μm). Artglass showed minimal change in thickness (30.20 ± 23.0 μm) compared to others. In the group of direct composite materials, Tetric experienced the lowest thickness change (31.0 ± 8.8 μm), while Charisma showed thickness variation of 65.3 ± 27.3 μm (Table 3; Fig. 2.). Among indirect composites, Vita Zeta LC showed the highest value of thickness variation (76.2 ± 31.3 μm; Table 4; Fig. 2).

Table 3. Average values and standard deviation of thickness and volume changes in direct resin based restorative materials.

Direct resin composite	Thickness variation (μm)	Standard deviation	Volume variation (mm^3)	Standard deviation
Tetric (Ivoclar Vivadent)	31.0	8.8	0.061	0.017
Kerr Herculite XRV (Kerr)	54.2	24.0	0.023	0.011
Charisma (Heraeus Kulzer)	65.3	86.2	0.089	0.019
Heliomolar (Ivoclar Vivadent)	86.2	32.6	0.091	0.026

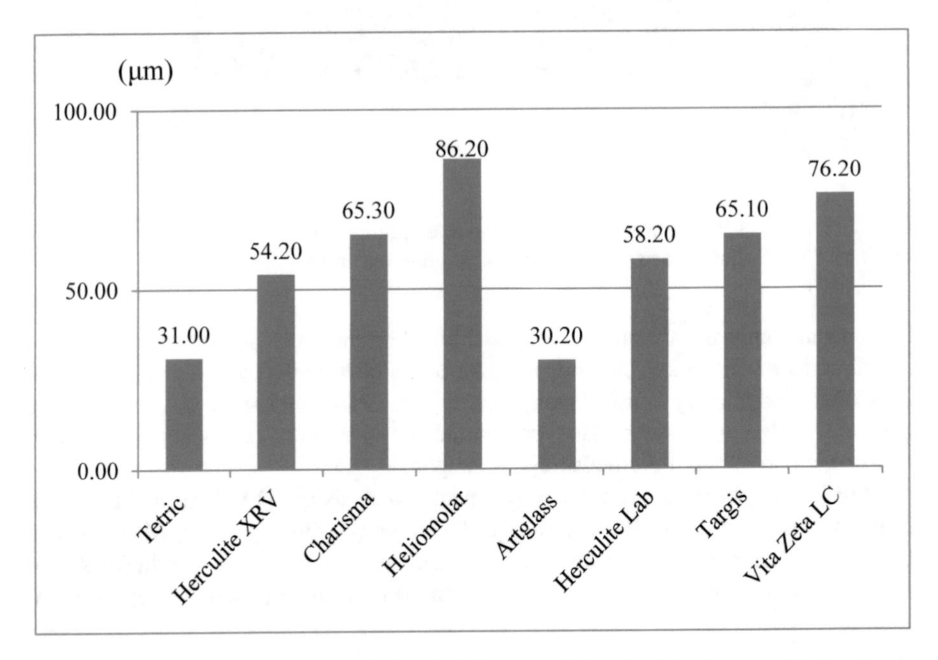

Fig. 2. Diagram of comparative analysis of thickness variation (μm) in direct and indirect resin composites.

Similar values of the thickness and volume changes were found in Kerr Herculite XRV (54.2 ± 24.0 μm; 0.023 ± 0.011 mm^3) and Kerr Herculte Lab (58.2 ± 24.9 μm; 0.029 ± 0.011 mm^3). Heliomolar showed the highest volume change compared to other direct composite resins (0.091 ± 0.026 mm^3). Tetric showed lower volume change (0.061 ± 0.017 mm^3) compared to Charisma or Heliomolar, but higher than Kerr Herculite XRV. Artglass changed its volume for 0.023 ± 0.011 mm^3, while Targis changed its volume for 0.0767 ± 0.031 mm^3. Vita Zeta LC had the lowest volume change (0.098 ± 0.031 mm^3) compared to the rest indirect resin composites tested in this study (Table 4; Fig. 3).

Table 4. Average values and standard deviation of thickness and volume changes in indirect composite resins.

Indirect resin composite	Thickness variation (μm)	Standard deviation	Volume variation (mm^3)	Standard deviation
Artglass (Heraeus Kulzer)	30.2	23.0	0.023	0.011
Kerr Herculite Lab (Kerr)	58.2	24.9	0.029	0.011
Targis/Vectris (Ivoclar Vivadent)	65.1	25.3	0.077	0.031
Vita Zeta LC (Vita)	76.2	31.3	0.098	0.033

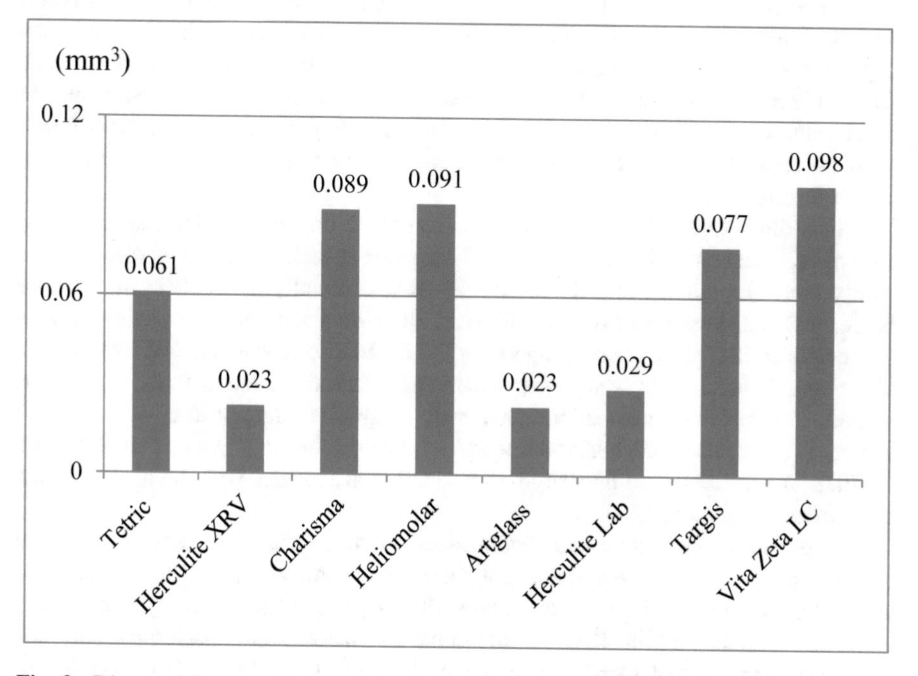

Fig. 3. Diagram of comparative analysis of volume variation (mm^3) in direct and indirect resin composites.

4 Discussion

Increase in demand for esthetics has led to the development of tooth-colored, metal-free restorations such as direct composite restorations, indirect composite inlays, and ceramic inlays or onlays [19]. Measurements of thickness and volume variations performed in this study can explain the resistance of tested composites resins to the applied occlusal loads. The changes in thickness and volume through years were followed by RPD's subsidence. Thus, RPD establishes closer contact to the supporting dental tissues and can change vertical dimension of occlusion. This is significant task of inlay/onlay and overlay materials especially considering patients with posterior related occlusal disbalance and bruxism. The ceramic inlays are more favorable restorations in these patients.

Although ceramic restorations are very resistant to compressive load [20] and have already proven themselves in many aspects (esthetics and long term clinical prognosis) they are still expensive, brittle, show tendency for fracture, poor tensile strength [21] and lead to wear of the opposing teeth. Teeth restored using ceramics inlay revealed increased crown stiffness while teeth under IRCs showed increased flexural strength of the crown [22]. IRCs was confirmed to be a good choice to ceramics for posterior teeth restorations [23]. Composite resin materials have shown a capability to absorb grate amount of compressive loading forces [24] and thus reduce occlusal forces by 57% more than dental ceramics [23]. This is associated with elastic property of organic matrix. It means that amortization of applied occlusal and para-occlusal forces were depended on elastic properties of applied restorative materials in a high percent. Clinical factor influences on the force transfer are the properties of dental tissues as substructures below composite restoration. Compared to porcelain and porcelain-fused-to-metal restorations, the transfer of masticatory forces is considerably less due to crown flexure [25, 26].

Thus, direct and indirect composite resins should be considered especially in the case of opposing natural teeth. However, therapeutic success and survival rate of direct composite restorations and IRCs is questionable especially in the case of presence harder materials in the opposing jaw like cast RPD, amalgam, metal-ceramic or metal-free-ceramics [26]. In this study opposing Co-Cr-Mo bowl was directed towards the direct and indirect resin composite specimens. This simulated the real conditions related to occlusion or mastication where resin composite filling or inlay occluded and accomplished contact with occlusal rest seat of clasp-retained removable partial denture (cRPD). The occlusal load intensity was inserted in accordance with the literature data on masticatory forces [17, 20].

The preparation technique and bonding agents can also affect longevity of composite restorations. In addition, inevitable polymerization shrinkage [27] should take into account as an adverse effect, although less when applied IRCs. Generally, during protocol of manipulation with IRCs an additional polymerization is performed and thus better activation of polymerization reactions [28–30]. Considering abovementioned, mechanical properties of composite material are crucial factors for long term clinical prognosis. Previously, it was concluded that the mechanical characteristics of composites depended on the resin monomer, filler (volume, weight and disposition), and coupling agents. Beside depth of polymerization and color stability, filler content is correlated with hardness, compressive strength, and stiffness. Increased filler loading and filler size leads to better mechanical performances of composite resin materials [31–33].

Results of this study are consistent to previous conclusions that great numbers of IRCs have advantages compared to direct composites, such as better mechanical performance and a significant reduction in polymerization shrinkage which is addressed to decreased volume changes [34]. However, considering the time dependence and storage within oral conditions, direct composites in some cases have shown increased mechanical performances compared to IRCs [35–37]. In clinical situation, shrinkage of IRCs is depended on shrinkage of composite cement.

This study confirmed that composite resin with higher filler loading and lower percent of organic matrix showed better mechanical properties compared to other composite resins. This was followed by decreased values of thickness and volume changes in Artglass. Also, Kerr Herculite and Tetric Ivoclar revealed slightly changes in volume and thickness, probably due to filler content and mechanical properties. It is known that the filler content increased the hardness values and tensile strength [38] which might be the reason for obtaining such results. Also, it may be argued about effect of resin composite restorative materials with higher filler loading and type of filler content on the long term prognosis of such teeth. This is due to improved mechanical properties decrease load amortization and thus induce higher stress and strain in periodontium [17].

Artglass is created basically as a restorative material for porcelain in PFMs metal-free veneers, inlays/onlays, crowns and bridges, or veneering of metal frameworks. Artglass consists of fine particles of hybrid composite material. Although, Artglass showed a higher degree of conversion (DC) and fracture toughness (FT) but a lower flexural modulus (E), hardness and wear resistance compared to Charisma considering the same curing methods [39]. Hence, in this study Charisma had more than twice a time higher values of thickness and volume variation. This indicates that Charisma was more sensitive to compressive forces and suffered from grater volume changes after loading. Reason for this may be lower values of compressive strength in Charisma compared to Artglass or Tetric. This fact should be also applied when explain a reason for higher values of thickness change in Vita Zeta LC and Targis/Vectris. In addition, higher percent of filler and size of irregular particles might influence on better mechanical properties in Artglass compared to others. Thus, an additional cure and the increased volume of inorganic fillers improved flexural strength and elastic modulus in IRCs [19]. For clinicans it is important that Artglass or Targis revealed higher fracture resistance of teeth restored with these IRCs compared to others composites [40]. This fact may exclude clinical failure of IRC restorations [41].

Regarding volume changes, Artglass showed minimal change in volume compared to the start up volume. The reason could be the presence of TEGDMA in the composite matrix. The presence of this monomer in the basic composition of Vita LC is associated with decreased flexural strength and increased modulus of elasticity of this material [42]. The characteristic flexibility of TEGDMA allows the creation of a dense and flexible polymer network that increases the composite elastic deformation [43]. As expected, the micro fine composite indicated poor mechanical properties and the lowest values of filler concentration [26]. The morphological characteristics of the fillers must also be considered, since they have been shown to be determining factors in the filler loading and the material strength [44]. This is important when interpreting the results of Heliomolar testing and should be considered when explain thickness change in Heliomolar specimens since mechanical behavior of composites depended on to their filler vol% [45]. High values of both, thickness and volume changes was found in Heliomolar probably due to presence of large amount of organic matrix and lower percentage of filler content. Thus increased polymerization shrinkage and lower mechanical properties were expected [4, 13, 40]. Considering Kerr Herculite XRV, higher value of thickness change is associated with lower hardness value. On the contrary, Kerr Helculite Lab has lower value of compressive strength and therefore experienced higher value of compared to initial stage of thickness [2, 13, 14]. However both products of Kerr showed less volume changes (Table 2) compared to other composites which is addressed to total filler loading and presence of barium silicate glass in their composition [1, 2, 7–9].

The nature of the filler, its' volume, how it is obtained and how much is added significantly affects the mechanical properties of the restoration material. As it is seen in Table 2, filler particles are irregularly shaped in all resin composite specimens employed in this study except Vita Zeta LC and Heliomolar where round particles are incorporated in organic matrix. In addition, the type of incorporated filler is not the same in all tested specimens. As previously mentioned, the physical and mechanical properties of the organic matrix are associated with filler particles incorporated in the organic phase [46]. This may be a reason for greater volume and Heliomolar changes during mechanical testing compared to other specimens. The filler reduces the thermal expansion coefficient and overall curing shrinkage, provides radio-opacity, improves handling and improves the aesthetic results [47]. Composites are usually composed of different forms of filler particles dispersed in a matrix phase [48]. Hybrid, micro filled, nano filled composites and IRCs have high percentage of filler particles. Micro filled composites have 37%–40% volume filler loading, whereas nano filled composites have 60% volume filler loading [49]. The aggregates and nanoparticles of the filler with volume up to 79.5% lead to high resistant to load due to better compression resistance [50].

Thus the use of IRCs may be recommended to patients with compromised periodontium for contouring tooth surfaces or implant supported therapy due to stability and less contraction during polymerization [51]. However, poor bonding between organic matrix and inorganic fillers and decreased wear resistance, marginal gap, micro leakage, high risk to bulk fracture, and adhesive failure are disadvantages when restored posterior teeth [14, 15].

Glass fiber reinforcement system in Targis/Vectris did not show satisfactory results considering mechanics. As it is known, fiber reinforced system composed of whether glass or polyethylene, as the commonly used fibers used for fabrication composite resin materials in dentistry. Fibers are important to stop a crack and thus improve resistance of composite. The resin matrix links fiber and influence on their geometrical orientation [52]. However, it was shown that fiber reinforcement did not improve mechanical properties considering experimental conditions which confirmed previous clinical evidence [53].

5 Conclusions

This study was conducted to investigate the thickness and volume changes in 8 types of resin composites often used in oral rehabilitation and clinical prosthodontics. It was found that all tested specimens indicated thickness and volume variations. This means that all observed specimens changed their thickness and volume compared to dimensional values before mechanical testing. Thus after mechanical testing composite specimens showed lower thickness and volume dimensions.

Further investigations should be conducted on randomized controlled experiments with long term follow-up to explain mechanical behavior of direct and indirect composite restorations. An association between mechanical properties of composites and their filler volume, and impact of the resin matrix should also be considered and investigated to reveal the real influence of these factors on mechanical properties of resin composites. In this experimental study some, but not all indirect composite specimens showed superior mechanical performance to direct composite specimens, highlighted Artglass as the material of choice viewed from mechanical aspect.

The following conclusions are drawn:

- Mechanical testing of selected composite specimens found no evidence regarding advantages IRCs over direct composite specimens due to some but not all IRCs showed higher values of thickness and volume changes;
- Regarding thickness of tested direct resin composite specimens, the highest resistance to dynamic vertical load showed Tetric while Herculite XRV, Charisma and Heliomolar showed lower resistance to dynamic load, respectively.
- The thickness of tested IRCs have shown minimal change in Artglass while changes in Kerr Herculite Lab, Targis, and Vita Zeta LC have shown decreased values, respectively.
- Considering volume variation in tested direct resin composite specimens, the value of the highest resistance to dynamic vertical load showed Tetric while Herculite XRV, Charisma and Heliomolar showed lower resistance to dynamic load, respectively.
- Volume variation of tested indirect resin composite specimens showed maximum changes in Vita Zeta LC specimens. Less change in volume was detected in Artglass, Herculite and Targis, respectively from the lowest to the highest value.
- The minimal thickness change including all specimens was found in Artglass probably due to high filler content, particle size and total filler content. Tetric was on the second place with specific composition.
- The lowest value of volume variation was found in both, Herculite XRV and Artglass. Barium glass and irregular shaped particles improved volume stability of Herculite and Artglass.
- The greatest changes in both, thickness and volume was found in Heliomolar probably due to its' composition.

References

1. Taira, Y., Hatono, H., Tokita, M., Sawase, T.: Thickness and surface structure of a ceramic layer created on three indirect resin composites with aerosol deposition. J. Prosthodont. Res. **54**, 168–172 (2010)
2. Brown, D.: The status of indirect restorative dental materials. Dent. Updat. **25**, 23–34 (1998)
3. Jackson, R.D., Morgan, M.: The new posterior resins and a simplified placement technique. J. Am. Dent. Assoc. **131**, 375–383 (2000)

4. Loguercio, A.D., Reis, A., Mazzocco, K.C., Dias, A.L., Busato, A.L., Singer, J.M., et al.: Microleakage in class 2 composite resin restorations: total bonding and open sandwich technique. J. Adhes. Dent. **4**, 137–144 (2002)
5. Leinfelder, K.F.: New developments in resin restorative systems. J. Am. Dent. Assoc. **128**, 573–581 (1997)
6. Alves, P.B., Brandt, W.C., Neves, A.C.C., Cunha, L.G., Silva-Concilio, L.R.: Mechanical properties of direct and indirect composites after storage for 24 hours and 10 months. Eur. J. Dent. **7**(1), 117–122 (2013)
7. Goldstein, R.E.: Sistemas adhesives de los composites. En: Goldstein RE. Odontología estética, vol. I, pp. 289–352. stm Editores, Barcelona (2002)
8. Bowen, R.L.: Properties of a silica-reinforced polymer for dental restorations. J. Am. Dent. Assoc. **66**, 57–64 (1963)
9. De la Macorra, J.C.: La contracción de polimerización de los materiales restauradores a base de resinas compuestas. Odontol. Conserv. **2**, 24–35 (1999)
10. Mitrovic, A., Mitrovic, N., Maslarevic, A., Adzic, V., Popovic, D., Milosevic, M., Antonovic, D., Thermal and mechanical characteristics of dual cure self-etching, self-adhesive resin based cement. In: Experimental and Numerical Investigations in Materials Science and Engineering, vol. 54, pp. 3–15. Springer (2018)
11. Millich, F., Jeang, L., Eick, J.D., Chappelow, C.C., Pinzino, C.S.: Elements of light-cured epoxy based dental polymer systems. J. Dent. Res. **77**, 603–608 (1998)
12. Tilbrook, D.A.: Photocurable epoxy-polyol matrices for use in dental composites I. Biomaterials **21**, 1743–1753 (2000)
13. Manhart, J., Kunzelmann, K.H., Chen, H.Y.: Mechanical properties of new composite restorative materials. J. Biomed. Mater. Res. **53**, 353–361 (2000)
14. Burke, F.J., Watts, D.C., Wilson, N.H., Wlson, M.A.: Current status ans rationale for composite inlays and onlays. Br. Dent. J. **70**, s269–s273 (1991)
15. Xu, H.H.: Dental composite resins containing silica-fused ceramic single-crystalline whiskers with various filler levels. J. Dent. Res. **78**, 1304–1311 (1999)
16. Tanasić, I., Tihaček-Šojić, Lj., Mitrović, M., Milić-Lemić, A., Vukadinović, M., Marković, A., Milošević, M.: An attempt to create a standardized (reference) model for experimental investigations on implant's sample. Measurement **72**, 37–42 (2015)
17. Tanasić, I., Milic-Lemic, A., Tihacek-Sojic, Lj., Stancic, I., Mitrovic, N.: Analysis of the compressive strain below the removable and fixed prosthesis in the posterior mandible using a digital image correlation method. Biomech. Model. Mechanobiol. **11**(6), 751–758 (2012)
18. Tanasić, I., Tihaček Šojić, Lj., Milić-Lemić, A.: Strain visualization of supporting tissues rehabilitated using two different types of removable partial dentures. Vojnosanitetski pregled. https://doi.org/10.2298/sarh170725181t
19. Manhart, J., Scheibenbogen-Fuchsbrunner, A., Chen, H.Y., Hickel, R.: A 2-year clinical study of composite and ceramic inlays. Clin. Oral. Investig. **4**, 192–198 (2000)
20. Tihacek Sojic, Lj., Milic Lemic, A., Tanasic, I., Mitrovic, N., Milosevic, M.: Compressive strains and displacement in a partially dentate lower jaw rehabilitated with two different treatment modalities. Gerodontology **29**(2), e851–e857 (2012)
21. Chabouis, F.H., Faugeron, S.V., Attal, J.P.: Clinical efficacy of composite versus ceramic inlays and onlays: a systematic review. Dent. Mater. **29**(12), 1209–1218 (2013)
22. Magne, P., Belser, U.C.: Porcelain versus composite inlays/onlays: effects of mechanical loads on stress distribution, adhesion, and crown flexure. Int. J. Periodontics Restor. Dent. **23**(6), 543–555 (2003)
23. Nandini, S.: Indirect resin composites. J. Conserv. Dent. **13**(4), 184–194 (2010)
24. Ciftçi, Y., Canay, S.: The effect of veneering materials on stress distribution in implant-supported fixed prosthetic restorations. Int. J. Oral Maxillofac. Implants **15**, 571–582 (2000)

25. Tanasić, I., Tihaček Šojić, Lj., Milić-Lemić, A.: Biomechanical interactions between bone and metal-ceramic bridges composed of different types of non-noble al loys under vertical loading conditions. Mater. Technol. **48**, 337–341 (2014)
26. Su, N., Yue, L., Liao, Y., Liu, W., Zhang, H., Li, X., Wang, H., Shen, J.: The effect of various sandblasting conditions on surface changes of dental zirconia and shear bond strength between zirconia core and indirect composite resin. J. Adv. Prosthodont. **7**(3), 214–223 (2015)
27. Mitrović, A.D., Tanasić, I.V., Mitrović, N.R., Milošević, M.S., Tihaček-Šojić, Lj.Dj., Antonović, D.G.: Strain determination of self-adhesive resin cement using 3D Digital Image Correlation Method. Srp Arh Celok Lek (2017). https://doi.org/10.2298/SARH170530176M
28. Santana, I.L., Lodovici, E., Matos, J.R., Medeiros, I.S., Miyazaki, C.L., Rodrigues-Filho, L. E.: Effect of experimental heat treatment on mechanical properties of resin composites. Braz. Dent. J. **20**, 205–210 (2009)
29. Soares, C.J., Pizi, E.C., Fonseca, R.B., Martins, L.R.: Mechanical properties of light-cured composites polymerized with several additional post-curing methods. Oper. Dent. **30**, 389–394 (2005)
30. Borba, M., Della Bona, A., Cecchetti, D.: Flexural strength and hardness of direct and indirect composites. Braz. Oral. Res. **23**, 5–10 (2009)
31. Fujishima, A., Ferracane, J.L.: Comparison of four modes of fracture toughness testing for dental composites. Dent. Mater. **12**, 38–43 (1996)
32. Miyazaki, M., Oshida, Y., Moore, B.K., Onose, H.: Effect of light exposure on fracture toughness and flexural strength of light-cured composites. Dent. Mater. **12**, 328–332 (1996)
33. Kim, K.H.: The effect of filler loading and morphology on the mechanical properties of contemporary composites. J. Prosthet. Dent. **87**, 642–649 (2002)
34. Dietschi, D., Scampa, U., Campanile, G., Holz, J.: Marginal adaptation and seal of direct and indirect Class II composite resin restorations: an in vitro evaluation. Quintessence Int. **26**, 127–138 (1995)
35. Da Fonte Porto Carreiro, A., Dos Santos Cruz, C.A., Vergani, C.E.: Hardness and compressive strength of indirect composite resins: effects of immersion in distilled water. J. Oral Rehabil. **31**(11), 1085–1089 (2004)
36. Cesar, P.F., Miranda, W.G., Braga, R.R.: Influence of shade and storage time on the flexural strength, flexural modulus, and hardness of composites used for indirect restorations. J. Prosthet. Dent. **86**(3), 289–296 (2001)
37. Reich, S.M., Petschelt, A., Wichmann, M., Frankenberger, R.: Mechanical properties and three-body wear of veneering composites and their matrices. J. Biomed. Mater. Res. A **69**(1), 65–69 (2004)
38. Chung, K.H.: The relationship between composition and properties of posterior resin composites. J. Dent. Res. **69**, 852–856 (1990)
39. Freiberg, R.S., Ferracane, J.L.: Evaluation of cure, properties and wear resistance of Artglass dental composite. Am. J. Dent. **11**(5), 214–218 (1998)
40. Soares, C.J., Martins, L.R., Pfeifer, J.M., Giannini, M.: Fracture resistance of teeth restored with indirect-composite and ceramic inlay systems. Quintessence Int. **35**(4), 281–286 (2004)
41. Manhart, J., Chen, H.Y., Hamm, G., Hickel, R.: Review of the clinical survival of direct and indirect restorations in posterior teeth of the permanent dentition. Oper. Dent. **29**(5), 481–508 (2004)
42. Asmussen, E., Peutzfeldt, A.: Influence of UEDMA, BisGMA and TEGDMA on selected mechanical properties of experimental resin composites. Dent. Mater. **14**(1), 51–56 (1998)
43. Sideridou, I., Tserki, V., Papanastasiou, G.: Study of water sorption, solubility and modulus of elasticity of light-cured dimethacrylate-based dental resins. Biomaterials **24**(4), 655–665 (2003)

44. Adabo, G.L., Cruz, C.A.S., Fonseca, R.G., Vaz, L.G.: The volumetric fraction of inorganic particles and the flexural strength of composites for posterior teeth. J. Dent. **31**(5), 353–359 (2003)
45. Ikejima, I., Nomoto, R., McCabe, J.F.: Shear punch strength and flexural strength of model composites with varying filler volume fraction, particle size and silanation. Dent. Mater. **19**(3), 206–211 (2003)
46. Hervás-García, A.: Composite resins. A review of the materials and clinical indications. Medicina oral, patologia oral y cirugia bucal **11**(2), E215-20 (2006)
47. Labella, R., Lambrechts, P., Van Meerbeek, B., Vanherle, G.: Polymerization shrinkage and elasticity of flowable composites and filled adhesives. Dent. Mater. **15**, 128–137 (1999)
48. Azeem, R.A., Sureshbabu, N.M.: Clinical performance of direct versus indirect composite restorations in posterior teeth: a systematic review. J. Conserv. Dent. **21**, 2–9 (2018)
49. Lu, H., Lee, Y.K., Oguri, M., Powers, J.M.: Properties of a dental resin composite with a spherical inorganic filler. Oper. Dent. **31**, 734–740 (2006)
50. Geraldi, S., Perdigao, J.: Microleakage of a new restorative system in posterior teeth. J. Dent. Res. **81**, 1276 (2003)
51. Leinfelder, K.F.: Indirect posterior composite resins. Compend. Contin. Educ. Dent. **26**, 495–503 (2005)
52. Van Heumen, C.C., Kreulen, C.M., Bronkhorst, E.M., Lesaffre, E., Creugers, N.H.: Fiber reinforced dental composites in beam testing. Dent. Mater. **24**, 1435–1443 (2008)
53. Behr, M., Rosentritt, M., Latzel, D., et al.: Fracture resistance of fiber-reinforced vs. non-fiber-reinforced composite molar crowns. Clin. Oral Invest. **7**, 135 (2003)

Surface Modification of Dental Materials and Hard Tissues Using Nonthermal Atmospheric Plasma

Jovana N. Stasic and Vesna Miletic(✉)

University of Belgrade School of Dental Medicine, Rankeova 4, Belgrade, Serbia
vesna.miletic@stomf.bg.ac.rs

Abstract. Research of nonthermal atmospheric plasma (NTAP) for dental applications has been increasing in recent years. This paper presents a literature review of potential use of NTAP for treatment of surfaces of dental materials and hard dental tissues. The aim of NTAP interaction with dental materials and tissues is surface modification for stable and durable material-to-material or material-to-tissue bonds. Reactive particles in NTAP and various mixtures of gasses increase hydrophilicity of material surface, which is known to be hydrophobic in implants, ceramics or dental composites, with or without roughness changes. Adhesion of cells to implant surface was shown to improve after NTAP treatment, thereby promoting successful osseointegration. Bonding ceramic materials to the prepared surfaces of teeth or fiber/metal posts was shown to improve after NTAP treatment. Hard dental tissues achieve primarily micromechanical bonds with composite materials using dental adhesives. Increased organic content in the form of collagen fibrils and residual water pose a problem for achieving adequate and long-term adhesive-dentin bonds. This problem has not been solved with current adhesive application protocols. It was recently shown that application of NTAP improves the hydrophilicity of dentin surface and changes its polarity, which can contribute to better distribution of adhesive resin and deeper penetration into the hybrid layer. Previous studies pointed to similar or better initial adhesive bonds with dentin. However, adhesive-dentin bonds are subject to degradation in the long-term also after NTAP treatment suggesting the need for further optimization of NTAP for application on dentin.

Keywords: Nonthermal atmospheric plasma · Implants · Ceramics · Adhesive · Dentin

1 Introduction

Recent advancements in various fields of technology and material science broaden potential applications in the dental field in terms of new and modified materials and/or interactions with human tissues. The primary purposes of dental materials are to restore function and esthetics of dental and oral structures lost due to disease or trauma. Materials commonly used for such purposes are (1) dental composites, for restoration of hard dental tissues, (2) ceramics, for oral rehabilitation in more extensive situations

N. Mitrovic et al. (Eds.): CNNTech 2019, LNNS 90, pp. 119–138, 2020.
https://doi.org/10.1007/978-3-030-30853-7_8

from large crown destructions of a single tooth to multiple missing teeth and (3) implants, for replacement of missing teeth.

Despite large chemical and structural differences of these materials, what they all have in common is the inevitable interaction with human tissues, whether it be hard dental tissues, enamel and dentin, in the case of composites and ceramics or bone tissue in the case of implants.

1.1 Challenges in Bonding Composite and Ceramic Materials to Tooth Tissues

Dental composites bond with tooth tissues *via* adhesive systems, mostly organic resin mixtures, whose primary mechanism of adhesion is micromechanical interlocking [1], with secondary chemical bonding reported for certain recent adhesive systems [2]. Adhesives and composites form covalent bonds in methacrylate monomers.

Dental adhesives are applied to enamel and dentin following the so-called 'total-etch' or 'self-etch' protocols which refer to surface demineralization by phosphoric acid or partial demineralization by functional acidic monomers in adhesive [3]. During this process, adhesive penetrates the interprismatic pits in enamel or interfibrillar collagen network in dentin creating a so-called 'hybrid layer' with its unique properties [1, 2, 4, 5]. Enamel is largely accepted as a favorable substrate, able to form stable and durable bonds with dental adhesives. However, dentin remains a much greater challenge for long-term adhesion [3]. Adhesive resin, especially following the 'total-etch' protocol, does not fully encapsulate the exposed collagen fibrils and residual water lowers monomer-to-polymer conversion [6] leading to suboptimal polymerization within the hybrid layer and potential elution of unbound monomers [7]. Adhesive phase separation occurs due to the inability of hydrophobic monomers to penetrate as deep as the hydrophilic ones within the hydrophilic hybrid zone following acid etching [8]. Long-term adhesive-dentin bonding is hindered by biodegradation processes within the polymer network, mainly hydrolysis and fatigue [9] but also within the collagen network of the hybrid layer due to the activation and detrimental effect of matrix metalloproteinases [10].

Dental ceramics bond to tooth tissues *via* resin or glass-ionomer luting cements which also create micromechanical and chemical (resin) or chemical bonds (glass-ionomer) with tooth tissues. Interaction of ceramics with resin or glass-ionomer cements is micromechanical and requires previous surface treatment using chemical or mechanical methods [11, 12].

Self-adhesive resin cements directly bond to tooth tissue without an intermediary adhesive layer. Although intended to create a similar adhesion pattern as dental adhesives, self-adhesive resin cements have shown only limited micromechanical and chemical interaction with enamel and dentin [13]. Initially hydrophilic upon mixing for facilitated wetting of tooth tissues, cements quickly become hydrophobic as acidic groups interact with calcium from hydroxyapatite or metal oxides from inorganic fillers [14]. Bonding of resin and glass-ionomer luting cements to silica-based dental ceramic materials is traditionally based on surface etching with hydrofluoric acid and subsequent silanization with silane coupling agents. Conversely, high-alumina and zirconia ceramics cannot be adequately bonded to luting cements through acid etching or silanization as they do not contain a silica phase. Air-abrasion with alumina particles, silica-deposition treatment using silica-coated alumina particles, surface functionalization with primers such as 10-MDP or selective infiltration etching are used to modify ceramic surface for adhesion [11, 12, 15]. These surface treatment methods have a limited effect, especially on zirconia. Furthermore, long-term adhesion is challenged by hydrolysis of silanes and primers [15].

1.2 Challenges in Osseointegration of Dental Implants

Dental implants interact with bone tissue on a macromolecular and cellular level to achieve stable and long-lasting implant-to-bone anchorage. Implants are mostly designed from titanium/titanium alloy or more recently zirconia. Implant surface plays an important role in the process of osseointegration [16], along with other major factors being surgical approach, patient factors and physical impact [17, 18]. Direct bone-implant contact is a preferred type of tissue response instead of the formation of fibrous soft tissue capsule which is known to lead to implant failure [19].

Roughness may be achieved through etching, blasting, anodization or nano-coating. Strong acids, such as hydrofluoric, hydrochloric, sulfuric or nitric acid, produce micrometer-sized pits and pores. Reproducibility of acid etching at the nanometer range and potential adverse effect on implant mechanical properties are reported limitations of this method [19]. Roughness created by blasting with ceramic particles depends on the size of the particles, mainly alumina, but may leave implant surface contaminated by such particles, resistant to various cleaning methods [19]. Micro- and nano-sized pores are created by oxide layer dissolution on implant surfaces using electric current but this anodization process is cumbersome and multifactor-dependent [19].

Titanium and hydroxyapatite plasma spraying creates nano-coating on implant surfaces by projecting the respective particles through plasma sources at high temperatures. The disadvantages of this method include titanium ion release in the peri-implant bone tissue potentially associated with peri-implant disease [20] as well as coating delamination [21] or increased risk for peri-implant bone defects associated with thick hydroxyapatite coating [22].

Surface topography, especially micro- and nano-sized roughness, and surface energy have been recently suggested as synergistic factors modulating both primary contact of blood components as well as further cell and tissue response (cell adhesion, proliferation and differentiation) [18].

1.3 Nonthermal Atmospheric Plasma

Also known as the fourth state of matter, plasma is an electrically conducting medium containing positively and negatively charged particles, generated by ionization of gas atoms and molecules following an energy input, such as heating or electric current. Plasmas may be classified as "thermal" or "nonthermal" based on the relative temperature of electrons, ions, atoms and molecules [23].

Nonthermal atmospheric plasma (NTAP) is suitable for application to sensitive organic matter as it operates at atmospheric pressure and low temperatures [24]. An inert gas, such as helium or argon, is used to generate plasma with temperatures as low as room temperatures which is especially important for vital cells and tissues. Cell and tissue interaction with NTAP occurs via abundant reactive oxygen and nitrogen species [25]. The fact that NTAP is "filamentary" or "contracted" [25] may be observed as an advantage or a limitation, depending on the application purposes and the area of its intended/desired use. For example, this characteristic of NTAP is advantageous is dentistry as the areas of application are small and strictly localized, so a well-targeted plasma reactivity is highly desirable.

NTAP interaction with materials is limited to surface modification without affecting the bulk material [25, 26]. Several basic mechanisms of surface modification by NTAP have been developed for biomedical application: etching or ablation, thin dielectric film deposition, chemical and/or physical surface modification, activation or functionalization. These processes are mainly chemically-based whilst physical surface modification and etching are modulated by reactive particle "bombardment" of material surface [25]. The resulting micro- and nano-topography is often characterized by increased roughness coupled with increased wetting *i.e.* hydrophilicity of material surface [26]. Additionally, surface functionalization may be tailored in such a way to attract cell adhesion and proliferation [27] or deter bacterial attachment [28].

Various NTAP sources include plasma brush, plasma jet, plasma needle and plasma pencil (Fig. 1). All of these sources are suitable for dental applications. Parameters such as working gas, power input, time and tip-to-surface distance characterize NTAP treatments.

Fig. 1. An NTAP needle.

The **aim** of this study is to provide a brief literature review on potential use of NTAP in surface modification of dental materials and hard dental tissues for improved material-to-tissue or material-to-material adhesion.

MEDLINE, Scopus and Web of Science databases were searched using the following keywords: "nonthermal atmospheric plasma" or "non-thermal atmospheric plasma", "dental", "implant", "ceramic", "zirconia", "enamel", "dentin". Also "nonthermal plasma" or "non-thermal plasma" search was followed by manual search in order to exclude plasma treatment under different pressure conditions and include only those at atmospheric pressure. Key NTAP parameters relevant for surface modification in this field are summarized and systematically presented. These parameters may serve as an indication for experimental NTAP applications as no standardization attempts have been made so far.

2 Surface Characteristics Relevant for Adhesion

Wetting explains the ability of a liquid to adhere to a solid surface. It is expressed as a contact angle between the liquid and the solid (θ) which is defined as the liquid-solid interface and the tangent line of the curve at the contact point of solid, liquid and gas [29] (Fig. 2).

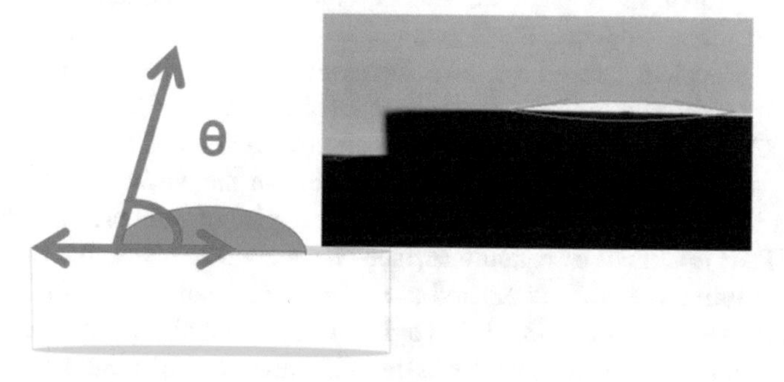

Fig. 2. A schematic of a contact angle (θ) of a liquid drop on a solid substrate and a reference liquid drop on a dentin disk.

The two extremes $\theta = 0°$ and $\theta = 180°$ indicate perfect wetting and non-wetting, respectively. Other degrees of wetting are expressed with a range of contact angles between 0° and 180°. It is generally accepted that contact angles of $0° < \theta < 90°$ indicate high wetting whilst those of $90° \leq \theta < 180°$ indicate low wetting.

When referring to water, wetting is also expressed as hydrophilicity, with low contact angles of water on hydrophilic and high contact angles on hydrophobic surfaces. Superhydrophilic surfaces exhibit water contact angles below 10° [29]. Hydrophilicity is dependent on interfacial energy between solid, liquid and gas and may be modified by changes in surface roughness as well as surface free energy (SFE) [29].

SFE is determined by intermolecular interactions that may be polar and apolar. Polar interactions are based on hydrogen-bonding or electron acceptor-electron donor (Lewis acid-base) interactions between polar moieties whilst apolar interactions are known as Lifshitz-van der Waals interactions. SFE of a solid may be calculated using three refence liquids, two polar and one apolar of known SFE parameters (γ) and measured contact angles (θ) on the tested solid, according to van Oss-Chaudhury-Good formula [30]:

$$(1 + cos\theta)_{\gamma_L} = 2\left(\sqrt{\gamma_S^{LW}\gamma_L^{LW}} + \sqrt{\gamma_S^{+}\gamma_L^{-}} + \sqrt{\gamma_S^{-}\gamma_L^{+}}\right) \quad (1)$$

Polar and apolar, as well as Lewis acid-base interactions of the tested solid may be differentiated by solving the above formula three times for three reference liquids [30].

3 Potential Application of NTAP in Surface Modification of Dental Ceramics

NTAP treatment of dental ceramics and implants has a significant effect on the surface of these materials in terms of roughness, surface energy and wetting properties. Increased roughness, surface energy and wetting improve material-to-material and material-to-tissue adhesion. In the case of dental implants, prevention of bacterial adhesion and growth on implant surfaces is important for long-term implant success.

NTAP treatment induces the formation of carboxyl groups on the surface of feldspathic porcelain resulting in increased hydrophilicity [31]. Increased hydrophilicity was reported for yttrium-tetragonal zirconia (Y-TZP) following NTAP treatment [32, 33]. NTAP alone or in combination with alumina sandblasting or primer functionalization increases reactive oxygen species on the surface of Y-TZP ceramic, resulting in increased SFE and decreased contact angles of reference liquids [32]. The effect of NTAP treatment of ceramic surface was shown to be similar or higher bond strength with luting materials compared to acid-treated, silane-coated [31], sandblasted [32] or untreated controls [33, 34]. Higher bond strength of colored zirconia or feldspathic ceramic to resin cement was reported after NTAP treatment compared to untreated specimens [35, 36]. Conversely, no significant differences were found between NTAP-treated and ground zirconia [37], zirconia abraded with silica-coated aluminum-oxide particles [38] in terms of shear bond strength to resin cements. Furthermore, lower shear bond strength to composite resin was reported for NTAP-treated feldspathic porcelain compared to standard acid-etch/silane coated surface treatment [35]. The inability of NTAP to improve bond strength to resin cements was seen both initially and after artificial aging [37]. Although NTAP was able to increase hydrophilicity, there were no changes in micro-roughness of zirconia surface [38]. These conflicting findings indicate that NTAP effects may be regime-dependent, requiring careful optimization for specific ceramic materials.

NTAP-induced surface modification of zirconia was shown to be time- and resin-dependent. A time-limited effect of increased hydrophilicity of zirconia was also reported for periods of 5–15 h [39] or up to 72 h [40] post-NTAP treatment. A positive effect on increased shear bond strength of NTAP-treated zirconia was registered for self-cured, self-adhesive RelyX U200 whilst significantly lower bond strength compared to untreated control was reported for dual-cured, self-etching adhesive resin Panavia F2.0 [40]. These time- and resin-dependent effects of NTAP on zirconia indicate that a potential chair-side rather than production stage application would be a viable mean of ceramic surface modification, especially if confirmed for other types of ceramic materials.

Favorable differentiation and proliferation of osteoblasts on NTAP-treated zirconia was related to increased hydrophilicity and coating with zirconium oxide nanoparticles [41]. An opposite effect on S. mutans adhesion was also reported [40], which warrants further investigation for obvious clinical relevance.

NTAP parameters used in previous studies are summarized in Table 1.

Table 1. Key NTAP parameters for surface modification of ceramic materials.

Authors	Parameters of NTAP			
	Gas	Input power/Voltage/Power supply	Tip-to-surface distance	Treatment time
Cho 2011	He	15 kHz, 2 kV	0.5 cm	Not stated
Jha 2017	N_2+O_2	500 V, 13 mA	2 mm	3 min 5 min 10 min 20 min 30 min
Lee 2016	Ar	200 W	Not stated	600 s
Liu 2016	Ar+O_2	10 mA, 0.8 kV	5 mm	2 min 5 min
Lopes 2014	Ar	Not stated	10 mm	60 s
Park 2017	Ar	300 W	5 mm	30 s
Park 2018	Ar	10 W	5 mm	30 min
Pott 2018	Ar	2.45 GHz, 10 W	10 mm	30 s
VBF Jr. 2018	Ar	Not stated	10 mm	60 s
Valverde 2013	Ar	1.1 MHz, 2–6 kV peak-to-peak, 8 W system power	5 mm	10 s

4 Potential Application of NTAP in Surface Modification of Dental Implants

NTAP may be effectively used to create "bio-functional dental implants" by surface functionalization with calcium phosphate, siloxane or peptide coatings. These coatings may further immobilize drug delivery systems onto implant surfaces, including bis-phosphonates and simvastatin, for expression of bone morphogenetic protein-2 (BMP-2), an effective inducer of bone regeneration, or adhesive proteins at the soft tissue-implant interface [42]. NTAP treated and bioactive peptide coated implant surface exhibited a synergistic effect on surface micro-roughness and hydrophilicity of titanium surface compared to separate treatments [43]. Such titanium surface then improved adhesion and proliferation of human bone marrow derived mesenchymal stem cells [43]. Titanium implants treated with NTAP enriched with 1% O_2 exhibited superhydrophilic surface with water contact angles approximating 0° which facilitated attachment and growth of human osteoblast cells [44]. Hydrophilicity further increased with exposure time and O_2 concentration (120 s *vs.* 60 s *vs.* 30 s and 1% *vs.* 0.2% O_2) [44]. Changes in surface chemistry and hydrophilicity of TiO_2 implant surface upon nitrogen- and air-based NTAP treatment improved cellular activity (viability, attachment and differentiation). Nitrogen-based NTAP was associated with higher osteogenic gene expression than air-based NTAP as a result of lower carbon atomic percentage on implant surface [45]. NTAP may synergistically act with other implant surface modifications, such as sandblasted, large-grit, and acid-etched (SLA), which improved osteoblast cell adhesion and proliferation [46]. SLA is known for hydrophobic surface, which may be effectively modified with NTAP toward desired hydrophilicity [46].

Nitrogen-, air- and ammonia-based NTAP treatment of titanium or titanium alloy surfaces increases SFE and hydrophilicity without changes in micro-roughness, increasing adhesion of mesenchymal stem cells but reducing adhesion of S. sanguinis [47]. Surface resistance to bacteria was associated with the so-called 'carbon-cleaning' *i.e.* reduction in C-H and C = O groups on NTAP functionalized surfaces [47]. The same effect on the adhesion of S. sanguinis was reported in another study, also for nitrogen-NTAP treatment [48]. This treatment reduced hydrocarbon and increased oxidized carbon species on the surfaces of both smooth and rough titanium implants. Changes in surface chemistry along with superhydrophilic surfaces following NTAP treatment significantly reduced attachment of S. sanguinis on implant surfaces [48]. Other bacterials strains, gram-positive S. aureus and S. mutans as well as gram-negative K. oxytoca and K pneumoniae, were also effectively prevented from adhesion on titanium implant surfaces following NTAP treatment. Plasma effects were time-dependent and more pronounced against gram-negative bacteria [49]. This was explained by greater susceptibility of gram-negative bacteria than cellular wall to adverse effect of surface oxidation during NTAP treatment and increase in reactive oxygen and nitrogen species [49]. Inhibition of bacterial adhesion and proliferation of S. mutans and S. aureus on titanium implant surfaces was reported in another study and also related to surface chemistry changes and increased hydrophilicity of implant surfaces [50]. S. mutans mono-species biofilm as well as multi-species oral microbial biofilm from human donors were affected by various NTAP treatments, including different sources, treatment times and O_2 addition to argon-based NTAP [51]. It is more clinically relevant to test an effect on oral biofilms due to complexities in biofilm architecture and extracellular matrix. This is reflected in the finding that S. mutans was more adversely affected by O_2 addition to argon gas in contrast to the multi-species biofilm, probably due to the resistance to oxidative gases of a complex biofilm [51].

NTAP parameters used in previous studies are summarized in Table 2.

Table 2. Key NTAP parameters for surface modification of dental implants.

Authors	Parameters of NTAP			
	Gas	Input power/Voltage/Power supply	Tip-to-surface distance	Treatment time
Duske 2012	Ar Ar+O_2	2–3 W	5 mm	30 s 60 s 120 s
Giro 2013	Ar+compressed air	1.5 MHz, 2–6 kV peak-to-peak, system power 230 V, 65 W	Not stated	20 s
Jeong 2017	N_2	2.24 kV, 1.08 mA, 2.4 W of power	3 mm	10 min
Karaman 2018	Air dielectric barrier discharge plasma	1.5 kHz, 31.4 kV	2 mm	5 s 15 s 30 s 45 s 60 s
Koban 2011	Ar/Ar+O_2 Ar/Ar+O_2 Ar	N/A 37.6 kHz, 8.4 kV 40 kHz, 10 kV	7 mm 5 mm 15 mm	1 min 2 min 5 min 10 min
Lee 2013	N_2	15 kV, 13 mA	3 mm	2 min 10 min
Lee 2017	N_2 N+ammonia	15 kV, 13 mA, 60 Hz	3 mm	10 s
Lee 2019	Air gas	15 kV, 13 mA	3 mm	2 min 10 min
Seo 2014	NO_2	2.4 W	3 mm	2 min 10 min
Shon 2014	He	1.9 kV, 2.9 mA, 15 kHz	Not stated	60 s

In addition to cell adhesion and proliferation and surface resistance to bacteria of implant surfaces following NTAP functionalization, osseointegration was assessed over a period of several weeks in animal models. Calcium-phosphate coating coupled with argon-based NTAP changed surface chemistry of titanium implants by increasing the percentage of hydrocarbon and O, Ca and P which, in turn, improved bone-to-implant contact and bone area fraction occupancy [52]. Similarly, alumina-blasted/acid-etched implant surfaces coupled with argon-based NTAP increased SFE and early implant stabilization [53].

A rare study on NTAP effects on zirconia implants showed similarities to titanium/titanium alloy implants in terms of increased hydrophilicity and maintained micro-topography. A positive effect on osseointegration was expressed as higher bone-implant contact ratio and bone volume compared to untreated specimens [54].

5 Potential Application of NTAP in Surface Modification of Hard Dental Tissues

As expected, enamel is not frequently a subject of NTAP experimental treatments, probably because adhesive bonding to enamel is widely accepted as optimized following phosphoric acid surface treatment. Dentin is the substrate of interest for potential NTAP surface modification with the majority of studies focusing on dentin.

Superhydrophilic enamel and dentin surfaces may be achieved after only 30 s of treatment, although this effect was more pronounced on dentin [55, 56]. Further studies confirmed improved wetting of NTAP-treated dentin [57–62] (Fig. 3).

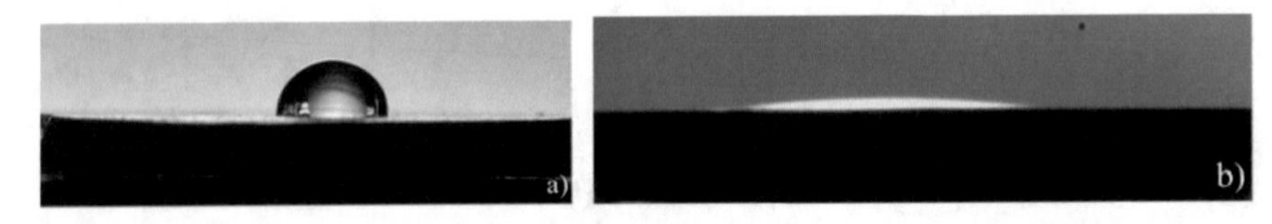

Fig. 3. Water droplet on (a) untreated dentin; (b) NTAP-treated dentin.

The addition of O_2 to the working gas may further increase NTAP-induced wetting properties [57] with 1% and 1.5% O_2 producing similar effects [62]. Longer exposure time increased dentin hydrophilicity (10 s *vs.* 30 s) [55, 63]. More pronounced hydrophilicity was seen on synthesized hydroxyapatite blocks than dentin [63], probably due to the complex nature of dentin involving certain micro-architecture of dentinal tubules, the presence of collagen fibers and water.

Surface changes in chemistry of NTAP-treated dentin indicate reduction in the atomic percentage of C [55, 57, 60, 61] and N [55, 60, 61] and increase in O, Ca and P [55, 57, 60]. These findings indicate the removal of adsorbed carbon and nitrogen species and disclosure of underlying inorganic calcium and phosphorus-containing hydroxyapatite with newly formed oxygen-containing reactive species.

NTAP parameters used in previous studies are summarized in Table 3.

Table 3. Key NTAP parameters for surface modification of dentin (§ enamel incl.)

Authors	Parameters of NTAP			
	Gas	Input power/Voltage/Power supply	Tip-to-surface distance	Treatment time
Ayres 2017	Ar	45 W[a]	10 mm	15 s
Ayres 2018	Ar	45 W[a]	10 mm	10 s 30 s
Chen 2013. §	Ar	5 W 10 W	2 mm	5 s 15 s 30 s 45 s
Dong 2013/2014/2015	Ar	2–3 W		30 s
Han 2014	He	Conventional: 2.4 kV, 2.5 mA, 8.0 kHz Pulsed: ~2 kV, 0.4 kHz, 5 voltage peaks	5 mm	30 s
Hirata 2015/2016	Ar	8 W	15 mm	30 s
Kim 2016	He	0.3 W	5 mm	20 s
Koban 2011	Ar Ar+O_2	Not stated	3 mm	10 s 30 s 60 s 120 s
Lehmann 2013. §	He He+H_2O_2 He+H_2O	2.45 GHz, average pulsed microwave power of 2 W, single pulse power of 250 W, pulse width 5 ms	2 mm	Not stated
Ritts 2010	Ar	5 W	Not stated	30 s 100 s 300 s
Stasic 2019	He He+O_2	1 W 3 W	2 mm 4 mm 8 mm	30 s
Šantak 2017. §	He	Not stated	1 cm	1 s 5 s 30 s 60 s 9 min
Zhang 2014	Ar	2–3 W	5-6 mm	30 s
Zhu 2018	He	Conventional: 67 kV, 17 kHz. Modified: 15 W	Not stated	5 s 10 s 15 s 30 s 45 s 60 s

[a]personal communication with author

SFE of dentin increased following various NTAP treatments [59, 62]. An increase in both polar and apolar components of SFE was reported for He-based NTAP [62] whilst only polar component increased after treatment with Ar-based NTAP [59]. The differences could be ascribed to different NTAP regimes and to different methods of calculation based on different reference liquids and formulas. Polar component of SFE was further analyzed in a recent study by Stasic et al. [62] showing that Lewis base increased whilst Lewis acid decreased after NTAP treatment. Lewis acid-base interactions are based on hydrogen bonding and electron donor-acceptor interactions, Lewis base being indicative of electron donor and Lewis acid of electron acceptor functionalities. Increase in Lewis base interaction sites indicates potential susceptibility for hydrogen bonding with adhesive functionalities during subsequent adhesive placement, contributing to adhesive-dentin bonding.

NTAP was shown to induce changes in the collagen fibrillar network in terms of fibrillar structure, mechanical and chemical characteristics, number and length of exposed collagen fibers [58, 63–65]. This may lead to better monomer interlocking within the hybrid layer [64], thicker hybrid layer with fewer structural defects and gaps at the dentin-adhesive-composite interface [66].

Improved hydrophilicity, SFE and changes in surface polarity may facilitate adhesive distribution on NTAP-treated dentin. However, it is somewhat unexpected that in wetting measurements, contact angles of actual adhesives are seldom measured. Stasic et al. [62] showed that, although the contact angles of reference liquids may be significantly lower on NTAP-treated than untreated dentin, the same does not necessarily hold for adhesive systems. Two universal adhesives (Single Bond Universal, 3M ESPE and Clearfil Universal Bond, Kuraray) showed material-dependent behavior. Whilst contact angles of both adhesives were significantly higher on phosphoric-acid etched than untreated dentin, Single Bond Universal showed slightly increased and Cleafil Universal Bond decreased contact angles following NTAP treatment compared to the self-etch protocol. These findings highlight the complex composition of adhesive mixtures and necessitate a more detailed investigation of adhesive-dentin interaction beyond measurement of contact angles of reference liquids.

The ultimate goal of any surface modification, NTAP included, of hard dental tissues is improved adhesive bonding. Indeed, higher bond strengths initially were reported for adhesives applied to NTAP-treated dentin [59, 63–71]. However, this initial positive effect on adhesive-dentin bond strength was not seen in the long-term for NTAP applied to previously untreated or acid-etched dentin [59, 72]. The results were aging-dependent *i.e.* simulated pulpal pressure resulted in higher adhesive bond strength to NTAP-treated than untreated dentin compared to direct water storage after 1 year [71]. Adhesive application protocol also affected NTAP effects in the long-term with 'etch-and-rinse' approach benefiting from 30 s NTAP treatment unlike the 'self-etch' approach which showed similar degradation to NTAP-treated dentin [63]. Similarly, a universal adhesive applied following the 'self-etch' protocol showed comparable results to NTAP-treated dentin after 1 year of storage [59].

Contrary to the previous findings, an increase compared to the initial bond strength was found for pulsed NTAP [69]. A decrease in adhesive bond strength to NTAP-treated dentin was reported after 2 months of storage, albeit maintaining higher values compared to untreated dentin both initially and after storage, suggesting a positive effect of NTAP surface modification [68].

These conflicting findings could be associated with differences in material composition, treatment parameters or storage conditions. Future research should focus on NTAP optimization for dentin application in order to achieve long-term benefit of improved wetting and reactivity induced by NTAP. Optimized regimes should exert no destructive or topographical surface changes, which was previously accomplished with different NTAP parameters [55, 57, 62, 71] (Fig. 4).

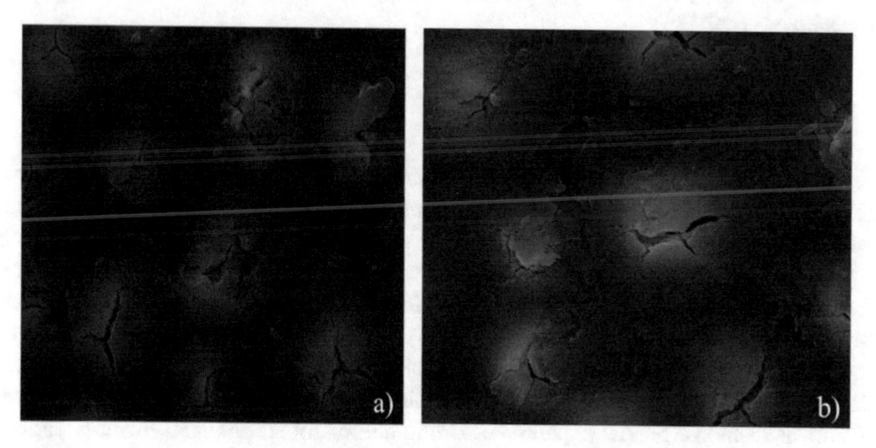

Fig. 4. A scanning electron micrograph of (a) untreated dentin; (b) NTAP-treated dentin.

Based on this literature review, it may be concluded that clinically relevant optimized NTAP treatment could be characterized by 30 s exposure time, 2–5 mm tip-to-surface distance and up to 3 W power.

6 Other Potential Applications of Nonthermal Atmospheric Plasma in Dentistry

Antimicrobial effect is one of the most prominent features of NTAP as studies reported on its effect against both planktonic [73] and biofilm-forming pathogens [74, 75]. Oral antimicrobial effects of NTAP were reported for dentin surfaces against multiple microbial strains, such as L. casei, S. mutans, C. albicans and E. coli [76]. Limited efficiency of NTAP was demonstrated in the case of intracanal infection, possibly due to the inability of current NTAP sources to act over longer distances [77]. NTAP modified with chlorhexidine was shown to improve antimicrobial effects of NTAP or chlorhexidine alone against E. faecalis endodontic biofilm [78].

NTAP may induce polymerization of resin-based adhesives with higher degree of conversion and without an adverse effect of water on such polymerization [79]. This finding is especially relevant for improved adhesive curing in the dentin hybrid layer known for residual water in the interfibrillar spaces within collagen network [1].

Plasma-induced polymerization of resin-based materials creates free radicals by energy transfer from collisions of excited particles (direct transfer) or photon irradiation (indirect transfer) [80].

NTAP may increase and accelerate bleaching efficiency of hydrogen peroxide [81, 82] and potentially replace carbamide/hydrogen peroxide with deionized water as a plasma-activated bleaching agent in non-vital teeth [83]. NTAP-assisted bleaching with carbamide peroxide lowers surface roughness of enamel compared to bleaching agent alone and reduces S. mutans adhesion [84]. Thermal tests inside the pulp chamber in vitro during such bleaching procedure exhibited a mild thermal challenge, not exceeding 37 °C [85].

7 Conclusions

NTAP is a powerful tool for surface modification of dental materials and hard dental tissues. Operating at low temperatures and atmospheric pressure, such plasma treatment appears safe for vital cells and tissues. The main effects of NTAP treatment include increased wetting and SFE with or without topographical changes in roughness. Surface chemistry is changed through interactions with abundant reactive oxygen and nitrogen species. The reported effects of NTAP surface modification include improved implant osseointegration and adhesive-dentin bonding. Conversely, NTAP has demonstrated a limited effect on ceramic-resin bonding compared to standard surface treatment methods. NTAP expresses a strong antimicrobial potential towards various planktonic and biofilm-forming microorganisms. NTAP optimization and standardization is required for application on dental materials and hard dental tissues.

Acknowledgment. This work was supported by research grant ON172007 from the Ministry of Education, Science and Technological Development, Republic of Serbia.

References

1. Pashley, D.H., Tay, F.R., Breschi, L., Tjaderhane, L., Carvalho, R.M., Carrilho, M., Tezvergil-Mutluay, A.: State of the art etch-and-rinse adhesives. Dent. Mater. **27**, 1–16 (2011). https://doi.org/10.1016/j.dental.2010.10.016. S0109-5641(10)00459-8 [pii]
2. Van Meerbeek, B., Yoshihara, K., Yoshida, Y., Mine, A., De Munck, J., Van Landuyt, K.L.: State of the art of self-etch adhesives. Dent. Mater. **27**, 17–28 (2011). https://doi.org/10.1016/j.dental.2010.10.023. S0109-5641(10)00466-5 [pii]
3. Miletic, V., Sauro, S.: Adhesion to tooth tissues. In: Miletic, V. (ed.) Dental Composite Materials for Direct Restorations, pp. 199–218. Springer, Cham (2018)
4. Nakabayashi, N., Pashley, D.H.: Hybridization of Dental Hard Tissues. Quintessence Publishing Co Ltd., Tokyo (1998)
5. Santini, A., Miletic, V.: Comparison of the hybrid layer formed by Silorane adhesive, one-step self-etch and etch and rinse systems using confocal micro-Raman spectroscopy and SEM. J. Dent. **36**, 683–691 (2008)

6. Santini, A., Miletic, V.: Quantitative micro-Raman assessment of dentine demineralization, adhesive penetration, and degree of conversion of three dentine bonding systems. Eur. J. Oral Sci. **116**, 177–183 (2008)
7. Miletic, V., Santini, A., Trkulja, I.: Quantification of monomer elution and carbon-carbon double bonds in dental adhesive systems using HPLC and micro-Raman spectroscopy. J. Dent. **37**, 177–184 (2009)
8. Ye, Q., Wang, Y., Spencer, P.: Nanophase separation of polymers exposed to simulated bonding conditions. J. Biomed. Mater. Res. B Appl. Biomater. **88**, 339–348 (2008)
9. De Munck, J., Van Landuyt, K., Peumans, M., Poitevin, A., Lambrechts, P., Braem, M., Van Meerbeek, B.: A critical review of the durability of adhesion to tooth tissue: methods and results. J. Dent. Res. **84**, 118–132 (2005)
10. Mazzoni, A., Scaffa, P., Carrilho, M., Tjaderhane, L., Di Lenarda, R., Polimeni, A., Tezvergil-Mutluay, A., Tay, F.R., Pashley, D.H., Breschi, L.: Effects of etch-and-rinse and self-etch adhesives on dentin MMP-2 and MMP-9. J. Dent. Res. **92**, 82–86 (2013). https://doi.org/10.1177/0022034512467034. 0022034512467034 [pii]
11. Mattiello, R.D.L., Coelho, T.M.K., Insaurralde, E., Coelho, A.A.K., Terra, G.P., Kasuya, A. V.B., Favarão, I.N., Gonçalves, L.d.S., Fonseca, R.B.: A review of surface treatment methods to improve the adhesive cementation of zirconia-based ceramics. ISRN Biomater. **2013**, 10 (2013). https://doi.org/10.5402/2013/185376
12. Della Bona, A., Borba, M., Benetti, P., Pecho, O.E., Alessandretti, R., Mosele, J.C., Mores, R.T.: Adhesion to dental ceramics. Curr. Oral Health Rep. **1**, 232–238 (2014). https://doi.org/10.1007/s40496-014-0030-y
13. Monticelli, F., Osorio, R., Mazzitelli, C., Ferrari, M., Toledano, M.: Limited decalcification/diffusion of self-adhesive cements into dentin. J. Dent. Res. **87**, 974–979 (2008). https://doi.org/10.1177/154405910808701012
14. Ferracane, J.L., Stansbury, J.W., Burke, F.J.: Self-adhesive resin cements - chemistry, properties and clinical considerations. J. Oral Rehabil. **38**, 295–314 (2011). https://doi.org/10.1111/j.1365-2842.2010.02148.x
15. Ozcan, M., Bernasconi, M.: Adhesion to zirconia used for dental restorations: a systematic review and meta-analysis. J. Adhes. Dent. **17**, 7–26 (2015). https://doi.org/10.3290/j.jad.a33525
16. Novaes Jr., A.B., de Souza, S.L., de Barros, R.R., Pereira, K.K., Iezzi, G., Piattelli, A.: Influence of implant surfaces on osseointegration. Braz. Dent. J. **21**, 471–481 (2010)
17. Chrcanovic, B.R., Kisch, J., Albrektsson, T., Wennerberg, A.: Factors influencing early dental implant failures. J. Dent. Res. **95**, 995–1002 (2016). https://doi.org/10.1177/0022034516646098
18. Rupp, F., Liang, L., Geis-Gerstorfer, J., Scheideler, L., Huttig, F.: Surface characteristics of dental implants: a review. Dent. Mater. **34**, 40–57 (2018). https://doi.org/10.1016/j.dental.2017.09.007
19. Le Guehennec, L., Soueidan, A., Layrolle, P., Amouriq, Y.: Surface treatments of titanium dental implants for rapid osseointegration. Dent. Mater. **23**, 844–854 (2007). https://doi.org/10.1016/j.dental.2006.06.025
20. Fretwurst, T., Nelson, K., Tarnow, D.P., Wang, H.L., Giannobile, W.V.: Is metal particle release associated with peri-implant bone destruction? An emerging concept. J. Dent. Res. **97**, 259–265 (2018). https://doi.org/10.1177/0022034517740560
21. Carrado, A., Perrin-Schmitt, F., Le, Q.V., Giraudel, M., Fischer, C., Koenig, G., Jacomine, L., Behr, L., Chalom, A., Fiette, L., Morlet, A., Pourroy, G.: Nanoporous hydroxyapatite/sodium titanate bilayer on titanium implants for improved osteointegration. Dent. Mater. **33**, 321–332 (2017). https://doi.org/10.1016/j.dental.2016.12.013

22. Madi, M., Zakaria, O., Kasugai, S.: Coated vs uncoated implants: bone defect configurations after progressive peri-implantitis in dogs. J. Oral Implantol. **40**, 661–669 (2014). https://doi.org/10.1563/AAID-JOI-D-12-00089
23. Fridman, A.: Elementary Plasma-Chemical Reactions. Plasma Chemistry, pp. 12–91. Cambridge University Press, Cambridge (2008)
24. Stoffels, E., Flikweert, A.J., Stoffels, W.W., Kroesen, G.M.W.: Plasma needle: a non-destructive atmospheric plasma source for fine surface treatment of (bio)materials. Plasma Sources Sci. Technol. **11**, 383 (2002)
25. von Woedtke, T., Reuter, S., Masur, K., Weltmann, K.D.: Plasmas for medicine. Phys. Rep. **530**, 291–320 (2013)
26. Liu, Y., Liu, Q., Yu, Q.S., Wang, Y.: Nonthermal atmospheric plasmas in dental restoration. J. Dent. Res. **95**, 496–505 (2016). https://doi.org/10.1177/0022034516629425. 0022034516629425 [pii]
27. Ohl, A., Schröder, K.: Plasma-induced chemical micropatterning for cell culturing applications: a brief review. Surf. Coat. Technol. **116–119**, 820–830 (1999)
28. Bazaka, K., Jacob, M.V., Crawford, R.J., Ivanova, E.P.: Plasma-assisted surface modification of organic biopolymers to prevent bacterial attachment. Acta Biomater. **7**, 2015–2028 (2011). https://doi.org/10.1016/j.actbio.2010.12.024
29. Otitoju, T.A., Ahmad, A.L., Ooi, B.S.: Superhydrophilic (superwetting) surfaces: a review on fabrication and application. J. Ind. Eng. Chem. **47**, 19–40 (2017). https://doi.org/10.1016/j.jiec.2016.12.016
30. Van Oss, C.J., Chaudhury, M.K., Good, R.J.: Interfacial Lifshitz-van der Waals and polar interactions in macroscopic systems. Chem. Rev. **88**, 927–941 (1988)
31. Han, G.J., Chung, S.N., Chun, B.H., Kim, C.K., Oh, K.H., Cho, B.H.: Effect of the applied power of atmospheric pressure plasma on the adhesion of composite resin to dental ceramic. J. Adhes. Dent. **14**, 461–469 (2012). https://doi.org/10.3290/j.jad.a25688. 25688 [pii]
32. Valverde, G.B., Coelho, P.G., Janal, M.N., Lorenzoni, F.C., Carvalho, R.M., Thompson, V. P., Weltemann, K.D., Silva, N.R.: Surface characterisation and bonding of Y-TZP following non-thermal plasma treatment. J. Dent. **41**, 51–59 (2013). https://doi.org/10.1016/j.jdent.2012.10.002
33. Liu, T., Hong, L., Hottel, T., Dong, X., Yu, Q., Chen, M.: Non-thermal plasma enhanced bonding of resin cement to zirconia ceramic. Clin. Plasma Med. **4**, 50–55 (2016). https://doi.org/10.1016/j.cpme.2016.08.002
34. Lee, M.H., Min, B.K., Son, J.S., Kwon, T.Y.: Influence of different post-plasma treatment storage conditions on the shear bond strength of veneering porcelain to zirconia. Materials (Basel) **9** (2016). https://doi.org/10.3390/ma9010043
35. Cho, B.H., Han, G.J., Oh, K.H., Chung, S.N., Chun, B.H.: The effect of plasma polymer coating using atmospheric-pressure glow discharge on the shear bond strength of composite resin to ceramic. J. Mater. Sci. **46**, 2755–2763 (2011). https://doi.org/10.1007/s10853-010-5149-1
36. Park, C., Yoo, S.H., Park, S.W., Yun, K.D., Ji, M.K., Shin, J.H., Lim, H.P.: The effect of plasma on shear bond strength between resin cement and colored zirconia. J. Adv. Prosthodont. **9**, 118–123 (2017). https://doi.org/10.4047/jap.2017.9.2.118
37. Pott, P.C., Syvari, T.S., Stiesch, M., Eisenburger, M.: Influence of nonthermal argon plasma on the shear bond strength between zirconia and different adhesives and luting composites after artificial aging. J. Adv. Prosthodont. **10**, 308–314 (2018). https://doi.org/10.4047/jap.2018.10.4.308

38. Vilas Boas Fernandes Junior, V., Barbosa Dantas, D.C., Bresciani, E., Rocha Lima Huhtala, M.F.: Evaluation of the bond strength and characteristics of zirconia after different surface treatments. J. Prosthet. Dent. **120**, 955–959 (2018). https://doi.org/10.1016/j.prosdent.2018.01.029
39. Lopes, B.B., Ayres, A.P.A., Lopes, L.B., Negreiros, W.M., Giannini, M.: The effect of atmospheric plasma treatment of dental zirconia ceramics on the contact angle of water. Appl. Adhes. Sci. **2**, 17 (2014). https://doi.org/10.1186/2196-4351-2-17
40. Park, C., Park, S.-W., Yun, K.-D., Ji, M.-K., Kim, S., Yang, Y.P., Lim, H.-P.: Effect of plasma treatment and its post process duration on shear bonding strength and antibacterial effect of dental zirconia. Materials **11**, 2233 (2018)
41. Jha, N., Choi, J.S., Kim, J.H., Jung, R., Choi, E.H., Ryu, J.J., Han, I.: Osteogenic potential of non thermal biocompatible atmospheric pressure plasma treated zirconia: in vitro study. J. Biomater. Tissue Eng. **7**, 662–670 (2017). https://doi.org/10.1166/jbt.2017.1626
42. Yoshinari, M., Matsuzaka, K., Inoue, T.: Surface modification by cold-plasma technique for dental implants—Bio-functionalization with binding pharmaceuticals. Jpn. Dent. Sci. Rev. **47**, 89–101 (2011). https://doi.org/10.1016/j.jdsr.2011.03.001
43. Karaman, O., Kelebek, S., Demirci, E.A., Ibis, F., Ulu, M., Ercan, U.K.: Synergistic effect of cold plasma treatment and rgd peptide coating on cell proliferation over titanium surfaces. Tissue Eng. Regen. Med. **15**, 13–24 (2018). https://doi.org/10.1007/s13770-017-0087-5
44. Duske, K., Koban, I., Kindel, E., Schroder, K., Nebe, B., Holtfreter, B., Jablonowski, L., Weltmann, K.D., Kocher, T.: Atmospheric plasma enhances wettability and cell spreading on dental implant metals. J. Clin. Periodontol. **39**, 400–407 (2012). https://doi.org/10.1111/j.1600-051X.2012.01853.x
45. Seo, H.Y., Kwon, J.S., Choi, Y.R., Kim, K.M., Choi, E.H., Kim, K.N.: Cellular attachment and differentiation on titania nanotubes exposed to air- or nitrogen-based non-thermal atmospheric pressure plasma. PLoS ONE **9**, e113477 (2014). https://doi.org/10.1371/journal.pone.0113477
46. Lee, E.J., Kwon, J.S., Uhm, S.H., Song, D.H., Kim, Y.H., Choi, E.H., Kim, K.N.: The effects of non-thermal atmospheric pressure plasma jet on cellular activity at SLA-treated titanium surfaces. Curr. Appl. Phys. **13**, S36–S41 (2013). https://doi.org/10.1016/j.cap.2012.12.023
47. Lee, J.H., Jeong, W.S., Seo, S.J., Kim, H.W., Kim, K.N., Choi, E.H., Kim, K.M.: Non-thermal atmospheric pressure plasma functionalized dental implant for enhancement of bacterial resistance and osseointegration. Dent. Mater. **33**, 257–270 (2017). https://doi.org/10.1016/j.dental.2016.11.011
48. Jeong, W.S., Kwon, J.S., Lee, J.H., Uhm, S.H., Ha Choi, E., Kim, K.M.: Bacterial attachment on titanium surfaces is dependent on topography and chemical changes induced by nonthermal atmospheric pressure plasma. Biomed. Mater. **12**, 045015 (2017). https://doi.org/10.1088/1748-605X/aa734e
49. Lee, M.J., Kwon, J.S., Jiang, H.B., Choi, E.H., Park, G., Kim, K.M.: The antibacterial effect of non-thermal atmospheric pressure plasma treatment of titanium surfaces according to the bacterial wall structure. Sci. Rep. **9**, 1938 (2019). https://doi.org/10.1038/s41598-019-39414-9
50. Yoo, E.M., Uhm, S.H., Kwon, J.S., Choi, H.S., Choi, E.H., Kim, K.M., Kim, K.N.: The study on inhibition of planktonic bacterial growth by non-thermal atmospheric pressure plasma jet treated surfaces for dental application. J. Biomed. Nanotechnol. **11**, 334–341 (2015)

51. Koban, I., Holtfreter, B., Hubner, N.O., Matthes, R., Sietmann, R., Kindel, E., Weltmann, K. D., Welk, A., Kramer, A., Kocher, T.: Antimicrobial efficacy of non-thermal plasma in comparison to chlorhexidine against dental biofilms on titanium discs in vitro - proof of principle experiment. J. Clin. Periodontol. **38**, 956–965 (2011). https://doi.org/10.1111/j.1600-051X.2011.01740.x
52. Giro, G., Tovar, N., Witek, L., Marin, C., Silva, N.R., Bonfante, E.A., Coelho, P.G.: Osseointegration assessment of chairside argon-based nonthermal plasma-treated Ca-P coated dental implants. J. Biomed. Mater. Res. A **101**, 98–103 (2013). https://doi.org/10.1002/jbm.a.34304
53. Teixeira, H.S., Marin, C., Witek, L., Freitas Jr., A., Silva, N.R., Lilin, T., Tovar, N., Janal, M. N., Coelho, P.G.: Assessment of a chair-side argon-based non-thermal plasma treatment on the surface characteristics and integration of dental implants with textured surfaces. J. Mech. Behav. Biomed. Mater. **9**, 45–49 (2012). https://doi.org/10.1016/j.jmbbm.2012.01.012
54. Shon, W.J., Chung, S.H., Kim, H.K., Han, G.J., Cho, B.H., Park, Y.S.: Peri-implant bone formation of non-thermal atmospheric pressure plasma-treated zirconia implants with different surface roughness in rabbit tibiae. Clin. Oral Implant Res. **25**, 573–579 (2014). https://doi.org/10.1111/clr.12115
55. Chen, M., Zhang, Y., Sky Driver, M., Caruso, A.N., Yu, Q., Wang, Y.: Surface modification of several dental substrates by non-thermal, atmospheric plasma brush. Dent. Mater. **29**, 871–880 (2013). https://doi.org/10.1016/j.dental.2013.05.002. S0109-5641(13)00119-X [pii]
56. Lehmann, A., Rueppell, A., Schindler, A., Zyla, I.M., Seifert, H.J., Nothdurft, F., Hannig, M., Rupf, S.: Modification of enamel and dentin surfaces by non-thermal atmospheric plasma. Plasma Process. Polym. **10**, 262–270 (2013)
57. Koban, I., Duske, K., Jablonowski, L., Schroeder, K., Nebe, B., Sietmann, R., Weltmann, K. D., Hubner, N.O., Kramer, A., Kocher, T.: Atmospheric plasma enhances wettability and osteoblast spreading on dentin in vitro: proof-of-principle. Plasma Process. Polym. **8**, 975–982 (2011)
58. Dong, X., Chen, M., Wang, Y., Yu, Q.: A mechanistic study of plasma treatment effects on demineralized dentin surfaces for improved adhesive/dentin interface bonding. Clin. Plasma Med. **2**, 11–16 (2014). https://doi.org/10.1016/j.cpme.2014.04.001
59. Hirata, R., Teixeira, H., Ayres, A.P., Machado, L.S., Coelho, P.G., Thompson, V.P., Giannini, M.: Long-term adhesion study of self-etching systems to plasma-treated dentin. J. Adhes. Dent. **17**, 227–233 (2015). https://doi.org/10.3290/j.jad.a34138
60. Šantak, V., Vesel, A., Zaplotnik, R., Bišćan, M., Milošević, S.: Surface treatment of human hard dental tissues with atmospheric pressure plasma jet. Plasma Chem. Plasma Process. **37**, 401–413 (2017). https://doi.org/10.1007/s11090-016-9777-3
61. Zhu, X., Guo, H., Zhou, J., Zhang, X., Chen, J., Li, J., Li, H., Tan, J.: Influences of the cold atmospheric plasma jet treatment on the properties of the demineralized dentin surfaces. Plasma Sci. Technol. **20**, 044010 (2018). https://doi.org/10.1088/2058-6272/aaa6be
62. Stasic, J.N., Selakovic, N., Puac, N., Miletic, M., Malovic, G., Petrovic, Z.L., Veljovic, D. N., Miletic, V.: Effects of non-thermal atmospheric plasma treatment on dentin wetting and surface free energy for application of universal adhesives. Clin. Oral Investig. **23**, 1383–1396 (2019). https://doi.org/10.1007/s00784-018-2563-2
63. Ayres, A.P., Freitas, P.H., De Munck, J., Vananroye, A., Clasen, C., Dias, C.D.S., Giannini, M., Van Meerbeek, B.: Benefits of nonthermal atmospheric plasma treatment on dentin adhesion. Oper. Dent. **43**, E288–E299 (2018). https://doi.org/10.2341/17-123-L
64. Zhang, Y., Yu, Q., Wang, Y.: Non-thermal atmospheric plasmas in dental restoration: improved resin adhesive penetration. J. Dent. **42**,1033–1042 (2014). https://doi.org/10.1016/j.jdent.2014.05.005. S0300-5712(14)00139-0 [pii]

65. Zhu, X.M., Zhou, J.F., Guo, H., Zhang, X.F., Liu, X.Q., Li, H.P., Tan, J.G.: Effects of a modified cold atmospheric plasma jet treatment on resin-dentin bonding. Dent. Mater. J. **37**, 798–804 (2018). https://doi.org/10.4012/dmj.2017-314
66. Dong, X., Ritts, A.C., Staller, C., Yu, Q., Chen, M., Wang, Y.: Evaluation of plasma treatment effects on improving adhesive-dentin bonding by using the same tooth controls and varying cross-sectional surface areas. Eur. J. Oral Sci. **121**, 355–362 (2013). https://doi.org/10.1111/eos.12052
67. Ritts, A.C., Li, H., Yu, Q., Xu, C., Yao, X., Hong, L., Wang, Y.: Dentin surface treatment using a non-thermal argon plasma brush for interfacial bonding improvement in composite restoration. Eur. J. Oral Sci. **118**, 510–516 (2010). https://doi.org/10.1111/j.1600-0722.2010.00761.x
68. Dong, X., Li, H., Chen, M., Wang, Y., Yu, Q.: Plasma treatment of dentin surfaces for improving self-etching adhesive/dentin interface bonding. Clin. Plasma Med. **3**, 10–16 (2015). https://doi.org/10.1016/j.cpme.2015.05.002
69. Han, G.J., Kim, J.H., Chung, S.N., Chun, B.H., Kim, C.K., Seo, D.G., Son, H.H., Cho, B.H.: Effects of non-thermal atmospheric pressure pulsed plasma on the adhesion and durability of resin composite to dentin. Eur. J. Oral Sci. **122**, 417–423 (2014). https://doi.org/10.1111/eos.12153
70. Kim, J.H., Han, G.J., Kim, C.K., Oh, K.H., Chung, S.N., Chun, B.H., Cho, B.H.: Promotion of adhesive penetration and resin bond strength to dentin using non-thermal atmospheric pressure plasma. Eur. J. Oral Sci. **124**, 89–95 (2016). https://doi.org/10.1111/eos.12246
71. Ayres, A.P., Bonvent, J.J., Mogilevych, B., Soares, L.E.S., Martin, A.A., Ambrosano, G.M., Nascimento, F.D., Van Meerbeek, B., Giannini, M.: Effect of non-thermal atmospheric plasma on the dentin-surface topography and composition and on the bond strength of a universal adhesive. Eur. J. Oral Sci. **126**, 53–65 (2018). https://doi.org/10.1111/eos.12388
72. Hirata, R., Sampaio, C., Machado, L.S., Coelho, P.G., Thompson, V.P., Duarte, S., Ayres, A.P., Giannini, M.: Short- and long-term evaluation of dentin-resin interfaces formed by etch-and-rinse adhesives on plasma-treated dentin. J. Adhes. Dent. **18**, 215–222 (2016). https://doi.org/10.3290/j.jad.a36134. 36134 [pii]
73. Puač, N., Miletić, M., Mojović, M., Popović-Bijelić, A., Vuković, D., Miličić, B., Maletić, D., Lazović, S., Malović, G., Petrović, Z.L.: Sterilization of bacteria suspensions and identification of radicals deposited during plasma treatment. Open Chem. **13**, 332–338 (2015). https://doi.org/10.1515/chem-2015-0041
74. Miletić, M., Vuković, D., Živanović, I., Dakić, I., Soldatović, I., Maletić, D., Lazović, S., Malović, G., Petrović, Z.L., Puač, N.: Inhibition of methicillin resistant staphylococcus aureus by a plasma needle. Cent. Eur. J. Phys. **12**, 160–167 (2014). https://doi.org/10.2478/s11534-014-0437-z
75. Sladek, R.E., Filoche, S.K., Sissons, C.H., Stoffels, E.: Treatment of Streptococcus mutans biofilms with a nonthermal atmospheric plasma. Lett. Appl. Microbiol. **45**, 318–323 (2007). https://doi.org/10.1111/j.1472-765X.2007.02194.x
76. Rupf, S., Lehmann, A., Hannig, M., Schafer, B., Schubert, A., Feldmann, U., Schindler, A.: Killing of adherent oral microbes by a non-thermal atmospheric plasma jet. J. Med. Microbiol. **59**, 206–212 (2010). https://doi.org/10.1099/jmm.0.013714-0
77. Schaudinn, C., Jaramillo, D., Freire, M.O., Sedghizadeh, P.P., Nguyen, A., Webster, P., Costerton, J.W., Jiang, C.: Evaluation of a nonthermal plasma needle to eliminate ex vivo biofilms in root canals of extracted human teeth. Int. Endod. J. **46**, 930–937 (2013). https://doi.org/10.1111/iej.12083
78. Du, T., Shi, Q., Shen, Y., Cao, Y., Ma, J., Lu, X., Xiong, Z., Haapasalo, M.: Effect of modified nonequilibrium plasma with chlorhexidine digluconate against endodontic biofilms in vitro. J. Endod. **39**, 1438–1443 (2013). https://doi.org/10.1016/j.joen.2013.06.027

79. Chen, M., Zhang, Y., Yao, X., Li, H., Yu, Q., Wang, Y.: Effect of a non-thermal, atmospheric-pressure, plasma brush on conversion of model self-etch adhesive formulations compared to conventional photo-polymerization. Dent. Mater. **28**, 1232–1239 (2012). https://doi.org/10.1016/j.dental.2012.09.005
80. Epaillard, F., Brosse, J.C., Legeay, G.: Plasma-induced polymerization. J. Appl. Polym. Sci. **38**, 887–898 (1989)
81. Santak, V., Zaplotnik, R., Milosevic, S., Klaric, E., Tarle, Z.: Atmospheric pressure plasma jet as an accelerator of tooth bleaching. Acta Stomatol. Croat. **48**, 268–278 (2014). https://doi.org/10.15644/asc47/4/4
82. Claiborne, D., McCombs, G., Lemaster, M., Akman, M.A., Laroussi, M.: Low-temperature atmospheric pressure plasma enhanced tooth whitening: the next-generation technology. Int. J. Dent. Hyg. **12**, 108–114 (2014). https://doi.org/10.1111/idh.12031
83. Celik, B., Capar, I.D., Ibis, F., Erdilek, N., Ercan, U.K.: Deionized water can substitute common bleaching agents for nonvital tooth bleaching when treated with non-thermal atmospheric plasma. J. Oral Sci. **61**, 103–110 (2019). https://doi.org/10.2334/josnusd.17-0419
84. Nam, S.H., Ok, S.M., Kim, G.C.: Tooth bleaching with low-temperature plasma lowers surface roughness and Streptococcus mutans adhesion. Int. Endod. J. **51**, 479–488 (2018). https://doi.org/10.1111/iej.12860
85. Nam, S.H., Lee, H.J., Hong, J.W., Kim, G.C.: Efficacy of nonthermal atmospheric pressure plasma for tooth bleaching. Sci. World J. **2015**, 581731 (2015). https://doi.org/10.1155/2015/581731

Thermo-Analytical Characterization of Various Biomass Feedstocks for Assessments of Light Gaseous Compounds and Solid Residues

Milena M. Pijovic[1], Bojan Jankovic[1(✉)], Dragoslava Stojiljkovic[2], Milos B. Radojevic[2], and Nebojša Manic[2]

[1] Department of Physical Chemistry, Institute of Nuclear Sciences "Vinča", University of Belgrade, Mike Petrovića Alasa 12-14, P.O. Box 522, 11001 Belgrade, Serbia
bojan.jankovic@vinca.rs
[2] Faculty of Mechanical Engineering, Fuel and Combustion Laboratory, University of Belgrade, Kraljice Marije 16, P.O. Box 35, 11120 Belgrade, Serbia

Abstract. Thermo-analytical characterization of selected biomasses (agricultural waste and wood biomass feedstock) through the pyrolysis process was performed under dynamic conditions. Slow pyrolysis (carbonization) regime (with a heating rate below 50 °C min^{-1}) was selected because it favours the residual solid (bio-carbon/bio-char) in the higher yields (change in the surface area of bio-char with pyrolysis conditions was dependent on the type of biomass feedstock). Comparison of results and discussions related to obtained percentage pyro char yields from thermo-chemical conversion of biomass feedstocks were generated from simultaneous thermal analysis (STA) (TGA-DTG-DTA apparatus). The analysis of gaseous products of pyrolysis was carried out using mass spectrometry (MS) technique. Releasing of the light gaseous compounds (mainly CO, CO_2, CH_4, and H_2 non-condensable gases) was monitored simultaneously with TGA measurements. Discussion related to this issue was performed from the aspect of the syngas production, as well as the versatility of selected biomasses in the gasification process where the various gasifying agent may be in use.

Keywords: Thermo-analytical characterization · Biomass · Pyrolysis · Mass spectrometry · Syngas

1 Introduction

The possibilities for utilizing biomass for energy needs are significant from many aspects. Uncertainty and insecurity in energy supply, the effects of environmental pollution, and the alarming effects of the climate change we experience in recent years indicate the necessity to consider replacement of fossil fuels and use alternative energy sources [1]. Accordingly, agricultural and environmental sectors are essential which

N. Mitrovic et al. (Eds.): CNNTech 2019, LNNS 90, pp. 139–165, 2020.
https://doi.org/10.1007/978-3-030-30853-7_9

will affect the effects of energy sector restructuring and the change from non-renewable to renewable energy sources. Also, agriculture and forestry have to offer biomass as a renewable source of energy, while in the field of environment, results of the implementing this instrument will likely be noticeable [2, 3].

Biomass represents the key renewable energy source (RES) in Serbia, with a share of 61% in the total technical potential (RES), of which 63% is agricultural and the rest is forest biomass [4]. Biomass is a renewable source that is nowadays intensively utilized and participates with around 10% in the primary energy balance [5, 6]. However, generally wood-based biomass in the form of heating wood for heating in households or, to a smaller amount, for combustion in heating boilers is used. The production of energy from agricultural biomass (harvesting and residual residues, mussels, waste from the slaughterhouse and food industry, etc.) is very uncommon, although by doing these additional benefits are obtained in terms of lowering the waste from agricultural production, which has been very intensive for water and land pollution in Serbia [7]. Agricultural biomass can be used either for direct combustion in heat generation furnaces, either as a raw material for the production of biogas which is later used as a fuel in plants for combined production heat and power (so-called CHP plants), in which case it is possible to use significant amounts of waste from farms or from the food industry, or slaughterhouses. Certain types of biomass, as well as waste edible oils, can be processed with suitable technological processes into bio-ethanol, or bio-diesel that can be subsequently mixed with a motor oil of crude oil origin [8, 9].

In order to keep the biomass permanently the renewable energy source, its use must be done in a sustainable manner, which implies appropriate management of forest resources and the way of using agricultural land. It is especially important to use as much as possible the waste from agricultural production as well as to carry out activities for more intensive use of agricultural land of a low category, as well as other marginal lands for production of biomass for energy purposes [10]. Regardless of how energy is produced, the biomass chain is extremely long. It begins with the cultivation of biomass, which involves the processes of harvesting, collecting, loading, transporting, storing, transshipping, crushing, processing or transformation, combustion preparations and ultimately only combustion of the appropriate biomass/bio-fuel form in power plants [11]. All these processes are working, organizing, and most of them capitally very intense. Due to the large investment and operating costs of the plant, the production of electricity from the biomass must be subsidized.

Despite the incentive measures, there are many limitations to the development of the biomass market for energy production. On the supply side, these barriers include insufficiently efficient collection and distribution of available biomass, logistical problems, underdeveloped mechanisms for long-term and safe supply, while on the demand side the factors that make it difficult to initiate investment are high initial costs and high-risk perception by banks to approve loans for the construction of plants using biomass. Also, investors are faced with numerous problems in the process of building facilities arising from the inconsistency of construction procedures, difficulties in connection with the connection of the plant to the network, insufficient knowledge of new technologies, etc. [12].

Pyrolysis is the fundamental chemical reaction process that is the precursor of both the gasification and combustion of solid fuels, and is simply defined as the chemical changes occurring when the heat is applied to a material in the absence of oxygen. Pyrolysis, which is one of the thermo-chemical conversion methods, converts biomass into valuable products, namely, solid char (or more correctly a carbonaceous solid) (bio-char), liquid and gas product yields, and compositions of which depend on the pyrolysis conditions. The study of pyrolysis is gaining increasing importance, as it is not only an independent process, it is also a first step in the gasification or combustion process but in these processes, it is followed by total or partial oxidation of the primary products. It is important to differentiate pyrolysis from gasification. Gasification decomposes biomass to syngas by carefully controlling the amount of oxygen present. Pyrolysis is difficult to precisely define, especially when applied to biomass. Namely, the older literature generally equates pyrolysis to carbonization, in which the principal product is a solid bio-char. Later definitions, besides obtaining the bio-char (solid) product, also include the consideration of light gaseous products and bio-oils which are main products. Pyrolysis produces a mixture of products under the three phases. The different types of pyrolysis processes differ in the type of product that is searched. The pyrolysis of biomass is a promising route for the production of solid (bio-char), liquid (tar, water and other organics, such as acetic acid, acetone and methanol) and gaseous products (H_2, CO_2, CO, CH_4). These products are of interest as they are possible alternate sources of energy [13].

There are many factors affecting pyrolysis product yields such as pyrolysis atmosphere, final temperature, particle size, heating rate, the geometry of apparatus and the initial amount of the sample. However, it is known, that higher final temperatures (600 °C and above) favor the gas formation, and relatively lower temperatures (400 °C and below) favor bio-char formation. The heating rate is also an important parameter for product yields. Conventional pyrolysis, in other words the slow pyrolysis (≤ 50 °C min^{-1}) is applied for thousands of years when the char-coal production is aimed. Generally equal amounts of the products are formed when the biomass pyrolyzed slowly. On the other hand, the much faster heating rates (≥ 1000 °C s^{-1}) are preferred for the fast pyrolysis where the higher yields of bio-oil can be produced [14].

The overall aim of this study was to (*i*) investigate whether the origin and composition of selected biomass feedstock does the influence the yield of gaseous products after slow pyrolysis, (*ii*) elucidate the effect of heating rate on the characteristic pyrolysis temperatures which were linked with various thermal transformation regions of biomass components (such as cellulose, hemicelluloses and lignin), and (*iii*) examine the impact of structural differences among tested biomass feedstock on abundance of gaseous products during the slow pyrolysis. Based on the STA-MS measurements of the monitored process for two different biomass feedstocks, the additional characterization assessment of the possible application of other evolved gaseous components, emphasizing on the possible syngas production, was also discussed.

2 Materials and Methods

2.1 Preparation of the Samples

Two biomass samples were tested for the research presented in this manuscript. The pistachios shell represent the agricultural waste and spruce wood as woody biomass. The pistachio shells were received from domestic sales sources, while sprigs of spruce wood which were used in measurements are collected from the forested habitats of the spruce woods located in the Republic of Serbia. Since pistachio shells are not consumed, and have a composition favorable for bio-char synthesis, they are potentially excellent candidates for activated carbon feedstocks, as well as for production of bio-fuel to partially power the bio-char production. While several research groups have identified the potential use of pistachio shells as a bio-fuel and bio-char precursor, there is no systematic evaluation of what conditions maximize the bio-gas production, as well as resulting bio-char yields obtained under the conditions applied in this work. Pistachio shells were washed with tap water and then distilled water. Pistachio shells were dried in an oven at 70 °C for 4 h. A kitchen blender was used for size reduction of dried shells. Ground samples were sieved to a particle size below 250 μm to prevent mass and heat transfer limitations.

The wood sample studied in this research is collected at Stara Planina mountain in the southeast part of Serbia. According to defined solid biomass classification, the spruce wood sample belongs to woody biomass and represents the softwood species which is dominantly widespread in central and southern Serbia. The tested sample was prepared for the experimental tests by considering of application the standard test procedures prescribed for biomass wood samples. These standard test procedures also set the test methods for obtaining data of proximate and ultimate analysis for both tested samples presented in this study [15].

2.2 STA – MS Measurements

The STA was performed in order to fully characterize the selected biomass samples, and clearly identified the main components of the gaseous products and obtained bio-char.

All tests were performed on simultaneous thermal analyzer NETZSCH STA Jupiter 449 F5 coupled with quadruple mass spectrometer QMS Aeolos 403D. Even though the QMS can scan ions up to a.m.u of 300, all measurements were of bar-graph type, scanning between the a.m.u of 10 and 50, because the most of discussed gasses are in that range. If the range was extended up to the instrument's limit, molar masses that are of interest for this research could not be scanned for long enough time so the acquired signal for respective masses would not be clear or strong enough. Because of that, each mass was scanned for 0.2 s, which means that the new scan cycle started every 8 s. That way, a sufficient amount of cycles – measured points for each mass, was obtained, so the representative curve could be made.

TG analysis (alongside with DTA (differential thermal analysis) scanning) was performed in order to obtain mass change curve [% vs. T] and derivative thermogravimetry (DTG) curve [% min^{-1} vs. T], so the MS curves can be analyzed more conveniently. Since it is known that diffusivity of evolved gas is a lot higher in the inert atmosphere when using helium as a carrier gas or the nitrogen, so, the argon was chosen as a carrier gas. Besides, in this way, lower carrier gas flow is required in order to acquire MS curves of acceptable intensity. Accordingly, 50 ml min^{-1} of argon was streaming as a purge gas in addition to 20 ml min^{-1} of argon as a protective gas – protects the highly sensitive internal balance (10–7 g) and maintains constant conditions around it so the measurement error for the TG curve is smallest possible. Each sample was heated at three heating rates β = 5, 10 and 20 °C min^{-1} from room temperature up to 850 °C, so the biomass can go through all the steps of decomposition, which ends at the mentioned temperature. Although, sample temperature controller (STC) was not used during measurements so the sample could not reach exactly 850 ° C, but only 815 °C. Even though, the final temperature of 800 °C is still considered fine, because during lignin decomposition period, sample mass changes negligibly from 800 °C to 900 °C [16]. It was chosen not to use STC in order to get a more linear sample temperature heating profile. Sample mass was in the range of 10.0 ± 0.5 mg, while the argon was of the highest purity available – N5.0, so that 10 ppm of oxygen cannot disturb the pyrolysis process. Different heating rates have been chosen to monitor the release period of certain gases and to compare the influence of heating rate.

In this paper, following mass numbers were analyzed: 12, 15, 16, 17, 18, 27, 28, 32, and 44, which in combination represent methane, water, ethane, ethylene, carbon monoxide, oxygen and carbon dioxide. Those gasses can be described with these molar masses because they are contained in them either as molecular ions or fragments. In the case of biomass, overlapping of gas releases is present and so certain molar masses can represent more than one gas.

3 Results and Discussion

3.1 Proximate/Ultimate Analysis of Tested Biomasses

The proximate and ultimate analysis of tested biomass (Pistachios shells and spruce wood) samples were presented in Tables 1 and 2 respectively. The performed analysis was done in accordance with EN ISO 18125 standard and was carried out by courtesy of Faculty of Mechanical Engineering (University of Belgrade) in fuel and combustion laboratory of the same research institution.

According to results showed in Table 1, the pistachio shell presents promising applicability in thermo-chemical conversion (even in a more complex process, such as combustion) due to lower moisture content (below 10.0%) and low ash content ($\sim$1.30%). Pistachio shell was characterized by a high amount of volatile matter (VM) (75.49%), which is in accordance with results by Ebeling and Jenkins [17] which they reported that walnut, pistachio and almond shells have 78.28%, 82.03% and 73.45% volatile matter, respectively. As volatile matter (VM) represent the carbon, hydrogen and oxygen components (usually a mixture of short and long chain hydrocarbon) in the biomass that turns to vapor when heated. Comparing these results with ones established for spruce wood (Table 2), it can be observed that for spruce wood sample, the slightly higher moisture content, but very low ash content (ash is the

Table 1. Proximate and ultimate analysis for Pistachios shell sample.

Proximate analysis (wt%)		Ultimate analysis[b] (wt%)	
Moisture	7.53	C	50.19
Volatile matter (VM)	75.49	H	6.30
Fixed carbon (FC)	15.68	O[c]	41.16
Ash	1.30	N	0.68
HHV (MJ kg^{-1})	18.22	S	0.22
LHV[a] (MJ kg^{-1})	16.76	H/C	1.50
Fuel ratio[d]	0.21	O/C	0.62

[a]Calculated according to EN ISO 18125:2017.
[b]On a dry basis.
[c]By the difference.
[d]Fuel ratio: the ratio of FC to VM (FR = FC/VM).

Table 2. Proximate and ultimate analysis for spruce wood sample.

Proximate analysis (wt%)		Ultimate analysis[b] (wt%)	
Moisture	9.43	C	47.22
Volatile matter (VM)	76.07	H	6.11
Fixed carbon (FC)	14.10	O[c]	43.27
Ash	0.40	N	0.08
HHV (MJ kg^{-1})	16.36	S	0.01
LHV[a] (MJ kg^{-1})	14.88	H/C	1.54
Fuel ratio[d]	0.18	O/C	0.69

[a]Calculated according to EN ISO 18125:2017.
[b]On a dry basis.
[c]By the difference.
[d]Fuel ratio: the ratio of FC to VM (FR = FC/VM).

non-combustible component of biomass) and high VM content were presented, as in the case of pistachio shell sample. Therefore, the lower the fuel's ash content was, the higher its calorific values (see HHV value in Tables 1 and 2), and the higher the VM content was (Tables 1 and 2) [18]. High VM content biomass samples are desirable for the pyrolysis process, since these are more reactive and easily devolatilized. Both biomass samples have approximately equal amount of fixed carbon (FC) (the percentage variation is $\Delta = 1.58\%$) and these values are in the range typically for FC present in the biomass feedstock. These results suggest that actual biomass samples can be good for bio-char production since the amount of bio-char which can be produced by an energy feedstock is limited by the amount of FC in the feedstock. Compared their HHV values, it can be identified that pistachio shell has a slightly higher value of HHV, which means that pistachio shell should be easier and much better decomposed (endured thermal transformation) than the spruce wood. Considering all above physical characteristics of tested biomass feedstocks, it seems that because of the high VM which were found in the samples, suggests the high potential of pistachio shell and spruce wood for *energy production* by pyrolysis and gasification [19]. It is an established fact that the lower ratio of H/C and O/C show the greater energy content of the material [20]. Hence, it seems that used biomasses (spruce wood and pistachios shell) are suitable for the pyrolysis.

Considering the ultimate analysis of biomass samples, the carbon content (C %) for pistachio shell reaches percentage above 50% (50.19%), while for spruce wood this value is slightly lower (47.22%) (Tables 1 and 2). Carbon is the most important constituent of biomass and represents the major contribution to the overall heating value followed by oxygen and then hydrogen. When carbon is not decomposed completely during thermal transformation, it leads to emissions of unburned gases, especially carbon monoxide. Higher oxygen concentrations (O %) in both samples, is responsible for their increased heating values. However, due to the low hydrogen content and high oxygen content of biomass samples, their pyrolysis is not efficient for the production of aromatic hydrocarbons. However, the significantly higher oxygen content may lead to the oxygenated pyrolysis (mainly liquid) products, but the much lower heating values than those of fossil fuels. Accordingly, removal of oxygenated species during or after the pyrolysis reactions is necessary to obtain a higher fuel grade product. Furthermore, both samples contain very low amounts of nitrogen and sulphur (Tables 1 and 2) compared with coal, which contains 1.40 and 1.70 wt% of nitrogen and sulphur, respectively. If the biomass itself or the pyrolysis products derived from the biomass are burnt for energy, the amounts of nitrogen oxides and sulphur oxides given off will be much lower than when burning fossil fuels. It is therefore beneficial to the environment when using the biomass for energy production. Considering both biomass feedstocks used in this work, it can be observed that their heating values are lower than those for fossil fuels, due to their high oxygen contents (Tables 1 and 2).

On the other hand, it is interesting to note that besides lowered sulphur contents, the pistachio shell has still higher percentage of S than ones present in spruce wood (which almost drops at zero), so, this is important for from environmental and technical point of view because some S derivatives are important atmospheric contaminants and negatively affect the installation. The SO_2 emissions result in acid rain, causing health, environmental and material damages, and it is transformed in the atmosphere to small particles that can have a very long life-time. Bearing this in mind, for forming of SO_2 and its subsequent emissions, however, there is a reduced probability for these events in our studied samples.

3.2 TG-DTG-DTA Analysis of Pistachio Shell and Spruce Wood Pyrolysis

Figure 1 shows the experimentally obtained TG-DTG and DTA curves of spruce wood pyrolysis under the slowly heating rate conditions (β = 5, 10 and 20 °C min^{-1}) in an argon (Ar) atmosphere. It can be seen from Fig. 1, that the moisture in the spruce wood was removed at up to 150 °C, where the main decomposition reactions of spruce wood started around 200 °C, and had a rapid incline to around 800 °C.

a

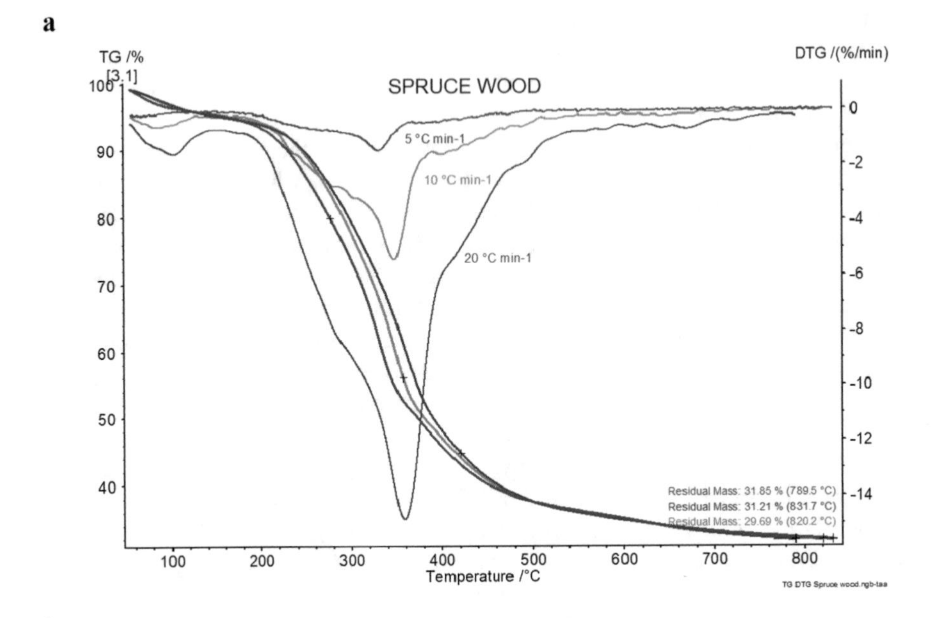

b

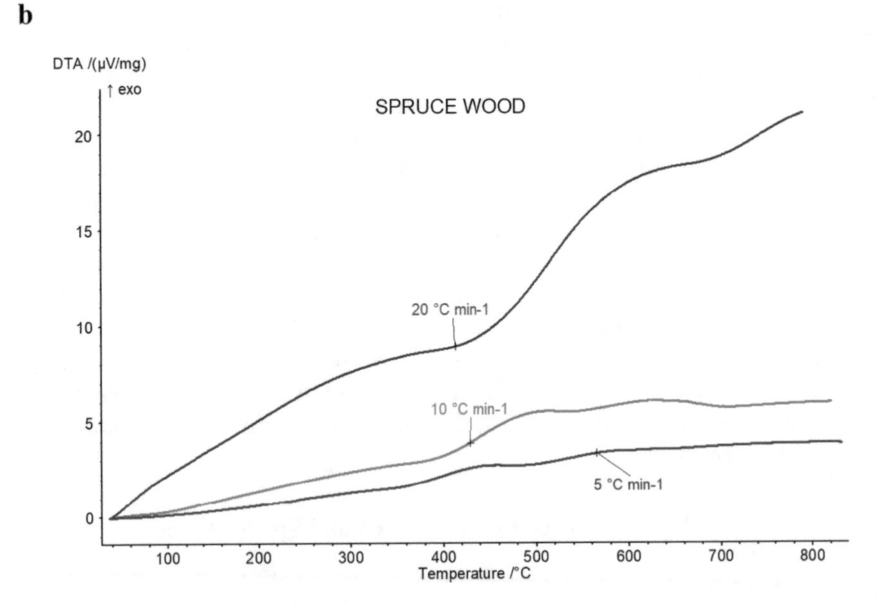

Fig. 1. TG-DTG (**a**) and DTA (**b**) curves of spruce wood pyrolysis in an argon atmosphere at different heating rates (5, 10 and 20 °C min^{-1}).

The main mass loss (~52.5%) includes temperature range of 200–500 °C which represent the active pyrolysis (devolatilization) zone, and in this zone, the decomposition reactions of three distinct primary components of wood (hemicelluloses, cellulose and lignin) occur. The disruption of bond linkages of wood bio-molecules caused by thermally-induced events, leads to the main products of slow pyrolysis process, which are the solid residue (bio-char) and inorganic gases and light hydrocarbons, where the global scheme of this process in accordance with Grønli et al. [21] proposal for softwoods (spruce wood belongs to softwoods) pyrolysis, includes four-component reaction scheme as:

$$extractives \rightarrow bio-char_{(e)} + volatiles$$
$$hemicelluloses \rightarrow bio-char_{(hc)} + volatiles$$
$$cellulose \rightarrow bio-char_{(c)} + volatiles$$
$$lignin \rightarrow bio-char_{(l)} + volatiles$$

Scheme 1. The slow pyrolysis reaction scheme for softwoods, which can be applied in this work.

As a four component in Scheme 1, it can be observed the presence of extractives decomposition, where extractives can be defined as non-structural wood constituent and usually represents the minor fraction in the wood sample. However, various parts of the same tree, e.g. stem, branches, roots, bark and needles, differ markedly with respect to both their amount and composition of extractives. The extractives comprise both inorganic and organic components. The inorganic components measured as ash seldom, while organic components are usually aliphatic and alicyclic compounds, phenolic compounds and other compounds. The composition of extractives varies widely from species to species, and the total amount of extractives in a given species depends on wood growth conditions. However, for spruce wood, the content of extractives does not go beyond 2.0% [22]. The decomposition reactions of extractives frequently take place in the lower temperatures region, between 200–260 °C in the case of softwoods [23].

The devolatilization rate in active pyrolysis zone strongly depends on the applied heating rate, where abruptly increases from 5 to 20 °C min^{-1}, which is clearly seen from DTG plots in Fig. 1. The release of gases is significantly accelerated with the increase in the heating rate reaching its maximum at the highest heating rate magnitude (Fig. 1a). Table 3 lists the values of characteristic pyrolysis reaction temperatures related to spruce wood and pistachio shell biomass samples.

The largest mass loss where the maximum rate of the spruce wood pyrolysis process is achieved was found at around 347.0 °C regarding all the heating rates. The maximum (peak) decomposition temperature for spruce wood pyrolysis increases with an increase of heating rate (Table 3; $T_{max(1)}$). This stage (which occurs immediately after moisture removal) can be attributed to the decomposition of hemicelluloses and cellulose. The actual stage was characterized by the inner inclination of DTG curves around 290/300 °C which further propagates into a well-defined peak, which width depends on the applied heating rate (Fig. 1a). Next stage starts at around 360 °C and ends about 500 °C, where this stage is characterized by smaller mass loss (Fig. 1a).

Table 3. The initial temperature ($T_{initial}$), the "shoulder" decomposition temperature (T_{sh}), the maximum (peak) decomposition temperature (T_{max}) and final temperature (T_{final}) values, related to slow pyrolysis processes of spruce wood and pistachio shell biomass samples.

Sample	Spruce wood				
Temperatures	$T_{initial}$ (°C)	$T_{max(1)}$ (°C)	$T_{max(2)}$ (°C)	T_{sh} (°C)	T_{final} (°C)
β (°C min^{-1})					
5	204.95	335.15	–	394.32	827.33
10	207.75	347.72	–	396.90	830.47
20	210.10	358.05	–	407.21	835.85
Sample	Pistachio shell				
Temperatures	$T_{initial}$ (°C)	$T_{max(1)}$ (°C)	$T_{max(2)}$ (°C)	T_{sh} (°C)	T_{final} (°C)
β (°C min^{-1})					
5	230.15	269.06	328.82	398.90	811.55
10	235.31	282.16	333.98	406.65	826.94
20	240.45	295.07	349.67	414.59	827.94

The considered stage can be attributed to the additional, the high-temperature cellulose decomposition, as well as the residual lignin decomposition, which was characterized by the appearance of the "shoulder" at T_{sh} and shifted with an increase of the heating rate value (Table 3; T_{sh}) [24–26]. Further, above 500 °C, the final stage takes place, with very low mass losses considering all heating rates, where the temperature exceeds of 700 °C, the TG curves become almost constant up to final temperatures (Table 3; T_{final}). This may be attributed to the residual mass which resembles the ash content in the spruce wood sample.

However, the indicated residual mass at the very end of the process represents the formed bio-char, where its percentage amounts are shown in Fig. 1a, for every considered heating rate. For the slow pyrolysis conditions applied in this work, in the case of spruce wood biomass, the mean bio-char yield of 30.92% was obtained at 800 °C. However, from above-established results, it can be concluded that the temperature range of 300–500 °C is more favorable for pyrolysis of spruce wood. This range was quite visible at DTA curves of spruce wood pyrolysis shown in Fig. 1b, and it is characterized by endothermic effect which is enlarged as the heating rate increases. As temperature increases beyond 600 °C, the some exothermic effect occurs, which corresponds to the bio-char formation (Fig. 1b). It is obvious that influence of heating rate is more emphasized for active pyrolysis process where the most of gaseous products release, but this effect is less pronounced in "passive" pyrolysis process which encompasses reactions during the decomposition of residual lignin bio-molecule fragments, above 500 °C.

It should be emphasized that in the case of a uniform rate of temperature rise conditions, the wood particles are required to be less than a fraction of a millimeter.

In the thermogravimetric analysis (TGA), this small size factor, is not a difficulty, except that the pyrolysis is no longer realistic. This is because in real applications: (*i*) evaporating moisture creates high pressures resulting in Darcy flow of steam and volatiles in the thick porous wood, (*ii*) tar decomposes when passing through porous char, and (*iii*) the heating rate is both locally high and low so that all four regimes of the wood pyrolysis are involved in a single wood slab. So, the four parallel reaction schemes for softwood pyrolysis, such as spruce wood (Scheme 1) represent actually, the competing reactions to produce bio-char and non-condensing gases of the wood sample that originated from primary decomposition reactions, while the secondary reaction of tar decomposition was excluded. These include possible reactions of CO_2, CO, and H_2O with char to form CO gas that converts to CO_2 during "glow" combustion (bio-char can burns in glowing combustion after the pyrolysis step, if some of the liberated oxygen was present in the reaction system, as a consequence of executive number of decomposition reactions involving hemicelluloses, cellulose and lignin components). Figure 2 shows the experimentally obtained TG-DTG and DTA curves of pistachios shell pyrolysis under the slowly heating rate conditions (β = 5, 10 and 20 °C min^{-1}) in an argon (Ar) atmosphere.

It can be observed from Fig. 2, that pyrolysis behavior of pistachio shell biomass is somewhat different from the pyrolysis behavior in the case of spruce wood biomass. The first significant difference occurs in the appearance of two DTG peaks within the temperature range of 250–385 °C in an active pyrolysis zone (Fig. 2a), which are located at temperatures $T_{max(1)}$ and $T_{max(2)}$ (Table 3), respectively. The first and the second DTG peaks attributable to $T_{max(1)}$ and $T_{max(2)}$ correspond to hemicelluloses and cellulose decompositions, respectively, where the first DTG peak is clearly visible and well-shaped defined, unlike for the case of spruce wood pyrolysis, where this is exposed as "artificial shoulder" (Figs. 1 and 2a). No clear DTG peak appears for the spruce wood sample. This is a consequence of the elevated content of hemicelluloses in pistachio shell sample ($\sim$50.0% and consists mainly of xylose) [27], while in the spruce wood sample, the content of hemicelluloses is significantly reduced ($\sim$11.0% as xylan) [28]. The content of the hemicelluloses directly dictates forming of two distinct decomposition regions, one attached to hemicelluloses, and other attached to another biomass component, *viz* cellulose. So, the main reaction zone includes the following temperature ranges as follow approximately: (a) ΔT_{I} = 250–325 °C and (b) ΔT_{II} = 325 – 450 °C. The former is due to early decomposition of hemicelluloses, whereas the later is determined by degradation of both cellulose and lignin [29, 30]. This behavior is clearly reflected by the values of $T_{max(1)}$ and $T_{max(2)}$ (where $T_{max(2)} > T_{max(1)}$ (Table 3) for pistachios shell sample), since that cellulose and lignin decompose at the higher temperatures as compared to hemicelluloses. It should be pointed out that the rate of hemicelluloses decomposition during pistachio shell pyrolysis expressed by the first DTG peak at $T_{max(1)}$ (Fig. 2a and Table 2) is accelerating rapidly by increasing of the heating rate.

So, the heating rate value may control the contribution of the reactions which can be involved during hemicelluloses, cellulose and lignin decompositions, creating the large number of different decomposition products, such as H_2, CH_4, CO, H_2O and CO_2 as the *major* gaseous decomposition products. Since on the elevated content of hemicelluloses in pistachios shell sample, among these gaseous products, carbon monoxide, vapor, and carbon dioxide can be evolved in much larger extent than methane, comparing with same products for the spruce wood sample. Furthermore, hemicelluloses and cellulose can be easily decomposed under the miscellaneous pyrolysis conditions, since these oxygen rich constituents show high thermal reactivity under decomposing conditions. This is because, some volatiles release after moisture

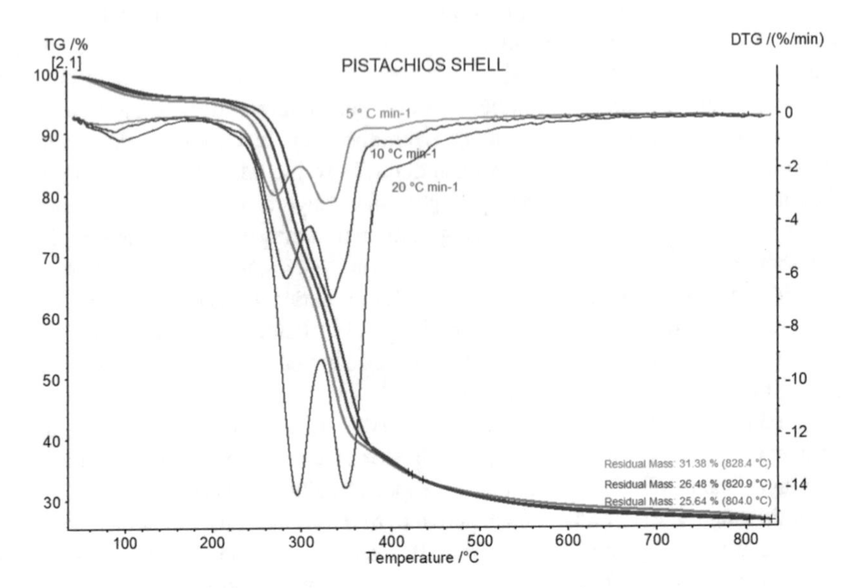

b

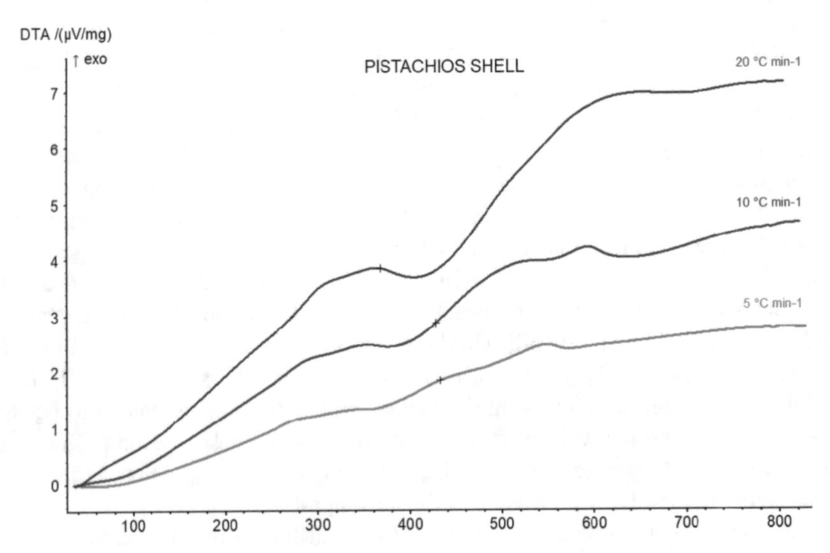

Fig. 2. TG-DTG (**a**) and DTA (**b**) curves of pistachios shell pyrolysis in an argon atmosphere at different heating rates (5, 10 and 20 °C min^{-1}).

removal (Fig. 2a) as a result of the disintegration of hemicelluloses, and just after that cellulose is begun to disintegrate, giving further volatiles as temperature increases. Of the major constituents of biomass feedstock, lignin is not as thermally reactive as the hemicelluloses and cellulose. Although lignin also began to decompose while hemicelluloses and cellulose were decomposing, the peaks of hemicelluloses and cellulose (Fig. 2a and Table 3) masked its peaks at lower temperatures.

Namely, the lignin pyrolysis behavior is expressed through the appearance of the "shoulder" at higher temperatures in comparison with that presented in the case of spruce wood sample (Table 3, T_{sh} values at different heating rates), but its prolonged further as a 'tailing' section especially at higher temperatures, after completion of the previous peaks of other two components, which is not observed in the case spruce wood pyrolysis (Fig. 1a). The tailing reaches high temperature regions up to 800 °C, where the lignin decomposition includes a much wider temperature interval, than ones which can be identified for spruce wood sample. This indicates that the lignin content in these biomass samples also varies and is not the same in percentage. Similar to spruce wood, taking into account a small particle sizes and the heating rate, the initial and final decomposing temperatures, the temperature of maximum mass loss rate are increase, when heating rates increase (Table 3). For the pistachios shell pyrolysis, the mean bio-char yield of 27.83% was obtained at 820 °C. Under given experimental conditions, it can be expected, that woody biomass probably gives a higher bio-char yield from lignin decomposition at lowered and subsequently residual lignin decomposition at elevated temperatures, while in the case of waste biomass, the gases evolved products are favored with higher yields, in comparison to bio-chars yield, primarily originated from hemicelluloses and cellulose decomposition reactions in the main temperature region of 250–325 °C. This can be identified from presented DTA curves shown in Fig. 2b, where deeper and more intense endothermic effects were observed and encompasses reaction regions where pistachios shell lignocellulosic components obey decomposition transformations. In the later stage, above 500/510 °C, the exothermic effect is detected since in this region, the bio-char is formed.

The endothermic heat region which was present for pistachios shell slow pyrolysis is a more intense and raises with a heating rate magnitude intensity, which is quite different from the one that is present in the case of spruce wood slow pyrolysis, which largely depends on the percentage distribution of components within complex chemical structure of tested biomass samples. For both samples, the results show that with an increasing of the heating rate, TG-DTG curves are shifted to higher temperatures.

For spruce wood, at low heating rate, the shape of DTG curve was narrow and short but then becoming broader and higher at the higher heating rates (Fig. 1a).

The shifting effect was reflected in the characteristic temperatures (Table 3), where the differences in $T_{max(1)}$ values from changing the heating rate from 5 to 10 °C min^{-1} are as follows: $\Delta T_{max(1)}$ = 12.57 °C (spruce wood), and $\Delta T_{max(1)}$ = 13.10 °C (pistachio shell), whereby this difference becomes enlarged, when the change occurs to the highest heating rate value.

Those behaviors were most probably because of the increased thermal lag, where at a given temperature, the higher heating rate implies that the material reaches the temperature in a shorter time. This phenomenon can be explained by combined effects of the heat and mass transfer at the various heating rates as well as by the kinetics guidance during the decomposition process.

However, the observed difference does not exceed 15 °C considering both tested samples, which was typical for using the conventional (micro lab-scale) TGA apparatus. In a given cases, as far as possible the lower heating rates with the small dimensions of biomass particles are strongly recommended, in order to facilitate the better heat transfer between the surroundings and the inside of tested samples, where the less energy is needed for this realization.

3.3 MS Spectra and Gas Evolved Analysis

Figure 3a and b shows the MS spectra of methane and its charged fragment during primary and secondary pyrolysis of spruce wood and pistachio shell samples monitored at heating rate of $\beta = 10$ °C min^{-1}.

a

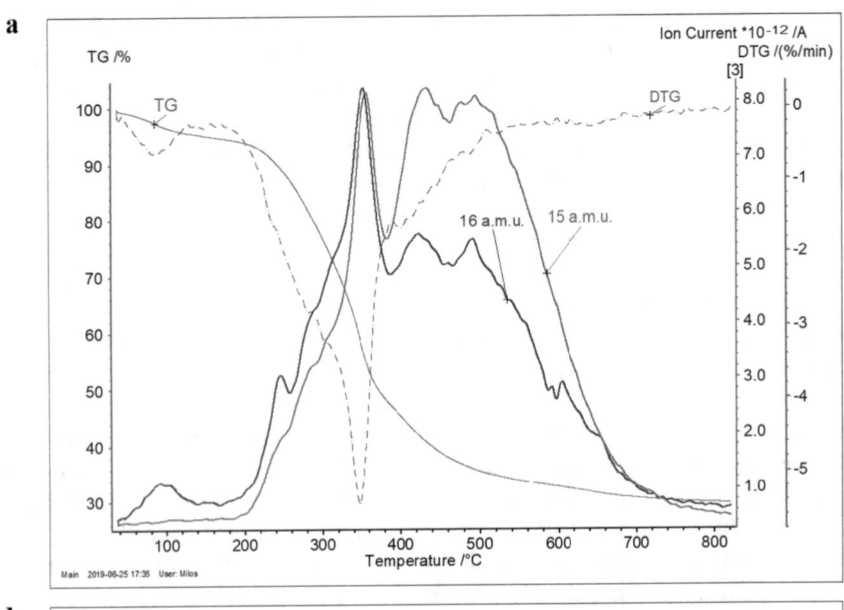

b

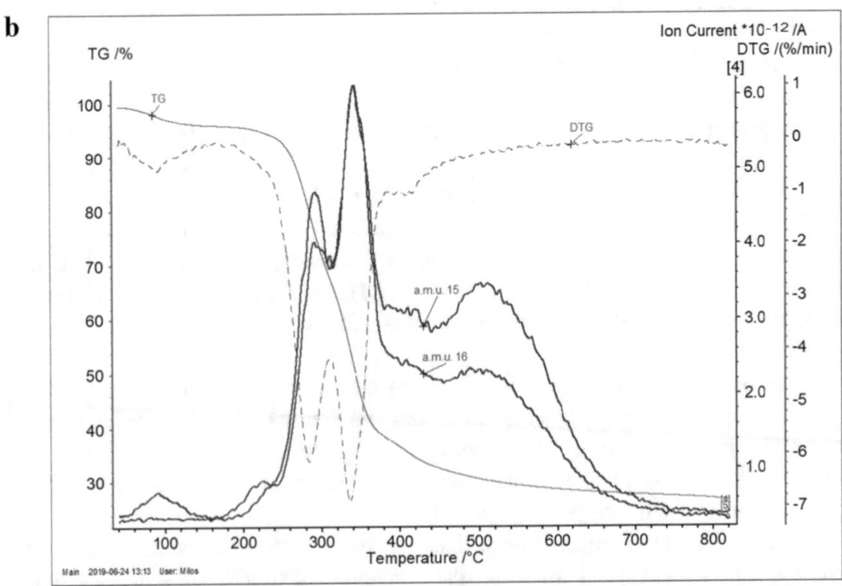

Fig. 3. The MS spectra of methane (CH_4) (16 a.m.u.) and its charged fragment (CH_3^+) (15 a.m.u.) together with simultaneous presentation of TG-DTG curves at 10 °C min^{-1}, during primary and secondary pyrolysis of (**a**) spruce wood and (**b**) pistachio shell samples.

The methane represents the supply fuel gas together with H_2 to the fuel cells for power generation. Considering results presented in Fig. 3, it can be observed that MS top peaks for CH_4 and CH_3^+ for both samples are overlap at around 350 °C, which means that largest production of methane comes from *cellulose pyrolysis*. Major deviations among MS spectra related to CH_4 releasing occur at higher temperature above 400 °C for spruce wood than for pistachio shell.

However, the second release regime for the methane takes place around 500 °C that originated from lignin decomposition reactions. The contribution of methane gas releases is higher in considered temperature interval for spruce wood sample, than in the case of pistachio shell sample. In both cases, beyond 500 °C, the production of methane in both cases is constantly reduced until the end of the process. The production of methane (CH_4) is more vigorous within the primary pyrolysis reactions than its production in the later phase of the process, taking into consideration the secondary pyrolysis reactions that occur at the higher temperatures (Fig. 3). So, for temperature higher than 500 °C, the residues obtained from the different constituents present a similar structure, which evolves towards a more condensed polyaromatic form by releasing CH_4 which is probably followed by releasing of the hydrogen [31].

Figure 4a and b shows the MS spectra of carbon monoxide and carbon dioxide as well as charged fragment during primary and secondary pyrolysis of spruce wood and pistachio shell samples monitored at heating rate of $\beta = 10$ °C min^{-1}.

It can be observed from Fig. 4, that in both cases, the MS spectra very well followed the corresponding DTG features during entire pyrolysis processes. In the case of pistachio shell pyrolysis (Fig. 4b), the hemicelluloses and subsequently (in a greater extent), the cellulose, are the main sources for the production of CO and CO_2. The amount of releases gases is much higher for pistachio shell pyrolysis than in the case of spruce wood pyrolysis (Fig. 4a), where the cellulose is the main biomass component in the woody sample that generates carbon oxides. For the spruce wood pyrolysis, in the high temperature region, at around 500 °C, CO and CO_2 gases were generated by the secondary cracking reactions of lignin (the residual lignin decomposition reactions). It should be noted that the increased amount of gas volatiles is a consequence of decreasing the particle size, especially in the case of fine particles (<1 mm). Also, increasing the temperature and (or) the heating rates supports the gas production while minimizing the production of solid residue. Considering MS results in Fig. 4, it should be noted that in the case of pistachio shell sample, no major reaction exists above 500 °C (which is not the case for spruce wood sample; Fig. 4a). According to this result, it can be said that choosing pyrolysis temperature as 500 °C is appropriate when obtaining of bio-char by physical activation. Obviously, for pistachio shell pyrolysis, we may expect that gas yield should increase, while the bio-char yield should decrease, by an increasing the temperature from 300 to 700 °C. The reactions responsible for the release of volatile compounds from lignin amorphous polymer are mostly due to the instability of the propyl chains, of some linkages between monomer units and of the methoxy substituents of the aromatic rings. After this step, responsible for the main release of primary volatiles, a charring process which consists in re-arrangement of bio-char skeleton in a polycyclic aromatic structure occurs. The volatile compounds released by these re-arrangement reactions are mostly low-weight incondensable gases [32].

a

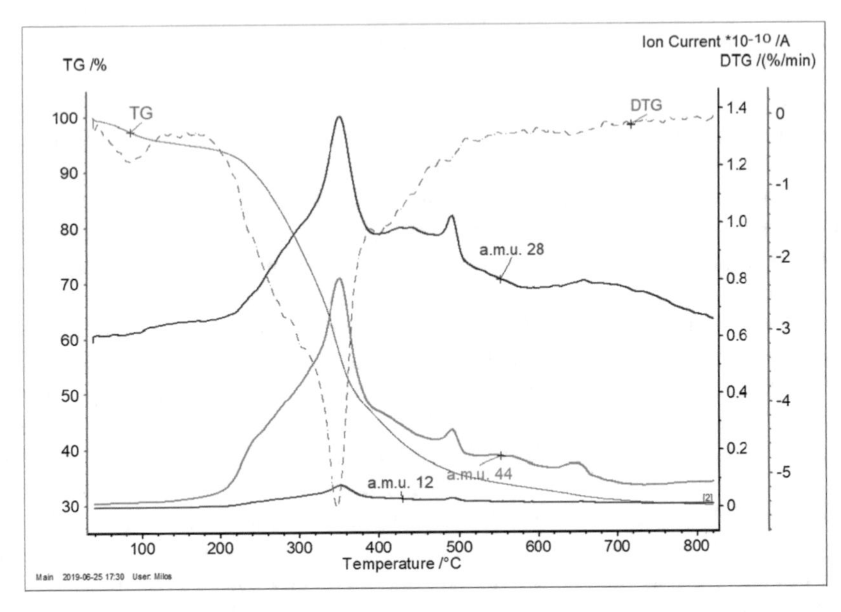

b

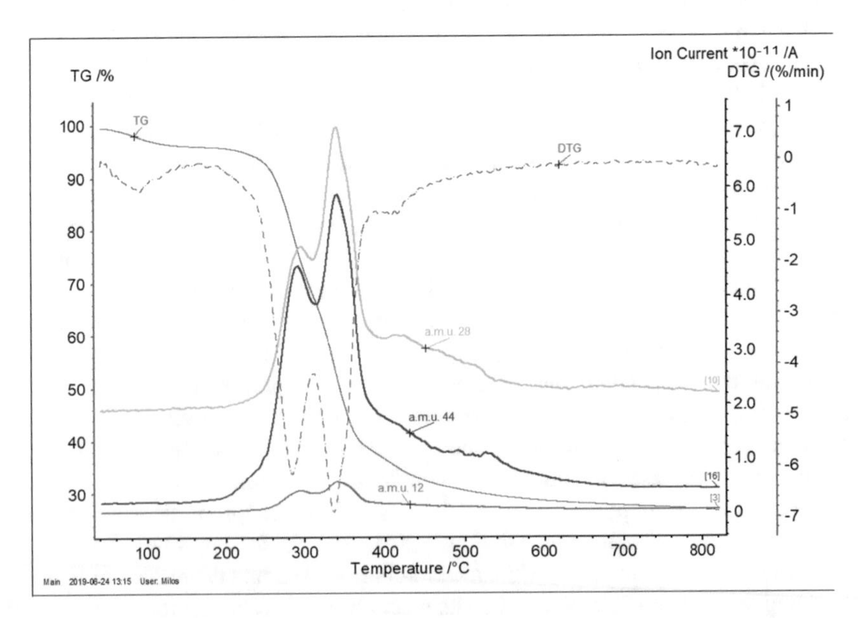

Fig. 4. The MS spectra of charged fragment (C_2^+) (12 a.m.u.), carbon monoxide (CO) (28 a.m.u.), and carbon dioxide (CO_2) (44 a.m.u.), together with simultaneous presentation of TG-DTG curves at 10 °C min^{-1}, during primary and secondary pyrolysis of (**a**) spruce wood and (**b**) pistachio shell samples.

In the case of wood and woody biomasses (such as our spruce wood sample), the generation of CO and CO_2 in the temperature range of 180–400 °C, occurs when the carbon C_γ is implied in the carbonyl or carboxyl functions, the same reaction is likely to take place, at the temperatures which are slightly higher [33]. For temperature higher than 450°C, most of the initial bonds between monomer units have been broken. Only the more stable such as the phenyl linkage 5-5 and the ether linkage 4-O-5 are still present [34, 35]. The main short substituents of the aromatic rings are $-CH_3$ or –OH. The bio-char becomes more and more aromatic and most of the evolved compounds are incondensable gases. At higher temperatures, $-CH_3$ substituents usually disappears between 500 and 600 °C [36], and this can be reflected through the local maximum on the MS spectrum for methane formation approximately between 500 and 550 °C (Fig. 3).

In addition, the rupture of the ether bonds 4-O-5 of lignin in spruce wood biomass can be effectively supposed to be a source of CO production [37].

Figure 5 shows the cumulative MS spectra of pyrolytic gases during slow pyrolysis of spruce wood (a) and pistachio shell (b) biomass samples at medium heating rate of 10 °C min^{-1}. The obtained MS spectra profiles obtained for tested samples are typical for lignocellulosic material originated from plant-based biomass [38].

In both cases, it could be observed that the signal intensity of the various gaseous products increases gradually with the temperature, indicating that the pyrolysis process of biomass particles was generally an endothermic reaction. The increase in the temperature accelerated the formation of the volatiles.

However, the occurrence of the individual gases and their representation expressed through the intensity of the MS signals is not the same for a given studied samples, differing in the temperature ranges and most likely at the time of their retention.

At the beginning of the process, the gas release is not obvious, mainly because of the internal heat transfer and the formation of precursors, such as the active cellulose, for further cracking reactions [39]. Up to 150/180 °C, the only strong MS signal originates from H_2O during dehydration step, where more vapor is released in the case of spruce wood thermal drying, since that this sample contains higher moisture content (Table 1). As the pyrolysis process of these two samples progressed, the releasing rate of CO, CO_2, and CH_4 went up to the maximum and then fell (except for spruce wood), ending the reaction up to 800 °C. However, under the same pyrolysis temperature, the tendency to form *double peaks* was observed in the H_2 releasing curve of spruce wood at the pyrolysis temperatures from 350 °C to 450 °C while not visible in that curve of pistachio shell (Fig. 5). In addition to the different peak shapes, the time required for complete releasing of H_2 from spruce wood is longer than for pistachio shell. This indicated that the reaction mechanism varied depending on the structural differentiation. The *double peaks* in H_2 releasing curve of spruce wood could be associated with hydrogenation and poly-condensation reaction of aromatic structures of the organic compound, especially the lignin. However, it should be emphasized that MS signals for all gaseous species in the case of pistachio shell pyrolysis, are more profiled and more symmetric in comparison for the same, in the case of spruce wood pyrolysis (Fig. 5a and b).

a

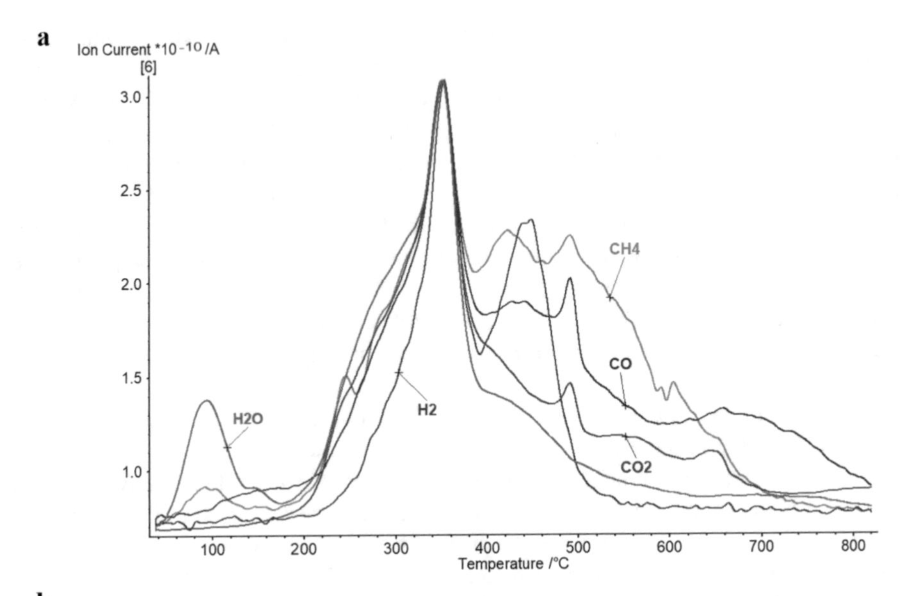

b

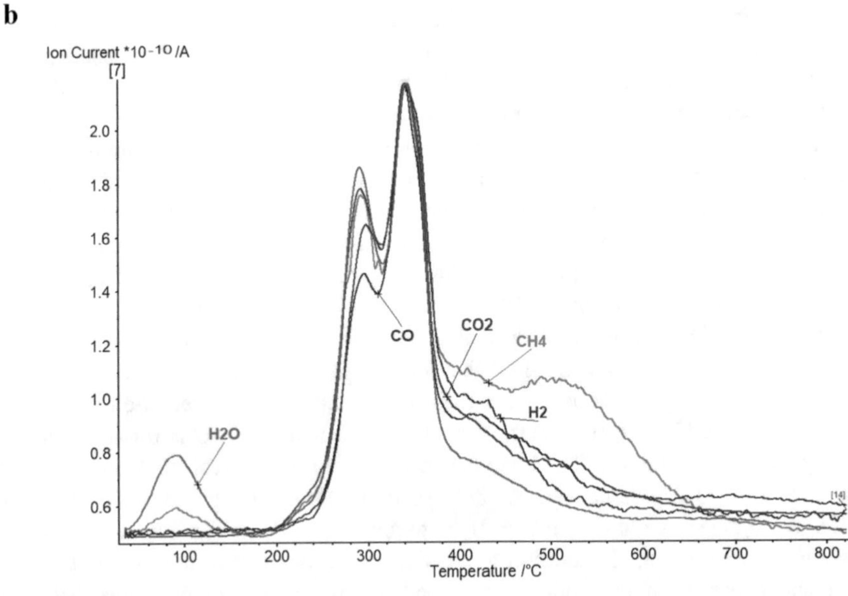

Fig. 5. The MS spectra of pyrolytic gases during slow pyrolysis (10 °C min^{-1}) of spruce wood (**a**) and pistachio shell (**b**) biomass samples. For evolved gases which were monitored (H_2O, CO, CO_2, CH_4 and H_2), the corresponding mass spectrometry signals are clearly indicated.

Furthermore, the formation of gas species from spruce wood demonstrated more sensitivity to the pyrolysis temperature changes than pistachio shell, especially for H_2 and CH_4. Unlike spruce wood, for the pistachio shell, the MS features for CO and CO_2 are more “ordered” with a very pronounced match with DTG curves, whereby high-profile peaks can be connected with elevated sensitivity of the rates of formation of these gases to free radical debris concentration. Previous researchers proposed that the pyrolysis process consisted of a group of complex primary or recombination reactions, and the free radical mechanism appeared to have a significant impact on the formation characteristics of gas products [40, 41]. Since that gas evolved MS profiles for CO and H_2 for pistachio shell pyrolysis, occur in a much narrower temperature range, and in the lower temperature position values at several target points (at 300 °C, 340 °C and 410 ° C) in comparison with the same issues related to spruce wood pyrolysis, it follows that reaction temperature and possible highest content of metal catalyst in pistachio shell sample, govern the process in a positive direction towards increasing the yield of more target products, such as CO and H_2 (the *syngas* positive coupling production). In the case of spruce wood pyrolysis, there is a negative feedback in the production of CO and H_2, regarding its "dissipative" character of MS signals (Fig. 5). However, it can be supposed that an increase of temperature will lead to an increase in the syngas fraction, but this can be strongly dependent on the used biomass feedstock types. On the other hand, increase or decrease in syngas fraction also depends on the applied pyrolysis regime (the slow, medium or fast pyrolysis regimes) and the effect of residence time. Namely, the slow (conventional) pyrolysis favors the decomposition of biomass feedstocks toward syngas fraction. The low temperature favors the production of methane other than hydrogen for both processes, while high temperature favors the production of hydrogen. In the case of spruce wood pyrolysis, at the higher temperatures, the volatile cracking reactions are promoted, which increase the production of non-condensable gases such as CO and CO_2 (Fig. 5a).

It should be noted that despite all the studies conducted to evaluate the effect of temperature on the yield of bio-char and non-condensable gases, there is ambiguity in the definition of the temperature of pyrolysis. Unfortunately, in both laboratory scale and industrial scale pyrolysis equipment, it is not possible to directly measure the temperature of the biomass particles, and it is generally assumed to be equal to the reactor temperature, or in some cases it can be estimated by mathematical modeling. However, these assumptions, in most cases, are far removed from reality, with mismatches (thermal lag) of temperatures as high as 100 °C. For slow pyrolysis conditions with small particle sizes (<1 mm), this assumption is valid due to the higher heating times, where it is easy to achieve thermal equilibrium between the biomass particles and the surroundings. In general, the temperature lags are greater when heating is undertaken at high temperatures or high heating rates. The slower rate of heating prevents that cellulose and hemicelluloses undergo the depolymerization reactions, but maximizing the volatiles residence time inside the biomass particle, and also prevents the volatiles cracking. So, in this case, the non-condensable gases are liberated quickly increased their abundance, and promotes the charring process. In the case of spruce wood pyrolysis, the production of CO for $T \geq 500$ °C must be due to the rupture of the oxygenated substituents, following the mechanisms similar to those involved during the lignin charring process. This explains higher bio-char amount obtained for spruce wood, in comparison to pistachios shells (see Fig. 1). The formation of H_2 at 450 °C is mainly due to dehydrogenation reactions [42] during the formation of a more condensed structure (mainly in the case of spruce wood pyrolysis). Some vapo-gasification reactions could also be responsible for production of hydrogen [42]. Some reactions occurring during the charring process of the hemicelluloses begin at lower temperatures than for the conversion of lignin and cellulose. For instance, the H_2 production is initiated at 450 °C during hemicelluloses conversion against 500 °C for r the other components. So, this phenomenon can be explained by the catalytic effect of the minerals [43], present in the high content in hemicelluloses, which can be probably related to the behavior of pistachio shell sample (see earlier discussion in sub-section "TG-DTG-DTA analysis of pistachio shell and spruce wood pyrolysis"). Considering higher ash content of pistachio shell sample (Table 1), the ash composition may consists the alkaline and alkaline earths oxides, such as K_2O, CaO, MgO as well as Fe_2O_3 [44], which can promote the hydrogen production from pistachios shells (present inorganic compounds in the ash can affects the conversion process; they have possibility of melting at low temperatures, interfering in the heat and mass transfer phenomena during pyrolysis process). Catalysts which can promote H_2 production and adjust the gas composition for downstream applications (e.g., Fischer–Tropsch synthesis) are CaO and Fe_2O_3 [45, 46], whereby these catalysts are presented in pistachios shells ash content [44].

3.4 The Gas Products Yield and Biomass Evaluation for Syngas Valorization

Several types of biomass feedstock exist on the market that can be valorized energetically. Amongst them are wood and agricultural wastes, which are well-suited for syngas production because of their abundant supply coupled with their currently low recycling rates and good fuel properties [47]. For waste-derived syngas, a major technical challenge must be overcome in order to achieve a wider market penetration, which is linked to the development of improved and cheaper syngas cleaning methods in order to meet the increasingly strict specifications of various end-uses devices and environmental emissions regulations. In this sub-section, based on previously established results, we performed the gas products yield analysis through considering possibilities of selected biomasses in their application for *syngas* production. In this context, regarding the initial biomass composition, lignocellulosic material with *high lignin* content, tends to increase *bio-char yields*, while lignocellulosic material with *high cellulose* and *hemicelluloses* content increases *gas yields*. Table 4 lists the lignocellulosic composition of spruce wood and pistachio shells biomasses reported in the open literature.

Table 4. The lignocellulosic composition (cellulose, hemicelluloses and lignin) for spruce wood and pistachio shells [48–51].

Biomass[a]	Cellulose (%)	Hemicelluloses (%)	Lignin (%)
Spruce wood	42.0	6.3	28.3
Pistachio shells	42.7	29.9	13.5

[a]% on dry basis

From data presented in Table 4, it can be seen that pistachio shells contain a higher percentage of cellulose and hemicelluloses components, and lower percentage content of the third component, the lignin, in comparison with wood biomass. The spruce wood is characterized with very low content of hemicelluloses, while the content of lignin is much higher than ones present in the pistachio shells biomass (Table 4). These data can be useful for preliminary prediction of solid and gas products yields after thermo-chemical conversion (pyrolysis) of tested biomass samples. As a presumption, the agricultural waste (pistachio shells) may be suitable for favoring the gaseous products, while the wood biomass (the spruce wood) may be suitable for enhanced bio-char production. Figure 6 shows the main gas products yields (including H_2, CH_4, CO and CO_2) obtained from slow pyrolysis process of pistachios shell (a) and spruce wood (b) biomass feedstock.

It can be observed from Fig. 6 that both samples exhibit approximately equal yields of hydrogen (H_2) and carbon dioxide (CO_2), but pistachios shell gives much higher yields of methane (CH_4) and carbon monoxide (CO) compared to the spruce wood. In general, if we consider the overall gas yields of both tested biomass samples, it seems that pistachios shells is more energetically favorable for syngas production and for future, the power-to-gas commercial exploitation. It can be assumed that in the case of spruce wood, eventually presented alkali salts tend to catalyze pyrolysis reactions to favor CO_2 production over the CO, so that leads to reduced yield in CO gas product (Fig. 6). So, the pistachios shell biomass feedstock would be much more convenient to perform gasification in the presence of oxygen or steam as oxidant agent, for the power and heat generation. This consideration only applies to permanent gases as shown in the diagrams in Fig. 6. Considering actual results and ones related to DTA and MS features, it can be assumed that spruce wood biomass can be used for bio-char gasification with carbon dioxide [52], especially at temperature equal or higher of 650 °C, since that susceptible released volume of CO_2 gas can be identified from MS spectra shown in Fig. 4a. From our perspective, for an even greater increase in the hydrogen concentration in syngas fraction of pistachios shell pyrolysis (Fig. 6), it can be suggested the adding some of alkaline earth metal oxides (such as CaO) into the prepared blend. Namely, CaO can affect the improving the H_2 yield and has the most significant effect on hemicelluloses (Table 4). Adding CaO can cause increasing the H_2 yield, through its generation from tar-cracking, reforming, and char-decomposition reactions at the elevated temperatures. With adding of metal catalyst, the significant amount of H_2 can be obtained even within the primary pyrolysis stage, where the water-gas shift reaction is driven toward H_2 formation. However, the supposed metal catalyst can also increase the rate of the bio-char decomposition and lowered its onset temperature. So, it can be suggested that production of H_2 from waste biomass sample which consists elevated content of hemicelluloses over the course of slow pyrolysis, could be modeled by summing the hydrogen evolutions from separate biomass components in their relevant proportions for the future planned research. Considering results showed in Fig. 6, the H_2/CO ratios were obtained, and their values are as follow: H_2/CO = 0.93 for pistachios shell, and H_2/CO = 1.31 for spruce wood, respectively. The lower H_2/CO ratio for pistachios shell indicates that produced syngas is richer in CO. So, in the gasification process, if the more oxygen is added, the more CO_2 will be produced, which can cause the raises the bed temperature inside gasifier, which in turn boosts the Boudouard reaction ($C + CO_2 \rightarrow 2CO$, endothermic) to produce more CO. On the other hand, regarding the influence of steam, an increase in H_2O concentration can

a

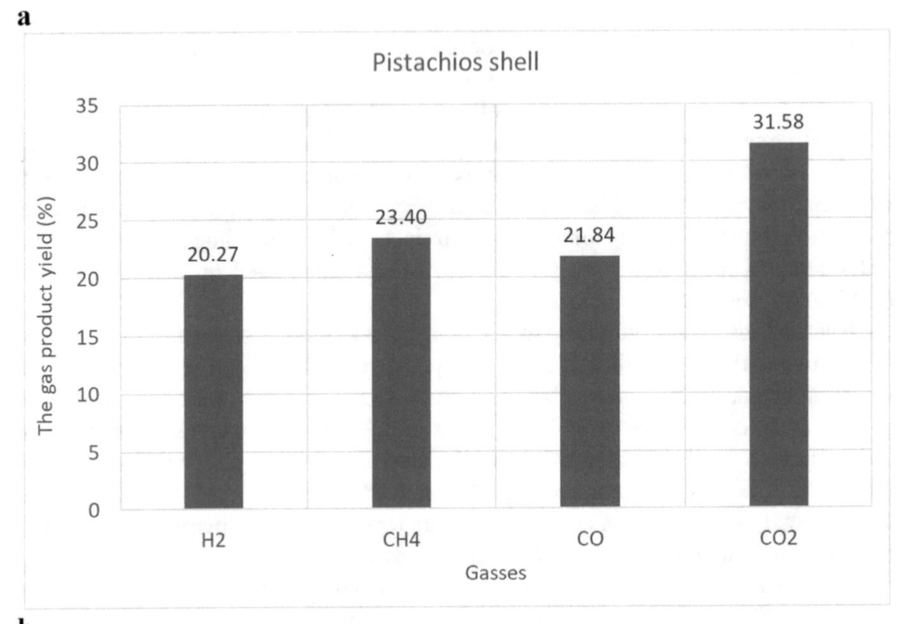

b

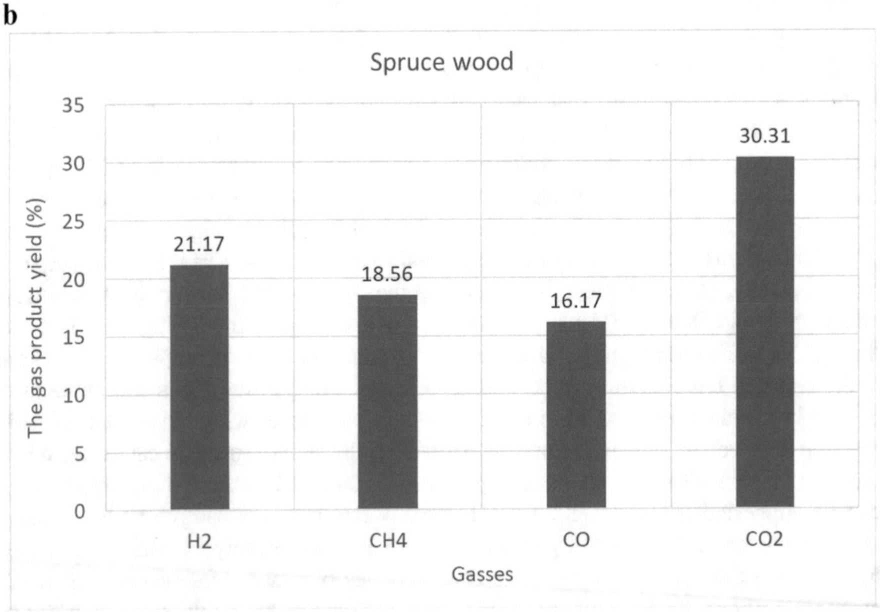

Fig. 6. The gas products yields obtained by slow pyrolysis process of pistachios shell and spruce wood.

result in the increase of H_2/CO ratio (the H_2/CO ratio for spruce wood is equal to 1.31, which is higher than ones for pistachios shell). More H_2O in the gasifier boosts the steam gasification reaction ($C + H_2O \rightarrow CO + H_2$, endothermic), thereby producing more CO and H_2. Much of this CO produced is consumed by the WGS (water-gas shift) reaction to favor H_2 production ($CO + H_2O \leftrightarrow CO_2 + H_2$, exothermic). It should be noted that slightly higher CO_2 yield obtained for pistachios shell (Fig. 6) lowers its H_2/CO ratio, in comparison with the spruce wood. Increasing the CO_2 concentration actually lowers the syngas quality, because the net increase of CO is small and H_2 production is further decreased. When the CO_2 is used as gasifying agent, its concentration volume must be smaller, if we use it for controlling syngas composition. So, for the quality improvement of syngas composition in the case of pistachios shell biomass pyrolysis, the use of metal oxides as a catalyst to increase the concentration of hydrogen in the syngas fraction is strongly recommended.

4 Conclusion

Pyrolysis of two selected biomass samples (pistachios shell and spruce wood) has been performed in simultaneous thermal analysis (STA) measurements, together with gas evolved analysis monitored by mass spectrometry (MS) analysis. TGA-DTA together with MS analysis provides an insight regarding the link between biomasses decomposition behaviors and their gaseous fractions. Overall, TG-DTG and gases profiles obtained suggest effective pyrolysis temperatures at which different fractions of considered biomass actively respond to the experimental conditions. Besides, this analysis also gives useful data on the specific temperatures, where the various heterogeneous reactions occur throughout the pyrolysis of studied biomasses inside the TGA apparatus.

It was found that the wood biomass (spruce wood) exhibits higher yield of bio-char (30.92%) than the agricultural waste biomass (pistachios shells) (27.83%), given that the lignin decomposition contributes in larger extents in the elevated bio-char formation, for the wood biomass feedstock. It was found that for both samples, the largest production of methane originates from cellulose pyrolysis at a temperature around 350 °C. The second release pathway for the methane takes place around 500 °C which originated from lignin decomposition reactions. It was found that the production of CH_4 is more vigorous within primary pyrolysis reactions than its production in the later phase of the process, taking into consideration secondary pyrolysis reactions which occur at higher temperatures. For pistachio shell pyrolysis, hemicelluloses and cellulose are main sources for production of CO and CO_2. The amount of releases gases is much higher for pistachio shell pyrolysis than for spruce wood pyrolysis. It was established that the increased amount of gas volatiles is a consequence of decreasing the particle size, where increasing the temperature supports the gas production while minimizing the production of solid residue. It was proposed that for pistachio shell pyrolysis, the gas yield may increase, while bio-char yield can decrease, if the pyrolysis temperature increases from 300 to 700 °C. Based on different MS peak shapes, it was found that time required for complete releasing of H_2 from spruce wood is longer than for pistachio shell. This indicates that the reaction mechanism varies and depends on the structural differentiation. The appearance of double MS peaks for H_2 releasing in the case of spruce wood pyrolysis is associated with hydrogenation and poly-condensation reaction of aromatic structures of the organic compound, especially the lignin. It was found that low temperature favors production of CH_4 other than hydrogen for both studied processes, while high temperature favors production of H_2. In the case of spruce wood pyrolysis, at higher temperatures, volatile cracking reactions are promoted, which increase production of non-condensable gases such as CO and CO_2.

Based on the overall gas yields analysis of both samples, it was established that pistachios shells is more energetically favorable for syngas production and for power-to-gas commercial exploitation. In this study, for the quality improvement of syngas composition in the case of the pistachios shell pyrolysis, the use of metal oxides as a catalyst to increase the concentration of hydrogen in the syngas fraction is strongly recommended. Also, in the actual research, we discuss possible gasification conditions which can be applied in the case of the use of tested biomass samples for power and heat generation.

Acknowledgement. Authors would like to acknowledge the financial support of the Ministry of Education, Science and Technological Development of the Republic of Serbia under the Projects 172015, 172045 and III42010.

References

1. Nordås, R., Gleditsch, N.P.: Climate change and conflict. Political Geogr. **26**(6), 627–638 (2007)
2. Stoeglehner, G., Narodoslawsky, M.: Implementing ecological foot printing in decision-making processes. Land Use Policy **25**(3), 421–431 (2008)
3. Upreti, B.R., van der Horst, D.: National renewable energy policy and local opposition in the UK: the failed development of a biomass electricity plant. Biomass Bioenergy **26**(1), 61–69 (2004)
4. Karakosta, C., Doukas, H., Flouri, M., Dimopoulou, S., Papadopoulou, A.G., Psarras, J.: Review and analysis of renewable energy perspectives in Serbia. Int. J. Energy Environ. **2**(1), 71–84 (2011)
5. Cherubini, F.: The biorefinery concept: using biomass instead of oil for producing energy and chemicals. Energy Convers. Manage. **51**(7), 1412–1421 (2010)
6. Hosseini, S.E., Wahid, M.A.: Hydrogen production from renewable and sustainable energy resources: promising green energy carrier for clean development. Renew. Sustain. Energy Rev. **57**, 850–866 (2016)
7. Dodić, S.N., Vučurović, D.G., Popov, S.D., Dodić, J.M., Ranković, J.A.: Cleaner bioprocesses for promoting zero-emission biofuels production in Vojvodina. Renew. Sustain. Energy Rev. **14**(9), 3242–3246 (2010)
8. Naik, S.N., Goud, V.V., Rout, P.K., Dalai, A.K.: Production of first- and second-generation biofuels: a comprehensive review. Renew. Sustain. Energy Rev. **14**(2), 578–597 (2010)
9. Balat, M., Balat, H.: Recent trends in global production and utilization of bio-ethanol fuel. Appl. Energy **86**(11), 2273–2282 (2009)
10. McKendry, P.: Energy production from biomass (part 1): overview of biomass. Biores. Technol. **83**(1), 37–46 (2002)
11. Wolfsmayr, U.J., Rauch, P.: The primary forest fuel supply chain: a literature review. Biomass Bioenergy **60**, 203–221 (2014)
12. Caillé, A., Al-Moneef, M., de Castro, F.B., Bundgaard-Jensen, A., Fall, A., de Medeiros, N. F., Jain, C.P., Kim, Y.D., Nadeau, M.J., Testa, C., Teyssen, J.: Deciding the future: energy policy scenarios to 2050. World Energy Council (2007)

13. Demirbas, A.: Biorefinery technologies for biomass upgrading. Energy Sources Part A **32**(16), 1547–1558 (2010)
14. Sharma, R.K., Wooten, J.B., Baliga, V.L., Lin, X., Chan, W.G., Hajaligol, M.R.: Characterization of chars from pyrolysis of lignin. Fuel **83**(11–12), 1469–1482 (2004)
15. Alakangas, E.: European standards for fuel specification and classes of solid biofuels. Solid Biofuels for Energy, pp. 21–41. Springer, London (2011)
16. Janković, B., Manić, N., Stojiljković, D., Jovanović, V.: TSA-MS characterization and kinetic study of the pyrolysis process of various types of biomass based on the Gaussian multi-peak fitting and peak-to-peak approaches. Fuel **234**, 447–463 (2018)
17. Ebeling, J.M., Jenkins, B.M.: Physical and chemical properties of biomass fuels. Trans. ASAE **28**(3), 898–902 (1985)
18. Van Loo, S., Koppejan, J.: The Handbook of Biomass Combustion and Co-firing, pp. 1–464. Earthscan Publications Ltd., London (2008). ISBN 9781849773041
19. Tonbul, Y.: Pyrolysis of pistachio shell as a biomass. J. Therm. Anal. Calorim. **91**(2), 641–647 (2008)
20. McKendry, P.: Energy production from biomass (part 1): overview of biomass. Bioresour. Technol. **83**, 37–46 (2002)
21. Grønli, M.G., Várhegyi, G., Di Blasi, C.: Thermogravimetric analysis and devolatilization kinetics of wood. Ind. Eng. Chem. Res. **41**(17), 4201–4208 (2002)
22. Yang, G., Jaakkola, P.: Wood Chemistry and Isolation of Extractives from Wood, pp. 3–47. Saimaa University of Applied Sciences, Literature study for BIOTULI Project (2011)
23. Rinta-Paavola, A., Hostikka, S.: A model for pyrolysis and oxidation of two common structural timbers. In: 1. Forum Wood Building Baltic 2019, pp. 1–10 (2019)
24. Varma, A.K., Mondal, P.: Pyrolysis of pine needles: effects of process parameters on products yield and analysis of products. J. Therm. Anal. Calorim. **131**(3), 2057–2072 (2018)
25. Radojević, M., Janković, B., Jovanović, V., Stojiljković, D., Manić, N.: Comparative pyrolysis kinetics of various biomasses based on model-free and DAEM approaches improved with numerical optimization procedure. PLoS ONE **13**(10), e0206657 (2018). https://doi.org/10.1371/journal.pone.0206657
26. Janković, B.: The pyrolysis process of wood biomass samples under isothermal experimental conditions - energy density considerations: application of the distributed apparent activation energy model with a mixture of distribution functions. Cellulose **21**, 2285–2314 (2014)
27. Sasaki, C., Kurosumi, A., Yamashita, Y., Mtui, G., Nakamura, Y.: Xylitol production from dilute-acid hydrolysis of bean group shells, In: Mendez-Vilas, A. (ed.) Microorganisms in Industry and Environment. From Scientific and Industrial Research to Consumer Products, pp. 605–613. World Scientific Publishing Co. Pte. Ltd., Singapore (2011). ISBN-13 978-981-4322-10-2
28. Laine, C.: Structures of hemicelluloses and Pectins in wood and pulp. Dissertation for the degree of Doctor of Science, Helsinki University of Technology, Department of Chemical Technology, Oy Keskuslaboratorio, Espoo, Finland, pp. 1–63 (2005). ISSN 1457-6252
29. Peters, B.: Prediction of pyrolysis of pistachio shells based on its components, hemicellulose, cellulose and lignin. Fuel Process. Technol. **92**(10), 1993–1998 (2011)
30. Janković, B.: The comparative kinetic analysis of Acetocell and Lignoboost® lignin pyrolysis: the estimation of the distributed reactivity models. Bioresour. Technol. **102**(20), 9763–9771 (2011)
31. Collard, F.-X., Blin, J.: A review on pyrolysis of biomass constituents: mechanisms and composition of the products obtained from the conversion of cellulose, hemicelluloses and lignin. Renew. Sustain. Energy Rev. **38**, 594–608 (2014)
32. Liu, Q., Wang, S., Zheng, Y., Luo, Z., Cen, K.: Mechanism study of wood lignin pyrolysis by using TG-FTIR analysis. J. Anal. Appl. Pyrol. **82**, 170–177 (2008)

33. Jakab, E., Faix, O., Till, F., Székely, T.: Thermogravimetry/mass spectrometry study of six lignins within the scope of an international round robin test. J. Anal. Appl. Pyrol. **35**(2), 167–179 (1995)
34. Nakamura, T., Kawamoto, H., Saka, S.: Pyrolysis behavior of Japanese cedar wood lignin studied with various modeldimers. J. Anal. Appl. Pyrol. **81**, 173–182 (2008)
35. Faravelli, T., Frassoldati, A., Migliavacca, G., Ranzi, E.: Detailed kinetic modeling of the thermal degradation of lignins. Biomass Bioenergy **34**, 290–301 (2010)
36. Sharma, R.K., Wooten, J.B., Baliga, V.L., Lin, X., Chan, W.G., Hajaligol, M.R.: Characterization of chars from pyrolysis of lignin. Fuel **83**, 1469–1482 (2004)
37. Wang, S., Wang, K., Liu, Q., Gu, Y., Luo, Z., Cen, K., Fransson, T.: Comparison of the pyrolysis behavior of lignins from different tree species. Biotechnol. Adv. **27**(5), 562–567 (2009)
38. Manić, N.G., Janković, B.Ž., Stojiljković, D.D., Jovanović, V.V., Radojević, M.B.: TGA-DSC-MS analysis of pyrolysis process of various agricultural residues. Thermal Sci. (2018). https://doi.org/10.2298/TSCI180118182M
39. Wang, S.R., Liu, Q., Liao, Y.F., Luo, Z.Y., Cen, K.F.: A study on the mechanism research on cellulose pyrolysis under catalysis of metallic salts. Korean J. Chem. Eng. **24**(2), 336–340 (2007)
40. Wu, Z., Yang, W., Li, Y., Yang, B.: Co-pyrolysis behavior of microalgae biomass and low-quality coal: products distributions, char-surface morphology, and synergistic effects. Bioresour. Technol. **255**, 238–245 (2018)
41. Janković, B., Manić, N., Dodevski, V., Popović, J., Rusmirović, J.D., Tošić, M.: Characterization analysis of poplar fluff pyrolysis products. Multi-component kinetic study. Fuel **238**, 111–128 (2019)
42. Widyawati, M., Church, T.L., Florin, N.H., Harris, A.T.: Hydrogen synthesis from biomass pyrolysis with in situ carbon dioxide capture using calcium oxide. Int. J. Hydrogen Energy **36**(8), 4800–4813 (2011)
43. Boumlin, S., Broust, F., Bazer-Bachi, F., Bourdeaux, T., Herbinet, O., Ndiaye, F.T., Ferrer, M., Lede, J.: Production of hydrogen by lignins fast pyrolysis. Int. J. Hydrogen Energy **31**(15), 2179–2192 (2006)
44. Okutucu, Ç.: Pyrolysis of Pistachio Shell. Master of Science Thesis, EGE University Graduate School of Natural and Applied Sciences, Department of Chemistry, Barnova-Izmir, September 2010, pp. 1–68 (2010)
45. Hao, Q.L., Wang, C., Lu, D.Q., Wang, Y., Li, D., Li, G.J.: Production of hydrogen-rich gas from plant biomass by catalytic pyrolysis at low temperature. Int. J. Hydrogen Energy **35**, 8884–8890 (2010)
46. Yang, H.P., Yan, R., Chen, H.P., Lee, D.H., Liang, D.T., Zheng, C.G.: Pyrolysis of palm oil wastes for enhanced production of hydrogen rich gases. Fuel Process. Technol. **87**, 935–942 (2006)
47. EN 14961-1. Solid biofuels - Fuel specifications and classes - Part 1: General requirements, 2010
48. Hakkila, P.: Utilization of Residual Forest Biomass, pp. 1–516. Wood Science. Springer, Heidelberg (1989)
49. Kim, D.: Physico-chemical conversion of lignocellulose: inhibitor effects and detoxification strategies: a mini review. Molecules **23**(2), 309 (2018). https://doi.org/10.3390/molecules-23020309

50. Shafiei, M., Karimi, K., Taherzadeh, M.J.: Pretreatment of spruce and oak by N-methylmorpholine-N-oxide (NMMO) for efficient conversion of their cellulose to ethanol. Bioresour. Technol. **101**, 4914–4918 (2010)
51. Salasinska, K., Polka, M., Gloc, M., Ryszkowska, J.: Natural fiber composites: the effect of the kind and content of filler on the dimensional and fire stability of polyolefin-based composites. Polimery **61**(4), 255–265 (2016)
52. Chen, S., Meng, A., Long, Y., Zhou, H., Li, Q., Zhang, Y.: TGA pyrolysis and gasification of combustible municipal solid waste. J. Energy Inst. **88**, 332–343 (2015)

Application of 3D Printing in the Metamaterials Designing

Boris Kosic[1(✉)], Aleksandra Dragicevic[1], Zorana Jeli[1], and Gabriel–Catalin Marinescu[2]

[1] Faculty of Mechanical Engineering, University of Belgrade, 11000 Belgrade, Serbia
bkosic@hotmail.com
[2] Faculty of Mechanics, University of Craiova, Craiova, Romania

Abstract. In the last couple of years, 3D printing has become one of the most popular manufacturing techniques in designing of the new prototype parts, mechanisms and machines. Available and relatively cheap 3D printing techniques allows fast manufacturing of different complicate prototypes (rapid prototyping) by using plastics as base material, like FDM. In the most cases, these prototypes are used only for functionality testing and design analyzing, yet that is a small part of the possibilities that 3D printing can give in parts design. For example, 3D printing can be utilized for creation of the cheap personalized limb prosthesis, toward specific needs of each patient. One of the biggest advantages of the method is production of the complex geometric shapes, done layer by layer, which provides a new level of freedom in part design and significantly impact their final characteristics, since conventional techniques of parts manufacturing have many limitations. By modifying the inner geometric structure of the parts produced from regular material, new properties of the engineered part are obtained, with different behavior, which the starting material didn't have. This, engineered materials are named metamaterials. Using the metamaterials it is possible to produce parts with properties that doesn't appear in nature. This paper will present how new parts, mechanisms, and machines can be designed and manufactured using metamaterials and 3D printing.

Keywords: 3D printing · Metamaterials · Internal structure · Smart design

1 Introduction

Manufacturing technology can be divided into two main groups: first is called subtractive and the second is called additive. First manufacturing technology represents a conventional technique where the part is made by removing excess material with different machines tools. The second technology is called additive manufacturing, this method was utilized by casting parts in various molds form different types of metal, plastics, but in modern days with the development of computers and modern algorithms for numerical control (CNC) new technology has emerged in the form of 3D printing [1].

In additive manufacturing (especially with 3D printers), parts can be optimized in such a way that the consumption of material is reduced to a minimum. Many different

N. Mitrovic et al. (Eds.): CNNTech 2019, LNNS 90, pp. 166–183, 2020.
https://doi.org/10.1007/978-3-030-30853-7_10

parts can be created as a hollow shell and in most cases it is enough to create mesh support in the form of different geometrical structures to achieve necessary structural integrity. This designing methodology can significantly reduce overall part mass and material consumption which further reduce the price. Those structures can be different geometrical shapes such as triangles, rectangles, sizes of those geometrical figures is defined by part functionality (to make stronger part infill should be denser) (see Fig. 1).

Fig. 1. Different densities of 3D printing infill [2].

1.1 3D Printing

Three-dimensional printing (3D printing) has been developed in the 1970s [1]. As manufacturing technology, 3D printing has been used for more than 40 years, and for most of this period it was only used for industrial manufacturing processes [1]. In the last couple of years, 3D printing has become popular because several companies have developed affordable 3D printers with prices less than 5000$ [3]. These printers use cheap plastic such as ABS or PLA and have good enough precision to create many different parts and devices.

3D printers have drawn the attention of the public a few years ago when CAD model for making a plastic gun that can't be detected by metal detectors on airports [4]. This gun was produced by a cheap 3D printer that was bought for less than 8000$ and its files with 3D models were downloaded almost 100,000 times before they were taken down [4].

Modern 3D modeling and simulation software tools have revolutionized designing processes in such a way that is possible to test many different ideas in the simulated environment until the final design is selected. Engineering software solution each day become more and more powerful but still final functionality testing must be conducted on real prototypes. As the industry grows, it became necessary to produce fast prototypes (rapid prototyping) which will enable to find design flaws and correct them in the next iterations (see Fig. 2). 3D printing enables relatively cheap fast production rapid prototyping especially when prototypes have complex geometry. Big impact 3D printing has on communication between people involved in the designing process because looking at real physical model can't match with software visualization or image on paper [5].

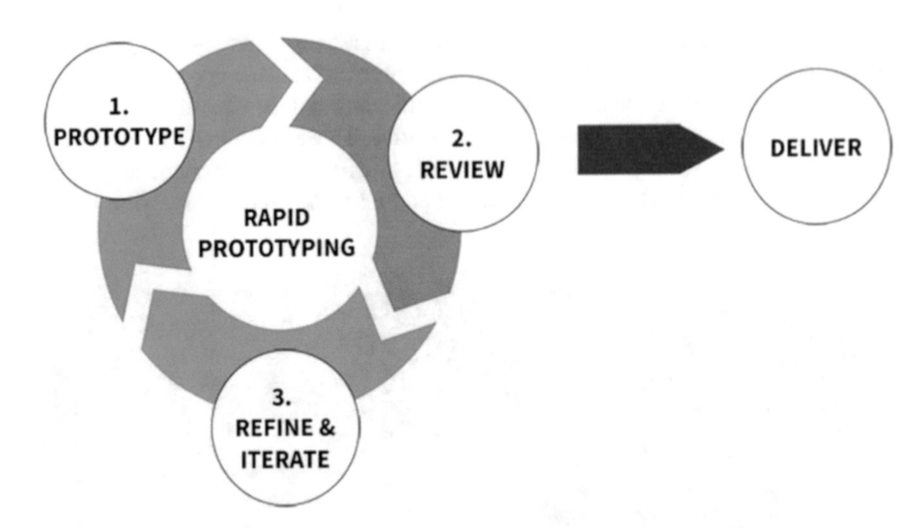

Fig. 2. Rapid prototyping processes [5]. 3D printing enables cheap testing prototypes for refinement of initial design to remove all mistakes that can impact the final product.

It is estimated that the global 3D printing market in 2017 was worth 7.01 billion dollars [6]. The flexibility of technology and possibility for application almost in every industry, supports its further research and development, and it is expected that 3D printer market will grow more than 16.5% in next seven years (see Fig. 3) [6].

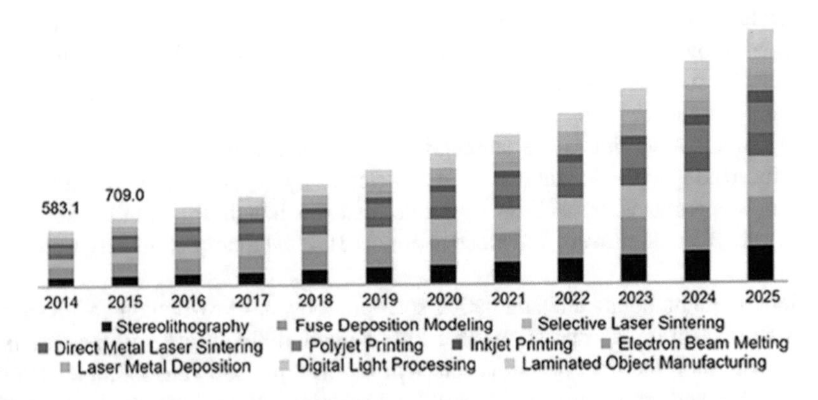

Fig. 3. 3D printing market sizes for each year form 2014 up to 2025 in millions USD [6]. Different colors represent different 3D printing types.

3D Printing Techniques. Part manufactured with 3D printing is created a layer by layer from CAD models. The process to obtain 3D printed parts can be summarized in several steps (see Fig. 4) [7].

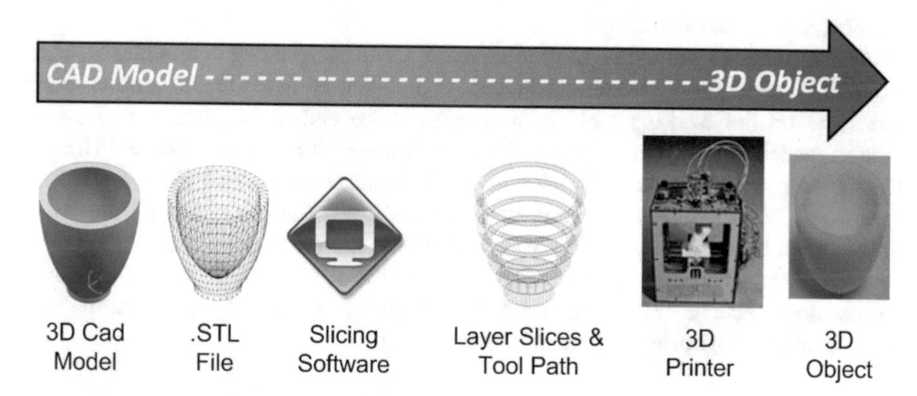

Fig. 4. 3D printing techniques [7]. First, the 3D model is created using some modeling software then that modes are converted into a.stl file which is imported into slicing software. Slicing software split the part into layers and create instruction for machines for each layer. Those instructions are sent to the machine which prints part.

- Model of part is created in some 3D modeling software
- That model is then divided into layers using slicing software
- And then printed with a 3D printer to obtain the final part.

There is a lot of different 3D printing techniques that use different materials and different processes to make parts layer by layer, but three most used are stereolithography (SLA), fused deposition modeling (FDM) and selective laser sintering (SLS), other techniques can be found in literature but there are no much differences how part is created compared to those three.

- Stereolithography - SLS is the first commercialized 3D printing technology. SLS printers utilize curable polymer resin where laser beam traces shapes of a part layer on its surface. When laser goes over resin surface emitted light beam harden thin layer of polymer, after finishing one layer platform moves and laser beam start to draw next slice on the top of the previous one. This method is common for all 3D printing techniques and only difference is the method for layer hardening [8].
- Fused deposition modeling - FDM method uses thermoplastic extrusion where the plastic filament is through a heated nozzle. The nozzle is heated over plastics melting temperature and liquefied plastic is deposited on the printing bed, layer by layer to form part [8].
- Selective laser sintering - SLS process uses fine powder sintering to print parts, where powder particles are fused together with a laser beam. A thin layer of powder is deposited on the printing bed and then laser beam trace part layer, when one layer is completed a new layer of powder is spread on top of that layer and sintering process start again. This process can be used to print almost any material that can be produced as a fine powder, those materials can be different types of plastics and metals [8].

Applications of 3D Printing

In the last couple of years, 3D printers and other CNC have started to miniaturize which led to the production of desktop machines. As the name suggests the desktop machines can stand on the working table in researches office and made relatively cheap 3D printed models and machined parts. Cheap and robust techniques such as FDM and SLA have become a mandatory part of any laboratory equipment. With the help of almost any 3D modeling software, new fixtures and mechanism can be created to further improve laboratory equipment functionality. 3D printing is an essential part of any new development but in last year's it started to use in the serial manufacturing process where special "3D printed farms" are deployed (see Fig. 5). Those farms use a large number of 3D printers to print the same part.

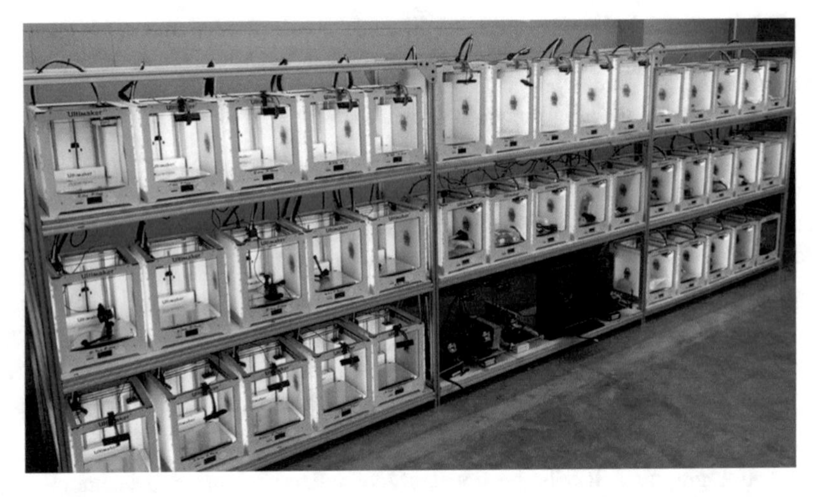

Fig. 5. 3D printing farm [9]. This is the cheapest way to produce small batches of plastic parts because their numbers cannot cover prices for mold production.

In a specific area of industry where a lot of similar parts and devices, that have small changes in size and geometry, only viable option to produce fast and cheap parts is 3D printing is. This application is most useful in prosthesis production where one artificial limb can cost more than a new car. Those prices go up even more if children need artificial limb because prosthesis must be changed as the child grows (see Fig. 6). On the other side when the prosthesis is made from metal parts to reduce overall mass they can be made from different tubings, in this case, 3D printer parts can be used as esthetic cover (see Fig. 7).

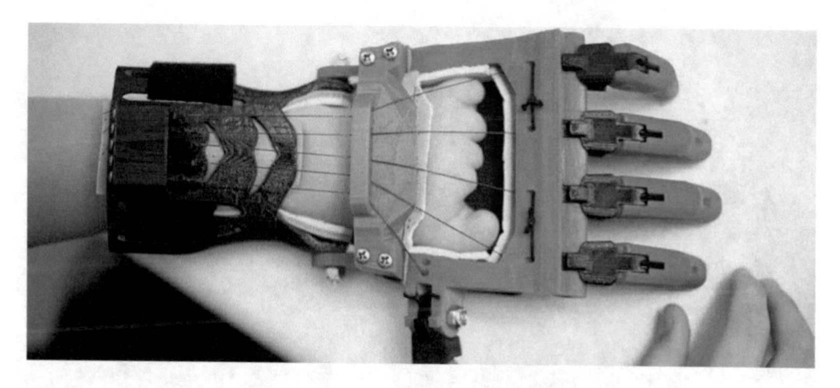

Fig. 6. 3D printed hand prosthesis [10]

Fig. 7. Esthetic covering for metal prosthesis [11].

Almost all parts made in human history have simple geometrical shapes or its combinations (prism, sphere, and cylinder) but with introducing 3D printers and CNC technology parts with more natural and more complex shapes have become the new focus of researches, also, 3D printing is utilized for parts that cannot be made with conventional technique (see Fig. 8).

One of the most interesting 3D printing application that is developed in last years is the 3D printer for printed circuit boards (PCB) where the DragonFly Pro can print in same time metal traces for component connection and board from isolating polymers [13].

Fig. 8. Arc bicycle first 3D printed steel frame [12]. This frame is almost impossible to produce with conventional manufacturing techniques.

1.2 Metamaterials

From the popularity and availability, in last years, of additive manufacturing techniques especially 3D printing and their advantages in creating cheap pars with complex geometries, new designing methodologies for creating various parts have emerged. Those methodologies give possibilities not only to design part shape but also its inner structure which can have a significant impact on the part response in its exploitation. Those part with artificially altered material inner structure can obtain new features simply through the precise designing of their internal geometry. Depend on geometrical structures that are implemented into the part during design, some of its properties can be enhanced or completely changed. First artificial material with this characteristic is created in the year 2000, obtained material had a negative refractive index in the microwave band [14]. This material response is obtained by making structures, with specific geometry, that interacts with microwaves in such a way that material overall response is as it has a negative refractive index. It is necessary to emphasize that this new behavior of the material is achieved only by creating specific microstructures that interact with microwaves.

In recent years, beside metamaterials that can interact with electromagnetic waves researches created many different types of metamaterials, some of them can interact with acoustics waves, others can have enhanced mechanical properties, etc.

One of the first examples of how properly designed metamaterials can be utilized to get new parts is 3D printed door handle that is printed as one piece. Relative movement necessary to get door handle functionality is achieved by deformations of parts inner structure. These parts are engineered in such a way that some of the part inner structure is deliberately weakened to allow large deformation without breaking door handle [15].

Further research in metamaterials is based on new 3D printing materials and new geometries of internal structure. Bodaghi et al. have tested structures that have triangles and hexagons as basic cells. They created a 3D model, simulation and experimentally to analyze both types of basic cells in two orthogonal directions, testing was done with compressive and tension loads [16]. Those studies resulted that mechanical response of metamaterial significantly depends on basic cell geometry and type of loading.

All of the presented metamaterials have two-dimensional geometry (2D metamaterials) Xin Ren et al. developed analyzed 3D metamaterial that has negative Poisson's ration (auxetic metamaterial). 3D metamaterial was created by repeating a cuboid basic cell that has spheres on joints [17]. This type of material is interesting because the plastic used for its 3D printing does not have negative Poisson's ratio but this engineered structure has. In the paper, authors claim that metamaterial response can be controlled by adjusting geometrical parameters of basic cells (cell thickness).

Bonatti et al. [18] investigate what impact different geometrical shapes have on energy absorption, they created 3D printed metal truss lattices with 20% relative density, and different basic cels. In conclusion, stated that with changing in the geometrical characteristic of basic cells it is possible to reduce weight and improve energy absorption of structure up to 35%.

2 Methods for Metamaterial Designing

Development in the area of computer techniques for 3D modeling in the previous 20–30 years in many ways simplifies the design of the part especially those ones with complex geometry. In the field of any type of engineering (mechanical, electrical, chemical…) simulation have become an indispensable part of the designing process. Well optimized process and part can be obtained through intensive analyzing and simulating of working conditions with the aid of computers. In most cases, simulation can solve a lot of problems especially when testing on a physical model can lead to unnecessary costs. Bugaric et al. have shown that simulation can provide detailed insight into flaws on river cargo terminals. This simulation can be used to find bottlenecks and provide solutions for further optimization of existing equipment, and reduce operating cost by reducing port waiting times [19]. Popkonstantinovic et al. were investigated how simulation can be utilized to improve the education process in the field of mechanism synthesis [20]. They analyzed eclipses abacus that is used to predict the solar and lunar eclipses. This type of simulation can further improve student understanding of complex mechanisms working principles and how small changing mechanism geometry can affect predictions of lunar and solar eclipses.

An interesting result in the application of simulations on mechanism synthesis is presented in [21]. They have investigated walking mechanism with one degree of freedom which generates bipedal motion similar to real human motion. The mechanism is designed in such a way that its movement follows trajectory center of mass of the human body, those results were compared with data acquired from Vicon Motion Capture system, and they show almost perfect overlapping. Since no two human bodies are the same for best performances in rehabilitation walker must be tweaked for everyone who needs it, with results and mechanism synthesis presented in that paper is it possible, through simulation, to tweak its parameters for each person that will use it.

Popkonstantinovic et al. simulated balance spring that is used in timekeeper mechanisms (mechanical watches) [22]. Used spring is torsion spring and has relatively large deformation (±5.2 rad or more than ±1.5 turns), but despite large twisting angle deformation are elastic. In the paper, they proposed a new method of simulations using a large number of relatively small deformation until summarized displacement is equal to deformation in normal working conditions. Angular movement of ±5.2 rad is divided into 208 small angular displacements, and for each of those 208 steps, one static simulation is calculated. This method is utilized is such way that angular steps are small enough to ignore stiffness changes caused by spring deformation. Deformed spring form first step is then used as an initial model for a second, in third simulation spring form second step is used as the initial model, this process continues until all 208 steps are calculated. Result acquired from this simulation method are very close to the experimental result, this result also has shown it is possible to reduce the complexity of the simulation process using a lot of simpler one which result can be combined to get the final result.

As seen for literature simulations are irreplaceable in designing and analyzing processes, and they represent the cheapest way to test new designing decisions and reduce the number of possible solutions for the final part model. Simulations will be especially important for metamaterials design as a number of the different applications increase in future because for any application new metamaterials must be designed and optimized.

3 Problem Formulation

Jeli et al. [23] analyzed the ski binding mechanism geometry, a mechanism that fixes leg for the ski during skiing (see Fig. 9). This mechanism should withstand intense forces and shocks which can occur during skiing and it should provide easy locking and unlocking when a person needs to remove ski from its legs. This is accomplished with a mechanism that has only three parts and one spring that provides the force necessary to keep the mechanism in locking or unlocking position [23]. The functionality of the mechanism is accomplished with two different cylindrical surfaces on "Fulcrum" and locking/unlocking action "cover" slides over those two surfaces, also spring force act on a fulcrum. Besides those parts, the mechanism can include a screw and nut that are used for defining the value of force necessary for locking and unlocking. As can it be seen from [23] mechanism all parts are made from plastics except spring which is made from steel. Metamaterials properties open the possibility to replace a lot of different parts with new parts made from metamaterials such is spring in ski binding mechanism.

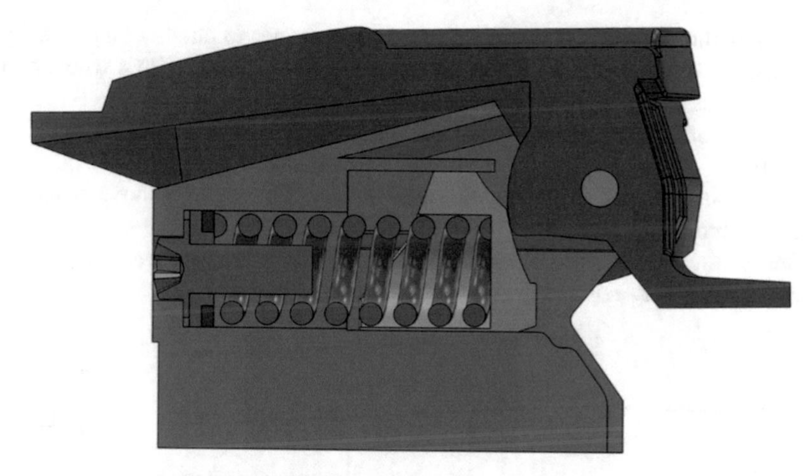

Fig. 9. Ski binding mechanism analyzed in [23]

From results of ski binding mechanism simulation and analysis of its movement it was determined that spring should compress at least 4 mm for a mechanism to work and should have 96 N/mm stiffness to work and analyzed mechanism form real ski. Outer diameter for spring is 18 mm and the uncompressed length was 50 mm. This spring is placed in housing with a chamber that has 20 × 20 mm crosssection. According to measured parameters new metamaterial spring needs to have features: 50 mm uncompressed length, 18 × 18 mm square base and it should be made from ABS, because it is most used plastics for 3D printing.

For metamaterial basic cell, rhombus shapes were chosen where rhombus was placed into 2 vertical columns which have 9 mm width. For testing purposes two 18 × 18 × 18 mm cubes were added, these cubes were used to attach necessary fixtures and forces for simulations. Because spring should move 4 mm within locking/unlocking action a minimal force, in the ideal case, that spring should produce is 384N. As in this case forces are one-directional different metamaterials geometries will be analyzed only in the direction of acting force. All modeling and simulations were done in SolidWorks 2018.

To optimize the metamaterial response for given parameters "design study" optimization tool from SolidWorks was used. The design study is a module/tool from SolidWorks software which is used for geometrical optimization of designed 3D models. This module is an upgrade on standard simulations and it allows to set up different geometrical parameters which will be changed until defined goals are met [24].

3.1 3D Modeling Workflow

The 3D model was created using the next workflow:

- The basic cell of metamaterials was modeled as a rhombus with filleted corners (R = 0.2 mm) to replicate real geometrical condition that can be achieved in manufacturing with 3D printers.

- Width of rhombus was 9 mm from one filleted corner to another and its height is chosen in such way that length of 50 mm should be covered with a whole number of rhombuses.
- These rhombuses represent a middle line of created metamaterial mesh (see Fig. 10).
- All sharp-edges was filleted with small radius 0.1 mm to remove singularities during simulation processes.
- After the creation of metamaterials two boxes 18 × 18 × 18 were added to create parallel faces between metamaterial should be pressed as spring.

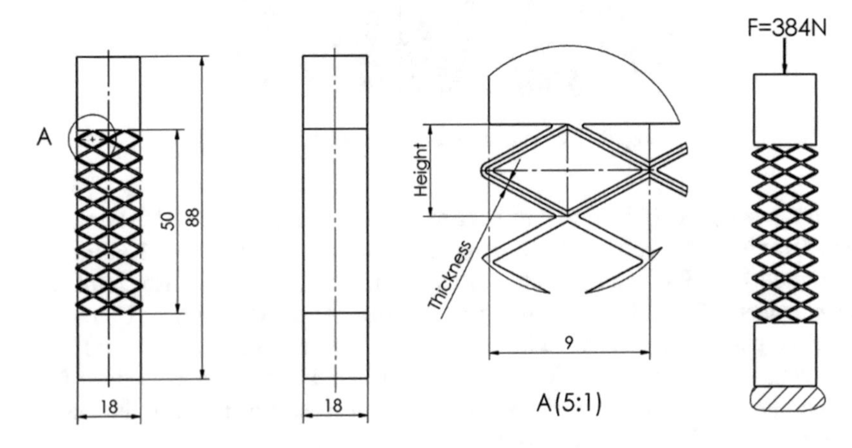

Fig. 10. Metamaterial size and shape (first two images), basic cell size (third image) and simulation set up (fourth image)

3.2 Simulation Workflow

Simulation is set up using the next workflow:

- Materials chosen for simulation was ABS with characteristics
  - Materials follow Hooke's law
  - Yield strength: 30 MPa
  - Tensile strength: 40 MPa
  - Elastic modulus 2000 MPa
  - Poisson's ration 0.3
  - Mass density 1020 kg/m3
- The model has planar geometry
- The model was fixed on the bottom (see Fig. 10)

- The used force was vertical and has value 384N because spring has a stiffness of 96 N/mm and it should be compressed 4 mm
- Mesh used in all simulations was "Standard mesh" with the characteristics (see Fig. 11):
  - Mesh type was planar 2D mesh
  - Maximum element size was 0.714 mm (Solidworks determines this value according to model geometry)
  - Minim element size was 0.0357 mm
  - The automatic transition was allowed
- To analyze different geometries design study was created with two changing parameters parameters
  - Basic cell thickness was changed between 0.5 and 1 mm in steps of 0.1 mm
  - Cell height was changed in such a way that cell numbers are increased in steps of 5 (10 cells, 16 cells, 20 cells and 25 cells) except for 16 cells because it is not possible to divide 50 mm with 15 cells and get cell size with a finite number of digits (see Fig. 12).
- As a result of each simulation two parameters was recorded
  - Maximal stress on the model
  - Maximal displacement to compare it with the length for which spring is compressed during locking/unlocking mechanism
- To calculate the resulting stress Von-Mises hypothesis was used.

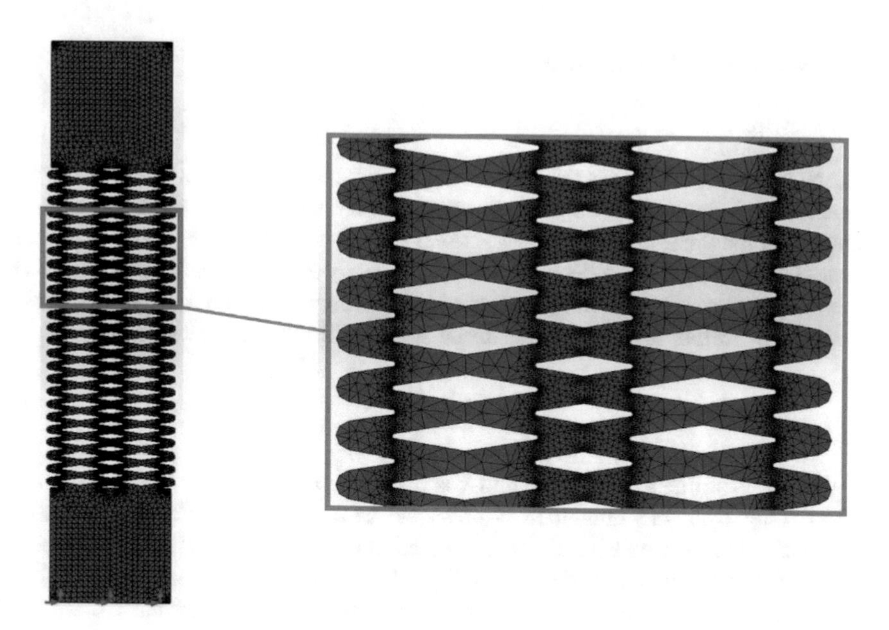

Fig. 11. Mesh used for simulations

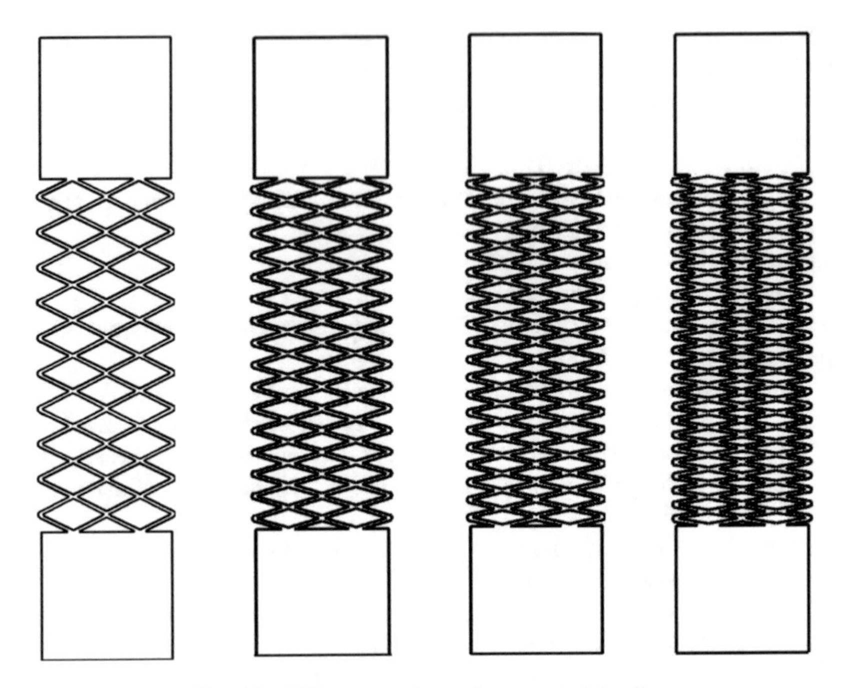

Fig. 12. Different numbers of metamaterial cells

4 Results and Discussion

4.1 Results

For each of four cell sizes six thicknesses analyzed which produces 24 simulations was created. For all simulations only two parameters were recorded: maximal stress on the material and maximal displacement. Maximal stress was followed to find is it possible to manufacture part from ABS material for a given load of 384N. To determine how much metamaterial is deformed under constant force maximal displacement were recorded. For further analysis metamaterial spring stiffness was calculated as:

$$k = \frac{F}{x} \tag{1}$$

Where k is stiffness, F is force and x is displacement.

The result from the simulation is presented as a graph, and where stresses, displacement and calculated stiffness are presented as a function of the number of cells of metamaterial in the direction of load (see Figs. 13, 14 and 15). On each graph there are multiple lines which represent results for different thicknesses of the basic cells.

4.2 Discussion

As expected different basic cells (size, number and shape) have a significant influence on metamaterial behavior which is can be compared with the result from the literature [16–18, 25]. Stress graphs show (see Fig. 13) that with an increase in cell density stress across the material decreases, this can be explained in two ways: first, stress is dependent on the density of metamaterial, as the density of metamaterial mesh increase the stiffness of structure also increase because of the volume of material which transfers load in inner structure is increased. This is supported by displacement graphs where the also maximal displacement of material are decreasing with increase with the number of cells, this conclusion can be confirmed with the result from [25] where is analyzed the impact of metamaterials thickness and cells size on overall metamaterial response on the constant force.

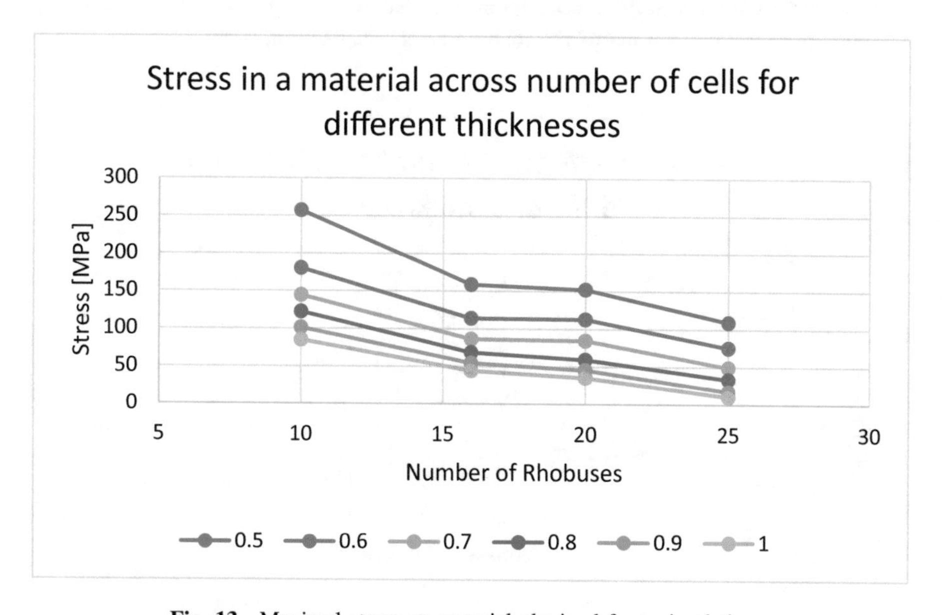

Fig. 13. Maximal stress on material obtained from simulations.

When diagram for stress (see Fig. 13) and displacement (see Fig. 14) are compared it was noticed that the change in stress between 10 and 16 basic cells is much higher than between 16 and 20 but if it is looked at the displacement diagram it can be noticed that the difference between those 3 cell numbers is almost linear. This result indicates that when the large displacement is divided into a lot of small-displacements across a large number of metamaterial cells, which can reduce deformation of the material and overall stress on inner structure. This means that large deformation of the whole structure is distributed to all metamaterials cell evenly. This is most visible on thinner cells (0.5 mm and 0.6 mm) as the thickness of cells increases the difference between stresses, as numbers of cells increases, converges to a linear relationship which confirms the first part of stress decrease. At 25 cells stress and displacement further decrease with higher slope (See Figs. 13 and 14) and this can be explained that metamaterial is dense enough that there is no more space to spread deformation across cells and most of the deformation is carried out by compressive deformation of ABS.

Finally stiffness of the structure is calculated according to Eq. (1) and the result is shown as graphs which shows exponential growth in material stiffness as the number and thickness of cells increase (see Fig. 15). Such behavior is a consequence of load is transfer through the structure of the metamaterial. A smaller number of cells and smaller thickness resulted that the deformation is more evenly spread across metamaterial structure (see Fig. 16) where the cell wall is flexed. As a number of cell increase and its thickness the deformation is concentrated in lines in such a way that ABS compression takes most of the load and not cell flection.

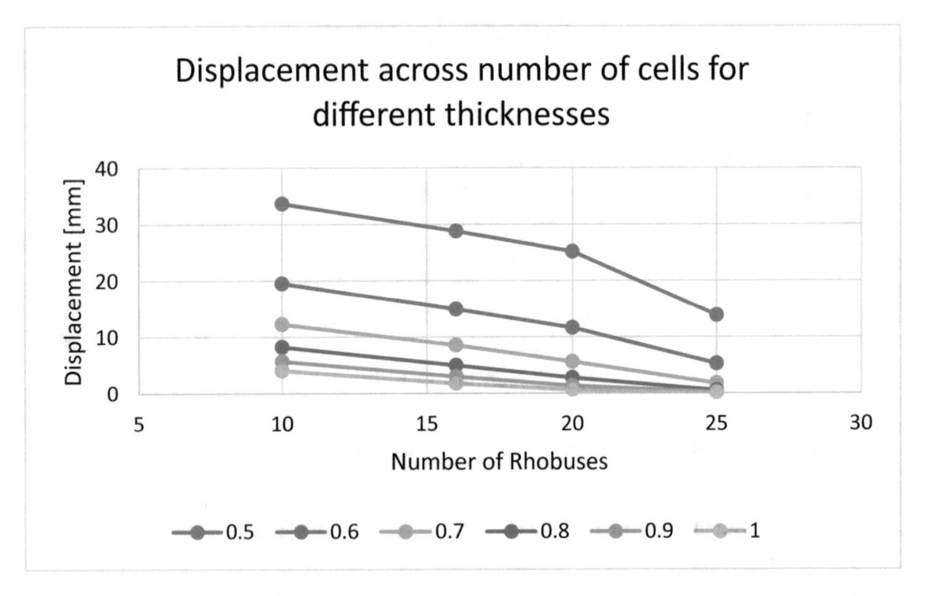

Fig. 14. Maximal displacement of the complete structure under the load of 384N for different cell numbers and thicknesses.

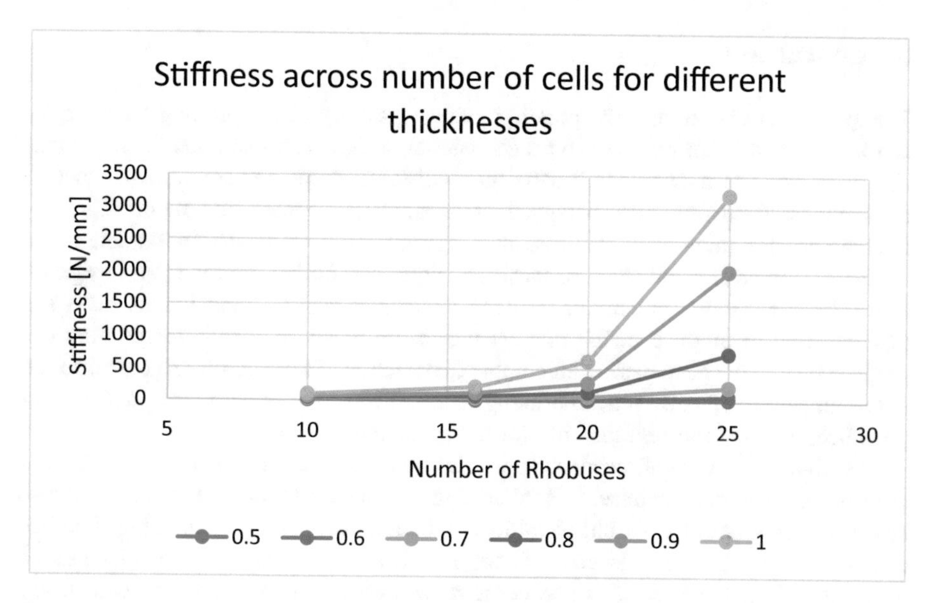

Fig. 15. Stiffness of whole structure calculated for the force of 384N and obtained displacement results from the simulation.

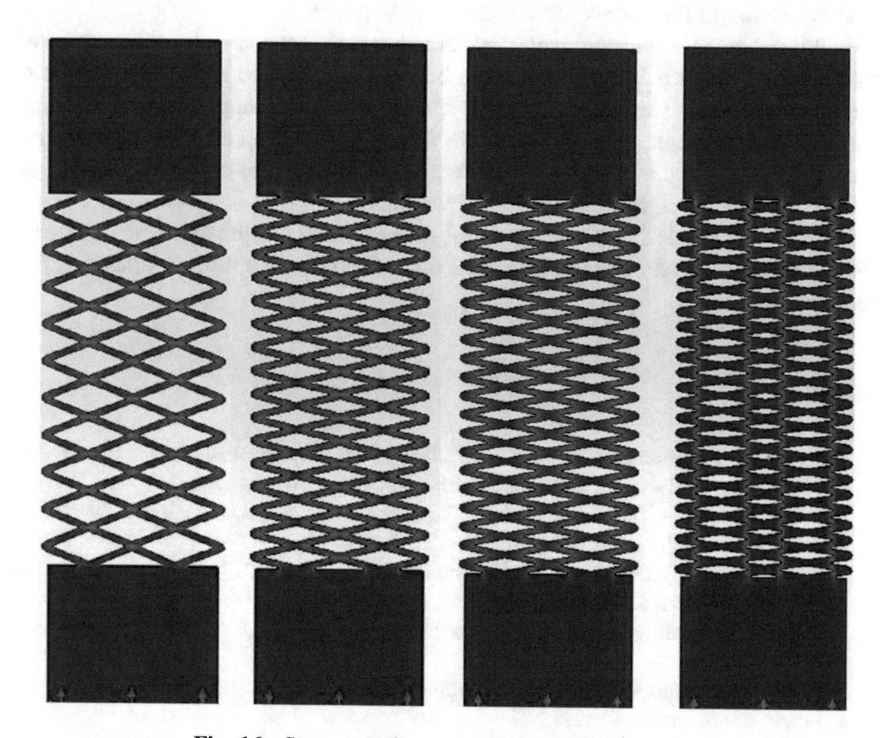

Fig. 16. Stress transfer across metamaterial structure.

5 Conclusion

This paper has shown how the popularity and versatility of 3D printing technologies can have a significant influence on metamaterial design. The main idea of metamaterials is to produce materials and parts that will fulfill numerous condition in specific applications. From engineer perspective metamaterials create one more degree of freedom in designing new devices, now special attention is drawn on inner part structure geometry. Application of these principles can further improve part design for a defined application, which can result in lower part cost by reducing its mass and material used to produce it. 3D printers which use plastic materials for printing can produce relatively cheap parts with complex geometries. Those complex geometries do not significantly influence the 3D printing process even complex parts can have a lower price because they use less material for the same application.

As obtained results showed metamaterial number of cells can have a significant influence on material response. It is interesting that metamaterials changed its response based on a number of cells but also this can be a consequence of changing rhombus diagonal length, but all this effect is negated with an increase in the thickness of metamaterial cells and when its number increase over a certain value. This conclusion opens possibilities to find some universal parameters which can be a starting point for metamaterial design. Such parameters can be surface or volume that metamaterials cover compared to the surface or volume of bulk material.

In this paper were analyzed is it possible to produce metamaterial spring which will replace spring in ski binding mechanism, but no one of tested metamaterials did not give a useful result. This indicates that simple geometrical shapes such are triangles squares octagons etc. may have limitations what can be achieved with them in metamaterial design, definitely more complex geometries as shown in [14, 17, 18] can produce metamaterials with completely new features.

Definitely smart design of metamaterial geometry can create parts with completely different properties and features from parts that are conventionally manufactured from same materials.

References

1. Krūgelis, L.: 3D printing technology as a method for discovering new creative opportunities for architecture and design. Landscape Archit. Art **13**(13), 87–94 (2018)
2. 3D Printing Infill – The Basics Simply Explained. https://all3dp.com/2/infill-3d-printing-what-it-means-and-how-to-use-it/. Accessed 21 June 2019
3. 2013 3D Printer Comparison Guide. https://newatlas.com/2013-3d-printer-comparison-guide/30187/. Accessed 21 June 2019
4. Bogue, R.: 3D printing: the dawn of a new era in manufacturing? Assembly Autom. **33**(4), 307–311 (2013)
5. Design better and faster with rapid prototyping. https://bigrep.com/applications/engineering-and-rapid-prototyping/. Accessed 21 June 2019
6. 3D Printing Market Size, Share & Trends Analysis Report by Component, by Printer Type, by Technology (FDM, Stereolithography), by Software, by Application, by Vertical, by Region, and Segment Forecasts, 2018–2025. https://www.grandviewresearch.com/industry-analysis/3d-printing-industry-analysis. Accessed 21 June 2019

7. Campbell, T., Williams, C., Ivanova, O., Garrett, B.: Could 3D Printing Change the World? Technologies, Potential, and Implications of Additive Manufacturing. Atlantic Council (2011)
8. All 10 Types of 3D Printing Technology in 2019. https://all3dp.com/1/types-of-3d-printers-3d-printing-technology/. Accessed 21 June 2019
9. Start your 3D printing lab in school – with a free Ultimaker! https://ultimaker.com/en/blog/25484-start-your-3d-printing-lab-in-school-with-a-free-ultimaker. Accessed 21 June 2019
10. D-Printed Prosthetics: Star Wars and Iron Man Inspire Bionic Hands. https://www.autodesk.com/redshift/3d-printed-prosthetics/. Accessed 21 June 2019
11. 3D Printed Prosthetics. https://www.forbes.com/sites/tjmccue/2014/08/31/3d-printed-prosthetics/#78b160833b45. Accessed 21 June 2019
12. Arc Bicycle: the world's first 3D printed steel bicycle frame. https://qeprize.org/createthefuture/arc-bicycle-worlds-first-3d-printed-steel-bicycle-frame/. Accessed 21 June 2019
13. 3D Printed Electronics Created on the DragonFly Pro System. https://www.nano-di.com/image-gallery-of-3d-printing-on-the-dragonfly-pro. Accessed 21 June 2019
14. Padilla, W., Basov, D., Smith, D.: Negative refractive index metamaterials. Mater. Today **9** (8), 28–35 (2006)
15. Ion, A., Frohnhofen, J., Wall, L., Kovacs, R., Alistar, M., Lindsay, J., Lopes, P., Chen, H.T., Baudisch, P.: Metamaterial mechanisms. In: 29th Annual Symposium on User Interface Software and Technology, Tokyo, Japan, 16–19 October 2016
16. Bodaghi, M., Damanpack, A.R., Hu, G.F., Liao, W.H.: Large deformations of soft metamaterials fabricated by 3D printing. Mater. Des. **131**, 81–91 (2017)
17. Ren, X., Shen, J., Tran, P., Ngo, T.D., Min Xie, Y.: Design and characterization of a tuneable 3D buckling-induced auxetic metamaterial. Mater. Des. **139**, 336–342 (2018)
18. Bonatti, C., Mohr, D.: Large deformation response of additively-manufactured FCC metamaterials: From octet truss lattices towards continuous shell mesostructures. Int. J. Plast. **92**, 122–147 (2017)
19. Bugaric, U., Petrovic, D., Jeli, Z., Petrović, D.: Optimal utilization of the terminal for bulk cargo unloading. Simul. Trans. Soc. Model. Simul. Int. **88**, 1508–1521 (2012)
20. Popkonstantiovic, B., Miladinovic, L.j., Obradovic, R., Jeli, Z., Stojicevic, M.: The Eclipses Abacus, the mechanical predictor of the solar and lunar eclipses. Simul. Trans. Soc. Model. Simul. Int. **95**(6), 499–507 (2018)
21. Stojićević, M., Stoimenov, M., Jeli, Z.: A bipedal mechanical walker with balancing mechanism. Tehnicki Vjesnik (Technical Gazette) **25**(1), 118–124 (2018)
22. Popkonstantiovic, B., Obradovic, R., Obradovic, M., Jeli, Z., Stojicevic, M.: Geometrical and mechanical characteristics of deformed balance spring obtained by simulation study. Simul. Trans. Soc. Model. Simul. Int. **92**(11), 981–997 (2018)
23. Jeli, Z., Kosic, B., Stojicevic, M., Berdic, S.: 3D modeling and analysis of ski binding mechanism. In: Uhl, T. (ed.) Advances in Mechanism and Machine Science, pp. 579–587. Springer, Cham (2019)
24. Kosic, B., Jeli, Z., Stoicevic, M., Popkonstantinovic, B., Dragicevic, A.: Metamaterial usage in design of bilateral prothetic legs. In: 8th International Conference of Engineering Graphics and Desing, Craiova, Romania, 15–17 May 2019
25. Kosic, B., Stoicevic, M., Jeli, Z., Popkonstantinovic, B., Duta, A., Dragicevic, A.: 3D analysis of different materials geometry and simulation of metamaterial usage. FME Trans. **47**, 349–354 (2019)

Manufacturing Process and Thermal Stability of Nanophotonic Soft Contact Lenses

Aleksandra Mitrovic[1](✉), Dragomir Stamenkovic[2], Dejana Popovic[3], and Aleksandra Dragicevic[4]

[1] University Union "Nikola Tesla", Faculty of Information Technology and Engineering, 11000 Belgrade, Serbia
aleksandramitrovic1926@gmail.com
[2] University of Belgrade, Faculty for Special Education and Rehabilitation, Belgrade, Serbia
[3] University of Belgrade, Vinča Institute of Nuclear Sciences, Belgrade, Serbia
[4] University of Belgrade, Faculty of Mechanical Engineering, Belgrade, Serbia

Abstract. Hydrogels have peculiar physical and chemical properties and therefore, are used in a variety of biomedical applications including drug delivery agents, prosthetic devices, the repair and replacement of soft tissues, contact lenses, etc. Consequently, investigation of mechanical, physical and chemical properties is crucial in biomedical application of hydrogels. Poly (2-hydroxyethyl methacrylate) (pHEMA), as a biocompatible hydrogel, was first hydrogel used for making soft contact lenses. Many researches have been modified pHEMA with the aim of improving its properties. Application of nanotechnology is one of the possible solutions for improving the characteristics of this biocompatible hydrogel. In this paper, polyhydroxyethyl methacrylate was used as standard material for soft contact lenses (SL 38). This material was incorporated with fullerene C_{60} (SL38-A), fullerol $C_{60}(OH)_{24}$ (SL 38-B) and fullerene metformin hydroxylate $C_{60}(OH)_{12}(OC_4N_5H_{10})_{12}$ (SL 38-C), respectively. Three new nanophotonic soft contact lenses were made. The main goal of this research was to develop appropriate process parameters for soft contact lens micro-turning. Also, studying the thermal decomposition of standard soft contact lens, pHEMA, as well as three new nanophotonic soft contact lenses was one of the main objectives. Results have shown that manufacturing process of nanofotonic soft contact lens is considered to be a micro-turning process regarding the cutting depth and tool nose ratio. Thermal stability of all three nanofotonic soft contact lenses was significantly improved comparing to the standard soft contact lens. Still, further research needs to be done so these nonophotonic soft contact lenses could find practical application in the field of biomedical engineering.

Keywords: Nanophotonic soft contact lenses · pHEMA · Hydrogel · Fullerenes · Production process · Thermal analysis

N. Mitrovic et al. (Eds.): CNNTech 2019, LNNS 90, pp. 184–199, 2020.
https://doi.org/10.1007/978-3-030-30853-7_11

1 Introduction

Hydrogels belong to the class of water swollen three-dimensional polymeric networks in which hydrophilic macromolecular chains are chemically and/or physically cross-linked. It is generally known that hydrogels are soft, flexible and with low surface friction values. Hydrogels could be biocompatible due to inherent hydrophilic nature [1, 2]. These type of polymers possess pelicular physical and chemical properties that make them useful for various applications [3–7]. Therefore, hydrogels are used in a variety of biomedical applications including drug delivery agents, prosthetic devices, the repair and replacement of soft tissues and contact lenses [1, 8–16].

The orientation of hydrophobic and hydrophilic parts of the macromolecules presents one of the most important influences on hydrogels surface characteristics [1]. It is known that amorphous regions are hydrophilic but crystalline regions are hydrophobic. Crystallinity has positively affect on the mechanical properties and solubility of polymers. Characteristics of polymers are influenced by hydrophobic and hydrophilic composition. The polymer hydrogels have good dissolution resistance in the body. This process will happen only if the polymer hydrogels overcome the heat and entropy factors associated with the crystalline regions during dissolution [2, 3, 17]. Investigation of swelling, mechanical and thermal properties but also morphology is crucial in biomedical application of hydrogels. Further, it is outstanding to carry out a great number of in vitro and in vivo studies in accordance with specific biomedical applications of such materials [18–20].

As it is mentioned, contact lenses are one of the most important applications of hydrogels. A revolution in the world of contact lenses occurred when it became possible to produce thinner lenses with high oxygen transmission. Soft contact lenses (SCLs) are made from hydrophilic polymers which easily and quickly absorb water to the stage of equilibrium. Degree of swelling for SCLs is defined by the external stimuli such as temperature, pressure, pH and etc. The appearance of such materials has initiated a number of research and development of physiologically more perfect soft contact lens materials [3, 21–28].

Poly(2-hydroxyethyl methacrylate) (pHEMA), as a biocompatible hydrogel, was used first for making soft contact lens. HEMA-derived hydrogels present around 22% of the total soft contact lens market [29]. Likewise, 2-hydroxyethyl methacrylate continues to be the most commonly used hydrophilic monomer for SCLs [10]. In the case of pHEMA, the -methyl group is hydrophobic whereas the hydroxyethyl group is hydrophilic. Several different surface studies have concluded that the pHEMA gel–air interface is hydrophobic [1]. The pore size and morphology of the pHEMA hydrogels is controlled by the water content in the initial polymerization mixture. Since a water droplet was placed on the pHEMA hydrogel surface, noticeable contact angle hysteresis has appeared. The soft contact lens biocompatibility characteristics are affected by the surface wettability properties of the hydrogel due to the contact between the soft contact lenses and the tear film [4, 8]. These hydrogels were composed of hydrophilic monomers that allowed interaction with water due to electrochemical polarity. Within the polymer network, these hydrogels could hold a high percentage of water. The hydrogels were also a flexible and oxygen-permeable class of material. All these factors have improved the comfort, wear time and oxygen permeability of soft contact lenses. Therefore, many SCL derivatives based on HEMA was synthetized. Development increased, since the oxygen permeability of HEMA hydrogels wasn't enough for extended SCL wear.

Soft contact lenses progress is significant today, however more requirement is demanded. Contact lenses ameliorated the quality of patients life, with two most common uses today - cosmetic reasons or for corrective vision in place of traditional spectacles [30]. Number of people developing various eye problems such as glaucoma, myopia and other eye conditions is increasing. Therefore, more effective treatments are required and new and improved soft contact lens materials are necessary [31]. Numerous sources investigate the hydrogel synthesis, bioavailability, eye-fitting, manufacturing [5–7, 9–16].

The hydrogel mechanical properties can be improved by using a cross-linking molecule. Ethylene glycol dimethacrylate (EGDMA) is s one of the most commonly used cross-linker and has two functional groups that allow the covalent bonds formation between two individual polymer chains. Hence, the polymer mass significantly increases and its ability to form a gel network improves. It is known that the crosslinks reduce the polymer-chain motion and affect swelling and oxygen transport [32]. For designing a soft contact lens for a particular application, balance must be achieved between these parameters.

To achieve better characteristics of SCL materials, it is crucial to understand the hydrogel properties. Depending on the patient's requirements, the properties of HEMA-based hydrogels may vary [33]. Tranoudis et al. showed that high water content hydrogels didn't always influence on poor mechanical properties. However, different intermolecular forces based on different chemical compositions complicate the comparing between the different polymer systems. Numerous polymer properties regarding structure could disclose more factors affecting the hydrogel mechanical properties. At the moment, HEMA-hydrogel structure modification is in progress regarding new lenses, innovations and applications.

Many researches have been modifying pHEMA to improve its properties [9, 10, 18, 28] and numerous studies tried to develop new soft contact lenses with better physical, chemical and mechanical properties [11–16].

The basic hydrogel structure has not changed due to the main properties important to contact lenses (e.g. oxygen permeability, water content, wettability, etc.). Modern HEMA hydrogel lenses have improved through hydrogel structure modification using different techniques (e.g. encapsulation, grafting). The modification materials were different - nanoparticles, anti-microbial agents and surfactants. A product based on HEMA hydrogel lens incorporated with silver nanoparticles (AgNPs) has been on the UK market. Other researchers studied the impact of HEMA hydrogel contact lens composition on AgNP properties [34]. Jung et al. studied common contact materials due to the ability to modify conformation with nanoparticle incorporation into the contact lens [35].

Having this in mind, it is important to develop new materials that would enhance soft contact lens properties, especially after manufacturing its surface. The SCL surfaces quality, both base and frontal is very important since it is placed directly on the eye surface.

Application of nanotechnology is one of the possible solutions to improve material characteristics. Nanotechnology includes imaging, measuring and modeling at scales that previously couldn't be approached. As a field of interest, nanotechnology includes production, characterization, the design and application of structures, devices and systems by controlling shape and size at nano scale [36]. The interest for material development of polymer-nanoparticles/nanocomposites focusing on structure-property relationships is increasing. Also, polymer-nanocomposites durability and degradability are crucial areas of research for polymer industry.

So far, many researchers have been investigating fullerene incorporation to the polymer structure [9, 10, 36–40]. Fullerene C_{60} and its derivatives can be used in contact lens material technology as they show strong electron affinity and can easily enter nucleophiles reactions. Powdered fullerene is a good optical limiter but has low water solubility and poor transparency.

Fullerenes mixed with polymers present a group of new optical filters with good properties (e.g. easy fabrication, predictable wavelength tuning, excellent performance stability etc.). Low water solubility is one of the main disadvantages of C_{60}, and for that, fullerenes are functionalized with polar groups (–OH and –COOH) [37, 40]. Fullerene enables better electromagnetic properties of transmitted light [39] that is convenient to the eye and optic nerve. SCLs with incorporated fullerene would have multiple potential uses.

The final step in a soft contact lens development is the method of manufacturing the new material. SCL manufacturing is an evolving field, with constantly improved manufacturing methods. The specific SCL manufacturing methodologies include operating conditions, speed of lathes, tip-edge parameters, and exact chemical recipes, amongst a huge number of parameters. For researchers, the lack of this information leads to difficulties in the evaluation process. However researchers can still analyze the products to help shape their research direction. There are concerns about the mechanical properties differences in the common HEMA-derived hydrogel lenses based on the manufacturing technique [41]. The polymerization and manufacturing method can cause the differences in the hydrogel properties. Polymerization steps during manufacturing affect the soft contact lens materials properties and therefore could improve with proper monitoring. The manufacturing- or process-driven development is likely implemented for economical or environmental reasons, or both.

Materials are an important consideration for both cast- and spin-molding processes. The mold surface and polymer solution interaction will affect the SCL surface finish. Ultra-precision and nano-manufacturing is a research area that could be applied to SCLs to enhance the material physical and optical properties [42–44]. These techniques had successful implementation in a number of areas, including biomedical device manufacturing, optics, and more [45]. The control of wettability can be achieved by using machining techniques [46]. Finally, there is a manufacturing cost that would also needs to be considered.

From all of the above mentioned, it should be emphasized that physical and mechanical properties of the SCL material could be changed after the fullerene incorporation. Also, it should be emphasized that physical and mechanical properties of the material for SCL could be changed after the incorporation of fullerenes.

In this paper, polyhydroxyethil methacrylate was used as standard material for soft contact lenses (SL 38). This material was incorporated with fullerene C_{60} (SL38-A), fullerol $C_{60}(OH)_{24}$ (SL 38-B) and fullerene metformin hydroxylate $C_{60}(OH)_{12}(OC_4N_5H_{10})_{12}$ (SL 38-C) respectively. Three new nanophotonic materials for soft contact lenses were obtained.

The aim of this study was to develop appropriate lathing process i.e. process parameters for soft contact lens micro-turning. Also, the scope of this work consisted in studying the thermal decomposition of standard soft contact lens, pHEMA, likewise three new nanophotonic soft contact lenses.

2 Materials and Methods

The materials used in this research have been provide as the result of the contract of scientific, technical and business cooperation between companies Optix (Belgrade, Serbia) and Soleko (Milano, Italy) with the University of Belgrade, Faculty of Mechanical Engineering. Polymerization of the new nanophotonical materials for soft contact lenses had performed according to the technology production laboratories at Soleko company (Milan, Italy).

Soft contact lenses are mostly produced using free-radical polymerization. This process is facile. Catalysts and other expensive reagents are not neccessary. Likewise, unwanted/unreacted chemicals can be removed from the material on post-fabrication cleaning processes [46].

The basic and nanophotonic materials for soft contact lenses were obtained using radical polymerization of 2-hydroxyethyl methacrylate and fullerene, fullerol and fullerene metformin hydroxylate, respectively. Derivatives of fullerene C_{60}, samples SL38-B and SL38-C, were previously dissolved in water. In the polymerization reaction, 2-hydroxyethyl methacrylate, HEMA, (Sigma Aldrich, $\geq 99\%$) was used as a monomer, ethylene glycol dimethacrylate (Sigma-Aldrich, 98%) was used as the cross-linking agent, fullerene C_{60} (MER Corporation, USA, $\geq 99\%$), and as an initiator benzoyl peroxide (Sigma-Aldrich, 75%) was used. From the obtained materials, nanophotonic soft contact lenses were made in the company Optix (Belgrade, Serbia).

2.1 Manufacturing of the Soft Contact Lenses

There are three methods for manufacturing soft contact lenses: cast-molding, lathe cutting or spin casting. Tested SCLs, SL38 as basic and SL38-A, SL38-B and SL38-C as nanophotonic SCLs were manufactured by lathe-cutting method. This process uses a special lathe for cutting an anhydrous block of material into the required shape. To obtain a soft contact lens, it was necessary to hydrate this intermediate.

As mentioned, soft contact lenses and nanophotonic soft contact lenses were synthesized using polymerization. Semifinished products were obtained in the process that was used to produce soft contact lenses. The term semifinished product implies working process on the frontal surface of the soft contact lens. Base surface that was produced using casting was already processed. Semifinished products represented casting.

Casting was placed on auxiliary accessories in the form of a acron i.e. casting beam and they make preliminary part. This part was placed in the feed rod. The machining parameters were set in software and cutting i.e. turning of the casting was done on CNC lathe ((Toric Lamda Polytech, LTD, UK). Figure 1 shows the turning of the working part of frontal surface for the soft contact lens. The soft contact lens had a moulded basic surface of the working part in an air operated collet (Fig. 1b).

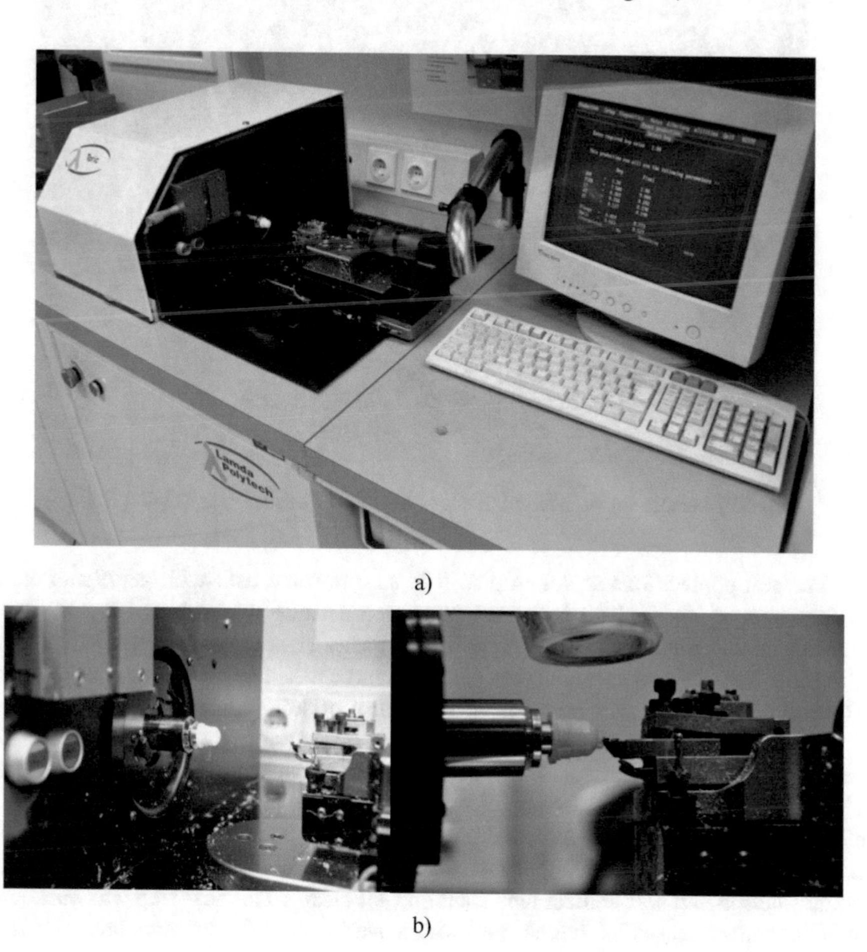

a)

b)

Fig. 1. (a) CNC lathe; (b) Turning of the casting (Optix, Zemun)

Working piece was sent to the polishing of the frontal surface. Polishing was done using the polishing paste for the soft contact lenses (CONTAPOL 1, Contamac, UK). The workpiece was put into a machine for removing lens from the casting (Lamda, UK). Thickness of the workpiece was measured from the center, CT (central thickness) with comparator (Messzenge, Germany), after the product was removed from the machine for removing lens from the casting. Furthermore, polishing the edges of the soft contact lens (Fig. 2) was done on the machine for polishing edges (Nissen, UK).

Rotational speed during the polishing process was n = 100 rev/min. The polishing paste contained aluminum oxide particles (Al_2O_3 grains) sized from 0.3 μm to 0.5 μm. The duration of polishing was within the interval 5 s–20 s.

Fig. 2. Machine for polishing the edges of the soft contact lens (Optix, Zemun)

The soft contact lens was then placed in an ultrasonic bath with petroleum on the temperature of 70 °C. The increased temperature value was selected due to easier cleaning of the contact lens from the polishing paste. The contact lens was placed in an ultrasonic bath for two seconds, then removed and cleaned using the medical gasoline. The magnifying glass was used for visual control of soft contact lenses.

Next step was hydration of the soft contact lens in a buffer solution for 24 h. The composition of the buffer solution included: sodium chloride BP or USP 8.5 g, boric acid BP or USP 1.0 g and Borax (di-sodium tetra borate) BP or USP 0.1 g in 1 l of distilled or deionized water. Acceptable pH value of the solution was from 7.20 to 7.40. After the hydration process, the soft contact lenses were cleaned with buffer solution, saline solution and soft contact lens solution (All clean). The next step was packaging of soft contact lenses in bottles and steam sterilization in an autoclave (HALLO SEKURIKLAV, Germany) at 121 °C, for 55 min.

2.2 Termogravimetric Analysis

Physical property of a substance is determined as a function of temperature using thermal analysis. This refers to the sample that is subjected to a controlled temperature program.

This group of techniques can measure changes of mass or changes in energy of a sample of a substance. Thermal analysis techniques have different applications: determination of changes in chemical composition, determination of purity, determination of phase changes, etc. [47].

Samples SL38, SL38-A, SL38-B and SL38 -C were tested using termogravimetric and differential thermal analysis. Simultaneous TG/DTA analysis was performed on an STD Q600 instrument (TA Instruments, USA) in the range from room temperature to 600 °C, in the platinum pots. The heating rate was 20 °C min^{-1}, and the weight of the samples was between 5 and 10 mg. The atmosphere in the furnace was ultrapure nitrogen, and the gas flow through the furnace was 100 cm^3 min^{-1} in all experiments.

3 Results and Discussion

The contact lens manufacturing process is considered to be a micro-turning process regarding the cutting depth and tool nose ratio. The micro turning process parameters are shown in Table 1 for the SCL base surface and in Table 2 for the SCL frontal surface.

Table 1. Process parameters for the soft contact lens - base surface

Soft contact lens – turning parameters of the base surface	
Number of revolutions of the main spindle – rotational speed [rev/min]	8500
Speed angle of the accessory motion of the rough turning – angular feed rate for rough turning [degrees/s]	7.000
Speed angle of the accessory motion of the fine turning – angular feed rate for fine turning [degrees/s]	4.500
Cutting depth for rough turning – nose radius for rough turning [mm]	0.400
Cutting depth for fine turning – nose radius for fine turning [mm]	0.200

Table 2. Process parameters for the soft contact lens - frontal surface

Soft contact lens – turning parameters of the frontal surface	
Number of revolutions of the main spindle – rotational speed [rev/min]	8000
Speed angle of the accessory motion of the rough turning – angular feed rate for rough turning [degrees/s]	8.000
Speed angle of the accessory motion of the fine turning turning – angular feed rate for fine turning [degrees/s]	4.000
Cutting depth for rough turning – nose radius for rough turning [mm]	1.000
Cutting depth for fine turning – nose radius for fine turning [mm]	0.050

Since the surface molecules and their orientation affect the final stage of surface quality, conformation states of polymers forming soft contact lens surface were transformed during the last step of the production [46]. This process could present a complex issue. It is important to emphasize that choosing the optimal machining parameters significantly influences surface quality of soft contact lenses and optical, medical and patient-comfort functionality. This paper should contribute to studies concerning these issues.

HEMA belongs to group of a high-water content, oxygen-permeable material. Hydrogel composed of only HEMA as monomer contains around 38% water. Water content of typically hydrogels is 20–80% depending on the comonomers. HEMA's contact lenses usually have suitable wetting properties due to highly polar properties. This means that SCLs are comfortable [46]. The oxygen permeability of hydrogels is property that makes this materials convenient for longer wear [48].

In our previous work [10] it was demonstrated that AFM measurements evaluated surface quality of new nanpohotonic soft contact lenses. AFM as nondestructive method, enable testing on the same sample by varying parameters. It was showed that SL38-C exhibited a more acceptable performance compared to SL38-A and SL38-B and base material SL-38. It is known that machined surface quality is determined by its roughness. From obtained results it was concluded that incorporation of fullerene's derivate into comercial SCL improved the surface roughness of new nanophotonic soft contact lenses.

All four samples for soft contact lenses: SL-38, SL-38A, SL-38-B and SL-38-C were tested using thermogravimetric (TG) and differential thermal analysis (DTA). Before thermogravimetric and differential thermal analysis, samples were kept at room temperature in desiccator in the standard boxes for soft contact lenses.

Thermogravimetric analysis is the most used thermal decomposition test. The procedure includes the following: the sample is brought up to the desired temperature and sample weight is monitored during the thermal decomposition process. It is common to expose the sample to a linearly increasing temperature at a predetermined rate of temperature rise since in practice, it is impossible to bring the sample up to the desired temperature before significant thermal decomposition occurs.

Regarding the differential thermal analysis, a sample and a reference inert material were exposed to the same linear temperature program. Sample and reference inert material had approximately the same heat capacity. Their temperatures were measured and compared. The temperature of the sample lagged behind the reference material so the thermal decomposition of the sample was endothermic.

Results of TG/DTA analysis of the tested samples SL38, SL38-A, SL38-B and SL38-C in nitrogen atmosphere are presented in Figs. 3, 4, 5 and 6, respectively. According to the TG and DTA curves the mass loss less than 4.00% in the interval from room temperature until 180 °C was noticed: 3.57% to 177 °C (SL38), 1.62% to 170 °C (SL38-A), 2.20% to 130 °C (SL38-B) and 1.85% to 144 °C (SL38-C). The mass loss can be explained through the water loss due to relatively poor connection in all samples, since dehydration process started at room temperature and ends at maximum of 177 °C. Dehydration process was followed by the expected endothermic effects. Characteristic values of endothermic DTA peaks were: 85, 112, 76 and 76 °C for SL38, SL38-A, SL38-B i SL38-C, respectively (Figs. 3, 4, 5 and 6).

After dehydration, the samples start degradation process. In this phase the mass lost is about 98% and TG-curves have steep slope. Complete decomposition of the all samples occurs bellow 500 °C and final temperatures of decomposition are marked in

Figs. 3, 4, 5 and 6. In all samples endothermic process of degradation was observed at DTA-curves at a very similar temperatures, in the range of 392–401 °C.

It can be considered that all tested samples were completely decomposed at about 500 °C and rest mass for all samples was 0% (regardless to some TG-curves with negative value of the rest mass, it is considered to be in the area of permitted deviation for derived measurements).

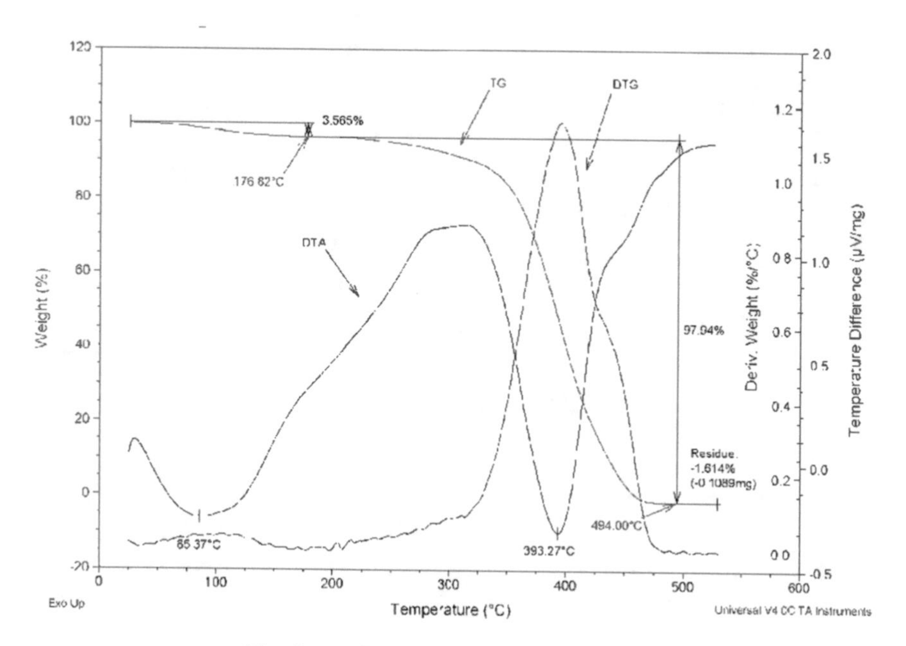

Fig. 3. TD/DTA curve of the sample SL38

Comparison of TG-curves for all tested samples is shown in Fig. 7. Based on the comparison, it can be seen that TG curves had a very uniformly course and showed almost the same weight loss at similar temperatures. Shaperd DTA curves also conformed this conclusion since two endothermic processes (dehydratation and decomposition) were carried out in all samples during the thermal decomposiotion, as it is shown in Figs. 3, 4, 5 and 6. Considering the different composition of the soft contact lenses samples, it should be noted that although small, these differences in the TG-curves are clearly visible and probably have significant infuence for the choice of materials of desired properties:

Samples SL38 and SL38-A showed almost the same thermal behavior. It should be noted that sample SL38 contained a higher moisture content compared to the sample SL38-A. SL38 had the highest mass fraction of water (3.57 wt%) compared to the all other samples. After the dehydration process, sample SL38-B was rapidly decomposed. This fact indicates that the sample B-SL38 was least stable since it decomposed at the lowest temperature unlike the temperature of decomposition of the rest anhydrous samples.

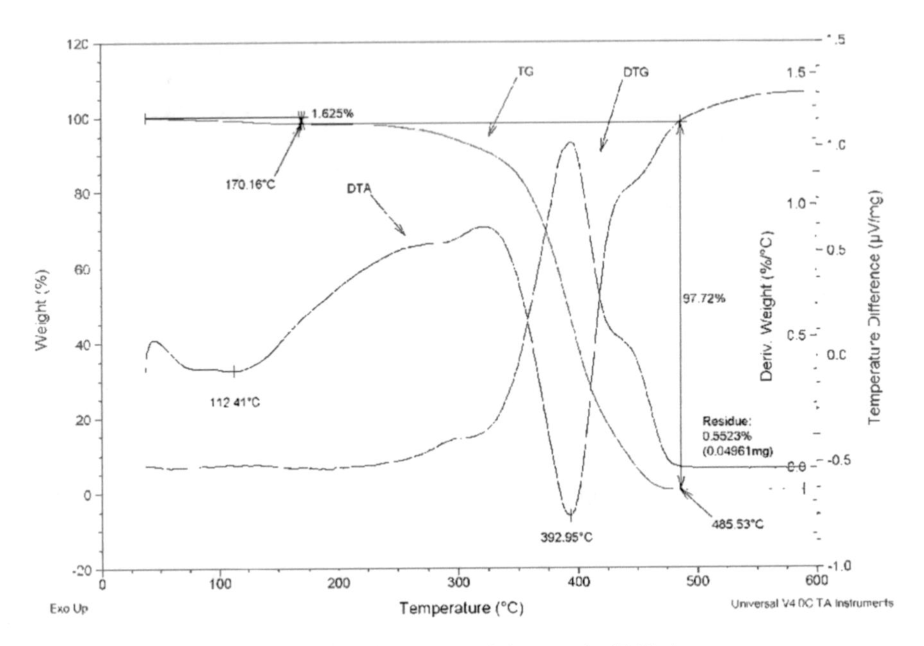

Fig. 4. TG/DTA curve of the sample SL38-A

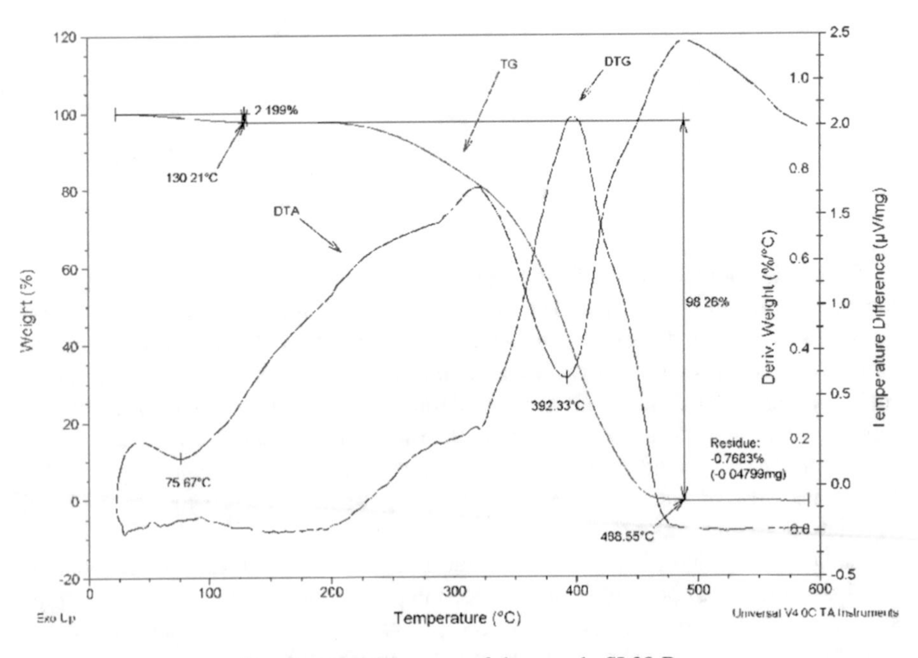

Fig. 5. TG/DTA curve of the sample SL38-B

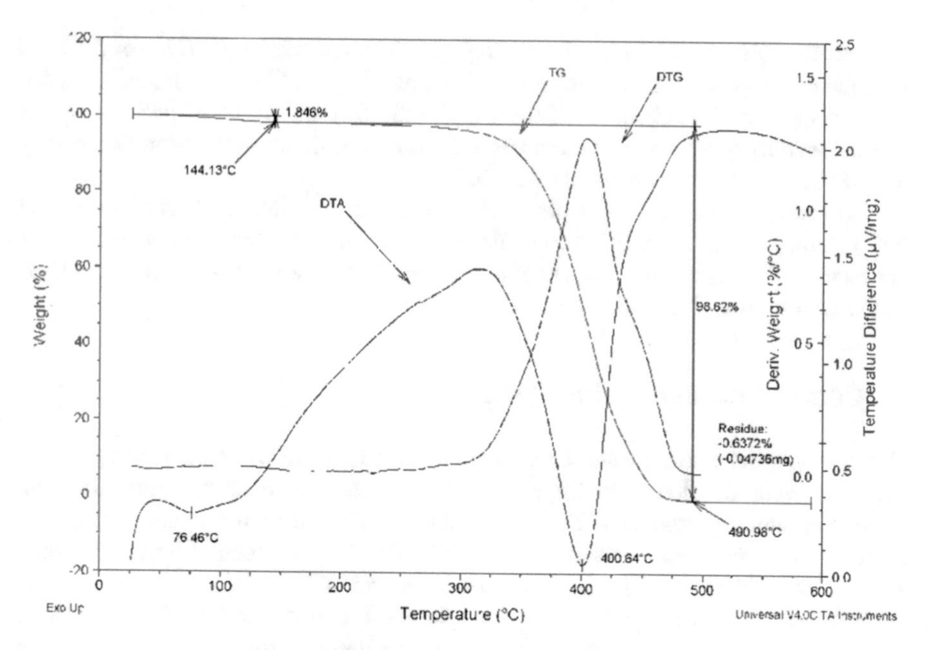

Fig. 6. TG/DTA curve of the sample SL38-C

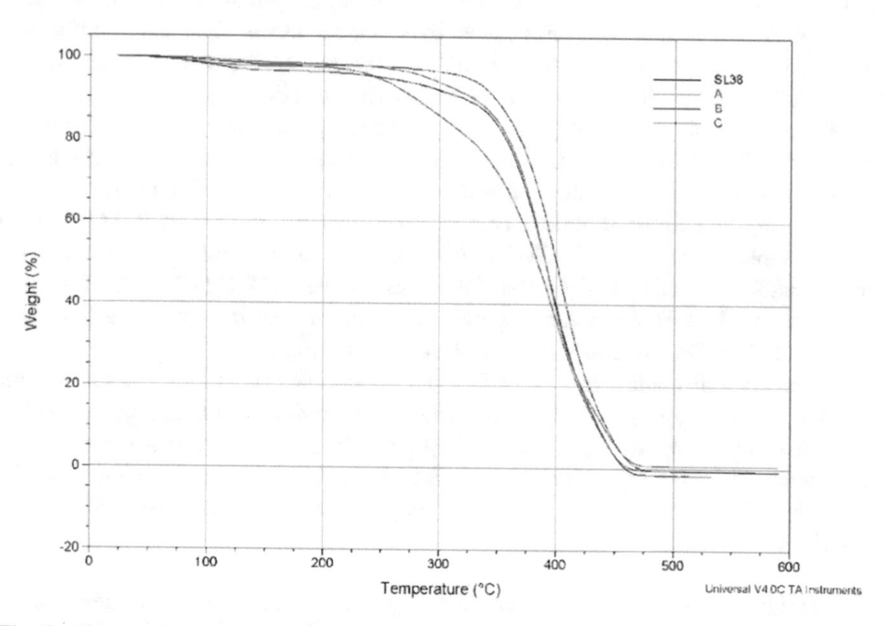

Fig. 7. Comparation of the results of TG analysis of samples SL38, SL38-A, SL38-B and SL38-C.

Decomposition temperature of the sample SL38-B was about 100 °C lower (150 °C) than the decomposition temperature of the sample SL38-C (237 °C). Relatively horizontal plateau on the TG-curves of the samples SL38-A and SL38 –C indicated better thermal stability compared to the samples SL38 and SL38-B in a temperature range of up to 250 °C. The most stable sample was SL38-C.

TG-curve of anhydrous sample SL38-C showed the highest weight loss (about 99%) during its decomposition, and therefore the greatest endothermic effect was presented in the degradation process of the sample. This endothermic effect was noticed as the largest area of the DTA-peak (Fig. 7).

4 Conclusions and Future Direction

New nanophotonic soft contact lens materials based on HEMA monomer and nanomaterials were developed to improve surface roughness after recommended and existing machining processes. Fullerene, fullerol and methformin hydroxylated fullerene) were incorporated into commercial material for soft contact lenses. Thermal stability analysis was also performed using TGA and DTA.

Furthermore, this research should contribute to the biomedical field, especially in the development of superior materials and also to improve soft contact lens existing manufacturing process. New manufacturing processes are one of the crucial factors for improving the material specifications. This is encouraging progress for designing new, improved materials and will only keep evolving as novel chemical modification techniques and materials. The application of new manufacturing methods can provide other solutions to specific challenges for soft contact lenses.

The methods of thermal analysis showed nearly identical thermal behavior of the samples SL38 and SL38-A. SL38-B sample had fastest degradation process at the lowest temperature (150 °C) after dehydration unlike the decomposition temperature of the remaining anhydrous samples. This indicated that the sample SL38-B was least stable. Samples SL38-A and SL38-C have a higher thermal stability compared to the samples SL38 and SL38-B. The most stable was sample SL38-C (237 °C). This result confirmed that two of three nanophotonic soft contact lens materials had improved thermal stability compared to commercial material for SCL.

In general, soft contact lens materials can be modified, thus they are suitable for improving manufacturing processes. Advances in precision manufacturing or modification of surface properties can help in improving physico-chemical and mechanical characteristic. In summary, future soft contact lens materials will continue to push the boundaries of materials sciences in order to adapt to the needs of a growing SCL-using population.

Acknowledgment. The authors are grateful to Optix (Belgrade, Serbia) for providing the material used in this study. This research was supported by Ministry of Education, Science and Technological Development of Republic of Serbia under Projects III45009 and TR35031.

References

1. Ketelson, H.A., Meadows, D.L., Stone, R.P.: Dynamic wettability properties of a soft contact lens hydrogel. Colloids Surf. B **40**, 1–9 (2005)
2. Peppas, N.A., Huang, Y., Torres-Lugo, M., Ward, J.H., Zhang, J.: Physicochemical foundations and structural design of hydrogels in medicine and biology. Annu. Rev. Biomed. Eng. **2**, 9–29 (2000)
3. Peppas, N.A., Bures, P., Leobandung, W., Ichikawa, H.: Hydrogels in pharmaceutical formulations. Eur. J. Pharm. Biopharm. **50**, 27–46 (2000)
4. Tranoudis, I., Efron, N.: Water properties of soft contact lens materials. Eye Contact Lens **27**, 193–208 (2004)
5. Mitrović, A., Munćan, J., Hut, I., Pelemiš, S., Čolić, K., Matija, L.: Polymeric biomaterials based on polylactide, chitosan and hydrogels in medicine, biomaterials in clinical practice. In: Advances in Clinical Research and Medical Devices, pp. 119–147. Springer, Heidelberg (2017)
6. Kalagasidis Krušić, M., Milosavljević, N., Debeljković, A. (Mitrović, A.). Üzüm, Ö.B., Karadağ, E.: Removal of Pb2 + ions from water by poly (acrlymide-co-sodium methacrylate) hydrogels. Water Air Soil Poll. **223**, 4355–4368 (2012)
7. Milosavljević, N., Debeljković, A. (Mitrović, A.), Kalagasidis Krušić, M., Milašinović, N., Üzüm, Ö.B., Karadağ, E.: Application of poly(acrlymide-co-sodium methacrylate) hydrogels in copper and cadmium removal from aqueous solution. Environ. Prog. Sustain. Energy **33**, 824–834 (2014)
8. Baker, M.V., Brown, D.H., Casadio, Y.S., Chirila, T.V.: The preparation of poly(2-hydroxyethyl methacrylate) and poly{(2-hydroxyethyl methacrylate)-co-[poly(ethylene glycol) methyl ether methacrylate]} by photoinitiated polymerization-induced phase separation in water. Polymer **50**, 5918–5927 (2009)
9. Debeljković, A.D. (Mitrović, A.D.), Matija, L.R., Koruga, Đ.L.J.: Characterization of nanophotonic soft contact lenses based on poly (2-hydroxyethyl methacrylate) and fullerene. Hemijska industrija **67**, 861–870 (2013)
10. Mitrovic, A., Bojovic, B., Stamenkovic, D., Popovic, D.: Characterization of surface roughness of new nanophotonic soft contact lenses using lacunarity and AFM method. Hemijska industrija, 2406–0895 (2018)
11. Debeljković, A. (Mitrović, A.), Veljić, V., Šijacki – Žeravčić, V., Matija, L., Koruga, Đ.: Characterization of materials for commercial and new nanophotonic soft contact lenses by Optomagnetic Spectroscopy. FME Trans. **42**, 89–93 (2014)
12. Mitrović, A.D., Stamenković, D., Conte, M., Mihajlović, S.: Study of the optical power of nanophotonic soft contact lenses based on poly (2-hydroxyethyl methacrylate) and fullerene. Contemp. Mater. **V-1**, 151–160 (2014)
13. Mitrović, A.D., Miljković, V.M., Popović, D.P., Koruga, Đ.: Mechanical properties of nanophotonic soft contact lenses based on poly (2-hydrohzethil methacrylate) and fullerenes. Struct. Integrity Life **16**, 39–42 (2016)
14. Miljković, V.M., Mitrović, A.D., Stamenković, D., Popović, D.P., Koruga, Đ.: Monte carlo simulation of light transport through lens. Struct. Integrity Life **16–1**, 125–130 (2016)
15. Stamenković, D., Kojić, D., Matija, L., Miljković, Z., Babić, B.: Physical properties of contact lenses characterized by scanning probe microscopy and optomagnetic fingerprint. Int. J. Mod. Phys. B **24**, 825–834 (2010)
16. Tomic, M., Bojovic, B., Stamenkovic, D., Mileusnic, I., Koruga, D.: Lacunarity properties of nanophotonic materials based on poly(methyl methacrylate) for contact lenses. Mater. Technol. **51**, 145–151 (2017)

17. Kaifeng, L., Timothy, C.O.: Poro-viscoelastic constitutive modeling of unconfined creep of hydrogels using finite element analysis with integrated optimization method. J. Mech. Behav. Biomed. Mater. **4**, 440–450 (2011)
18. Tomic, S.Lj., Micic, M.M., Dobic, S.N., Filipovic, J.M., Suljovrujic, E.H.: Smart poly(2-hydroxyethylmethacrylate/itaconicacid)hydrogelsfor biomedical application. Radiat. Phys. Chem. **79**, 643–649 (2010)
19. Koetting, M.C., Peters, J.T., Steichen, S.D., Peppas, N.A.: Stimulus-responsive hydrogels: theory, modern advances, and applications. Mater. Sci. Eng. R. **93**, 1–49 (2015)
20. Enas, M.A.: Hydrogel: Preparation, characterization, and applications: a review. JAR. **6**, 105–121 (2015)
21. Brannon-Peppas, L., Harland, R.S.: Preparation and characterization of cross linked hydrophilic networks. In: Absorbent Polymer Technology, pp. 45–66. Elsevier, Amsterdam, (1990)
22. Gupta, P., Vermani, K., Garg, S.: Hydrogels: from controlled release to pH-responsive drug delivery. Drug Discov. Today **7**, 569–579 (2002)
23. Davis, K.A., Anseth, K.S.: Controlled release from cross linked degradable networks. Crit. Rev. Ther. Drug Carrier Syst. **19**, 385–423 (2002)
24. Safrany, A.: Radiation processing: synthesis and modification of biomaterials for medical use. Nucl. Instrum. Meth. Phys. Res. Sect. B-Beam Interact. Mater. Atoms **131**, 376–381 (1997)
25. Rosiak, J.M., Yoshii, F.: Hydrogels and their medical applications. Nucl. Instrum. Meth. Phys. Res. Sect. B-Beam Interact. Mater. Atoms **151**, 56–64 (1999)
26. Opdahl, A., Kim, S.H., Koffas, T.S., Marmo, C., Somorjai, G.A.: Surface mechanical properties of PHEMA contact lenses: viscoelastic and adhesive property changes on exposure to controlled humidity. J. Biomed. Mater. Res., Part A **67**, 350–356 (2003)
27. Kim, S.H., Opdahl, A., Marmo, C., Somorjai, G.A.: AFM and SFG studies of PHEMA-based hydrogel contact lens surfaces in saline solution: adhesion, friction, and the presence of non-crosslinked polymer chains at the surface. Biomaterials **23**, 1657–1666 (2002)
28. Xiang, Y., Chen, D.: Preparation of a novel pH responsive silver nanoparticle/poly (HEMA–PEGMA–MAA) composite hydrogel. Eur. Polymer J. **43**, 4178–4187 (2007)
29. Nichols, J.: Contact Lenses 2017. Contact Lens Spectrum, 20–25 (2018)
30. Bui, T.H., Cavanagh, H.D., Robertson, D.M.: Patient compliance during contact lens wear: perceptions, awareness, and behavior. Eye Contact Lens **36**, 334–339 (2010)
31. Kirchhof, S., Goepferich, A.M., Brandl, F.P.: Hydrogels in ophthalmic applications. Eur. J. Pharm. Biopharm. **95**, 227–238 (2015)
32. Seo, E., Kumar, S., Lee, J., Jang, J., Park, J.H., Chang, M.C., Kwon, I., Lee, J.S., Huh, Y.: Modified hydrogels based on poly (2-hydroxyethyl methacrylate) (pHEMA) with higher surface wettability and mechanical properties. Macromol. Res. **25**, 704–711 (2017)
33. Tranoudis, I., Efron, N.: Parameter stability of soft contact lenses made from different materials. Contact Lens Anterior Eye **27**, 115–131 (2004)
34. Shaynai-Rad, M., Khameneh, M., Mohajeri, S.A., Fazly Bazzaz, B.S.: Antibacterial activity of silver nanoparticle-loaded soft contact lens materials: the effect of monomer composition. Curr. Eye Res. **41**, 1286–1293 (2016)
35. Jung, H.J., Chauhan, A.: Temperature sensitive contact lenses for triggered ophthalmic drug delivery. Biomaterials **33**, 2289–2300 (2012)
36. Kumar, A.P., Depan, D., Tomer, N.S., Singh, R.P.: Nanoscale Particles for polymer degradation and stabilization—Trends and future perspectives. Prog. Polym. Sci. **34**, 479–515 (2009)
37. Giacalone, F., Martýn, N.: Fullerene polymers: synthesis and properties. Chem. Rev. **106**, 5136–5190 (2006)

38. Ahmed, R.M., El-Bashir, S.M.: Structure and physical properties of polymer composite films doped with fullerene nanoparticles. Int. J. Photoenergy **2011**, 1–6 (2011)
39. Riggs, J.E., Sun, Y.P.: Optical limiting properties of (60) fullerene and methano (60) fullerene derivative in solution versus in polymer matrix. J. Phys. Chem. A **103**, 485–495 (1999)
40. Peng, N., Leung, F.S.M.: Novel fullerene materials with unique optical transmission characteristics. Chem. Mater. **16**, 4790–4798 (2004)
41. Maldonado-Codina, C., Efron, N.: Impact of manufacturing technology and material composition on the mechanical properties of hydrogel contact lenses. Ophthalmic Physiol. Opt. **24**, 551–561 (2004)
42. Fang, F., Xu, F.: Recent advances in micro/nano-cutting: effect of tool edge and material properties. Nanomanuf. Metrol. **1**, 4–31 (2018)
43. Fang, F.Z., Zhang, X.D., Gao, W., Guo, Y.B., Byrne, G., Hansen, H.N.: Nano manufacturing—perspective and applications. CIRP Ann. Manuf. Technol. **66**, 683–705 (2017)
44. Zhu, L., Li, Z., Fang, F., Huang, S., Zhang, X.: Review on fast tool servo machining of optical freeform surfaces. Int. J. Adv. Manuf. Technol. **95**, 2071–2092 (2018)
45. Kang, C., Fang, F.: State of the art of bio implants manufacturing: part II. Adv. Manuf. **6**, 137–154 (2018)
46. Stephen, S., Musgrave, A., Fang, F.: Contact lens materials: a materials science perspective. Materials **12**, 261 (2019)
47. Martınez, G., Sanchez-Chaves, M., Marco Rocha, C., Ellis, G.: Thermal degradation behavior of 2-hydroxyethylmethacrylate–tert-butyl acrylate copolymers. Polym. Degrad. Stab. **76**, 205–210 (2002)
48. Laftah, W., Akos, N.I., Hashim, S.: Polymer hydrogels: a review. Polymer-plastics Technol. Eng. **50**, 1475–1486 (2011)

Engineering

Probabilistic Simulation of Incremental Lifetime Cancer Risk of Children and Adults Exposed to the Polycyclic Aromatic Hydrocarbons – PAHs in Primary School Environment in Serbia, Model Development and Validation

Rastko Jovanovic(✉) and Marija Zivkovic

Vinca Institute of Nuclear Sciences, Laboratory for Thermal Engineering and Energy, University of Belgrade, Belgrade, Serbia
{virrast,marijaz}@vinca.rs

Abstract. Polycyclic aromatic hydrocarbons (PAHs) are considered to be major air pollutants with a strong negative influence on human health. Many of them are toxic with high carcinogenic potential. Children and school staff spend a significant portion of daytime at schools, mostly indoors. Therefore, the hypothesis can be made that air quality significantly impacts their health. A health risk assessment, performed by calculating Incremental lifetime cancer risk (ILCR), was conducted in the framework of this study. Indoor and outdoor PAHs concentrations were measured in typical Serbian primary school. Total suspended particles (TSP) and gas-phase PAHs from the air were collected both inside the school building and in the outside school environment. Average indoor and outdoor PAHs concentrations were used to calculate benzo[a]pyrene equivalent (BaPeq) concentration. A significantly higher BaPeq was observed in the gas-phase than in the TSP, due to a high amount of low molecular PAHs present in the gas-phase. The measured BaPeq concentration values were fitted to the appropriate mathematical distribution and used as an input parameter for stochastic ILCR modeling. Different body weight and inhalation rate distributions were used for sampling during ILCR calculations. The performed sensitivity analysis showed that the two different recommended values of cancer slope factor had a major impact on the ILCR values. Based on this, it was decided to perform simulations using cancer slope factors for individual PAHs. The obtained ILCR values for both children and adults were greater than the allowed level, indicating high potential lung cancer risk. It may be concluded that it is necessary to improve indoor air quality in schools applying measures for lowering TSP PAHs with high carcinogenic potential.

Keywords: Polycyclic aromatic hydrocarbon · School population · Incremental lifetime cancer risk · Modeling · Risk assessment

N. Mitrovic et al. (Eds.): CNNTech 2019, LNNS 90, pp. 203–220, 2020.
https://doi.org/10.1007/978-3-030-30853-7_12

1 Introduction

Environmental pollution and its negative influence on human population health have been a source of the major concern of the scientific community in recent years. Rise in environment air pollution is associated with negative effects on human health including increased cancer risk, increased lung diseases, increased mortality rates, and ecosystems damage [1]. The economic sectors, which contribute to the environmental air pollution the most are transport, industrial, energy (mainly power plants), and waste management [2, 3].

Polycyclic aromatic hydrocarbons (PAHs) are organic compounds consisting of two or more benzene rings. PAHs are typical results of incomplete combustion of solid fuels such as coal, biomass, petrol, diesel, and wastes. The major PAHs anthropogenic sources include traffic, coal combustion for power generation in thermal power plants, domestic and industrial combustion of wood and other biomass, and waste treatment [4–6]. PAHs, present in the air, exist both in the gas phase and in the particulate matter (PM). The majority of PAHs bound in the particulates are found in fine fractions, with an aerodynamic diameter smaller than 2.5 μm ($PM_{2.5}$). Due to their small diameter, $PM_{2.5}$ PAHs behave similarly to gas molecules that can penetrate to the alveoli lung regions and even penetrate into the circulatory system [7].

PAHs have different toxic effects on human health, among which their carcinogenic potential is of great concern. Both the US Environmental Protection Agency (USEPA) and the World Health Organization (WHO) listed a number of PAHs as probable human carcinogens [8, 9]. Children are more vulnerable population group with regard to harmful effects induced by air pollution than adults. Compared to adults, children are more vulnerable to poor outdoor and indoor air quality due to their still developing physiological and immunological systems, and greater inhaled breath per unit mass [2, 5].

USEPA classified sixteen PAHs as priority pollutants, with some of these PAHs being carcinogen [10]. Considering the carcinogenic PAHs potential and large PAHs emissions it is essential to evaluate population, especially children, cancer risk levels caused by PAHs exposure. However, determining cancer risk levels from the available experiential measurements may be very challenging due to several reasons: the amount of experimental data are usually relatively limited, mechanism of PAHs influence on cancer development is very complex, interactions between different PAHs are not yet completely understood, etc… Thus, mathematical modeling is often employed, as an efficient tool in determining population cancer risk levels. In general, different classes of mathematical models can be employed for risk assessment. Stochastic analysis based on the Monte Carlo method [11] is the most suitable approach since it can take into account different, random, probabilities involved with lifetime risk levels calculations.

Gungormus et al. conducted a carcinogenic risk assessment due to PAHs exposure via inhalation and dermal routes. Exposure-risk probability distributions were calculated using Monte Carlo simulations. The results showed that risk due to dermal exposure is lower than risk due to inhalation exposure. It was estimated that more than 99% of the population is under higher than acceptable risk level in case of inhalation route, while this percentile is lower in case of the dermal route (at about 28%) [12].

Yu et al. performed an assessment of human incremental life cancer risk (ILCR) using Monte Carlo method. The obtained results showed that diet accounted for 85% low-molecular-weight PAH (L-PAH) exposure, while inhalation contributed to approximately 57% of high-molecular-weight PAH (H-PAH) exposure of the Beijing population. Moreover, it was shown that L-PAHs exist both in gaseous and particulate phase, while H-PAHs were determined mostly in the particulate phase. Thus, it is necessary to give more attention to particulate-phase PAHs, which have considerably high carcinogenic potential, in order to reduce cancer occurrence [13].

Han et al. performed health risk assessment due to PM bound PAHs of the elderly population. Monte Carlo based probabilistic simulations showed that more than 95% of the investigated population has higher than acceptable lung cancer risk. It was also shown that there were no significant differences in risk levels between male and female inhabitants [14].

In paper of Wang et al. PAHs from both gaseous phase and particulates were taken into account. ILCR calculations were performed based on a Monte Carlo simulation for different age and gender population groups. ILCR values, caused by total (gaseous phase and PM) PAHs, for all population groups were higher than the accepted, pointing out high cancer risk. The performed sensitivity analysis showed that the empirical cancer slope factor used in simulations has a significant impact on ILCR values [15].

Oliveira et al. characterized PAHs levels in ten primary schools in urban areas in Portugal. Only PM bound PAHs were analyzed in this study. It was determined that the main source of indoor PAHs is infiltration from the outdoor environment. It was also shown that the main source of PM PAHs is traffic, both from gasoline and diesel vehicles. The calculated risk levels were higher than those accepted, both by USEPA and WHO regulations. However, cancer slope factor values provided by USEPA and WHO recommendations led to significantly different ILCR values [5].

Hamid et al. showed results of probabilistic health assessment using Monte Carlo simulations of outdoor and indoor PAHs on the exposed population in two different cities (Rawalpindi and Islamabad) in Pakistan. The results showed that the estimated life cancer risk in an indoor environment is higher than in an outdoor environment. Incremental life cancer risk model also showed that the highest risk levels are associated with the ingestion route, then with the dermal route, and the lowest levels are associated with the inhalation route. It was identified that the main sources of PAHs emitted to the air originate from gasoline and diesel combustion from the city vehicles [16].

Li et al. performed work focused on vehicle exhaust PAHs (VEPAHs). Incremental life cancer risk model based on Monte Carlo simulations was used to evaluate the cancer risk for residents of Zhengzhou city in China. The results showed that although the majority of VEPAHs are suspended in the soil, VEPAHs concentration in the surrounding air is higher than allowed. The main exposure route in this work was the inhalation route, followed by dermal and ingestion routes.

Škrbić et al. performed experimental and modeling work to determine the influence of PAHs suspended in the street dust on human health in the urban area of Novi Sad city in Serbia. Authors used probabilistic analysis based on Monte Carlo simulations to assess the cancer risk for children and adult population taking into account three possible exposure routes (ingestion, dermal, and inhalation). The obtained ILCR values were higher than allowed with a median total cancer risk of 10^{-5}. It was concluded that high values of obtained ILCR should be of great concern as a potential source of future health problems [6].

The main aim of this work is to suggest novel ILCR probabilistic model for risk assessment based on Monte Carlo simulations. The model development is performed in the three phases. First, the same type of ILCR model, as found in the above-written literature review was programmed and applied on children and adult (school employees) population in typical Serbian primary school in the city of Zaječar. Influence of totally 16 different PAHs in the gaseous phase and in total suspended particles (TSP) was taken into account. Measured PAHs concentrations in outdoor and indoor environments were used to calculate benzo[a]pyrene equivalent (BaPeq) concentration. BaPeq is the main PAH for which cancer slope factor value, necessary for ILCR probabilistic models development, is derived. WHO recommended value for cancer slope factor was used as an input parameter for ILCR calculations in this step, as suggested in [5]. Then, the ILCR model was applied to the same population exposed to the same BaPeq concentration, using USEPA recommended cancer slope factor values, following the procedure described in [14]. The obtained ILCR probability dependences of BaPeq concentration differed for about two orders of magnitude using different (WHO and USEPA) suggested values.

Authors conducted a sensitivity analysis of all input parameters on final ILCR value in the second phase of this study due to significant differences in the calculated ILCR distributions. The performed analysis showed high ILCR model's dependency on the cancer slope factor value (distribution).

Based on these results it was decided to develop a novel probabilistic model based on Monte Carlo simulations for ILCR probabilities calculation. The main feature of the suggested model is that it uses individual PAHs' concentrations with corresponding cancer slope factor for each involved PAH. Recommended values for cancer slope factors were adopted from USEPA [17]. The developed model showed good performance in predicting ILCR probability levels. Moreover, repeated sensitivity analysis confirmed that the novel model is independent of cancer slope factors values. However, it should be noted that the obtained risk levels were lower than in the case of the two previously used ILCR models. This can be explained by the fact, that the suggested model uses individual values for cancer slope factor for each modeled PAH concentration. However, USEPA guidelines define these values only for certain PAHs with the highest carcinogenic potential. Therefore, some PAHs were not taken into account, mainly those with a high concentration in the gaseous phase, which have a major impact in the other models based on the BaPeq concentration.

The obtained ILCR values for both children and adults were higher than the significant level, indicating potential lung cancer risk. It may be concluded that it is necessary to improve indoor air quality in schools applying measures for lowering TSP PAHs with high carcinogenic potential.

2 Materials and Methods

2.1 Site Description

All in-situ measurements and sample collection used in this work are performed in the city of Zaječar, Serbia. Zaječar is a relatively small city located in the eastern part of Serbia, shown in Fig. 1. The city underwent major industrial development after the Second World War. Previously known as a mainly agricultural center, it also became the industrial center of the region with food and metal processing industry. It is currently populated by around 44000 inhabitants, and the district heating is implemented only in a small number of households (less than 20% of the total number). Majority of households located in the town area are heated by the individual heating stoves burning wood and coal as a fuel. This significantly increases air pollution, and PAHs emissions, during the heating season. Moreover, the city is heavily overpopulated by passenger vehicles, which also substantially contributes to total PAHs emissions.

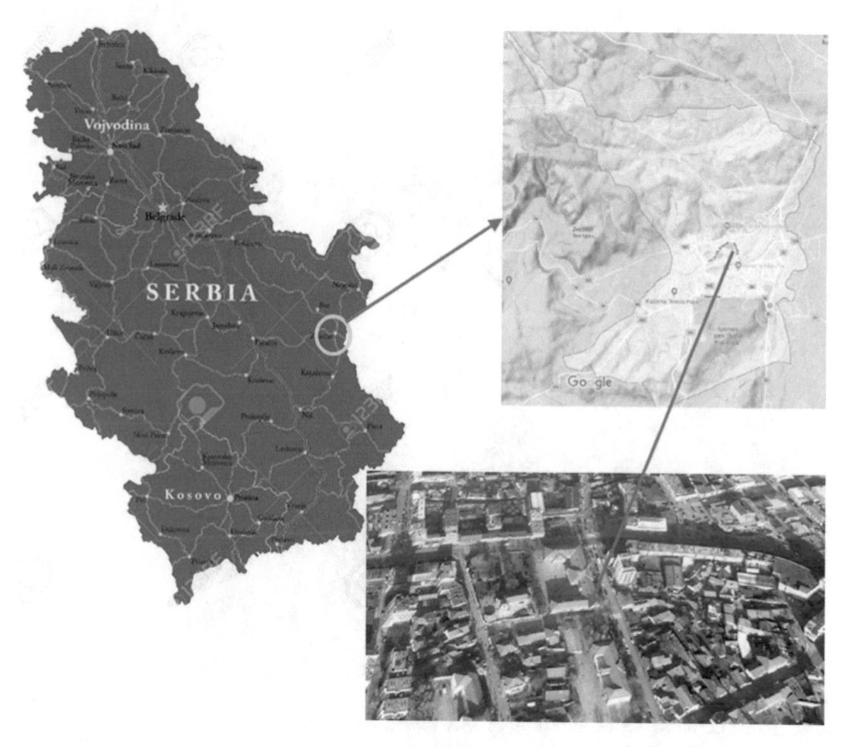

Fig. 1. Primary school – in-situ measurement site location, (43.902695 N, 22.285223 E) Zaječar, Serbia

2.2 Sampling Procedure and Sample Chemical Analysis

The sampling campaign was conducted simultaneously in both indoor and outdoor primary school environments during full ten working days. All samples were collected using the low volume reference sampler, Sven/Leckel Sven/LACKEL LVS3 (LVS) with size-selective inlets for TSP fraction for 24 h period. Totally four samplers were used, three were positioned inside classrooms and one was positioned in the schoolyard. The sampling flow rate of each sampler was 2.3 m^3/h (38 L/min). The air was drawn by LVS reference sampler through a quartz filter (Whatman QMA, 47 mm), to collect total suspended particles (TSP) and then through polyurethane foam (PUF) to collect pollutants in the gas phase. PUF plugs were cleaned before sampling in a Soxhlet extractor. Extraction was performed by acetone for 8 h and by a mixture of diethylether/hexane (JT Baker, HPLC analysis grade) with ratio 1/5 for 16 h. The filters were kept at 900 °C, before sampling, for 4 h so that all organic compounds could be removed and blank values reduced. Filters weight was measured before and after the sampling procedure for determination of collected TSP. The obtained weight difference was used to calculate the number concentration of the captured particles. Total air volume sucked in by the sampler was used to calculate the mass concentrations of TSP and gaseous phase PAHs.

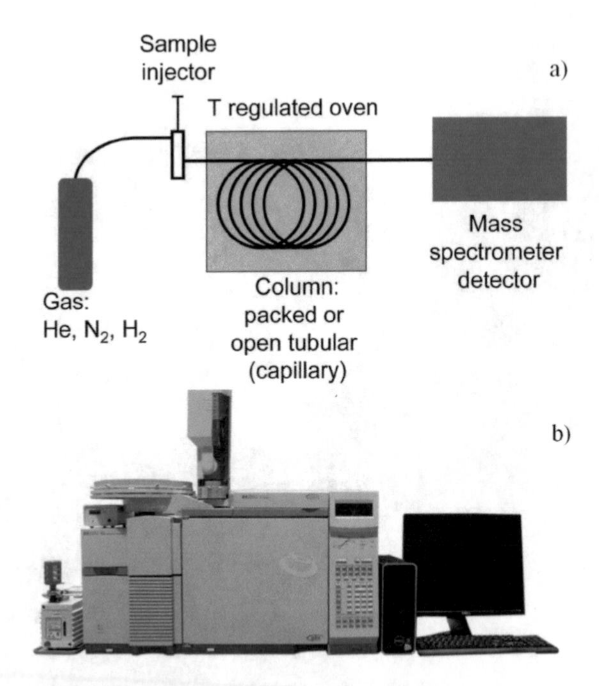

Fig. 2. (a) Schematic representation of gas chromatography mass spectrometry process [18], (b) Agilent GC 6890/5973 gas chromatograph mass spectrometer (taken from [19])

After the sampling, PUFs were extracted by the mixture of acetone and hexane (1:1 ratio) for 8 h at 90 °C using Soxhlet extraction. The filters were extracted using microwave extraction. Hexane/acetone (JT Baker, HPLC analysis grade) mixture in proportion: 12.5 ml n-hexane/12.5 ml acetone was used as a solvent during extraction in accordance with EPA 3546 standard. After the extraction was finished, the solvent

Table 1. Individual PAHs measured in this study

No.	Compound name and abbreviation	Skeletal formula	Molecular weight [$gmol^{-1}$]
1	Naphthalene (Nap)	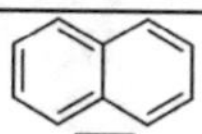	128
2	Acenaphthylene (AcPy)	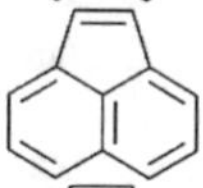	152
3	Acenaphthene (AcP)	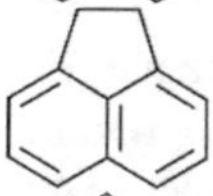	154
4	Fluorene (Flu)	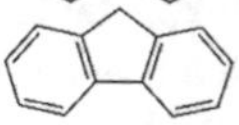	165
5	Phenanthrene (PA)	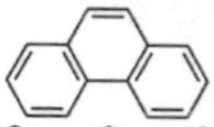	178
6	Anthracene (Ant)	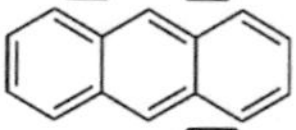	178
7	Fluoranthene (FL)	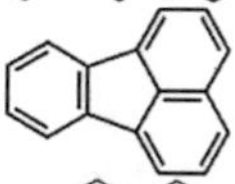	202
8	Pyrene (Pyr)	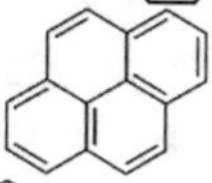	202
9	Benzo[a]anthracene (BaA)	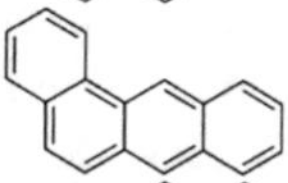	228
10	Chrysene (CHR)	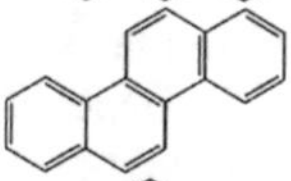	228
11	Benzo[b]fluoranthene (BbF)	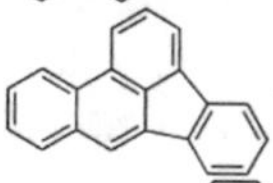	252
12	Benzo[k]fluoranthene (BkF)	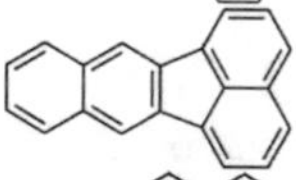	252
13	Benzo[a]pyrene (BaP)	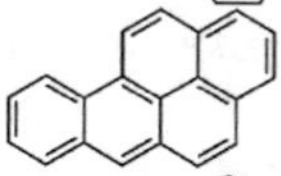	252
14	Dibenz[a,h]anthracene (DBA)	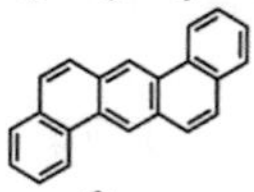	278
15	Benzo[g,h,i]perylene (BghiP)	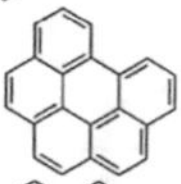	276
16	Indeno[1,2,3-c,d]pyrene (IND)	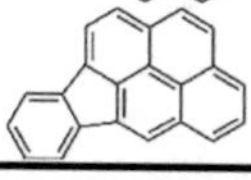	276

volume was reduced to 1 ml, using rotary evaporation under the reduced pressure of 55.6 kPa. N-hexane solution was then reduced to 0.25 ml under a nitrogen stream at room temperature. Known quantities of the internal standard were added to estimate the method recovery. PAHs were analyzed using gas chromatography coupled with mass spectrometry (GC-MS) using Agilent GC 6890/5973 gas chromatograph mass spectrometer with a DB-5 MS capillary column (30 m × 0.25 mm × 25 μm), shown in Fig. 2. PAHs analysis was performed according to EPA Compendium Method TO-13A) [20].

Totally, sixteen US EPA priority PAHs were determined: naphthalene (Nap), acenaphthylene (AcPy), acenaphthene (AcP), fluorene (Flu), phenanthrene (PA), anthracene (Ant), fluoranthene (FL), pyrene (Pyr), benz[a]anthracene (BaA), chrysene (CHR), benzo[b]fluoranthene (BbF), benzo[k]fluoranthene (BkF), benzo[a]pyrene (BaP), dibenz[a,h]- anthracene (DBA), benzo[g,h,i]perylene (BghiP) and indeno[1,2,3-cd]pyrene (IND), as shown in Table 1.

3 Mathematical Modeling

3.1 Risk Assessment Model

ILCR model is based on Monte Carlo simulations to quantitatively estimate the exposure risk of children and adults population groups in primary school caused by PAHs' exposure via the inhalation route. Input factors necessary to determine ILCR values have complex and sometimes highly skewed distributions. In order to overcome this challenge simpler, analytical, models are based only on the central values of input parameters (mean, median, etc…). However, ILCR values obtained by these simpler models provide very limited information. In order to have more detailed insight into ILCR probability distribution and its dependence on different input parameters, it is necessary to incorporate realistic input parameters distributions into the model. This is performed by fitting physical input data distribution to the corresponding statistical mathematical distribution, which will provide the best fit for that particular input variable.

Toxicity equivalence factors (TEFs) values relative to BaP are used to calculate the BaPeq concentrations of the measured PAHs. This is done by multiplying individual PAH concentrations by the corresponding TEF value. In this way, the carcinogenic potency of all measured PAHs can be calculated using the sum of all individual BaPeq. TEFs values suggested in the work of Nisbet and LaGoy were adopted for BaPeq calculations in this work [21].

Daily exposure inhalation dose is calculated using BaPeq total concentration distribution using Eq. (1):

$$E_j = \sum_{i=1}^{2} C_i \times IR_{ij} \times t_{ij} \tag{1}$$

Where i = 1, 2 corresponds to outdoor and indoor environments, j = 1, 2 corresponds to different population age groups: children and adults, E_j is daily exposure inhalation dose for the jth age group [mg·day^{-1}], C_i is BaPeq concentration in the ith environment [mg·m^{-3}], IR_{ij} is inhalation rate in the ith environment for the jth age group [m3·h^{-1}], and t_{ij} is daily exposure time span in the ith environment of the jth age group [h·day^{-1}].

ILCR is calculated using the following equation:

$$ILCR = \frac{CSF \times EF}{AT} \sum_{j=1}^{2} \frac{E_j \times ED_j}{BW_j} \quad (2)$$

Where $ILCR$ is an incremental lifetime cancer risk (dimensionless), CSF is inhalation cancer slope factor for BaP [kg·day·mg^{-1}], AT is averaging time adopted to 70 years for carcinogens (according to [22]), E_j is a daily exposure dose for the jth group [mg·day^{-1}], ED_j is the exposure duration for the jth group [year], and BW_j is body weight for the jth group [kg].

The novel probabilistic model calculates ILCR risk assessment based on cancer slope factors for the individual PAHs using their individual BaPeq concentrations, thus Eq. (1) becomes:

$$E_{jk} = \sum_{i=1}^{2} C_{ik} \times IR_{ij} \times t_{ij} \quad (3)$$

Where k = 1–11 is a number of individual PAH, E_{jk} is daily exposure inhalation dose for the jth age group due to exposure to the kth PAH [mg·day^{-1}], C_{ik} is BaPeq concentration of the kth PAH in the ith environment [mg·m^{-3}].

Total daily exposure inhalation dose is given by the equation:

$$E_j = \sum_{i=1}^{2} \sum_{k=1}^{11} C_{ik} \times IR_{ij} \times t_{ij} \quad (4)$$

The expression for the ILCR, based on individual cancer slope factors and their corresponding PAH concentrations, takes the following form:

$$ILCR = \frac{CSF \times EF}{AT} \sum_{j=1}^{2} \frac{\sum_{k=1}^{11} E_{j,k} \times ED_j}{BW_j} \quad (5)$$

It is important to note that even when the complexity of the combined different statistical distribution is minimized as much as possible, it is often impossible to determine the accurate distribution of the result. Thus, numerical calculations are a necessary computational tool for this task. Monte Carlo simulations are an efficient method for modeling and characterization of dependent variables distribution based on different distributions of the associated input independent variables. All input quantitates were sampled independently from their appropriate distributions, as shown in Table 2. Ultimately, statistical distributions of values of dependent variables (E and $ILCR$), which have frequency similar to the real, physical population, are obtained as a result of the performed Monte Carlo probabilistic simulation.

Totally one million Monte Carlo iterations were performed in order to obtain stable results. High model performance was achieved using the MATLAB package. Namely, the MATLAB package is very efficient in tensor and vector calculus, and also allows simple code vectorization. These features enabled to avoid "for" loops iterating over a sequence and consequently dramatically increased code performance, compared with different spreadsheet software packages commonly used for this kind of simulations.

Table 2. Input risk parameters considered as random variables for different age groups in different environments

Variable (measurement units)	Outdoor		Indoor	
	Value	Distribution type	Value	Distribution type
C_i [ng·m^{-3}] (TSP)	5.4572 ± 2.5756	Normal	3.5241 ± 1.9672	Normal
C_i [ng·m^{-3}] (gaseous phase)	81.6915 ± 62.3634	Normal	41.8942 ± 29.8485	Normal
IR [m^3·day^{-1}] (children)	9.98 ± 1.83	Log-Normal	9.98 ± 1.83	Log-Normal
IR [m^3·day^{-1}] (adults)	14.27 ± 2.37	Log-Normal	14.27 ± 2.37	Log-Normal
t_i [h·day^{-1}]	3	Constant	5	Constant
CSF [*kg·day·mg*$^{-1}$] (WHO based [9])	304.5	Constant	304.5	Constant
CSF [*kg·day·mg*$^{-1}$] (USEPA based [23])	3.14 ± 1.8	Log-Normal	3.14 ± 1.8	Log-Normal
AT [*days*]	25500	Constant	10	Constant
ED [*years*]	3	Constant	3	Constant
BW [*kg*] (children)	32.5 ± 7.1	Log-Normal	32.5 ± 7.1	Log-Normal
BW [*kg*] (adults)	65.3 ± 6.2	Log-Normal	65.3 ± 6.2	Log-Normal

3.2 Uncertainty Analysis

Model uncertainty arises from the estimation of both exposure and effects. Monte Carlo simulation introduces variability and uncertainty into ILCR calculations, giving a more realistic insight into the probability assessment. Multiple independent runs were performed with different number of Monte Carlo steps: 5e+04, 1e+05, 5e+05, 1e+06, and 2e+06. Stability of results was achieved for the case of one million Monte Carlo simulations.

3.3 Sensitivity Analysis

Sensitivity analysis was performed in order to determine the most important input parameters that contribute to the model's uncertainty. The sensitivity of each input variable relative to one another is determined calculating rank correlation coefficients between each input variable and output (ILCR) variable and then estimating the contribution of each input variable to the variance of the output variable, by squaring the output variable variance and normalizing to 1.

4 Results and Discussion

TEF values, PAHs concentrations, and calculated BaPeq values, fitted to the normal distribution are shown in Table 3 (outdoors) and Table 4 (indoors). PAHs concentration in the gaseous phase is much higher than in TSP. Gaseous PAHs account for 93.7% of total PAHs outdoors, and for 92.2% of total PAHs indoors. This is in good agreement with literature data, in which gaseous PAHs are frequently found in the significantly higher concentrations than TSP PAHs [2]. It is important to note that PAHs with 2–4 aromatic rings (Naphthalene, acenaphthylene, acenaphthene, fluorene, phenanthrene, anthracene, fluoranthene, pyrene, benzo[a]anthracene, and Chrysene) are found predominately in the gas phase. On the contrary, PAHs with more than five aromatic rings (Benzo[b]fluoranthene, benzo[k]fluoranthene, Benzo[a]pyrene, Dibenz [a,h]anthracene, Benzo[g,h,i]perylene, and Indeno[1,2,3-c,d]pyrene), which usually come from combustion of solid fuels, are almost completely bound in the TSP.

Table 3. Outdoor TEFs, PAHs concentration, and BaPeq concentration normal distributions

PAH Name	TSF [−]	TSP PAH [ng/m^3]		TSP BaPeq [ng/m^3]		Gaseous PAH [ng/m^3]		Gaseous BaPeq [ng/m^3]	
		Value	St. Dev.	Value	St. Dev.	Value	St. Dev.	Value	St. Dev.
Nap	0.001	0.0267	0.032	2.67e−5	3.2e−5	4730.11	3322.73	4.7301	3.3227
AcPy	0.001	0.22	0.1977	2.20e−4	1.98e−4	3436.73	2357.85	3.4367	2.3578
AcP	0.001	0.1467	0.1616	1.47e−4	1.62e−4	514.16	149.32	0.5142	0.1493
Flu	0.001	0.555	0.7062	2.78e−4	3.53e−4	1769.77	527.10	0.8849	0.2635
PA	0.001	0.095	0.0437	4.75e−5	2.19e−5	3819.12	575.00	1.9096	0.2875
Ant	0.01	0.0233	0.0103	1.17e−5	5.16e−6	752.02	157.78	0.376	0.0789
FL	0.001	2.0917	1.7555	1.05e−1	8.78e−2	983.45	158.82	49.1727	7.9409
Pyr	0.001	2.4217	1.995	2.42e−3	1.99e−3	830.04	150.58	0.83	0.1506
BaA	0.1	3.7783	2.2489	1.89e−2	1.12e−2	346.81	810.21	1.734	4.0511
CHR	0.01	4.7483	2.6415	1.42e−1	7.92−2	602.65	1458.70	18.0796	43.7611
BbF	0.1	4.1133	1.8642	4.11e−1	1.86e−1	0.01	0	0.001	0
BkF	0.1	3.5683	1.4407	1.78e−1	7.2e−2	0.01	0	0.0005	0
BaP	1	3.4133	1.7553	3.41	1.76	0.01	0	0.01	0
DBA	0.1	2.5833	0.9054	2.58e−1	9.05e−2	0.01	0	0.001	0
BghiP	1	0.7883	0.245	8.67e−1	2.69e−1	0.01	0	0.011	0
IND	0.01	2.975	1.0386	5.95e−2	2.08e−2	0.01	0	0.0002	0
ΣBaP				5.4572	2.5756			81.6915	62.3634

Table 4. Indoor TEFs, PAHs concentration, and BaPeq concentration normal distributions

PAH	TSF [−]	TSP PAH [ng/m^3]		TSP BaPeq [ng/m^3]		Gaseous PAH [ng/m^3]		Gaseous BaPeq [ng/m^3]	
		Value	St. Dev.	Value	St. Dev.	Value	St. Dev.	Value	St. Dev.
Nap	0.001	0.023	0.0205	2.3e−05	2.05e−05	1209.24	406.41	1.2092	0.4064
AcPy	0.001	0.1091	0.0451	1.09e−04	4.51e−05	838.09	448.40	0.8381	0.4484
AcP	0.001	0.1075	0.0719	1.08e−04	7.19e−05	254.8	41.69	0.2548	0.0417
Flu	0.001	0.0881	0.0495	4.4e−05	2.47e−05	1715.82	319.2	0.8579	0.1596
PA	0.001	0.1058	0.0279	5.29e−05	1.40e−05	3602.6	868.96	1.8013	0.4345
Ant	0.01	0.0288	0.0133	1.44e−05	6.63e−06	317.47	114.29	0.1587	0.0571
FL	0.001	0.5472	0.2271	2.74e−02	1.14e−02	470.41	92.3	23.5203	4.6148
Pyr	0.001	0.6146	0.2690	6.15e−04	2.69e−04	805.78	1216.65	0.8058	1.2167
BaA	0.1	0.79	0.4399	3.95e−03	2.2e−03	190.01	335.47	0.95	1.6773
CHR	0.01	1.0907	0.5943	3.27e−02	1.78e−02	229.42	391.05	6.8825	11.7314
BbF	0.1	2.2013	1.0887	2.20e−01	1.09e−01	6.26	12.32	0.6259	1.2318
BkF	0.1	1.9419	1.1772	9.71e−02	5.89e−02	6.35	11.96	0.3174	0.5978
BaP	1	2.2376	1.3094	2.24	1.31	3.66	7.23	3.66	7.2310
DBA	0.1	1.9649	0.9819	1.96e−01	9.82e−02	0.01	0	0.001	0
BghiP	1	0.6037	0.307	6.64e−01	3.38e−01	0.01	0	0.011	0
IND	0.01	2.1892	1.1167	4.38e−02	2.23e−02	0.01	0	0.0002	0
ΣBaP				3.5241	1.9672			41.8942	29.8485

Outdoors PAHs concentration is higher than indoors PAHs concentration, which is in agreement with the previous experimental studies performed in Serbian schools [24]. Higher outdoors PAHs concentrations are caused by the strong contribution of outdoor source, mainly from infiltration of ambient emissions.

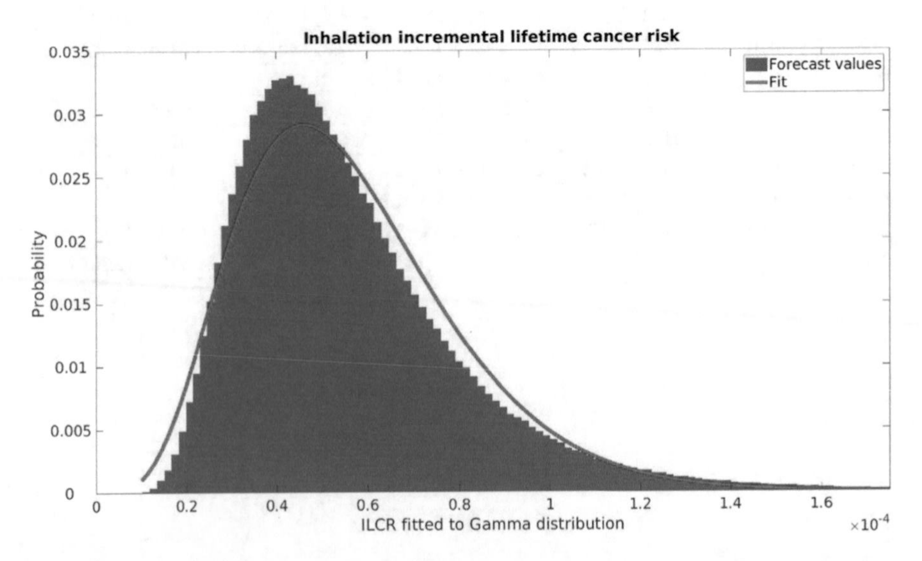

Fig. 3. ILCR probability histogram fitted with Gamma distribution, WHO slope factor

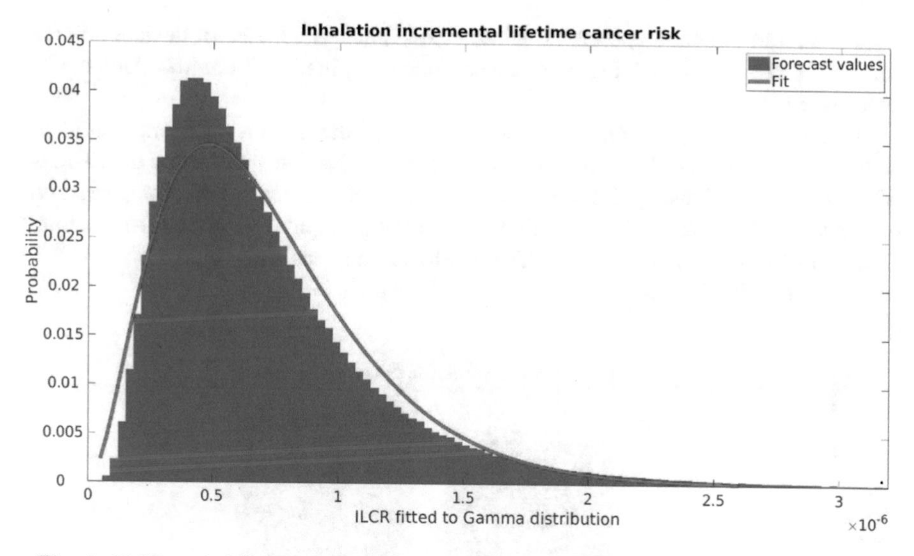

Fig. 4. ILCR probability histogram fitted with Gamma distribution, USEPA slope factor

5%	0.0032
50%	0.0334
95%	0.0836
Confidence interval	Standard Deviation

Fig. 5. Results of the uncertainty analysis, given in terms of uncertainty in distributional statistics of simulated exposure

Calculated ILCR probability histograms and fitted distributions for WHO and USEPA recommended cancer slope factors are shown in Figs. 3 and 4, respectively. ILCR values based on WHO recommended cancer slope factor, are very high with 95th percentile high above the general acceptable carcinogenic risk level of 1e−06. High calculated values pointed out further research in calculating ILCR risk using USEPA recommended cancer slope factor. In this case, the 95th percentile is at 1.3e−06 which is still higher than acceptable carcinogenic risk level of 1e−06. It may be concluded that both approaches (WHO and USEPA) showed higher ILCR probabilities than allowed. However, due to big difference in the above mentioned ILCR distributions calculated using different cancer slope factor values, uncertainty analysis was conducted in order to discover the cause of possible error in the applied models.

Results of the uncertainty analysis in the simulated inhalation exposure as a descriptive statistic for a chosen set of distribution percentiles and the mean are shown in Fig. 5. The same values were obtained for both WHO and USEPA based ILCR models. This can be explained by the uncertainty calculation procedure. The analysis takes into account changes in the output parameters depending on a number of Monte Carlo iterations while neglecting the direct contribution of the input parameters. Moreover, an uncertainty analysis was conducted for inhalation exposure, which does not incorporate cancer slope factor value, Eqs. (1) and (4). Standard deviation is 8.3% for 95th percentile indicating that uncertainties originating from the Monte Carlo procedure are small.

Uncertainty analysis confirmed the accuracy of the conducted Monte Carlo procedure. However, it did not provide an explanation for the differences in ILCR distributions obtained using different cancer slope factor values. Therefore, sensitivity analysis, which quantifies the contribution of different input distributions on the output (ILCR) distribution, is conducted. The results of the sensitivity analysis for USEPA recommended ILCR probability distribution is shown in Fig. 6.

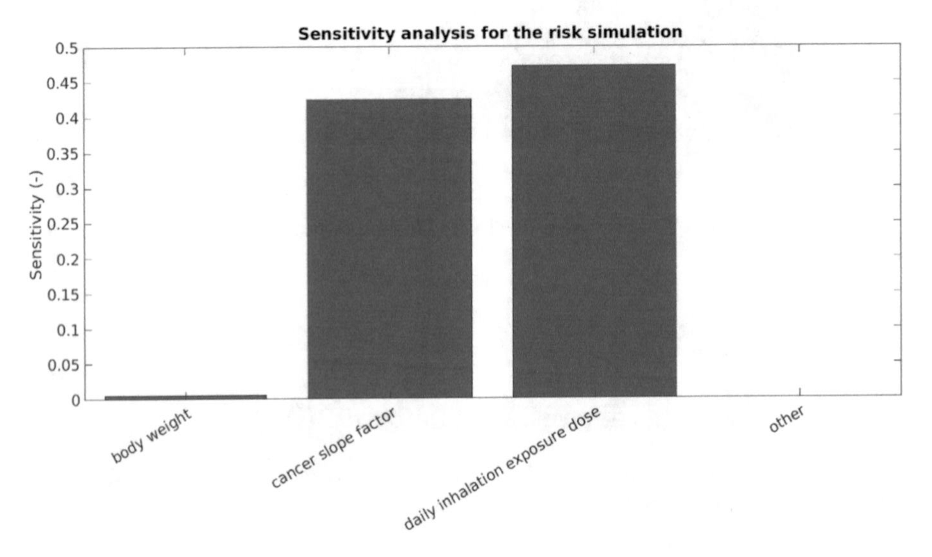

Fig. 6. Results of the sensitivity analysis on ILCR, based on USEPA slope factor

ILCR sensitivity analysis shows that cancer slope factor and daily inhalation exposure dose are the most influential variables on ILCR with relative contributions of 43% and 47.5%, respectively. ILCR was calculated using the individual cancer slope factors for the corresponding PAHs due to high influence of the used cancer slope factor input distribution on output ILCR values. The obtained histogram fitted with Gamma distribution calculated following this procedure is shown in Fig. 7. It can be seen that the obtained values fall into the accepted category (less than 1e−06). However, it should be noted that the individual cancer slope factors are defined by USEPA only for the PAHs with highest carcinogenic risk. Thus, it can be considered that the calculated probability ILCR distribution actually shows a very dangerous trend, since it is focused on the most health-threatening PAH compounds. Moreover, the conducted sensitivity analysis for the suggested novel probabilistic ILCR model showed that output ILCR distribution dependency of slope factor is eliminated.

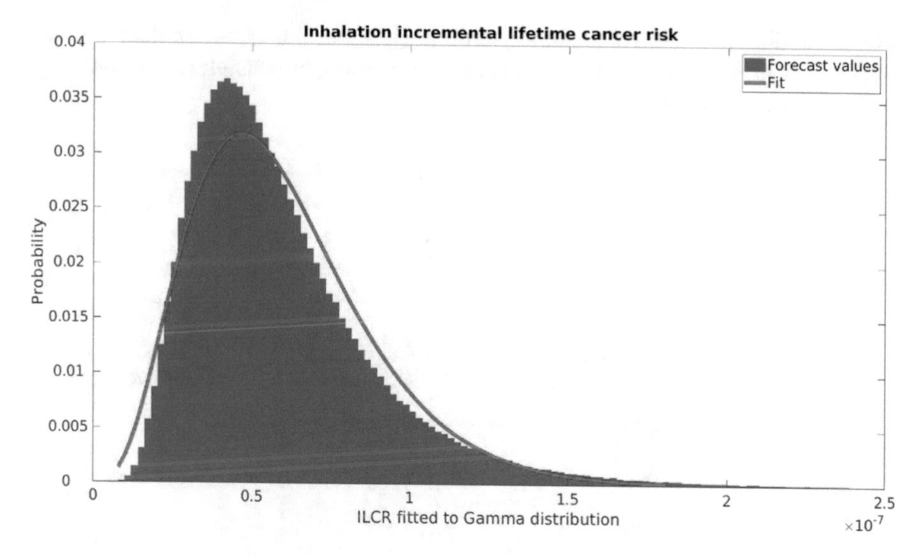

Fig. 7. ILCR probability histogram fitted with Gamma distribution, individual slope factors recommended by USEPA

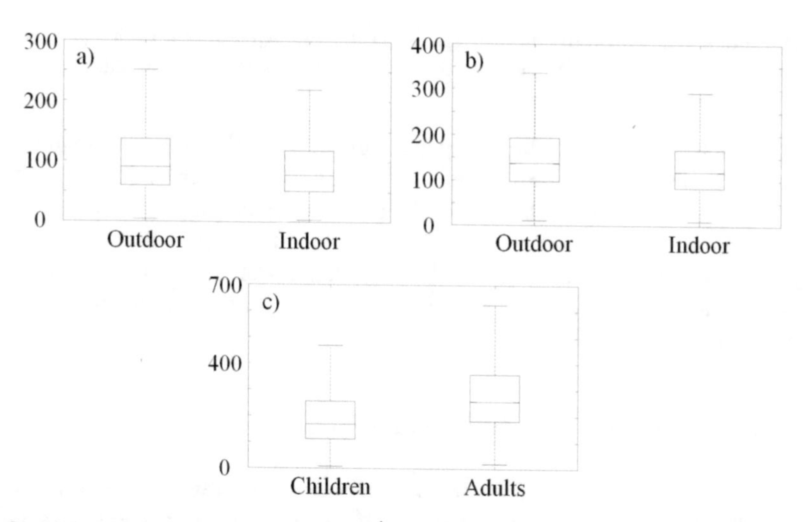

Fig. 8. B[a]Peq daily exposure doses [ng·d^{-1}]: (a) children exposure, (b) adults' exposure, and (c) total exposure (single value cancer slope factor based models)

BaPeq daily exposure for different population groups and in different environments, for models based on the single cancer slope factor is shown in Fig. 8. It can be seen that outdoor exposure is higher than indoor because of higher BaPeq outdoor concentrations, which overcompensate for shorter time spent at an outdoor environment. Adult population has higher daily inhalation dose mainly due to their higher inhalation rate compared with children inhalation rate concentration. Since the time spent inside the school building is 5 h/day compared with 3 h/day spent outdoors, considerably high indoor daily exposure doses require much attention, especially taking into account high BaPeq indoor concentrations.

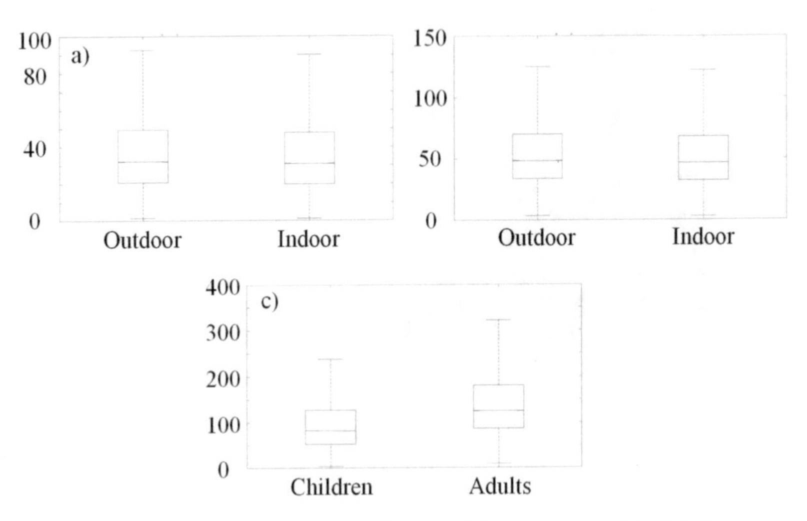

Fig. 9. B[a]Peq daily exposure doses [ng·d^{-1}]: (a) children exposure, (b) adults' exposure, and (c) total exposure (novel model based on the individual cancer slope factor values)

BaPeq daily exposure for different population groups and in different environments, for the novel model based on the individual cancer slope factor values is shown in Fig. 9. The suggested model showed similar trends as the existing models based on the single value cancer slope factor. The main difference is in lower maximum value of daily exposure, obtained using the newly suggested model. This is caused by the fact that individual values of cancer slope factor are derived only for the PAHs with the highest carcinogen potential.

5 Conclusions and Future Direction

Novel model based on probabilistic Monte Carlo simulations was developed to assess ILCR probabilities of school children and adult population in both outdoor and indoor environments due to their exposure to gas and TSP PAHs.

The main novelty of the proposed model is that it takes into account individual cancer slope factor values for the corresponding PAHs.

The results obtained with the existing models differ significantly depending on the value adopted for the cancer slope factor distribution. However, ILCR values distribution obtained for the two recommended cancer slope factors showed increased ILCR risk than allowed.

Uncertainty analysis was conducted and showed that difference in the modeled results does not originate from the Monte Carlo procedure.

The performed sensitivity analysis showed a high dependency of the output ILCR distribution of the input cancer slope factor.

The newly suggested model is not sensitive to the contribution of the input cancer slope factor values.

ILCR probabilities calculated by the newly suggested model are lower compared with those obtained using existing models. This is caused by the fact that the novel model takes into account only PAHs with the highest carcinogen potential.

ILCR probabilities obtained using the new model are considerably high and point out for the necessity for the higher efforts in health risk management in urban environments.

It is expected that the novel model can be further improved interpolating cancer slope factor values to include the all carcinogenic PAHs, which can be part of the future work.

References

1. Air quality in Europe - 2018 report, EEA Report No 12/2018. European Environment Agency, 1050 Copenhagen, Denmark (2018)
2. Oliveira, M., et al.: Children environmental exposure to particulate matter and polycyclic aromatic hydrocarbons and biomonitoring in school environments: a review on indoor and outdoor exposure levels, major sources and health impacts. Environ. Int. **124**(1), 180–204 (2019)
3. Jovanović, R., et al.: Mathematical modelling of swirl oxy-fuel burner flame characteristics. Energy Convers. Manag. **191**(1), 193–207 (2019)
4. Ravindra, K., Sokhi, R., Van Grieken, R.: Atmospheric polycyclic aromatic hydrocarbons: Source attribution, emission factors and regulation. Atmos. Environ. **42**(1), 2895–2921 (2008)
5. Oliveira, M., et al.: Polycyclic aromatic hydrocarbons in primary school environments: Levels and potential risks. Sci. Total Environ. **575**(1), 1156–1167 (2017)
6. Škrbić, B., Đurišić-Mladenović, N., Živančev, J., Tadić, Đ.: Seasonal occurrence and cancer risk assessment of polycyclic aromatic hydrocarbons in street dust from the Novi Sad city Serbia. Sci. Total Environ. **647**(1), 191–203 (2019)
7. Kim, K.H., Kabir, E., Kabir, S.: A review on the human health impact of airborne particulate matter. Environ. Int. **74**(1), 136–143 (2015)
8. United States Environmental Protection Agency. https://www.epa.gov/risk/regional-screening-levels-rsls-users-guide#toxicity. Accessed 31 May 2019
9. Air Quality Guidelines for Europe. 2nd edn. World Health Organization, Regional Office for Europe, Copenhagen, Denmark (2000)
10. Guidelines for Carcinogen Risk Assessment EPA/630/P-03/001F. www.epa.gov/sites/production/files/2013-09/documents/cancer_guidelines_final_3-25-05.pdf. Accessed 31 May 2019
11. Jovanović, R., et al.: Lattice Monte Carlo simulation of single coal char particle combustion under oxy-fuel conditions. Fuel **151**(1), 172–181 (2015)
12. Gungormus, E., Tuncel, S., Hakan Tecer, L., Sofuoglu, S.C.: Inhalation and dermal exposure to atmospheric polycyclic aromatic hydrocarbons and associated carcinogenic risks in a relatively small city. Ecotoxicol. Environ. Saf. **108**(1), 106–113 (2014)

13. Yu, Y., et al.: Risk of human exposure to polycyclic aromatic hydrocarbons: a case study in Beijing, China. Environ. Pollut. **205**(1), 70–77 (2015)
14. Han, B., et al.: Assessing the inhalation cancer risk of particulate matter bound polycyclic aromatic hydrocarbons (PAHs) for the elderly in a retirement community of a mega city in North China. Environ. Sci. Pollut. Res. **23**(20), 20194–20204 (2016)
15. Wang, T., et al.: Pollution characteristics, sources and lung cancer risk of atmospheric polycyclic aromatic hydrocarbons in a new urban district of Nanjing, China. J. Environ. Sci. (China) **55**(1), 118–128 (2017)
16. Hamid, N., et al.: Elucidating the urban levels, sources and health risks of polycyclic aromatic hydrocarbons (PAHs) in Pakistan: Implications for changing energy demand. Sci. Total Environ. **619–620**(1), 165–175 (2018)
17. Lloyd, A.C., Denton, J.E.: Air Toxics Hot Spots Program Risk Assessment Guidelines Part II Technical Support Document for Describing Available Cancer Potency Factors. California Environmental Protection Agency, Office of Environmental Health Hazard Assessment, Air Toxicology and Epidemiology Section, Oakland, California, USA (2005)
18. GNU Free Documentation License. https://www.gnu.org/licenses/fdl-1.3.html. Accessed 30 May 2019
19. Conquer Scientific. https://conquerscientific.com/product/agilent-technologies-6890-5973-gcms-system. Accessed 30 May 2019
20. Compendium of Methods for the Determination of Toxic Organic Compounds in Ambient Air. 2nd edn. U. S. Environmental Protection Agency, Research Laboratory Center for Environmental Research Information, Cincinnati, Ohio, USA (1999)
21. Nisbet, I.C.T., LaGoy, P.K.: Toxic equivalency factors (TEFs) for polycyclic aromatic hydrocarbons (PAHs). Regul. Toxicol. Pharmacol. **16**(3), 290–300 (1992)
22. Risk assessment guidance for Superfund Volume I: Human health evaluation manual, supplemental guidance “Standard default exposure factors”. Washington, USA (1991)
23. Chen, S.C., Liao, C.M.: Health risk assessment on human exposed to environmental polycyclic aromatic hydrocarbons pollution sources. Sci. Total Environ. **366**(1), 112–123 (2006)
24. Zivkovic, M., et al.: PAHs levels in gas and particle-bound phase in schools at different locations in Serbia. Chem. Ind. Chem. Eng. Q. **21**(1–2), 159–167 (2014)

Revitalization and Optimization of Thermoenergetic Facilities

Marko Ristic[1(✉)], Ivana Vasovic[2], and Jasmina Perisic[3]

[1] University of Belgrade, Institute Mihajlo Pupin, Volgina 15, 11000 Belgrade, Serbia
marko.ristic@pupin.rs
[2] Lola Institute, 11000 Belgrade, Serbia
[3] Union University Belgrade, 11000 Belgrade, Serbia

Abstract. Global demand for electrical energy is in constant growth because of the constant increase in the number of inhabitants and the economy growth. This paper presents the way the increasing demand for energy and the constant increase in the capacity of power plant production systems exacerbates the working conditions of the system parts and leads to the remaining working life reduction. Efficient operation of thermal power plants is the basis of economical production of electricity based on combustion of lignite. Ventilation mill as such is one of the basic equipment in thermal power plants which work has a significant impact on the level of energy efficiency. This paper presents the complexity of the optimization problem of the different thermal power plants and the system demands to use multidisciplinary research. However, any optimization of an energy plant, such as a ventilation mill, in this case, represents a unique approach to the problem of operating efficiency of the plant, and as such it asks for many new requirements and solutions. This paper also presents the way for solving this complex problem by using detail analyses of work of system of ventilation mill and dust channel by analysing the period of failures. This paper describes the following: the usages of CFD numerical simulation in FLUENT software for determining the working conditions in ventilation mill and dust channel; possibilities for use the process of coating damaged parts with wear resistance material and procedure for selection and coating on real working parts; analyses of damaged coated working parts of ventilation mill after testing in exploitation condition; data analysis of working parts before and after coating, time and time and resource saving. One of the main goals is to analyse the damaged coated working parts of ventilation mill after testing in exploitation condition.

Keywords: Energy saving · Thermal power plant · Failure · Coating · CFD

1 Introduction

Global demand for electrical energy is in constant growth because of the increasing population and the economy growth. There is no doubt that energy is an important area for every country, for the entire economy and society. Energy production must and should be part of the strategic plan for country and society in order to ensure the continuing development of society and industry. The analysis of all development

N. Mitrovic et al. (Eds.): CNNTech 2019, LNNS 90, pp. 221–249, 2020.
https://doi.org/10.1007/978-3-030-30853-7_13

scenarios for electrical energy production shows that general tendency is decreasing consumption of invested resources for electrical energy production [1–3]. Analysing the total world production of electrical energy, it can be observed that the dominate way of production is by burning fossil fuels (coil, nature gas oil and etc.). Comparing the economic development and undeveloped countries it can be noticed that development countries have bigger demand for electrical energy, which is very obvious, because of the development of industries and infrastructures, comparing to the underdeveloped countries. According to these data, the production and usage of the electrical energy are in direct function of economic development [1, 2]. Analysing the global demand for production of energy, it can be noticed that the most dominant source for obtaining electrical energy is by burning of fossil fuels [3, 4]. Considering the data obtained from the International Agency for Energetic (IAE), the percentage of fossil fuel in production of electrical energy are more than 70% with potential growth in the next twenty years Fig. 1. Production of electrical energy from fossil fuels loses its priority in relation to the production of electrical energy from other sources, such as nuclear energy and renewable sources. The world reserves of oil and natural gas are estimated to be in exploitation for at least 20 years, which is much less than coal reserves, which leads to the conclusion that new plants will continue to develop, as well as the existing technologies for using solid fossil fuels [1, 2].

Serbia is now in process of the industry rebuilding by opening new factories and increasing general production. This process has been caused by a constant increase in electrical energy usage and production. The production of electrical energy is increased without developing and building new energetic facilities but by extending the current capacities to the limits. By developing and improving all process in production electrical energy from excavation of coil across coil preparation up to the process of burning. Due to the constant increasing requirements for electrical energy, in the last few years there are some big investments in energy facilities, in the two biggest thermal power plants TPP Nikola Tesla and TPP Kostolac. In last few years Serbia has followed global tendency in building renewable energy capacities, building small hydro power plants, wind turbines, solar energy system, but their capacities are very small and do not have big impact in the process of producing the energy and fossil fuels will remain dominant source for production. Current process of obtaining the electrical energy is from the current facilities which work in the up level of capacity. This process is followed by decreasing quality of coil which leads to the system damages. This paper presents the usage of multidisciplinary research in identifying the problem in the system and the process of solving it [2–4].

In the Republic of Serbia, the dominant main source (fuel) for production of electrical energy is achieved by the exploitation and burning of coil (lignite). Thermal power plants have the continuous research process of finding more economical and efficient use of lignite, which can have a positive effect both locally and globally, bearing in mind that 40% of the world's global coal reserves make lignite [1, 6].

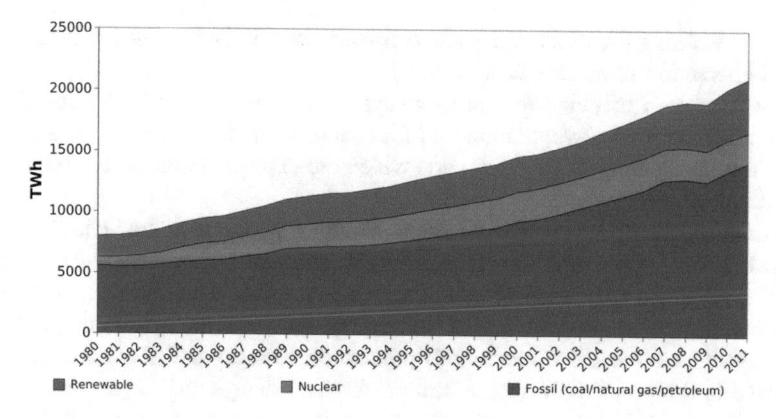

Fig. 1. Annual production of electrical energy in the world (per primary source of energy)

The most important deposits of coal in the Republic of Serbia are deposits of lignite (soft brown coal). Geological reserves of lignite in relation to the geological reserves of all types of coal in the Republic of Serbia go up to 97% [1, 2]. Projection of producing electrical energy from new and current facilities in the period up to 2030 is given in Fig. 2.

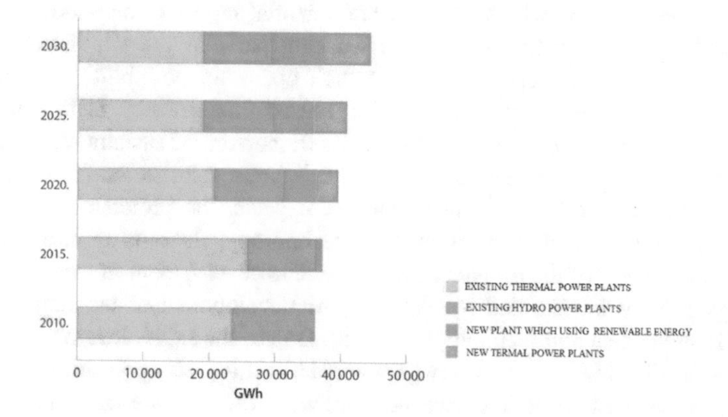

Fig. 2. Projection of electrical energy production in Serbia up to 2030

Efficient working of thermal power plants is based on the most economical production of electrical energy, obtained from combustion of lignite. Analysing whole system for preparation and combustion of coil, one of the most critical parts of the system for electrical energy production is ventilation mill.

When the materials and construction in mining and energetic industries are analysed, the most dominant damages are under the influence of numerous factors, such as: various mechanical stresses, elevated-low temperatures, compositions and atmospheric conditions, shapes and dimensions of a part or construction; structure and properties of material and quality of surface [5–8]. All those factors individually had effect on the remaining working life of materials and constructions but the biggest influence is in the mutual interaction of all this factors [8, 9].

The main aim of this paper is to analyse the possibility of increasing the wear resistance of the parts of ventilation mill for coal grinding in the power plants. Optimal technology for coating of working parts which are exposed to damages and wear will be selected by this analysis.

Working parts of the thermal energetic facilities, in this case the ventilation mill is exposed during exploitation mostly to intensive abrasive and erosive wear and also to impact loading at elevated temperatures, which lead to damages and materials lost, this processes shorten the working life of each components and the system in general. The consequences are the reduction in mill production capacity and its ventilation effects compared to the projected value, as well as frequent delays due to parts replacements that significantly affect the productivity, economy and energy-efficiency of the system.

The approach of this paper is to unite multidisciplinary research of ventilation mills of thermal power plants including a variety of theoretical, numerical, empirical and experimental methods [9–13]. The available literature shows the worldwide usage of numerical methods for different applications leading to optimization of the complex system [12]. This research continues the previous research which has been conducted for other working parts of ventilation mill. Impact plates of ventilation mill are already redesigned and it is confirmed that numerical simulation is a good tool for reducing the number of modal experiments and for more precise analysis of the process which occur during the complex flow which can indicate potential redesign methodology [14].

Researching and optimization of the working parts of the ventilation mill in a thermal power plant is a complex area, which opens up possibilities for observing problems from various aspects. The main goal of this research is to optimize the operation of the thermal power plant within the framework of sustainable development strategy, from the aspect of technological and economic valorisation [5, 12, 16]. In the countries with the developed industry, in recent years, the protection of surfaces of various working parts is designated as a key technology, since its application depends largely on the quality of the product or machine. The development of surface protection technology is very intense and is related to the development of techniques in many basic areas of science and technology [12–18]. Due to the rapid development, the gap between scientific research and development in relation to application in industry, especially in small enterprises, where there are no development sectors, increases, it is not possible to monitor such intensive development. Knowing these facts, industrially developed countries of the European Union support the transfer of knowledge in the field of the application of new protection technologies (protective layers) [15, 17].

The reliability and proper working life of the components of the thermal energetic facilities is ensured by the use of modern technologies and control methods in their design. It is achieved by selecting the appropriate material, which directly affects the efficiency of the thermal power plant and the lifetime of the working life [15].

Extreme working conditions (wear and cyclic loads) are just some of the problems in which present thermal power plant systems function. Due to inadequate exploitation, which is significantly influenced by the input parameters (the bad quality of coal and the increasing demands for energy), the working life of machines is decreasing. The trends in the development of new machines today are based on high productivity, causing a significant rise in loads, loads, speeds and operating temperatures [18]. As a result, problems with increased wear, corrosion, imbalance, vibration and finally fracture occur.

Fig. 3. Ventilation mill working parts after exploitation

Abrasiveness of coal, in addition to its physical properties, depends on its hardness and strength, the composition of mineral matter, dimensions and shape of coal particles (round, cuboid, sharp edges), the presence of various substances that increase or decrease its abrasiveness (pyrite, sand, clay) [19].

Within the coal dust preparation in the thermal power plant there are a series of processes that cannot be observed independently of each other. These are the grinding of coal in the ventilation mill, then the grinding of coal dust behind the impact wheel, drying the coal in the mill, as well as the wear of the working elements of the fan mill and its housing [12, 15]. During the process of grinding coal, it is necessary to use the appropriate energy to overcome the cohesive forces inside the particle, between its crystals and the cohesive forces within the larger crystals themselves. There are several basic modes of force effect on the particle being milled: crushing, cutting, impact and friction [12, 15].

Reducing the wear of the working parts caused by erosion can be achieved in ventilation mills by reducing the speed of particles movement, by selecting the appropriate materials from which the working parts of the mill are made, by applying coating layers to working parts. The angle of the particles and work piece collisions must be taken into account, as well as removing solid particles from the working fluid, which has a favourable effect on abrasive wear. With the increase in the diameter of the impeller mill, the erosive and abrasive wear of the working parts of the mill decreases beside the erosive [4, 12, 15].

Analysing the previous experimental research has indicated a linear dependence of abrasive wear resistance and mechanical properties of materials [4, 15–21]. Based on the mechanical properties, especially hardness, the behaviour of metals in the conditions of wearing can be predicted. However hardness is not the only characteristic that affects the wear resistance and wear in general. Parameters, which also affect the wear resistance, are the structure, shape, size and distribution of micro constituents in the coating [8–12, 15]. A final goal of this research lead to reduction of the expenses in at least one of the phases: from designing, through production to an operation, and maintenance [4, 12, 22–24].

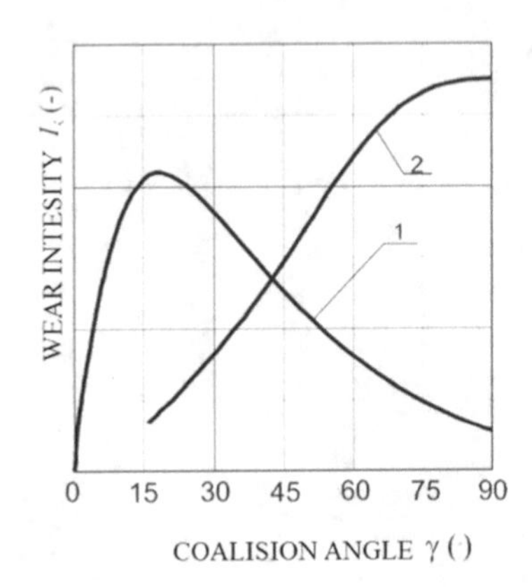

Fig. 4. Intensity of erosive wear, 1- tough materials, 2- brittle materials [4]

Multiphase flow of mixture in the ventilation mill consists of air, coal, sand and other mineral particles. Problem of the ventilation mill working parts wearing occurs during the high presence of sand in the mixture, whereas the percentage of the sand depends on the coal quality and cannot be influenced (Fig. 4).

Several revitalization methods are used for analysing and proposing the reconstruction and revitalization of ventilation mill. First, processes inside the ventilation mill and dust channel must be understood in order to analyse the results obtained by testing. Methods which are used in the processes of revitalization are CFD numerical analyses of flow inside the ventilation mill, thermography testing to see thermal process during the grinding process and, finally, according to the analysis results, the proposals for coating and reconstruction of the working part of the ventilation mill are obtained.

This research is very complex and it analyses various conditions and carries out several reconstructions of the ventilation mill. This paper presents only one reconstruction proposal in order to describe the usage of multidisciplinary research in this complex construction. This paper includes the results of testing only some working parts and working conditions in the ventilation mill. For increasing the working life of elements in the ventilation mill, the coating process is used. It uses the original working parts which are coated with material resistant to wearing. The present reliable coated surface treatment method, selected to improve the surface properties of working parts material with a coated material, gives to original working parts different characteristic. For the purpose of this research, the used filler material that has excellent resistance to wearing and oxidation is homogeneously deposited onto the surface of a substrate [12–15]. Process of revitalization represents a unique process, and that is a reason why in practice there are no standard procedures for defining and applying the most optimal technology of reparation by apply of coating or the assessment of remaining life of repaired components [4–12].

The selection of optimal coating procedures, filler materials and coating technology is done based on the results of numerical simulation, structural and mechanical analyses of samples from experimental coating model.

The character of the multiphase flow in the ventilation mill and mixing duct, where recirculation gases, pulverized coal, sand, abrasive materials and other materials are included, is directly related to the efficiency of the thermal power plants. The all particles interaction is analysed with the surface of working parts of ventilation mill in a two-phase gas–solid particle flow. In this paper the flow around house of ventilation mill around the element which gives the additional grinding and which is positioned on the house of mill is presented and analysed. Damages are indirectly determined based on the results of the flow numerical simulation; vectors of the velocity and distribution of particles in the house ventilation mill [12–15, 24–27].

Abrasive particles, reached with minerals (dominantly sand) of higher density than coal, significantly increase the wear of mill components, reducing the period between two repairs. The obstacle elements in house of ventilation mill (position is shown by arrows Fig. 2) of the ventilation mill under repair are shown in the Fig. 3. They are the elements that are the mostly exposed to wearing and they will be analysed in this paper. These obstacle elements are located at the house of ventilation mill between working wheel and house of ventilation mill and they can increase effect of grinding and generally the ventilation mill efficiency.

The obstacle elements increase the fineness of grinding of coal but they also reduce the ventilation capacity of the mill, therefore, it is necessary to remove the coal dust separator from the mill. In this way, the flow process of coal grinding in a ventilation mill with obstacle elements significantly reduces the wear of the working elements of the mill and especially the working wheel with the impact plates, extends working time of these mills and improves the process of coal dust drying.

Previous research indicates that increasing the number of obstacle elements increases the number of possible collisions of coal dust particles and in this way increases the efficiency of grinding. Also, the grinding efficiency can be influenced by increasing the dimensions of the obstacles as well as by reducing the distance between the working wheel and house of ventilation mill (reducing the average diameter of position of obstacles elements).

An example of the successful reconstruction of the combined ventilation mill DGS 100S, which is located at the TPP Nikola Tesla A in Obrenovac, is described in detail in the literature. [15, 28–31] The goal of the reconstruction is to achieve an increase grinding of coal. Reconstruction is done by incorporating three flat plates of rectangular cross-section with a width of 150 mm and a thickness of 15 mm. The plates were placed in the spiral of the ventilation mill house as an obstacle at the exit from ventilation mill to area of dust channel. After 900 h of working of the combined mill in the significant wear of working and obstacle elements, the grinding fineness is slightly increased and the humidity of coal powder is reduced, with the unchanged other working parameters of the ventilation mill. Further experiment has been carried out in order to increase the number of obstacles placed in the house of ventilation mill. The tests have been carried out with 9, 10, 11, 12, 14, 15, 24 and 25 obstacle elements. They are placed in a row with the impact wheel starting from the exit of ventilation mill to the dust channel, increasing their number to one point where whole 25 obstacles cover whole house of ventilation mill. By imposing obstacle elements, the ideal surface of coal powder increases relatively to the mill of slaughtering mill and reduces the available effort of the mill in proportion to the increase in the number of obstacles. [29–34] To simulate the process of ventilation mill working parts wearing, the sandblasting machines are used. Verification of analysed results is done by experimental testing performed at the samples in sandblasting machines chamber. Medium for simulating wear is quartz sand and the samples are positioned at the angle of 75°.

The infrared thermography has been used to analyse the ventilation mills in the thermal power plant Kostolac B. This paper presents the analyses of the experimentally obtained unsteady surface temperatures with the aim to justify application and convenience of the infrared (IR) thermography to order the measuring places of high temperatures. The measurements have been conducted by the industrial thermo-camera FLIR E40, as an affordable measuring technique for the early test phases.

2 Ventilation Mill with Dust Channel

System Ventilation mill and dust channel for grinding and delivering coal for burning is one of the main parts in the thermal power plants system that has a significant influence on the level of energy efficiency in the whole thermal power plant. The plant system consists of eight ventilation mills of EVT N 270.45, with a nominal capacity of 76 t/h of coal shown in. Mills are directly connected to the burner system consisting of four levels [1, 10].

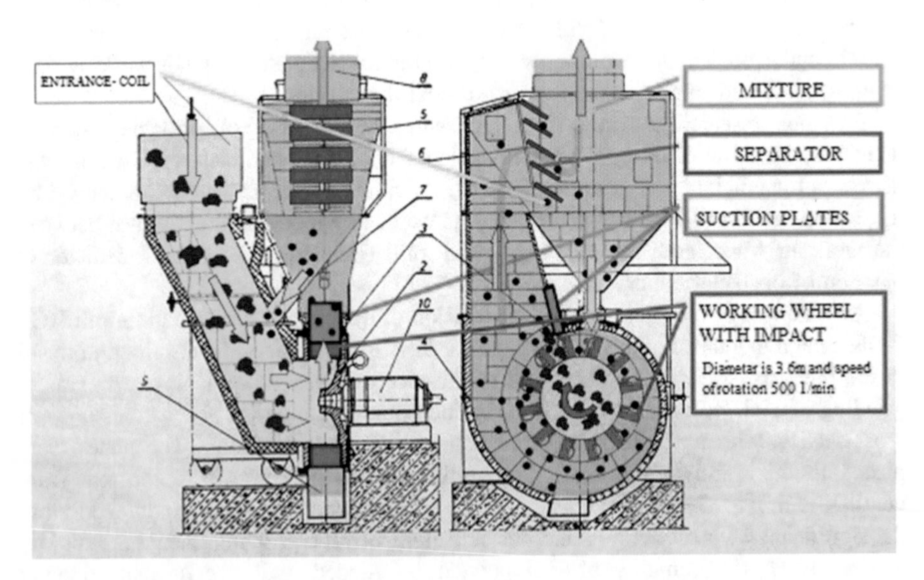

Fig. 5. Schematic shown of ventilation mill with dust channel [5].

The character of the multiphase flow in the ventilation mill and dust channel, where recirculation gases, pulverized coal, sand, abrasive materials and other materials are included, is directly related to the efficiency of the thermal power plants. Abrasive particles, reached with minerals (dominantly sand) of higher density then coal, significantly increase the wear of mill components, reducing the period between two repairs. During the process of grinding coal in ventilation mills, intensive wear occurs primarily on impact plates of the impeller and obstacles in slaughter, and then of other parts of the mill, such as its housing, the basic disk of the working wheel, the suction plate, etc., The wearing process depends on a number of factors, which relate to: the quality of the working fuel (coal), the geometry of the mill, and the material from which its vital parts are made, which cannot be fully comprehended, and determine their mutual dependence [5].

3 Numerical Simulation and Analyses of Results

The numerical simulation of the multiphase flow in this research present the first step in identifying possibilities for revitalization of the mill parts exposed to wear. By analysing the velocity vector and the distribution of flow mixtures, i.e. particles, it is possible to locate the critical points in the construction of the ventilation mill. Character of multiphase flow in the ventilation mill TE-Kostolac B, where components of air, coal dust, sand and other minerals particles participate, is directly related to the operation function of the ventilation efficiency of the mill, as well as to the wear process of vital parts. Changes in quality and flow of the mixture parameters directly affect changes in the combustion chamber and boiler plant. In addition to the efficiency of these changes, they are directly reflected in the emission of the combustion products.

The ANSYS FLUENT software package based on the finite volume method is used for numerical flow simulations. The Reynolds averaged Navier–Stokes partial differential equations of the turbulent multiphase flow in the mill are solved. A solution of the dispersed gas–solid flow is obtained, where granular phases are pulverized lignite and sand particles. The coal moisture is introduced as gas phase [1–6]. Because of the memory limitations and convergence behaviour on one side and geometry and flow complexity on the other side, the mixture model of the Euler–Euler approach is chosen. It is used geometry with obstacles in wall of ventilation mill, to see if these processes increase effect of grinding.

Numerical simulation of multiphase flow in a ventilation mill with obstacles elements is the first step in exploring the impact of obstacles on the wall of mill on the characteristics of the ventilation mill and the selection of the geometry of obstacles. For the purpose of this research numerical simulation is used to analyse flow in ventilation mill, for the two models and six geometries of with different position and angle of obstacles elements. Obstacles elements of mill have presented the plates on the wall of ventilation mill which have purpose to increase the effect of gridding. Location of these obstacles is shown in Fig. 6 The primary phase of the flow is recirculation gases, and the secondary phase of coal dust and sand particles. The first is the mixture model in the Eulerian approach, which gives the value of the static pressure at the exit of the aero mix channel, and the second Lagrangian approach (DPM), which injects solid phase particles and monitors their motion in the gaseous phase. The first generates the required geometry, that is, the shape of the impact plates and obstacles elements, as well as their position and mutual distance. In the second step, a network is generated in the numeric domain. The third step involves the choice of a multiphase model, the definition of primary and secondary phases, the definition of boundary and initial conditions, a turbulent model, numerical simulation to the convergence of the solution, and analyses the obtained results. It is applicable to very low loading and intermediate one for the Stokes number less than 1. The continuity, momentum and energy equations for the mixture, the volume fraction equations for each secondary phase and algebraic expressions for the relative velocities are solved in this model. In the considered multiphase flow the loading is very low, meaning that the gas phase influence the solid particles by turbulence and drag, but the particles have no influence on the gas carrier. The mixture model is as good as full Eulerian models in the cases where coupling between the phases is strong, or when the interphase laws are unknown. Also, computational expenses are very high for the Eulerian model due to a large number of nonlinear highly coupled transport equations.

The efficiency of grinding in ventilation mill depends primarily on the size of the kinetic energy that is passed on to coal particles in their collisions with impact elements of the impeller. In fan mills, collisions occur predominantly at the input edge of the impact plate at speeds approximately equal to the average velocity of the plate at the particle contact point in the range of 50–60 m/s. Further grinding takes place by friction of the particles on the impact surface when moving it towards the outlet edge of the plate. Crushed and dried coal particles, which passed through the impact plates of the ventilation mill, reach the absolute speed at an output of 80–100 m/s.

In order to increase the grinding efficiency of ventilation mills, especially in the part related to the subsequent grinding of coal behind the impact wheel, fixed immobilized obstacle enclosures around the house of ventilation mill in the form of obstacles are placed.

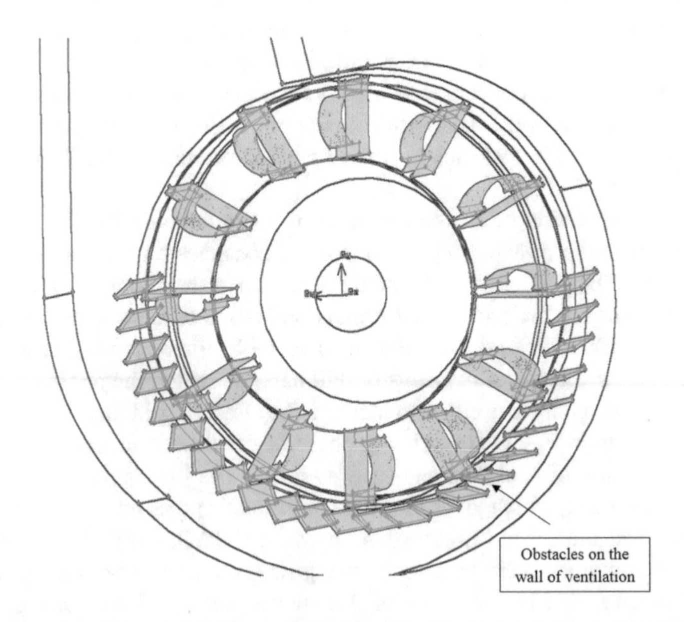

Fig. 6. (a) The geometric model of the ventilation mill with duct and channels to the entrance at the burner (arrows indicates locations of obstacles)

The first step in preparation for numerical simulation is geometrical modelling of the mill obtained from the several part drawings. The numerical modelling procedure of the multiphase flow in the ventilation mill is made up of two steps. The first step is geometry preparing and mesh generating. An unstructured tetrahedral grid is generated by consisting of 2996772 volume and 706444 surface elements. Geometric model of the ventilation mill with air mixture channel and impeller wheel with impact plate and suction plates are shown on. For the purpose of this research is used the complex analyses with two model of flow and four geometry models of obstacles. In this paper it will observe Eulerian analysis of multi-phase flow where is analysed static pressure on exit of ventilation mill. This experiment is analysed multiphase flow with three phases. The primary phase is recirculation gases, and secondary particles of coal dust and sand. In the numerical simulation of the multiphase flow, two models were used for each of the four geometries. The first is the model of a mixture in the Eulerian approach, which served to obtain the static pressure at the exit of the dust channel, and the second Lagrangian approach in which the motion of solid phase particles in the gaseous phase is monitored. This paper will show results obtained from Lagrange approach, it will present only one model of geometry which are done for this experiment. The total amount of coal which entering the mill is 16.72 kg/s. This value includes water and mineral substances. Numerical simulations were done under the condition that 40% of the total mass at the entrance to the mill are mineral substances, or our case sand.

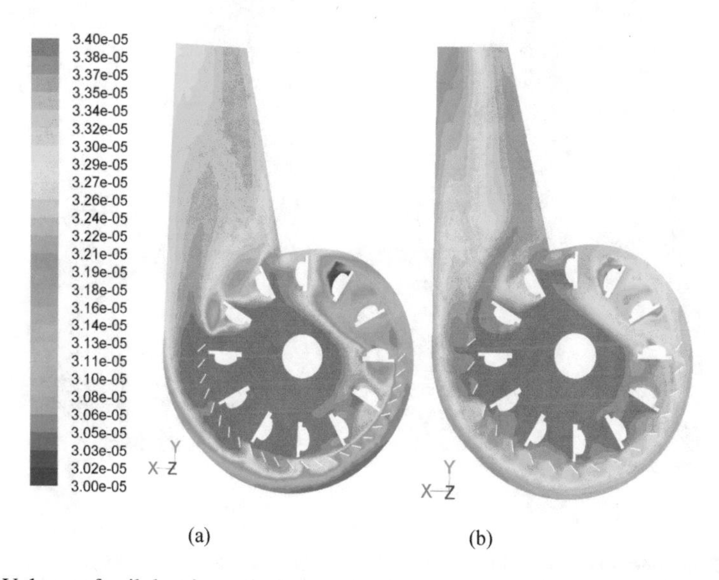

Fig. 7. Volume of coil dust in vertical plane for obstacles angle 45°, and central angle between obstacles (a) 8° and (b) 12°

The numerical results were obtained for the speed of the impeller circuit equal to 500 rpm. In this model, coal dust and sand are considered as secondary granular phases, whose movement has the greatest impact on the gravity force, the resistance force, and the coefficient of restitution in the collisions of the particles of these two phases. The diameter of the coil particles is 150 μm, while the size of particle sand was 300 μm. The volume fraction of coal dust at the entrance to the mill has a value of $3.44 \cdot 10^{-5}$, and sand $3.48 \cdot 10^{-5}$. In Fig. 7 it is shown the volume of coil dust in vertical plane for obstacles plane of 45° and two central angle. For the fixed angle of setting obstacles elements, the increase in the central angle, or the step between these elements, leads to a significantly evener distribution of coal dust in the zone working wheel of ventilation mill, in area between the obstacles, and in the dust channel. In Fig. 8 is shown the velocity field, which have low changes of intensity in vertical planes, and which is on different distance form entrance of mill. That is important for all geometries, and in Fig. 3 is given ventilation mill whit obstacles on the house on the angle 45° and distance between obstacles is 8° form the central angle of wheel and on the planes in the distance 0.5 m, 0.7 m and 0.9 m from mill entrance.

In Fig. 8 (a) and (b) it can observe three characteristic zones. First area is high velocity speed on upper part between working wheel of ventilation mill and house of mill, causes of this are high rotation speed of ventilation mill and also reduced space between hose of mill and working wheel caused by adding the obstacles on house of mill and increasing length of impact plates for 50 mm.

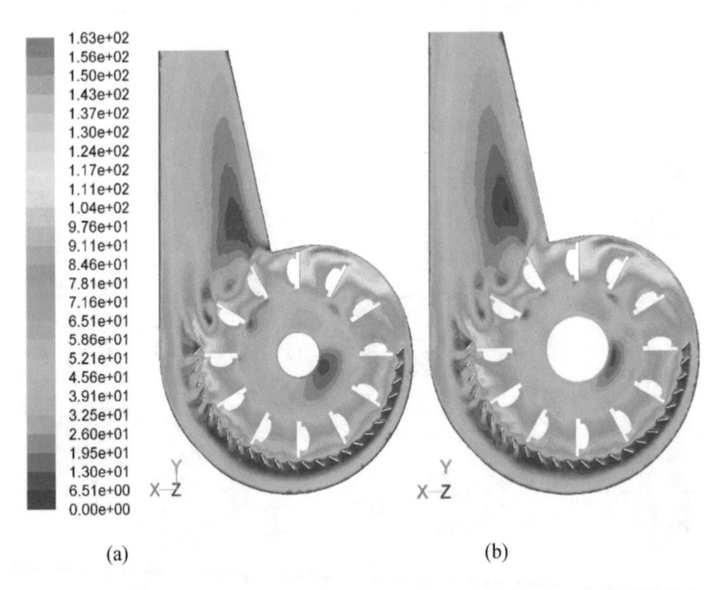

Fig. 8. Speed intensity of mixture in ventilation mill with obstacles with angle of 45° and steps of obstacle in 8° relative on central angle ventilation mill (a) obstacles on the distance 0.5°m from mill entrance (b) obstacles on the distance 0.7 m from mill entrance

Second area is very low speed in areas around the obstacles of ventilation mill. Third area is zone with very low speed in dust channel, close to place where distance between working wheel and house at least and place where done the suction of mixture.

Numerical simulation shows that maximal speed of mixture in whole ventilation mill is lower if is increased the angle of obstacles which is put in house of ventilation mill. Speed vector around the obstacles and on the surface of plates is shown in Fig. 9. Obstacles are representing barrier for mixture flow, which leads to intensive whirling and reduction of the speed of mixture between obstacles in ventilation mill. These effects become increasingly dominant in reducing the pitch of the obstacles. It can also be seen that the speed at the left side of the obstacles (upward path) at the entrance to the dust channel is considerably higher compared to the speeds on the opposite side of the obstacles (downhill).

Analysing the vector of relative speed on impact plates if ventilation mill shows that vector is oriented radial form centre. This numerical simulation has been done with original impact plates without coated layer in surface and results show that wear of plates is same as in exploitation condition. Speed intensity depend from position of impact plate and it is in interval from 50 m/s to 100 m/s.

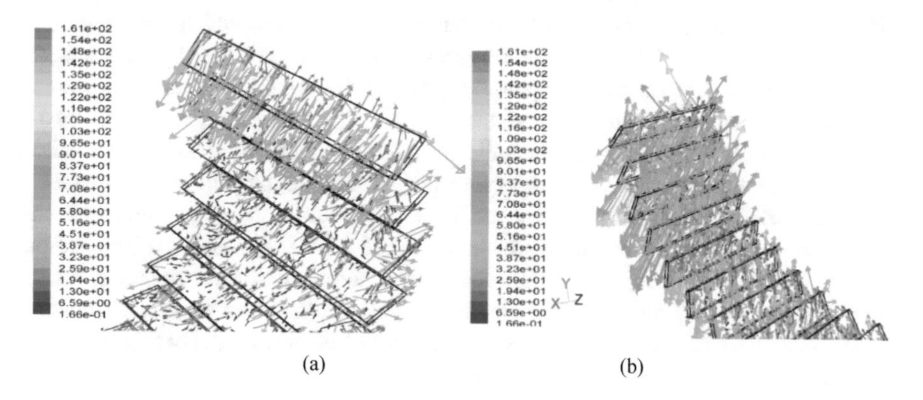

Fig. 9. Speed intensity of mixture in ventilation mill with obstacles with angle of 45° and steps of obstacle in 8° relative on central angle ventilation mill (a) Speed vector on the surface of obstacles on the right side of ventilation mill (b) Speed vector on the surface of obstacles on the left side of ventilation mill near dust channel

Obstacles in the house of ventilation mill lead to turbulence flow of mixture lead to decrease of pressure of mixture in relative to ventilation mill without obstacles show dependence of static pressure relative to angles of obstacles.

4 Thermography Testing

Analysing the temperature distribution in the ventilation mill and dust channel is done by thermography testing. This measurement has given another aspect in the operation the ventilation mill in the thermal power plant. Processes which occur during the grinding have increase the working temperatures and direct influences in the working life of components in the ventilation mill. This testing is essential for optimal selection process of revitalization of ventilation mill.

The ventilation mill heat balance is necessary in determining its degree of usefulness or fuel consumption. Thermal analysis of the ventilation mill in thermal power plant connects the calculation of the assembly plant for the preparation of the pulverized coal and the combustion chamber, in order to determine the recirculation rate and the temperature of the flue gases at its end, and controlling the thermal budgets of the other of the heating surfaces. The IR camera lens receives emitted radiation from three sources from its surroundings, as follows: from the observed object (this one is a function of the object's temperature), from the object's surroundings (reflected from the object), and the atmosphere (a grey body which partially absorbing, transferring and radiating the radiation) [20, 21]. The measurement of the temperature on the outside walls of the ventilation mill was done by a thermography camera. Heat transfer is causing the temperature rise in a zone of certain thickness of the heated wall, thus produce the density change in respect to the air at rest, far away from the heated wall.

Thermography camera recording was performed after the insulation was restored on the side walls of the house ventilation mil. On the thermography images, is shown scale with measured temperatures and selected temperature zones in which maximum temperatures are recorded. The measured temperatures depend on several parameters: the operating temperature of the multiphase mixture in the ventilation mill, the state of insulation in the ventilation mill and the ambient temperature. The analysis of the thermography Fig. 11 and the results indicates the temperature in homogeneity and the presence of insulation damage on the system of the ventilation mill - dust channel.

Herein, the emission from the test model is a function of its temperature corrected for the emittance of the surface and transmittance of the atmosphere. The various sources, including the objects in the close surrounding, emit their radiation energy which then is partially reflected from the test model. The reflected energy is a function of the temperature of the objects in the surrounding of the test model (a common one), reflectance and transmittance of the test model.

In this paper are presented results obtained from the global zones of the specific temperature intervals by, selecting the zones with the highest/lowest temperatures. Furthering, the measuring lines are selected in correspondence with ventilation mill superficial geometry, accounting the places with the expected higher level of turbulence in the flow (for example around the obstacles as are the fastening ribs, the flanges, the service openings, and the joints, *etc.*). Finally, for the series of temperature readings along with measuring lines, during the recording time, are reorganized, processed and compared.

4.1 Thermography Measurement of Ventilation Mill 6

In Fig. 10a is shown ventilation mill 6 which is part of this experiment. Main goal of thermography recording of Ventilation mill is to fully understand the process of wear and fully understand working condition of ventilation mill.

(a) ventilation mill (b) Front view (c) left view

Fig. 10. Measurement of temperature distribution by height of ventilation mill 6

Results obtained from recording ventilation mill 6 by thermography camera shows that working part of ventilation mill are exposed to elevated temperature Fig. 10b and c shows group thermograms for ventilation mill 6, measured in different high and positions. Measuring the temperature and distribution along the ventilation mill 6, from the different angles of observation, it observed a zone with elevated temperatures which goes up to 198 °C from the front side of ventilation mill which is shown in Fig. 10c. The area on the left front side of the ventilation mill it can observe damages of insulation layers. In this area temperature is the same as is inside of ventilation mill. The temperature fluctuations at the front of the ventilation mill do not show a pronounced dynamic, and it is possible to conclude the stabile distribution of temperature in time.

In Fig. 12 the setup for the temperature measurements is shown. The six measuring spots are ordered along the support part, Fig. 12, assuming the creep or no flow along struts. The five vertical, VL, and the twelve horizontal lines, HL, were selected by matching the locations of the maximal local temperatures and geometry details on the ventilation mill front.

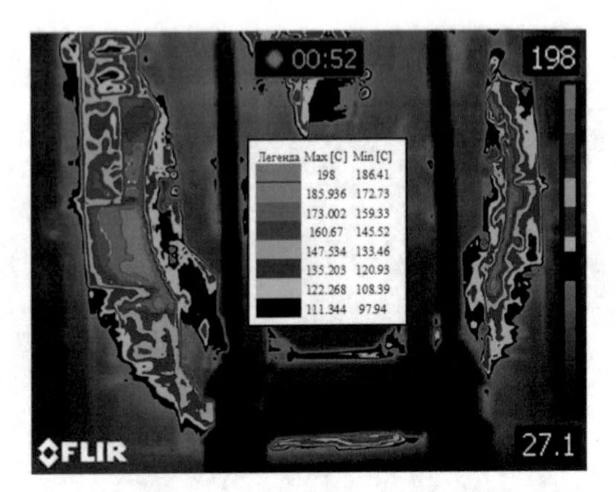

Fig. 11. Record of ventilation mill 6 front view

This experiment has wide range of records for purpose of this paper will be presented selected characteristic measurements. By analysing thermography, it can have observed changes colours of outer shell which is present high surface temperature. In mark zone temperature is between 173 °C–186 °C in the front of ventilation mill left and right along the flange with the dispenser. One more zone with higher temperatures from 186 °C–198 °C on the surface of ventilation mill near the flange. This zone is indicated in Fig. 12 as uninterrupted.

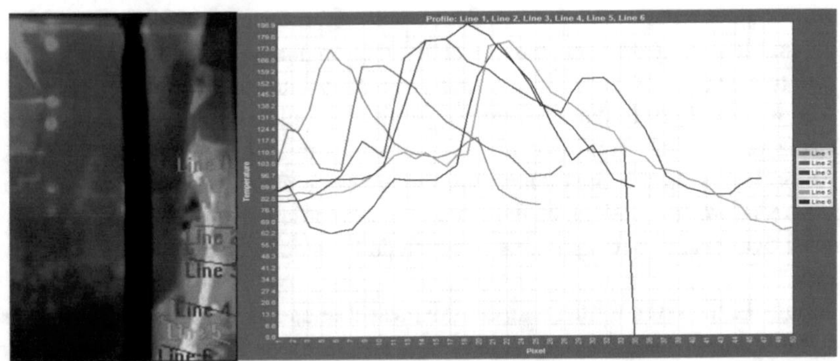

(Measuring line from right sides and temperature distribution according to measuring lines

Fig. 12. Thermography record front view

Figure 12 shows the temperature distribution at the front of ventilation mill in the flange area, whereby the recording is made from a smaller distance, from which there is a slight difference in the temperature values when comparing the results in Fig. 12. Results also confirms the presence of a high-temperature area at the front area of ventilation mill in area of flange and the presence of enlargement in the zone of the reinforcement rib on the left. The 186 °C–194 °C temperature range also confirms the distribution of the temperature along the measuring lines on the on the right, Fig. 12. By temperature distribution in real time it possible to estimate the temperature at all measuring lines. Variations in temperature during measuring time can be due to the part of uneven heat transfer between the isolation of ventilation mill and the outside environment, caused by the creation of forced air flow and partly due to the unevenness of the heat supply from the inside of the mill.

The research confirmed that it is necessary to periodically examine the temperature distribution on the outer walls of the ventilation mill with an IC thermography, because the obtained results allow the rapid, contactless determination of all temperature changes as well as on time interventions that would maintain and optimize the technological and functional performance of the thermoelectric power plants.

5 Modal Analyses of Coated Samples

Experimental modal analysis is done to select optimal coating technologies to increase working life of elements in ventilation mill. Speed and pressure from numerical simulation has identify parts of mill which are the most exposed to wear, also speed and angle of mixture which act on surfaces of working parts. The process of selection the optimal coating procedures, filer materials and coating technology which will be used in exploitation condition in later stage of experiment, it will be done based on the results of numerical simulation, structural and mechanical properties of samples from experimental coating model which is explained in this chapter. Experimental research indicates approximately linear dependence of abrasive wear resistance and mechanical properties of materials. Especially hardness can be predicted behaviour of metals in the conditions of wearing, based on the mechanical properties. The penetration of the abrasive particles and high hardness is inversely proportional to the hardness of the surface layers [3, 22, 30–32]. However, hardness is not the only characteristic that affects the wear resistance and wear in general. Parameters which also effect on the wear resistance beside hardness are the structure, shape, size and distribution of micro constituents in the coated layer [3, 12–27]. The wear resistance of a coated alloy depends on many other factors such as the type, shape and distribution of hard phases, as well as the toughness and strain hardening behavior of the matrix [4]. It is necessary to select an appropriate filler material which has optimal properties, great hardness and optimal resilience, which lead to reducing the wear. In previous research wide scale of filler materials resistance to wear is analysed [12] and for the purpose of this experiment, one group of filer material for model testing is selected and analysed potential implementation in the functional testing.

Modal testing of macrostructure has been done, diagrams of distributions of hardness have been made, zone of the surface layer and Heat affected zone (HAZ). Microstructural analyses were obtained with scanning electron microscopy (SEM) and with EDS analysis. The choice of filler materials and coating procedures which will potentially use and applied in the redesign of ventilation mill working elements it can be done from the results of this research. Optimization of ventilation mill performances is achieved by correlating the inevitable technological process and the state of the main working parts. Numerical simulation of multiphase flow, based on the available parameters, indicated the modes of failure system components and facilitates the removing of causes.

5.1 Experimental Procedure of Modal Testing

Modal testing is carried on the samples coated with adequate layer. Samples dimensions 200 × 200 × 15 mm, that are made of hot-rolled steel sheet S355J2G3 (ASTM 572), were prepared and used as substrates. In order to prevent the inclusion formation in the material deposits, prior to process of coating, the oxide layers were removed from the substrate surface by means of the grinding. Before coating, samples are preheated at Tp = 160–170 °C. Conventional process of coating (welding) was used: manual metal arc welding (MMA) and flux core wire welding (FCAW), plasma coating, and HVOF (High Velocity Oxy-Fuel) process. After process of deposition samples with coatings were air-cooled down to the room temperature.

Table 1. Materials and coating procedures

Coating material	Nominal chemical compositions (M*/C and M/Fe wt. % ratio)	Process hardfacing
PG6503	Ni-Cr-Bo-Si/60% WC	Plasma process
55586C	CoCrFeCW/WC	HVOF
4010 EC	Fe-Cr-C-Si/Ti 0.18 (7.0 and 0.6)	MMA
4395N	Fe-C-Cr-Ni-Mo-W-B	FCAW

The filer materials, used in this paper, were presented in Table 1, with coating procedures, signs and nominal chemical compositions and commercial name manufactured by Castolin Eutectic Co, Ltd, Vienna. The most important condition of processes of deposition is that have low degree of mixing and the manufacturing cost of deposition consumables. Coating is carried out by technological lists with defined parameters (power, voltage, welding speed, line energy, preheating of materials and coatings, and application method of coatings).

5.2 Microstructural Analysis and Hardness Test

Samples for structure and hardness analyses were carried out by the water jet on the plane perpendicular to the coating surface. The obtained cross-sections are ground with SiC abrasive papers down to P-1200 and polished with alumina suspensions down to 1 μm. Microstructural tests were carried out on polished samples etched with a solution of 3% Nital. The microstructures of coated specimens were observed by scanning electron microscope (SEM) and their chemical compositions were examined by energy-dispersive spectroscopy (EDS). The SEM-EDS analysis was creating electron imaging allowed different phase's morphologic description of the coatings and EDS compositional maps were used to qualitatively describe chemical variations in the microstructure. Also, the measurements of micro- hardness in the cross section of coated samples were made using a Vickers hardness tester HV 10.

5.3 Experimental Testing

The microstructures with chemical analysis of the coated samples are analysed; also hardness was evaluated. Structural tests were carried out on prepared and polished samples for purpose to scanning electron microscopy (SEM) with EDS. Also, the measurements of hardness in the cross section of coated samples were made by Vickers (HV10).

Table 2. Materials and coating g procedures

Coating material	Thickness of layer in mm	Degree of mixing, %	Hardness, HV10
PG6503	≈4.5	11	750–1049
55586C	≈0.5–0.7	10	644–701
4010 EC	≈5.2	20	887–1028
4395N	≈5	10	918–1091

The low degree of mixing (11% and 10%) is achieved in coated samples with filler material PG6503, 55586C and 4395. The maximum of hardness is achieved in samples coated with alloy 4395 (MMA) and 4010EC. Measurement of hardness distribution is done for all samples and range of values is presented in the Table 2. Values of hardness depend of point which is tested, coating layers is very tick and depend if the device possibilities. Process of measurement has been done in thru cross section of modal coated samples. The characteristic of coated layers and the variation in hardness can be explained by the distribution of the different phases along the depth of the coated deposit, as it was evident from microstructural investigation. Macrostructure and measurement of hardness distribution of the metalized sample cross section with a hard coating with filler material PG 6503 done with plasma metallization is presented in Fig. 5. Hardness vales of all samples are given in the Table 2. Diagram in Fig. 13 indicates that in near surface areas, points 7 and 13 has the highest values of hardens in the more than 1000 HV.

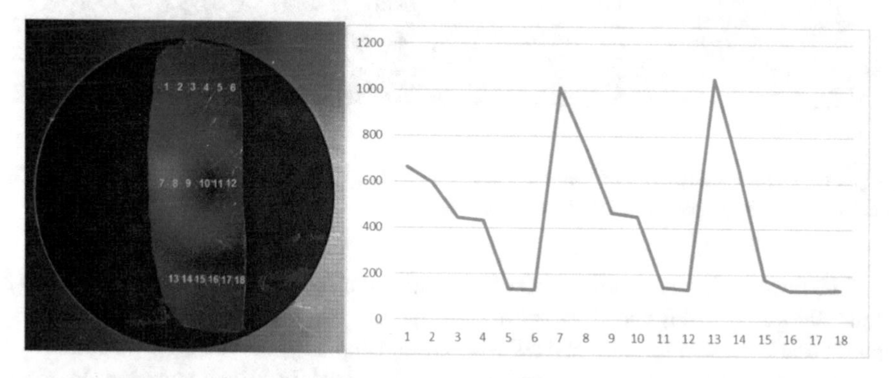

Fig. 13. Macrostructure and hardness distribution on the sample cross section done with filler material PG 6503.

Results shown in the Table 2 indicates that a filler material 4010 EC and 4395N, show relatively homogenous distribution of hardens and microstructure in the cross-section and much higher values of hardness relative to other two samples. Values of hardness in those two samples with filler materials 4010 EC and 4395N are higher in 20–40%. Besides that, hardness interval in cross section is higher for 10–80%. Those two samples have much higher values of degree of mixing base material and filer material because are done by conventional process of coating (welding). Values of hardness are different in fact that some filler materials have similar chemical compound. This is consequences of different coating procedures and technologies of applying. SEM micrographs with EDS of different phases of coated samples with filler material PG6503 are shown in Fig. 6. The EDS data, shown in Table 3, indicate on different matrix compositions and presence of carbides in coated layer.

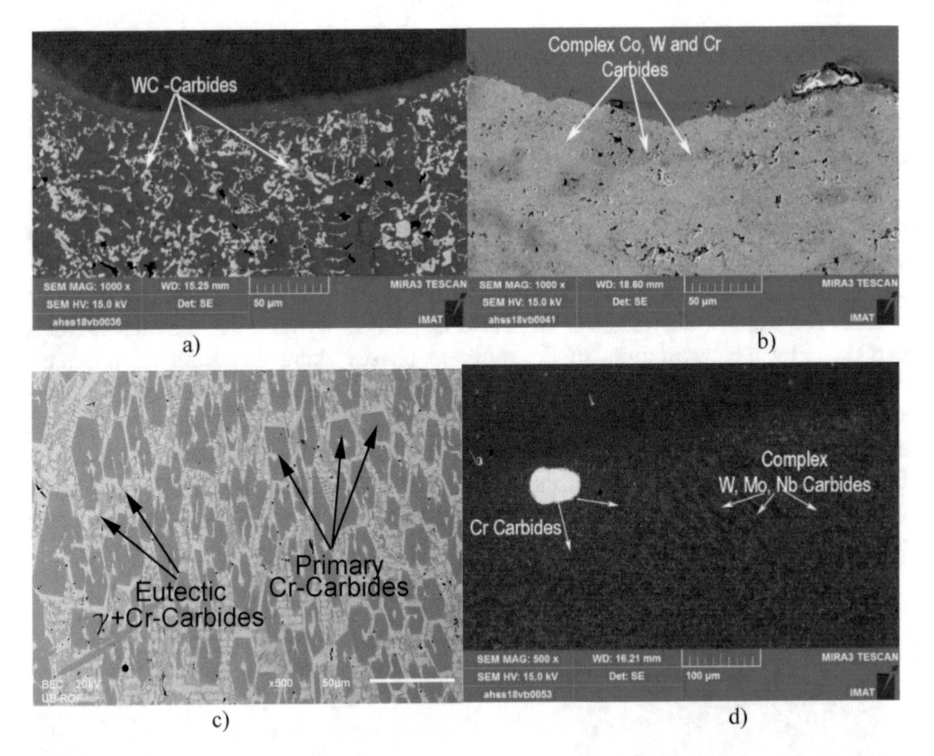

Fig. 14. The structures of the samples, (a) – PG6503, (b) 55586C, (c) 4010 and (d) 4395 hardfaced coatings (back-scattering electron images).

The near-surface structure of PG 6503 coating consisted of large WC grains point 2 (light phase, Fig. 14a) and EDS spectrum (Fig. 14b and c) embedded in the Ni-based matrix (dark phase, Fig. 15a) and EDS spectrum (Fig. 14d) in which Fe was dissolved in a major amount, whereas Si and B were dissolved in minor amounts. In many locations, the small, worm-like, and random-oriented WC particles were also observed.

The distribution of large WC particles (120 ± 31 μm) was non-uniform in coating's thickness direction but their presence was largest in the near-surface region of PG 6503 coating. During solidification achieves also complex carbides the near-eutectic structure and dominant presence of blade-like primary Cr-carbides, Cr-borides (Fig. 14a). In sample coated with PG 6503 is observed the presence of small chromium-based boride/carbide reinforcing particles and their short interparticle distance which strengthened more effectively the Ni-based matrix.

The Fig. 14b shows the structure of the material 55586C. Deposition of material shows homogenous structures with presence of porosity. Matrix of this coating is Fe base matrix with eutectic W carbides. Analysing the filler material 55586C it can be seen the presence of complex Co and Cr carbides. Coated sample show a homogeneous distribution of carbides with polygonal and spherical shape. During the process of deposition its created very complex carbides, dominantly built with W but it crates various spectrum of complex structures. Phases in Fig. 15 shows dominantly eutectic W-carbide and presence is around 16%. Other complex carbides are W-Co-Cr around 27%, Fe-Co-W around 19%, W-Co-Cr –Fe around 5% of presence.

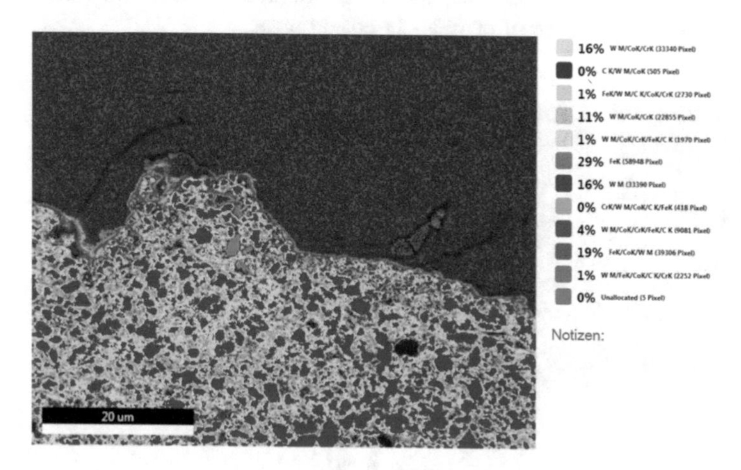

Fig. 15. Samples coated with filer material 55586C mapping of structures

Table 3. EDS chemical composition of the phases in alloys PG6503, 55586C, 4010 and 4395 in mass %

Samples	Phases	C	Fe	Co	Ni	Cr	Nb	Mo	W
PG6503 Ni-Cr-Bo-Si/60% WC	Matrix	2.62	64.29	/	24.74	0.35	/	/	7.28
	Carbides	4.33	40.15	/	12.91	0.42	/	/	42.20
55586C CoCrFeCW/WC	Matrix	5.51	74.97	3.91	/	2.66	/	/	12.94
	Carbides	8.10	2.62	14.85	/	5.58	/	/	68.85
4010 EC Fe-Cr-C-Si/Ti 0.18 (7.0 and 0.6)	Matrix	3.84	81.94	0.50	0.21	9.99	/	/	/
	Carbides	10.11	37.02	0.52	0.25	52.25	/	/	/
4395N Fe-C-Cr-Ni-Mo-W-B	Matrix	5.90	64.08	/	/	13.15	0.67	7.56	8.20
	Carbides 1	4.35	22.88	/	/	18.25	7.55	16.26	30.55
	Carbides 2	3.31	58.11	/	/	34.04	/	1.04	3.50

The structure of the obtained 4010 coatings comprises a larger rod-like primary Cr-carbides with the total absence of the blade-like type of morphology (Fig. 14c). These results are in agreement with previous research findings [12, 17, 35] in which the addition of carbon facilitates transition from blade-like towards the rod-like morphology. The rod-like morphology is superior when wear applications are considered. The increase of the Cr-carbides volume fraction is probably the result of the carbon and chromium addition [17, 18, 26, 35]. It has been point out that large primary carbides, which are identified to be (Cr, Fe)7 C3 (Table 3), are formed from the melt of the coating (hardfaced) electrodes and exhibit columnar growth with a hexagonal cross section. However, the samples coated with filler materials 4395 show a homogeneous distribution of carbides with polygonal and spherical shape (Fig. 16). The EDS date shown in Table 3 indicate that coated layer is phase with high content of W, Cr, Nb, Mo and C. Present carbides are the primary, special carbides type WC, MoC, NbC, complex carbides Fe3(W, Mo)3C and Cr-carbides [3, 12, 26]. As the special and complex carbide fraction increases, hardness and wear resistance improve [1–3].

SEM micrographs with EDS of different phases of coated samples PG6503 are shown in Fig. 16. The EDS data, shown in Table 3, indicate on different matrix compositions and presence of carbides in coated layer.

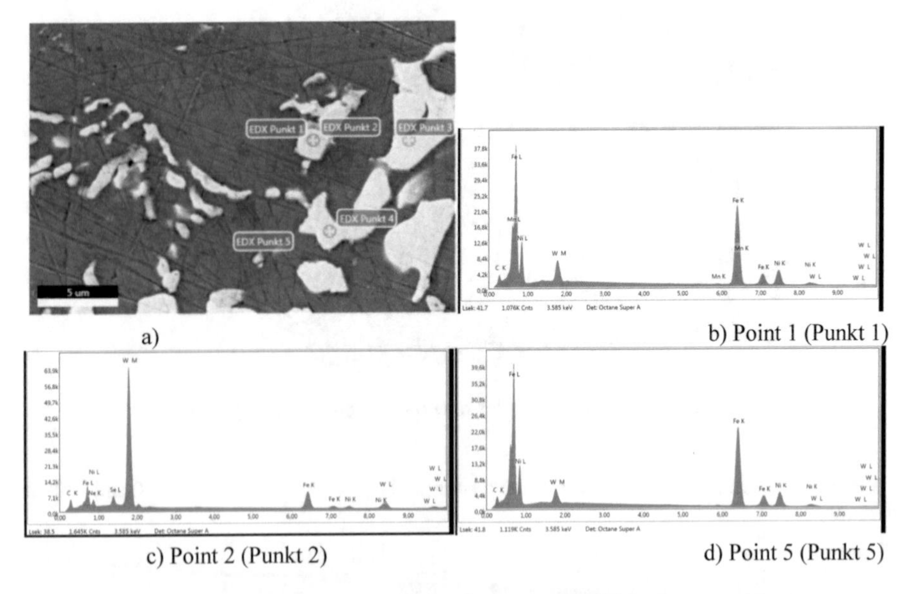

Fig. 16. SEM micrograph and EDS structure of the coated sample PG6503

6 Experimental Testing of Modal Samples

Wear of working parts in ventilation mill has been indicated by lost the material from the surface. Experimental testing which is set in this research has opportunity to simulate the working condition in ventilation mill in order to analyses behaviour filler material in exploitation condition. For simulation of working condition is used standard sandblasting chamber and machinery for material cleaning. Set of machinery is shown in Fig. 17. This research shows the analyse, positions of obstacles in the wall of ventilation mill (their angle, density) in order to increase the grinding efficiency and reduce the wear of working parts. The grinding efficiency behind the impact wheel will be greatest if the obstacles elements are placed administratively on the direction of movement of larger particles of coal dust. The angle at which coal particles leave the impact wheel depends largely on the coefficient of friction between these particles and the metal of the impact plate. For the friction coefficient of 0.6, the angle at which the carbon powder leaves the impeller is 29.5°, while for the coefficient of friction 0.8 this angle has a value of 25.7°. The angle at which the transport gas leaves the working wheel would be smaller than the angle under which coal particles are made, but this only applies if coal dust particles are of such a size that, due to inertia, they cannot fully follow the gases of the gaseous fluid. When the coil particles are small enough, as is the case are analyzed, the mixture of the gas and solid phase acts as a whole and the angle α under which the impeller is greater than the angle for the conveyor fluid and smaller than the angle for the charcoal powder. The term for the angle under which the obstacles elements should be placed so that they are governed on the path of the largest particles of coal dust is given in [1] as

$$\beta = arcsin\left(\frac{D_1}{D_0}\cos\alpha_u\right) \tag{1}$$

Where α_u is the angle at which coal particles leave the working wheel, D_o to the middle diameter of the position of obstacles elements on house ventilation mill, and D_1 is the diameter of the working wheel. By modifying the above term, provided that the mixture as a whole hits the barriers obstacles elements at a right angle and taking into account that D1/Do = 0.92, for the angle $\alpha = 25°$, the angle of setting the slaughtering obstacles $\beta = 54.3°$ is obtained, while for $\alpha = 20°$ it gets $\beta = 58.7°$. Such specific angular values β do not differ much from the value in which the numerical simulation gives the lowest value of the static pressure at the entrance to the dust channel.

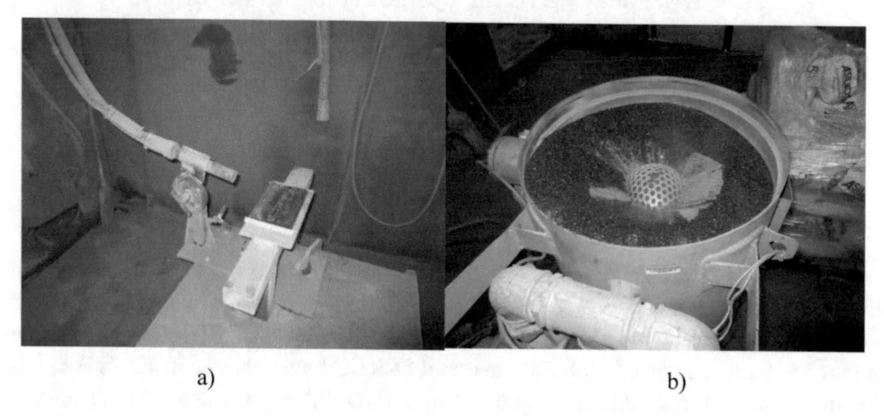

a) b)

Fig. 17. Machinery for cleaning metal in sandblastings chamber (a) experiment set; (b) sand for simulation wear

For the purpose of this experiment the analyses of coating materials are done for two different angles 20° for 75°. Those two defined angles are in coincidence with angles which are obstacles in house of ventilation mill are exposed to mixture flow. Those two angels are representing most critical wear expose of working parts on flow of mixture of coil, sand and other mineral mater.

It is necessary to done appropriate simulation of working condition in ventilation, exposing of working parts to the mixture of coil and mineral mater. The dominate wear in working of ventilation mill has been from sand which can be up to the 40%. For the simulation process and for the purpose of speed process of wear is used the sand and sandblasting machinery. The range of reusable and non-reusable blasting abrasives in different grain sizes, degrees of hardness and grain shapes is correspondingly broadly based.

Selected blasting abrasives, are frequently based on processed natural or artificial minerals, and are mostly used in mobile blasting systems, for instance on construction sites. ASILIKOS is a synthetic, mineral, environment-friendly blasting abrasive in accordance with ISO 11126-4N/CS/G. It consists of slag-tap granulate and can be used in accordance with the requirements of BGR 500 (Table 4).

Table 4. Asilikos sand technical details.

Name	Silicon dioxide	Aluminium oxide	Iron(III) oxide	Calcium oxide	Potassium oxide
Quantity	42–58%	23–32%	3–15%	2–8.5%	0.5–4.6%

6.1 Experimental Testing Modal Plates Are 75°

Mass reducing of modal test plates during the experimental test is shown in Table 5. Modal test plate are exposed to angle of sand flow 75°. Time of exposure to sand flow is similar for every group of filler material with small differences.

Table 5. Materials and coating g procedures for exposure angle 75°

Coating material	Time of exposure (s)	Material lost (kg)	Material lost (%)
PG6503	373	0.066	1.5
55586C	371	0.081	2.79
4010 EC	363	0.052	1.24
4395N	352	0.020	0.5

After test performance minimum material lost has sample with filler material 4395N for 0.020 kg which is shown in Fig. 10. Samples coated with filler materials 4010EC have 0.052 kg, PG6503 has 0.066 kg and 5586C has 0.81 kg which is shown in Fig. 18. Samples coated with filler material 55586C have material lost 2.79% of total volume deposited material on sample which is 0.09–0.1 kg. Generally, filler material 4395 show best wear resistance for this simulate wear, where is analysed mass lost and percentages of mass lost relative of volume deposit layers. But generally all filler materials except 55586C has god wear resistance on sand flow to angle of 75°. Percentages of material lost has some different values, in the Table 5 all results of material lost relative to volume of coated material are presented. Filler material 4395 shows minimum material and percentages are 0.5% total, relative it masses of coated sample. Deposit layer of filler material is between ~ 1.300 kg per sample.

Filler material PG 6503 has material lost 0.066 kg which is shown in Table 5, mass of coated layer is ~ 1.660 kg per sample and total material lost is 1.5%. Filler material 4010EC has material lost 0.052 kg which is shown in Table 5. mass of coated layer is ~ 1.450 kg per sample and total material lost is 1.2%. In Fig. 18 are shown all coated samples after testing in sandblastings machinery. On several samples is observed the damages of surface layers. The biggest damage is present in sample coated with filler material 55586C, it can see big hole from sand flow. Sample coated with filler material 4395N has small surface damages, but this material has minimum material lost. Other two samples coated with filler materials 4010EC and PG6503 show the minimum surface damages relative to other two samples. Those two samples have the bigger material lost then sample coated with filler material 4395N.

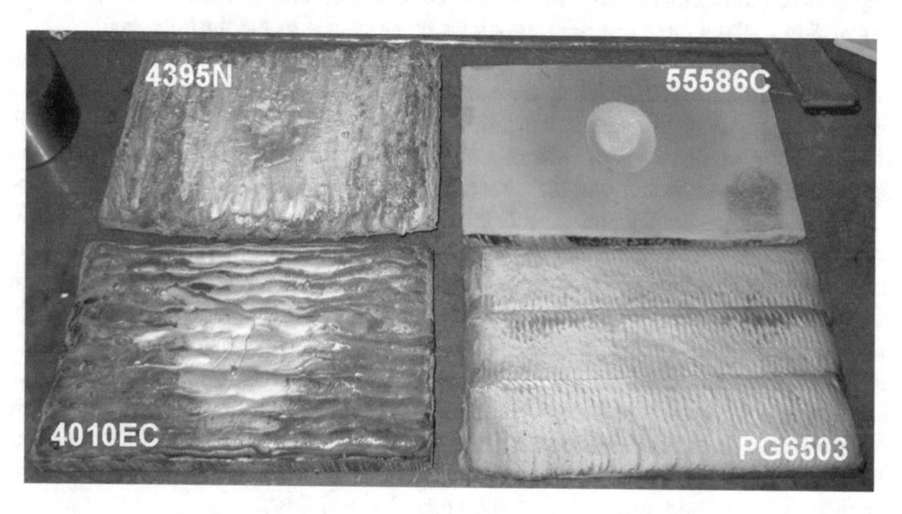

Fig. 18. Appearance of modal testing plate after experiment

7 Conclusion

There is no doubt that energetic was and will be area of special importance for the entire economy and society. If the supply and production of energy is made a stable, modern and well-organized sector, it is certain that this will mean good for the entire economy of the country. Conversely, if energy will not have given enough attention from the point of view of strategic planning, there is a certain weak position and poor prospects of the economy as a whole [2].

The numerical simulations of the multiphase flow are performed for a ventilation mill in the Kostolac B power plant. The results obtained by the numerical simulation clearly show that the CFD methods done in ANSYSS FLUENT provide detailed analyses of flow in ventilation mill with obstacles element in house. The results obtained in the simulations are using in determination of locations, which is most exposed to wear, due to sand and other mineral particles movement. Numerical simulation of the multiphase flow is performed for system ventilation mill and dust channel with recirculation gases as the primary phase and particles of coal dust and sand as secondary granular phases. A used model is the mixture in Euler's approach was in order to obtain a reliable static pressure at the exit of the dust channel, and the Lagrange approach to see how obstacles elements affect the movement of particles of different sizes. Several geometries of obstacles elements of rectangular cross-section are defined, and their setting angle and step are changed. For the central angle of 8° and 12° the position of obstacles elements was at 45°, 55°. In the mixture model, the coil dust particles are 150 μm and the sand particles are 300 μm. The volume fraction of coal dust at the entrance to the ventilation mill has a value of 3.44•10-5, and sand 3.48•10-5. When is present the fixed angle of obstacles elements, it is noted that increasing the step between the elements leads to a much evener distribution of coal dust in the area of the working wheel, between the obstacles elements, and in the dust channel.

Numerical analysis determines angles under which particles of mixture leave the house of ventilation mill. This paper analyses the appropriate angle of installation of obstacle elements in the ventilation mill object in order to increase the grinding efficiency. The most efficiency of ventilation mill grinding will be if the obstacles elements on the house are set directly in the direction of movement of larger parts of coal dust. The angle at which coal particles leave the house of mill depends largely on the friction coefficient between these particles and the impact plate. For the friction coefficient of 0.6, the angle at which the coal dust leaves the dust channel is 29.5°, while for the coefficient of friction 0.8 this angle has a value of 25.7°. The angle under which the mixture leaves the ventilation mill to dust channel would be smaller than the angle under which coal dust particles are made, but this only applies if coil dust is a critical size that, can movement together to the mixture inertia. When the coil dust in mixture are small enough, as in the case of analysis, the mixture of gas and solid phases behaves as a whole and the angle α under which the stroke of the wheel is greater than the angle for the conveyor fluid, and less than the angle for coal dust.

The thermography measurement indicates that working part of ventilation mill are working at elevated temperature which can increase damages during the work. This is very important for use appropriate coating material resistant to elevated temperature. The equipment and time costs as well as the efficiency of the IR thermography as a method for identification of turbulence zones are affordable and sufficiently well in the cases when other methods are not available or applicable. The IR thermography is intended and justified for the early-stage observations. Further, for results that are more precise the high-speed and high-resolution IR camera type is recommended. Experimental tests indicated that as the number of obstacles in the house of ventilation mill increase, the grinding process is better and remaining working life of the elements has been increased.

On the basis of the numerical simulation results and experimental tests of the modal testing, the selection of critical area, optimal coating technologies and filler materials for apply on working parts of ventilation mill will have carried out. Deep analysis of several parameters is used before selecting optimal filler materials and coating technologies for modal samples. By analysing the properties of wear resistance coatings (layers) of model samples, the results of numerical simulations, the characteristics of the available equipment and additional materials, as well as other presented results, the method of application of wear resistance coatings, coating materials and conditions for applying them has been carried out. More precisely, the following criteria were applied in the selection: working conditions, abrasion resistance, required finishing quality, permissible thickness, porosity of the layer and cohesion strength of the base and applied layer, characteristics of the basic material, shape and size of the work piece and costs of whole process. Wear resistance is determined by the size, shape, distribution and chemical composition of the carbides, as well as by the matrix microstructure. To evaluate the wear behavior of the different Fe, NI, Co-based coating alloys it is important to regard the macro hardness and the microstructure and to set them into relation. Coatings with lower hardness showed lower abrasive wear resistance, but the dependence (hardness vs. wear rate) was nonlinear.

After the applying alloys on modal samples, the simulation of wear conditions which occur on ventilation mill is done which is close to working conditions. For the purpose of this experiment the analyses of coating materials are done for angle for 75°. Selected angle are representing most critical wear expose of working parts on flow of mixture of coil, sand and other mineral mater.

Comparative, quantitative and visual, analyses have indicated that samples goes from 2.79% with filler materials 55586c to 0.5% of weight loss with filler materials 4395, and greater resistance to wear compared to the original plates with 3.9% weight loss. The experimental tests have pointed to the possibility using this filler materials and coating procedures can lead to wear reducing of working parts ventilation mill compared to the existing ones. Other two materials coated with filler materials 4010EC and PG6503 has weight lost from 1.21 to 1.5, but is very important to check weight of deposited layers where 4395N has smallest one. Analysing the weight of all plates it can be observed that plate with filler material 4395N have the minimum weight lost.

The application of this approach can reduce the number of possible repairs and extends the period between them and give significant economic effects. The experimental functional tests of revitalized mill wearing parts in real exploitation conditions show that the proposed modification, coating technologies and filer materials can give good results. Research results are applied in the thermal power plant of Kostolac B and can be used for other parts in facilities and in similar thermal energetic facilities. The application of appropriate reconstructions, supported by additional research, in other coal preparation system, can expect good results in terms of reducing the wear of the working elements of the ventilation mill and extending their lifetime, and thus facilitating coil exploitation with a smaller number of stops in exploitation.

Acknowledgements. This paper is the result of research within the project TR 34028, which is financially supported by Ministry of Education, Science and Technological Development of Serbia, Messer Tehnogas and PD TE – KO Kostolac.

References

1. Kozić, M., Ristić, S., Puharić, M., Katavić, B., Prvulović, M.: Comparison of numerical and experimental results for multiphase flow in duct system of thermal power plant. Sci. Tech. Rev. **60**, 39–47 (2010)
2. Kozic, M., Ristic, S., Katavic, B., Puharic, M.: Redesign of impact plates of ventilation mill based on 3D numerical simulation of multiphase flow around a grinding wheel. Fuel Process. Technol. FUPROC-03583, pages 14 (2012)
3. Blau, P.J.: Metals Handbook, volume 18, Friction, Lubrication and Wear, 9th edn., pp. 320–380. ASM, Metals Park (1993). ISBN 978-1-61503-163-4
4. Kazemipour, M., Shokrollahi, H., Sharafi, S.: The influence of the matrix microstructure on abrasive wear resistance of heat-treated Fe–32Cr–4.5C wt% hardfacing alloy. Tribol. Lett. **39**, 181–192 (2010)
5. Perković, B., Mazurkijevič, A., Tarasek, V.: Lj. Stević, reconstruction, and realization of the projected modernization of power block B2 in the TE Kostolac. Termotehnika **1**, 57–81 (2004)
6. Ristic, M., Prokic-Cvetkovic, R., Kozic, M., Ristic, S., Katavic, B.: Wear reducing of ventilation mill suction plates based on the multidisciplinary research. J. Balkan Tribol. Assoc. **21**(3), 493–513 (2015)
7. Shah, K.V., Vuthaluru, R., Vuthaluru, H.B.: CFD based investigations into optimization of coal pulverize performance: effect of classifier vane settings. Fuel Process. Technol. **90**, 1135–1141 (2009)
8. Bhambare, K.S., Ma, Z., Lu, P.: CFD modeling of MPS coal mill with moisture evaporation. Fuel Process. Technol. **91**(2010), 566–571 (2009)
9. Ristic, M., Prokic-Cvetkovic, R., Kozic, M., Ristic, S., Pavisic, M.: Numerical simulation of multiphase flow around suction plates of ventilation mill in the function of extending its remaning working life. FME Trans. **45**(4), 154–158 (2016). ISSN 1451-2092
10. Ristic, M., Radovanovic, L., Prokic-Cvetkovic, R., Otic, G., Perisic, J., Vasovic, I.: Increasing the energy efficiency of thermal power plant Kostolac B by the revitalization of ventilation mills. Energy Sources Part B (2015). https://doi.org/10.1080/15567249.2015.1014977. ISSN 1556-7257
11. Liu, C., Jiang, D., Chu, F., Chen, J.: Crack cause analysis of pulverizing wheel in fan mill of 600 MW steam turbine unit. Eng. Fail. Anal. **42**(2), 60–73 (2014)
12. Perisic, J., Radovanovic, L.J., Milovanovic, M., Petrovic, I., Ristic, M., Bugarcic, M., Perisic, V.: A brine mixing mobile unit in oil and gas industry—an example of a cost-effective, efficient and environmentally justified technical solution. Energy Sources Part A **38**(23), 3470–3477 (2016)
13. Ulmanu, V., Badicioiu, M., Caltaru, M., Zecheru, G., Draghici, G., Minescu, M., Preda, C.: Research regarding the hard-facing of petroleum gate valves by using high velocity oxygen fuel technology. J. Balkan Tribol. Assoc. (4), 551–558 (2010). ISNN 1310-4772
14. Perišić, J., Milovanović, M., Petrović, I., Radovanović, L., Ristić, M., Speight, G.J., Perišić, V.: Application of a master meter system to assure crude oil and natural gas quality during transportation. Pet. Sci. Technol. **36**(16), 1222–1228 (2018)
15. Ristic, M., Vasovic, I., Alil, A., Obradovic, J., Prokic -Cvetkovic, R.: Failure analysis of carbon steel screws under the service in the presence of corrosion environment. J. Balkan Tribol. Assoc. **21**(3), 493–51416 (2018). ISSN 1310-4772
16. Perisic, J., Milovanovic, M., Petrovic, I., Radovanovic, L., Ristic, M., Perisic, V., Vrbanac, M.: Modelling and risk analysis of brine mixing mobile unit operation processes. Energy Sources Part B **12**(7), 646–653 (2017)

17. Kovacevic, M., Lambic, M., Radovanovic, L., Kucora, I., Ristic, M.: Measures for increasing consumption of natural gas. Energy Sources Part B **12**(5), 443–451 (2017)
18. Gercekcioglu, E., Odabas, D., Dagasan, E.: Effect of impact angle on the erosive wear rate of dual-phase (DP) AISI 1020, AISI 1040, AISI 4140 and AISI 8620 steels. J. Balkan Tribol. Assoc. **15**(1), 25–35 (2009). ISNN 1310-4772
19. Buchely, M.F., Gutierrez, J.C., LéOn, L.M., Toro, A.: The effect of microstructure on abrasive wear of hardfacing alloys. Wear **259**, 52–61 (2005)
20. Šolar, M., Bregant, M.: Filler materials and application for repair welding. Weld. Weld. Struct. **51**(2), 71–77 (2006)
21. Bouzakis, K.-D., Michailidis, N.: Coatings' and other materials' mechanical properties considering different indenter tip geometries and calibration procedures. J. Balkan Tribol. Assoc. **1**, 10–18 (2008). ISNN 1310-4772
22. Wang, X., Hanb, F., Liu, X., Qua, S., Zoua, Z.: Microstructure and wear properties of the Fe–Ti–V–Mo–C hardfacing alloy. Wear **265**, 583–589 (2007)
23. Kirchganer, M., Badisch, E., Franek, F.: Behavior of iron-based hardfacing alloys under abrasion and impact. Wear **265**, 772–779 (2008)
24. Zhou, Y., Huang, Z., Zhang, F., Jing, S., Chen, Z., Ma, Y., Li, G., Ren, H.: Experimental study of WC–Co cemented carbide air impact rotary drill teeth based on failure analysis. Eng. Fail. Anal. **36**(2013), 186–198 (2013)
25. Hou, L.-F., Wei, Y.-H., Li, Y.-G., Liu, B.-S., Hua-Yun, D., Guo, C.-L.: Erosion process analysis of die-casting inserts for magnesium alloy components. Eng. Fail. Anal. **33**, 457–464 (2013)
26. Correa, E.O., Alcantara, N.G., Tecco, D.G., Kumar, R.V.: Development of an iron–based hardfacing material reinforced with Fe-(TiW)C composite powder. Metall. Mater. Trans. A **38**(5), 937–945 (2007)
27. Choo, S.-H., Kim, C.K., Euh, K., Lee, S., Jung, J.-Y., Ahn, S.: Correlation of microstructure with the wear resistance and fracture toughness of hardfacing alloys reinforced with complex carbides. Metall. Mater. Trans. A **31**(12), 3041–3052 (2000)
28. Ray, S.K., Nandi, P., Mukhopadhyay, M.S.: Development of chromium-rich hardfacing welded deposit using in situ carbothermic reduction. J. Mater. Sci. Lett. **11**, 1469–1470 (1992)
29. He, B., Zhu, L., Wang, J., Liu, S., Liu, B., Cui, Y., Wang, L., Wel, G.: Computational fluid dynamics based retrofits to reheated panel overheating of no. 3 boiler of Dagang Power Plant. Comput. Fluids **36**, 435–444 (2007)
30. Thermo investigation and analysis of boiler plant blocks B1 and B2 in Kostolac TE, PD Ltd. Production–technical sector (2014) (internal study)
31. Katavić, B., Jegdić, B., Prokolab, M., Prvulović, M., Budimir, S., Milutinović, Z.: Optimal parameters estimation by the analytical methods for the welding of the GS-36Mn5 steel. In: Congress Welding 2012 & NDT 2012/Proceedings of abstracts (2012). ISBN 987-86-82585-10-7
32. Katavić Boris, J.B., Odanović Zoran, H.N., Mladenović, M., Jaković, D., Ristivojević, M.: Prediction of optimal parameters of repair welding steel 13CrMo4-5 by analytical methods. Zavarivanje i zavarene konstrukcije **55**(3), 91–96 (2010)
33. Vencl, A., Gligorijević, B., Katavić, B., Nedić, B., Džunić, D.: Abrasive wear resistance of the iron and WC based hardfaced coatings evaluated with scratch test method. Tribol. Ind. **35**(2), 123–127 (2013)
34. Gligorijević, B.R., Vencl, A., Katavić, B.T.: Characterization and comparison of the carbides morphologies in the near-surface region of the single- and double-layer iron-based hardfaced coatings. Sci. Bull. Politehnica Univ. Timisoara Romania Trans. Mech. **57**(71), 15–21 (2012). ISSN 1224-6077

Integral Model of Management Support: Review of Quantitative Management Techniques

Sinisa M. Arsic(✉) and Marko M. Mihic

Faculty of Organizational Sciences, University of Belgrade, 154 Jove Ilica st, Belgrade, Republic of Serbia
sinisaars@telekom.rs, marko.mihic@fon.bg.ac.rs

Abstract. In this paper, an overview of existing methods and techniques of quantitative management has been elaborated, which can be used for the definition of an integral model for management support. A special focus is on the needs of management in terms of a gradual, scientific approach to the organization's management processes, appropriate methodologies and techniques, validation and verification tools for defined models. The subject of the research is an analysis of the scope and parameters necessary for defining an integral model for support of management in profit-oriented companies. In addition to analysis of the input parameters for the model and analysis of the output parameters from the model, the subject of the research includes defining the ways to implement the model in everyday business practice.

A more detailed analysis of a set of potential approaches for defining a quantitative model, with the help of the concept of "machine" and "statistical" learning. The above concepts are necessary for modelling the relationship between all previously analyzed parameters in order to achieve proposed goals. The choice of a reliable model defined by quantitative research should be carried out using the methods and techniques presented. Key advantage of using quantitative management is reflected in the maneuvering power of multiple functions that reflect real business processes, that is, in the correct modelling of real occurrences in business. In this way, the exploitation of all variables is enabled, which have a sufficiently high level of correlation with main business outcomes.

Keywords: Quantitative management · Model · Machine learning

1 Introduction

Modern research is increasingly concerning the use of statistical tools to successfully analyze extremely complex sets of parameters, i.e. modelling future outcomes, based on previous historical data (knowledge). Data-based can be applied in various areas (finance management, marketing, medicine), and is currently taking into account practical application, since it is increasingly defined as the core business intelligence tool in large corporations. Machine learning methods and techniques, along with the concept of analyzing large databases, create an indispensable part of the understanding of the usefulness of quantitative management, but also as an exploitation tool of model outputs, thus supplementing decision support system.

N. Mitrovic et al. (Eds.): CNNTech 2019, LNNS 90, pp. 250–264, 2020.
https://doi.org/10.1007/978-3-030-30853-7_14

In [1] was created the process of developing a model for predicting the company's entry into the problematic phase of existence, which implies three main components:

- Sample definition and data collection (design and description of parameters, design of sample size);
- Selection of methods and specific parameters (business indicators and others) necessary for defining a model for predicting future values of parameters of interest;
- Testing and confirming the credibility of the model in terms of significance and error in assessment (confidence interval).

Bearing in mind the existing theoretical and practical research, the purpose of this chapter is to present practical examples of the successful application of quantitative management in the function of management support.

The literature review paid much attention to potential consideration parameters in defining an integral model to support the management of profit-oriented companies, which could be used by the owner or top management members. It is necessary to provide the necessary and adequate level of quality of the defined model, and this can be achieved through methods and measures for verification and validation of the models, whose tools are available in the existing literature, but whose application is not possible every time or is in the author's focus.

2 Concepts of Statistical and Machine Learning

James [2] has defined key areas of statistical learning, on the basis of which mathematical models of quantitative management can be defined, based on the concept of statistical learning. Linear regression, linear regression assemblies, classification, "resampling" methods, methods of selection and regularization of the linear model are considered.

Du Jardin [3] analyzed existing models of classification, on the basis of which the forecasting of the values of parameters is carried out, over the long-term horizon. The prediction period was expected to pull down accuracy levels. Lately, we are considering models which take into account the company's life cycle in terms of most important business parameters and the company's bankruptcy forecast.

An overview of existing research over the past 50 years on the topic of concepts of applying statistical learning is given in the table below (Table 1).

Table 1. Evolution of existing concepts

Research	Focus of research	Tested parameters
Beaver [4]	Financial ratio indicators as basis for predicting business crisis	Classification methods
Von Stein and Ziegler [5]	Development of procedures for early warning of latent risks	Parameters available from balance sheets
Laitinen [6]	Dimensions of influence of financial condition on potential bankruptcy	Investment indicators

(*continued*)

Table 1. (*continued*)

Research	Focus of research	Tested parameters
Sheppard [7]	Effects of strategic survival factors of the company	Financial leverage
Mossman [8]	Ability for prediction of different models	Prediction power
Nam and Jinn [9]	Models for early warning of forthcoming crisis	Prediction momentum
Gepp and Kumar [10]	Prediction of bad business results based on statistic learning	z-test of bankruptcy regression analysis
Tian [11]	Research of relative importance of different bankruptcy predictors	Lasso regularization

The conclusion in most of the examined studies is that companies that are more likely to undergo organizational changes have a significantly higher risk of entering a problematic phase, and that the need for a quality model of prediction is more pronounced. However, Xu and Wang [12] directly link poor management and inadequate decision making with the company's failure, citing business efficiency as a good parameter of the forecast.

Lussier and Halabi [13] considered the key parameters for the forecast of business success (or failure) in their research. De Bock [14] analyzed the power and simplicity of interpreting the model for the prediction of business failure, citing an example where the output is defined based on a set of parameters that characterize the company's business profile, activities and business results in the previous period.

The model is adapted with a cubic "regression spline", and the final set of parameters for forecasting the company's entry into problems is generated with the help of "lasso" regularization, elastic network and "ridge" regression, with similar fitting results. Li and associates [15] used the classification and tree regression, and a 30-hold holdout method was used to evaluate the model, i.e. a method of allocating 30 parts of the sample from the entire test sample.

In order to adequately validate and verify the process of defining an integral model for management support, it is necessary to first determine the preconditions and ways through theoretical review of the literature on the subject of validation and verification. Collins [16] concludes that rarely can be found published research which fully describes the defined prediction model in terms of validation of results, thus resulting in limited possibilities for real use.

2.1 Measures of Verification of Prediction Model

The task of verification consists in determining whether the detected gap in the existing literature is directly related to the parameters defined in the model, and whether it is in correlation with the opinion of the examined managers (executive directors or owners of family companies). Also, verification should provide an answer as to whether the input parameters are adequately defined and specified, and whether their use will produce the desired effects.

The field of performance of the model verification measures is to test the consistency, completeness and overall behaviour of an integral model under very different circumstances. The aim of the implementation of measures is to remove all "blind" terminations, in order to provide a systematic, clearly defined outcome from the model, in any combination of entered initial parameters, with a clearly defined structure of relations between parameters.

An additional reason for the use of verification measures (and not exclusively statistical tests aimed at checking the accuracy of defined conclusions) are found in the previously identified shortcomings of existing model for forecasting business results, which have never experienced becoming a handy tool for supporting business process management. The assumption is that exclusive use of identical distribution for test set and training set is not a sufficient guarantee that the model is complete and consistent at the level that it can be trusted.

Goodfellow and Papernot [17] state in their research: "It is clear that testing the expected inputs is sufficient for the traditional application of the models defined by the process of machine learning, but the verification of those unusual inputs is also necessary to guarantee the quality of the model. Mechanical learning models can be easily reversed by checking with unusual inputs. Testing can no longer be sufficient for defects in such models, and in the future, higher quality verification techniques will be required to examine the full effectiveness of the model. Developing model quality guarantees, characterized by the verification of the level of confidence about input that is properly processed, is central to the future of machine learning, and will almost certainly be based on formal verification via verification".

2.2 Measures of Validation of Prediction Models

Unlike verification, the validation considers the accuracy of prediction using the defined model. Validation tools will serve to determine whether the defined model (input system, interactions and recommendations following the predicted outcomes) adequately represents all those parameters which are positioned as key in the literature review, as well as those parameters defined by the top management/owners of companies. Validation is achieved with adequate settings in model definition through adequate software support, as a continuous process which compares the possibilities for successful prediction before the existence of an integral model, and the possibility of a successful prediction of business results in the future, after influencing the impact of knowledge defined after implementation of the model.

In literature, there are standard measures of describing model associativity, distinguishing basic measures and measures of predictive association. According to Healey and Gerald Prus [18], the most important are:

- Kendall Tau-b, with the help of which one can analyze the absolute correlation measure between the ordinal parameters (positivity/negativity affects the direction of correlation);
- Spearman Rho, which can be used to analyze the data-based correlation coefficient, where the level of separation between the arrays is considered (approximation to level "1" indicates a higher level of correlation);

- Gamma, as a measure of symmetric association between two ordinal parameters (approximation to level “1” indicates a higher level of correlation);
- Lambda, as a measure of association that reflects the error of estimation of the dependent parameter, when the new independent parameter is assigned, in the analysis of the dependent parameter (level “0” means that the independent parameter does not contribute to the prediction of the dependent, while level “1” denotes a perfect correlation);
- Goodman Kruskal-Tau, on the basis of which the error of estimation of the dependent parameter is calculated (the value “1” denotes the complete association, and “0” is the complete absence of the same).

An additional way to validate the model is cross-validation-a statistical tool for evaluating the performance of a defined model, as well as a specific check of the contribution of parts of the empirical sample to the overall level (inaccuracy) of the defined model. Kohavi [19] lists the training sub-set and the validation sub-set as the main elements of each cross-validation set. Efron [20] supplemented with the usual methods for cross-validation, such as:

- holdout,
- leave-one-out cross-validation,
- k-fold cross validation.

In addition, the mentioned methods and measures for evaluating the integral model are not finite, so Ivanescu [21] lists the practical exit of the model output through the bootstrapping method, the adjusted R2 coefficient and the shrunken R2 coefficient.

It is important to emphasize that the processes of verification and validation consist of a number of extremely complex activities, and that the prerequisite for the success of the model will lie in reaching the level at which the planned measure of verification and validation is met. That is, to what extent it is possible to test the variability of the model over time (by adding or subtracting parameters from the model). Measures aiming to establish and further strengthen the discovered relationships in the model, which is one of the expected outcomes of the model. The application of verification and validation methods, can contribute significantly, to increasing the reliability of the defined model, which is the prerequisite for any practical application (beyond the control of the author of the model).

2.3 Parametrization of Assumptions Prior to Definition of a Model

For the purposes of operationalization of parameters necessary for adequate definition of an integral decision support model, it is necessary to collect information on parameter values from several different angles (source types) in order to confirm the relevance assumption. The key segments are: entry information about the company, organizational, strategic, financial and non-financial.

Specific parameters (variables) which can be observed for each segment can be a product of a qualitative research (organized with focus groups of interviews), where a researcher can identify the integral set of research parameters influencing the business results of the company.

Entry parameters describing the personal information of the company, usually consist of the following:

- Type of ownership - a parameter defining ownership relations, rights and obligations (joint stock companies, limited liability companies, entrepreneurs, partnerships, corporations, holdings);
- Region (city, village) - parameter that determines external business preconditions;
- Number of employees/annual income - a parameter determining the size of the company and its potential for generating additional revenue and new investments;
- The number of years since the founding of the company - the parameter that refers to the longevity of the company and the parameter that correlates with the level of maturity of the organization;
- Industry - a parameter that is determined by the specific industry sector within which the organization operates;
- Type of business (production, services, combined)

The table below contains a potential set of variables divided by segment, which can be discussed and broadened within a qualitative research.

Table 2. Parameters involved with each segment

Segment	Research	Parameters
Organization	[22–24]	Owner is the same person as the CEO - a parameter that identifies the owner with the chief manager, with the idea of understanding if there is a concentration of power in the image of the owner; CEO is the same person as the professional manager - similar to the previous parameter, with the exception that this parameter allows testing the differences between the owner (s) occupying the position based on previous experience, or based on the formal level of expertise (validated by the academic institution or certification body); The owner already had a company (which went bankrupt, or was sold) - a parameter which provides an insight into the history of previous endeavours of the owner, which can contribute to the future quality of decision-making; Level of corporate culture - a parameter that talks about the compliance of the mission and the vision of top management members and company owners
Strategy	[25, 26]	Impact of consumers - a parameter which defines to whom it is intended to offer products and services of a company, to a public or private owned company, operating in foreign or domestic market; Market competition - a parameter which defines the level of competition on the market where the company markets goods; Suppliers - parameter describing the length of the chain in the process of production or service delivery, as well as the diversity of the network of associates The share of online sales in total sales - a parameter that shows the level of supply representation in all cities and regions where there is no physical place for selling products and services; Level of innovativeness - the number of innovations in the phase of the idea (approved by management), which are not yet in the prototype stage or brought to commercialization;

(*continued*)

Table 2. (*continued*)

Segment	Research	Parameters
		Share of innovation projects - the level of investment in innovation in relation to the company's total revenues; Operations with foreign markets, in terms of exports and imports; Global horizon of planning (quarter, half or yearly, and longer midterm periods) - parameter referring to focus of managers, in terms of planning business ventures; Critical points (checkpoints): a composite parameter, defined in the form of the expected growth/declination. Critical points relate to the expected future market growth or entry into new markets, an increase in share in the existing market, or the acquisition of other operations
Financial-business	[27, 28]	Operating income and expenses, Receivables from sales of goods and services, Short-term liabilities Investments, Liabilities or receivables from suppliers
Non-financial	[29]	Human factor in decision-making (stress, motivation, intuition) Family members working in the company (in the case of a family company)

By evaluating individual, organizational and market performances, while taking into account the trends and institutional framework for the operations of a particular company, it is possible to use an integral model, which can be further upgraded, and which serves the management as a modern tool for everyday use. In addition, the defined quantitative model is also an excellent input for the continuous analysis of the trend by the company's life cycle phases, which can be used by individual users and organizations, which in different ways provide the resources necessary for the functioning of profit-oriented companies (institutions of the government, banks, investment funds, etc.).

3 Procedure for the Definition of an Integral Model Based on Statistical Learning

After the necessary pre-selection of variables (defined within the chapter Parametrization of assumptions prior to definition of a model) which allows a broader and more complete understanding of current business practice (clustering of organizations operating on the same market or within the same industry branch), it can be discovered that only after the parametrization activities, it is possible to approach the definition of a procedure which enables researchers and organizations to identify the life cycle of the organization, and, on the other hand, to create a framework for the business performance forecasting model, which are identified as key for strategic management of the company. The process thus created would be re-structured over time, so that it best responds to changes in business system trends. As the integral of the quantitative model for the forecast of business results has already been thoroughly elaborated, that is, after qualitative research has been confirmed "on the ground", the results of the model can be interpreted as part of various other analyzes and research.

The process of defining an integral model is presented with iterative steps, which should be undertaken by any researcher who wants to repeat the modelling process later. To understand the context and broaden the reader's readiness for the forthcoming description of the procedure for defining an integral model, it is desirable to consider the following authors and literature:

- James [2] was considering statistical learning as a way to understand complex data sets and seemingly unrelated trends;
- De Bock [14], who analyzed the application of the machine learning model to predict the collapse of business, with the balance between the power of the prediction model and the ability of the model to be interpreted as quality (2017);
- Hofmann and Klinkenberg [30], Rapid Miner as a data mining tool and business analytics support.

Below follows the iterative view of the steps by which it is possible to train and test the model for successful prediction of business results.

Step 1. Firstly, several process must be assessed, such as understanding the relationships between variables which contribute to the process of forecasting business results. Second, variables with a dual character, on the one hand, as variables in the modelling process, and, on the other hand, as outcomes of the forecast model, or previously unaddressed values of the variables. The basic task of the forecast model is to find the form/prediction function, which deviates minimally from the realistic function that maps the correlations between the research variables, and leads to the possibility of prediction the key dependent variables (business outcomes described in Table 2). Since it is necessary to provide a forecast of business results in order to achieve the main goal modelling, as well as to provide an understanding of the relationship between the variables, no component of the model should be a "black box", or, the model must be designed in order to provide an optimal level of interpretation, since predictability without interpretation is not sufficient.

Step 2. The next step involves identifying the most important variables, which contribute to the most realistic forecast of key business indicators. Also, this step involves the identification and analysis of key relationships (positive, negative) between the "outcome" of the forecast and variables which contribute to the overall forecast. Finally, it is also necessary to define the relationships between "outcomes" and individual variables, through linear equations, or by graphically representing more complicated relationships (scatter plots, etc.).

In order for the model to be trained based on the optimal level of flexibility, it is necessary to apply a nonlinear approach to modelling. That is, to actively define the balance between predictability and interpretability of model results, i.e. to manage the level of understanding (correlations) between "outcomes" and research variables will be more challenging or less challenging. Also, this step involves defining the experimental region of variables in order to eliminate the unreliability of the model in the case of entering values that are beyond the experimental region.

Step 3. The next step is the process of model training, that is, learning based on data, through a parametric model of statistical learning. The parameter model chosen is an estimate of the prediction function of the key business parameters, which, with the help of several parameters, contribute to the quality of the forecast, i.e., reduce the confidence interval of the predicted value of a certain outcome. The alternative consists of a non-parametric approach to machine learning based on data, which does not imply any assumption of the function of the prediction function, therefore it does not contribute to reaching the main goal of this research.

Step 4. Involves the creation of *n* decision trees and n simple linear regression (correlation of dependent variables based on only one independent variable), preparation of rules based on the analysis of individual decision trees, as well as all linear conditions. The functions (regression splines) for continuous events are estimated. Different prediction models are analyzed (linear regression, multiple regression). The process of training the model based on data was accompanied by an analysis of the model's flexibility (which has not yet been tested on previously unused data) in order to determine the optimal level of complexity of the model. Excessive tracking of "noise" in data leads to a higher level of flexibility, but also to an unsatisfactory level of interpretation of the model.

Step 5. Regarding access to statistical learning, supervised or unsupervised, the need conditioned by the main goal and research issues impose the use of both approaches. To be clearer, with the supervised approach, the specific value of the "outcome" of the prediction is defined within the confidence interval (95% or 99%) and the prediction interval (or deviation in the forecast). With an unsupervised approach, there is no outcome which can be formulated as the unknown, since the main goal is to understand the relationships between variables and the database classification/ segmentation, in order to move from the unmanaged to a manageable domain.

Step 6. Prior to the testing process on previously unintended data, it is necessary to separately analyze the effect of interaction (the influence of one or several parameters on the outcome) to see if some parameters have a greater impact with common observation than individually. Also, part of the same step is an analysis of multicollinearity, which aims to eliminate variables which do not contribute directly to the outcome. James (2013) claims that we should eliminate all variables without which it is possible to maintain the same (or similar) level of indicator R.

Step 7. The test data set (records of parameter values of individual companies which were not used in the training process of the model) will be provided by sharing the initial sample of *m* companies on the training and test set. Testing processes will measure the accuracy of the prediction (comparison of the ROC curve), as well as the percentage of errors of the first type ("false" predicted exactly) and the percentage of mistakes of another type ("false" predicted incorrectly). An estimation error is estimated on the test error. Approaches to evaluate model optimality (Cp, RSS, Adjusted R2 parameters) are taken into account and a decision is made about the model with optimal level of balance between predictability and the ability to interpret the results.

Step 8. This step consists of implementing approaches for regulation of the selected model using regression splines (Lasso and Ridge), to check whether it is possible to produce better predictability or better interpretation of the model. Certain regression coefficients, with the values of the parameters for the prediction, will be reduced or eliminated, in order to reduce the influence of some regression parameters.

Step 9. Approach model validation, i.e. retraining methods, to check the variability of the regression model due to taking into account data not previously used for model training, but also to demonstrate that the results of modelling are statistically significant, that is, to explain and present how the used sample can represent the necessary and sufficient level of information with the purpose of creating an integral model.

Step 10. The last step in the modelling process is the sensitivity analysis in terms of eliminating or adding a new variable of interest, at a later time (post validation period), which improves the criteria of optimal predictability. It is especially valuable to foresee a training moment based on the new set of prediction variables, as well as the implementation of the regularization process for the redefined version of the model.

Upon completion of the presented procedure, often within the preparation processes of scientific research papers, the results are tested in accordance with the set of imposed research hypotheses. Below is a summary of how concrete outputs are derived from a quantitative model, and how can these derivatives provide support in testing the main and auxiliary hypotheses.

4 Testing of Research Hypothesis Justifying the Model

The methods and techniques of quantitative management, in addition to the previously shown, have the potential to be of use in the scientific sense, so that they can also be used within processes of testing the underlying research hypotheses. Below is a possible demonstration of mathematical proof that tests a particular problem or gap in the existing literature.

- If slope/s are ALL equal to zero, then the unknown variable is not a linear function of $x_{1,}$ x_q;
- For our Hypothesis Test, our goal is to "Reject" the hypothesis "ALL Slope/s = 0";

If "ALL Slope/s = 0", then model would be no better than the mean value line for making predictions.

1. State The Null and Alternative Hypotheses.
   - Null Hypothesis = H_0 = All Slope/s = 0
   - Alternative Hypothesis = H_a = All Slope/s <> 0
     - "At Least One Slope is NOT Equal to Zero"
2. Set Level of Significance = "alpha".
   - Alpha = risk of rejecting H_0 when it is TRUE.
   - Alpha determines the hurdle for whether or not the test statistic just represents sample error or there is a true "statistically significant" difference (past the hurdle).
   - Alpha is used to compare against p-value. P-value <= Alpha, we Reject H_0 and accept H_a
   - Alpha is often 0.05 or 0.01.

- When testing the slope, when we get a statistically significant difference, it will mean:
  - It is reasonable to assume that at least one of the slopes is not zero.
  - It is reasonable to assume that there is a statistically significant relationship.

3. Rejection Rule:
   - "If p-value is less than our alpha, we reject H_0 and accept Ha, otherwise, we fail to reject H_0."
4. From Sample Data calculate the Test Statistic and then calculate the p-value of the Test Statistic.
   - Use F Test Statistic (and F Distribution (F Test Statistic, q, n − q − 1)) for Testing Overall Significance:
     - F Test Statistic is displayed with (1):

$$F = \frac{\mathrm{SSR}/q}{\mathrm{SSE}/(n-q-1)} = \frac{\mathrm{SSR}/(dfRegression)}{\mathrm{SSE}/(dfError)} = \frac{MSR}{MSE} \tag{1}$$

where:
F- test statistic;
SSR- sum of squared residuals;
MSR- mean squared residuals;
MSE- mean squared error;
q- number of predictor variables;
n- sample size;

- Use t Test Statistic (and t Distribution) for testing individual Slope and Y-Intercept:
  - t Test Statistic for Slope is displayed in (2):

$$\mathrm{t} = \frac{b_1}{\frac{s}{\sqrt{\sum (x_i - \bar{x})^2}}} \tag{2}$$

where:
t - t-test value;
b_1 - slope of the sample regression line;
x_i – predictor variable i^{th} value;
x_{mean} - mean value of all predictor variable values;
S - standard error of slope coefficient;

  - t Test Statistic for Y-Intercept is displayed with (3):

$$\mathrm{t} = \frac{b_0}{s * \sqrt{\sum (x_i^2)/(n * \sum (x_i - \bar{x})^2)}} \tag{3}$$

where:

t - t-test value;
b_0 - intercept of the sample regression line;
x_i – predictor variable i^{th} value;
x_{mean} - mean value of all predictor variable values;
n - sample size;
S - standard error of slope coefficient;

– p-value for t Test Statistic:
Two tailed t distribution (Absolute value (t Test Statistic), $n - q - 1$)

- Data Analysis Regression Output provides F & t Test Statistics, & p-values.
- The key is: If p-value is less than Alpha, Reject H_0 and Accept H_a.

If we reject H_0 and accept Ha it can be considered that the sample evidence suggests that at least one slope is not equal to zero. It is reasonable to assume that there is a significant relationship at the given level of significance. If x (independent variable) and y (dependent variable) are related, then the linear relationship explains a statistically significant portion of the variability in y over the Experimental Region. If we fail to reject H_0, it can be concluded that the sample evidence suggests that all slope/s are equal to zero. Finally, it is reasonable to assume that a significant relationship does not exist, at the given level of significance.

5 Conclusion

The concept of applying statistical (machine) learning for purposes of defining a model, which can support managers with decision-making processes, has gained growing importance in the field of quantitative management in recent years. Modern business trends indicate that European markets gradually gain on quantity and variety of quality due to the various influences and sources of capital, so that the economic growth of individual European economies is reflected through continuous development of business environment in each separate country.

It is very likely that the implementation of parallel (secondary) applications will be realized, during the process of defining an integral model, but that will be a completely new understanding of the problem. A good example was given by Wu [31], who implemented clustering techniques while exclusively observing companies which failed (went bankrupt), as opposed to most existing research analysing samples of companies who went bankrupt and who avoided bankruptcy. A logical conclusion can be formulated that effects of applying each model must be tested by validating and verifying the results of the modelling process, as well as by checking the adequacy of the selected modelling tools.

As one of the key constraints on the application of the scientific approach for modelling of management support processes, the fact is that there is an adequate number of predictive forecasting models internally developed in large companies in the world. Also, there can be found a significant number of vendors supplying predictive analytics tools, with a clear purpose of supporting companies in need. It is yet to be seen that in the next few years, predictive analytics tools (based on machine learning) will record biggest investment levels, even compared with business intelligence tools [32].

Such models represent a result of many years of work and a much larger population in which these models have been trained. However, due to business confidentiality issues and best practices – these models are in most cases inaccessible to the general (academic) public. Model improvements are only possible in accordance with publicly presented copies of the best world practice and the best examples from literature [33, 34]. This study follows accordingly with conclusions made by Stefanovic [35], who successfully analysed decision making process by integrally taking into account quantitative variables (financial, business) with human factor variables.

The future directions of the research should include the relationships of business parameters directly related to the company's business, with external parameters, which shape the lower and upper boundaries for the predicted overall outcome of key company performance indicators. These external parameters include macroeconomic trends, indicators of associated organizations, banks, interest rates, the impact of political instability or conflict escalation in the region. These possible extensions of the integral quantitative management model, would support further validation of the model, with a new dimension in terms of model sustainability over time.

References

1. Zopounidis, C., Dimitras, A.I.: Multicriteria Decision Aid Methods for the Prediction of Business Failure. Kluwer Academic Publishers, Dordrecht (1998). ISBN 978-1-4419-4787-1
2. James, G., Witten, D., Hastie, T., Tibshirani, R.: An Introduction to Statistical Learning (2013). ISBN 978-1-4614-7138-7
3. Du Jardin, P.: Dynamics of firm financial evolution and bankruptcy prediction. Expert Syst. Appl. **75**, 25–43 (2017)
4. Beaver, W.H.: Financial ratios as predictors of failure. J. Account. Res. **4**(3), 71–111 (1966)
5. Von Stein, J.H., Ziegler, W.: The prognosis and surveillance of risks from commercial credit borrowers. J. Bank. Financ. **8**, 249–268 (1984)
6. Laitinen, T.: Financial ratios and different failure processes. J. Bus. Financ. Account. **18**, 649–673 (1991)
7. Sheppard, J.P.: Strategy and bankruptcy: an exploration into organizational death. J. Manag. **20**, 795–833 (1994)
8. Mossman, C.E., Bell, G.G., Swartz, L.M., Turtle, H.: An empirical comparison of bankruptcy models. Financ. Rev. **33**, 35–53 (1998)
9. Nam, J.H., Jinn, T.: Bankruptcy prediction: evidence from Korean listed companies during the IMF crisis. J. Int. Financ. Manag. Account. **11**, 178–197 (2000)
10. Gepp, A., Kumar, K.: Business failure prediction using statistical techniques: a review. Financial distress Prediction. Research project (2010)
11. Tian, S., Yu, Y., Zhou, M.: Data sample selection issues for bankruptcy prediction. Risk Hazards Crisis Public Policy **6**, 91–116 (2015)
12. Xu, X., Wang, Y.: Financial failure prediction using efficiency as a predictor. Expert Syst. Appl. **36**(1), 366–373 (2009)
13. Lussier, R.N., Halabi, C.E.: A three-country comparison of the business success versus failure prediction model. J. Small Bus. Manag. **48**(3), 360–377 (2010)

14. De Bock, K.W.: The best of two worlds: Balancing model strength and comprehensibility in business failure prediction using spline-rule ensembles. Expert Syst. Appl. **90**, 23–39 (2017)
15. Li, H., Sun, J., Wu, J.: Predicting business failure using classification and regression tree: an empirical comparison with popular classical statistical methods and top classification mining methods. Expert Syst. Appl. **37**(8), 5895–5904 (2010)
16. Collins, G.S., de Groot, J.A., Dutton, S., Omar, O., Shanyinde, M., Tajar, A., Voysey, M., Wharton, R., Yu, L.M., Moons, K.G., Altman, D.G.: External validation of multivariable prediction models: a systematic review of methodological conduct and reporting. BMC Med. Res. Methodol. **14**(40), 2–22 (2014)
17. Goodfellow, I., Papernot, N.: The challenge of verification and testing of machine learning (2017). cleverhans-blog.com
18. Healey, J.F., Gerald Prus, S.: Statistics. a tool for social research. Nelson Education (2010). ISBN: 978-0-17-644253-8
19. Kohavi, R.: A study of cross-validation and bootstrap for accuracy estimation and model selection. In: Proceedings of International Joint Conference on AI, pp. 1137–1145 (1995)
20. Efron, B.: Estimating the error rate of a prediction rule: improvement on cross-validation. J. Am. Stat. Assoc. **78**, 316–331 (1983)
21. Ivanescu, A.E., Li, P., George, B., Brown, A.W., Keith, S.W., Raju, D., Allison, D.B.: The importance of prediction model validation and assessment in obesity and nutrition research. Int. J. Obes. **40**(6), 887–894 (2016)
22. Stosic, B., Mihic, M., Milutinovic, R., Isljamovic, S.: Risk identification in product innovation projects: new perspectives and lessons learned. Technol. Anal. Strat. Manag. **29**(2), 133–148 (2017). https://doi.org/10.1080/09537325.2016.1210121
23. Petrovic, D., Mihic, M., Stosic, B.: Strategic IT portfolio management for development of innovative competences. In: Tan, A., Theodorou, P. (eds.) Handbook on Strategic Information Technology and Portfolio Management, pp. 150–169. IGI Publishing. Information Science Reference, Hershey, USA (2008). ISBN 978-1-59904-687-7
24. Mihic, M., Petrovic, D., Obradovic, V., Vuckovic, A.: Project management maturity analysis in the serbian energy sector. Energies **8**(5), 3924–3943 (2015)
25. Arsic, S: Key factors of project success in family small and medium sized companies: the theoretical review. Manag. J. Sustain. Bus. Manag. Solut. Emerg. Economies **23**(1) (2018). ISSN 1820-0222
26. Arsic, S., Banjevic, K., Nastasic, A., Rosulj, D., Arsic, M.: Family business owner as a central figure in customer relationship management. Sustainability **11**(1) (2019). https://doi.org/10.3390/su11010077
27. Mihic, M., Todorovic, M., Obradovic, V.: Economic analysis of social services for the elderly in Serbia: two sides of the same coin. Eval. Program Plan. **45** (2014). https://doi.org/10.1016/j.evalprogplan.014.03.004. ISSN: 0149-7189
28. Mihic, M., Petrovic, D., Vuckovic, A.: Comparative analysis of global trends in energy sustainability. Environ. Eng. Manag. J. **13**(4), 947–960 (2014). Print ISSN: 1582-9596. eISSN: 1843-3707
29. Mihic, M., Arsic, S., Arsic, M.: Impacts of entrepreneurs' stress and family members on SMEs' business success in Serbian family-owned firms. J. East Eur. Manag. Stud. (2015). ISSN 0949-6181
30. Hoffmann, M., Klinkenberg, R.: RapidMiner: Data Mining Use Cases and Business Analytics Applications. Taylor & Francis Group LLC (2014). ISBN: 978-1-4822-0550-3
31. Wu, W.W.: Beyond business failure prediction. Expert Syst. Appl. **36**(2), 1593–1600 (2009)
32. Finances. https://financesonline.com/predictive-analysis/. Accessed 25 May 2019
33. Kim, C.: Quantitative Analysis for Managerial Decisions. Addison-Wesley Pub. Co., USA (1976). ISBN: 978-0201037395

34. Anderson, D.R., Sweeney, D.J., Williams, T.A., Wisniewski, M.: An Introduction to Management Science: Quantitative Approaches to Decision Making. Cengage Learning EMEA, London, UK (2009). ISBN: 978-0324202311
35. Stefanovic, M., Stefanovic, I.L.: Decisions. decisions— Paper presented at PMI® Global Congress 2005—North America. Toronto. Ontario. Canada. Newtown Square. PA: Project Management Institute (2005)

Critical Speed of Flexible Coupling - Determining with CAE Software

Elisaveta Ivanova(✉) and Tihomir Vasilev

Nikola Vaptsarov Naval Academy,
73 Vasil Drumev Street, 9026 Varna, Bulgaria
e.ivanova@naval-acad.bg

Abstract. The present article deals with FEM analysis of flexible coupling with rubber elastic element. Working parts of coupling are half of all cylinders made with carrying ring. Elastic elements are subjected to compression loading when transmitting torque. At certain load level, forces, acting on cylinders, cause their radial deformation. By default - metal fingers, with their shape, must limit radial deformation. However, in the process of operation, under torque and speed of rotation, it deforms and changes its overall size. Fingers press the loaded cylinders and deform them plastically at certain speed and load. As a result of this deformation, the elastic element loses a work stable and needs to be replaced. Elastic elements of couplings are made of a variety of polymeric materials whose properties have a significant impact on dynamic properties of coupling.

Radial deformation of the loaded cylinders is determined by FEA method using CAE software. Critical speeds are defined, resulting in pinching of rubber elastic element that depends on properties of the used material. Critical rotational speed increases with increase elastic modulus of the elastic element of 4 to 8 MPa and decreases with increased density of the material from 1080 to 1900 kg/m^3. Results show, that achieving maximum rotational speed for investigated flexible coupling without plastic deformation requires elastomer with a modulus of elasticity of $E \geq 6$ MPa and density $\rho \leq 1100$ kg/m^3. Results of these tests will be used to correct some of the geometric parameters of an elastic element or more suitable geometry of metal fingers to ensure the required coupling load.

Keywords: Flexible coupling · Rubber element · Critical speed · FEM analysis

1 Introduction

The elastic element provides the basic properties of the flexible coupling. The trouble-free, reliable and long-lasting operation of the flexible coupling depends on the properties of the elastomer from which the elastic element is made [1]. The elastic element in the various couplings is made of different polymeric materials.

The elastomers are formed from a mixture of basic polymer and other chemical substances. This mixture of components is designed to provide the necessary features to optimize product performance. Various additives, reinforcing agents, activators, plasticizers, accelerators, antioxidants, etc. are added to the elastomers. Many combinations

N. Mitrovic et al. (Eds.): CNNTech 2019, LNNS 90, pp. 265–274, 2020.
https://doi.org/10.1007/978-3-030-30853-7_15

can be made. The ISO 1629 standard identifies approximately 25 types of elastomers. In practice, the elastomer is evaluated by hardness, modulus of elasticity, resistance to various stresses, oxygen, oils, aging, etc. Hook's law can't be applied to the elastomers because the deformation of these materials is not directly proportional to the load applied. The properties of the elastomers are similar for those of the highly elastic materials. Elastomers have many characteristics: energy absorption, flexibility, high elasticity, long service life, moisture, pressure and heat protection, non-toxic properties, variable hardness, good dielectric properties etc. [2]. The characteristics of different brands of elastomers are investigated by standardized tests and are shown in Table 1.

Table 1. Characteristics of different brands of technical rubber and elastomers [3, 4]

Elastomer type	Hardness HSh A	Density ρ, g/cm^3	Tensile strength Rm, MPa	Modulus of elasticity E, MPa
SILICONE	60	1,2	7	4,5
POLYURETANE	60/95	1,25	28/45	4,5 -> 9,5
EPDM	65	1,35	7,5	6
NBR	70	1,4	5	7,5
SBR	65/70	1,5	4	6–7,5
VITON	73 ± 5	1,9	min 0,4	8,5

The deformation mechanism of the elastomers varies greatly in comparison with the materials commonly used in technical engineering. Hardness is the property of the material, which expresses the degree of resistance to deformation. The modulus of elasticity varies with the hardness change of the material (Fig. 1) [5, 6]. The fourth column of modulus of elasticity E in Table 1 is added according to Fig. 1.

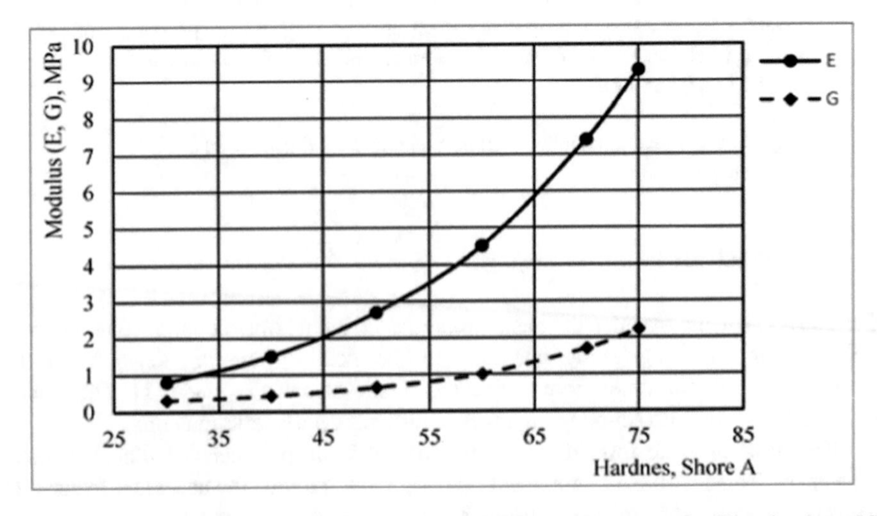

Fig. 1. Influence of Elastic (E) and Shear (G) moduli of the elastomer to the Shore hardness [6]

The influence of physico-mechanical properties of the elastic element on the deformation process of a variety of flexible couplings with a rubber elastic element was examined by simulations in the Solid Works Simulation environment [7–10]. The results show that the density of the material has a significant impact on the relative deformations and stresses in the elastic element. The elastic modulus of the rubber maximally influences the deformed coupling state at maximum load. Therefore, these two properties of the elastomer - modulus of elasticity of the rubber and density of the material are selected for investigation of the radial deformation of the elastic element (Table 1).

The elastic coupling loses its intended working capacity at a large deformation of the elastic element. As a result of this deformation, the elastic element loses a work capability and needs to be replaced. The nominal, maximum torque and permissible speed for each elastic coupling are specified in the standard. However, up to now, there is a lack of information about the critical speed of the elastic couplings.

The aim of the present article is to investigate the critical rotational speed of an elastic coupling with a rubber elastic element depending on the properties of the material used, by FEM analysis.

2 Methodology of the Research

The object of the study is an elastic coupling with a rubber elastic element type SEGE under BDS 16420-86 [11, 12]. The coupling tested with CAE software has a nominal torque T_n = 80 Nm. The maximum torque at short-term overload recommended by the standard is T_{max} = 200 Nm and permissible rotation speed n = 4500 min^{-1}. The three-dimensional model of the coupling, shown in Fig. 2, is prepared by Solid Works software.

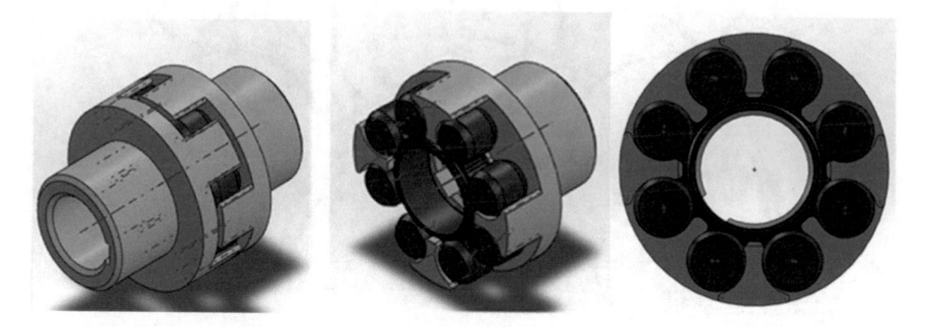

Fig. 2. CAD model of SEGE-type elastic coupling Tn = 80 Nm [8]

The elastic coupling has a normally loaded elastic element. The torque loads half of the cylinders in one direction of compression, and they deform. The other half of the cylinders is loaded at the reverse. In the presence of dynamic loads of sufficient amplitude, all cylinders are loaded. The design and operation of the coupling is described in detail in [8].

The elastic coupling SEGE under certain operating parameters, such as speed and load, loses its intended working capacity. The inspection of the elastic element shows plastic deformation of the cylinders (Fig. 3), which requires its replacement.

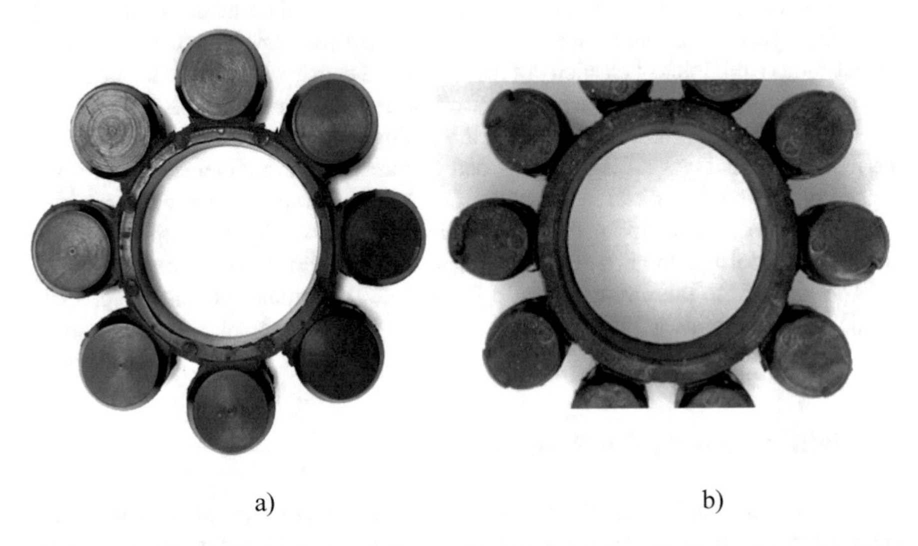

Fig. 3. Undeformed elastic element before work – (a) and deformed after replacement – (b).

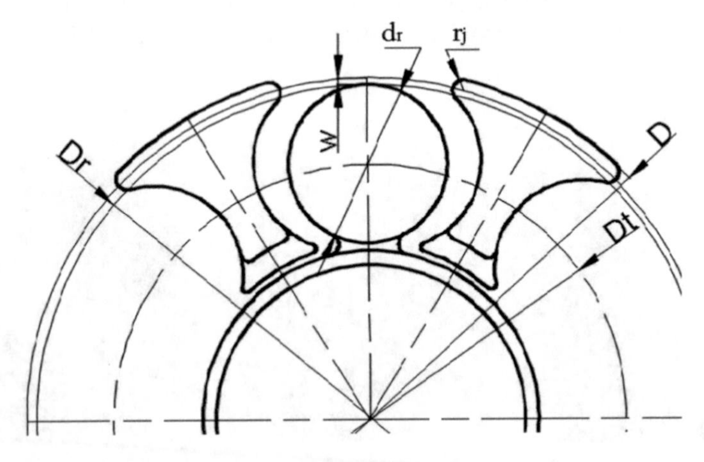

Fig. 4. Cylinder gripping pattern in the elastic element

During the operational process the elastic element deforms radially. The critical angular speed of rotation of the elastic coupling is associated with this radial deformation. In the coupling design, the metal fingers of standard size limit the radial deformation of the loaded cylinders (Fig. 4).

The maximum deformation for each cylinder, as shown in Fig. 4, can be calculated for a standard coupling using the formula (1) [13]:

$$w_{\max} = \frac{D - D_r}{2} - r_j \tag{1}$$

Where: D - outer diameter of the fingers of the metal semi-coupling; Dr - outer diameter of the elastic element; rj - rounding radius of the fingers.

The radial deformation of the SEGE test coupling with nominal torque 80 Nm must not exceed the following value, calculated by Eq. (1):

$$w_{\max} = \frac{110 - 105}{2} - 2 = 0.5\,\text{mm} = \Delta l \tag{2}$$

This radial deformation determines critical speed of rotation of the coupling.

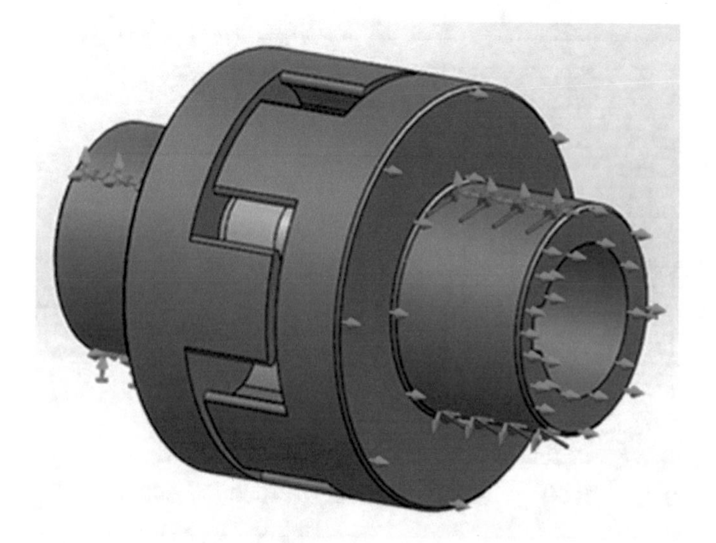

Fig. 5. CAD model of the SEGE elastic coupling with applied forces and reactions.

Two non-linear simulation tasks are prepared in Solid Works Simulation environment, for the elastic coupling shown in Fig. 5 [14]. The calculation model is explained in detail in [8, 9]. For the curve type of the material strengthening, a Mooney Rivlin type model with two constants was used [15]. The relationship between the Shore hardness of the rubber and the constants that determine the elastic properties of the rubber given in [16], are used. They are also compliant with the specific load on working elements of the coupling, defined by the Standard - maximum torque $T_{\max} = 200$ Nm and permissible rotation speed n = 4500 min^{-1}.

To study the radial deformation of the elastic element, depending on the properties of the elastomer, the elastic modulus of the rubber and the density of the material are selected from Table 1. Two simulations are performed.

3 Results and Analysis

During the first simulation the modulus of elasticity of the rubber is changed ($E = 4 \div 8$ MPa) with step of 1 MPa. The following parameters are also set - Poison coefficient for rubber - $\mu = 0{,}499$ and rubber density $\rho = 1080$ kg/m^3 [17]. The model is loaded through 100 steps, or 45 min^{-1} rotation speed for each. The centrifugal forces are applied in the center of the cylinders.

The obtained results are graphically presented in Fig. 6. The rotation speed leading to compression of the loaded cylinders from the metal fingers is considered critical at a permissible radial deformation $\Delta l = 0{,}5$ mm, calculated by formula (2).

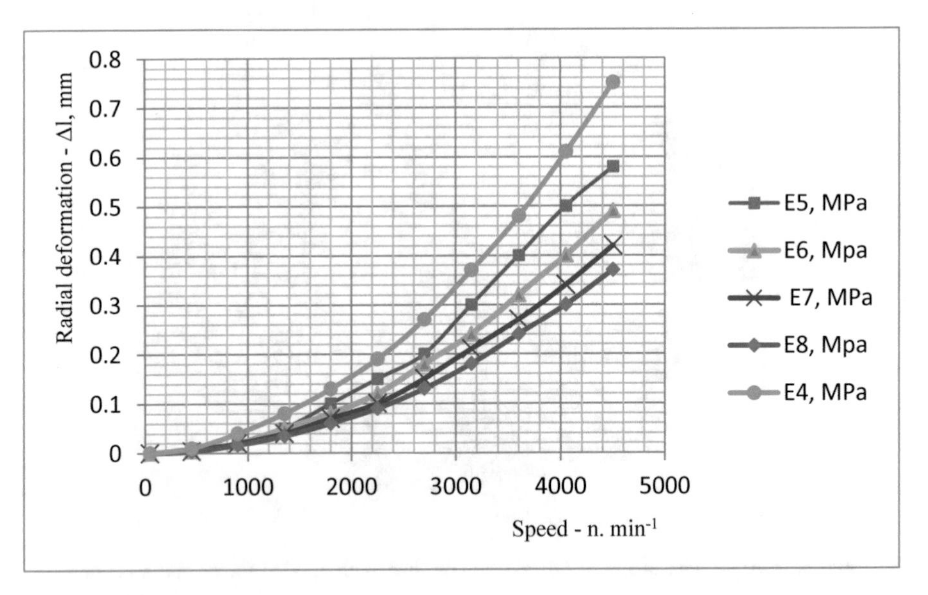

Fig. 6. Change of the radial deformation Δl depending on the rotation speed of the coupling for polymers with different elastic moduli.

Analysis of the results shows that the critical speed of rotation for a rubber with modulus $E = 4$ MPa is $n \cong 3600$ min^{-1}. While for a rubber with modulus $E = 5$ MPa the critical rotation speed is $n \cong 4000$ min^{-1}. At a higher rotation speeds, the elastic elements will be pinched. Therefore, the polymer of this standard coupling should have a modulus of elasticity higher than 6 MPa to reach a maximum permissible speed of rotation $n = 4500$ min^{-1}.

Considering these results, a modulus of elasticity of the rubber E = 6 MPa was adopted for the second simulation. During the second simulation the rubber density - ρ is changed (Table 1). A Poisson coefficient for rubber is set - μ = 0,499.

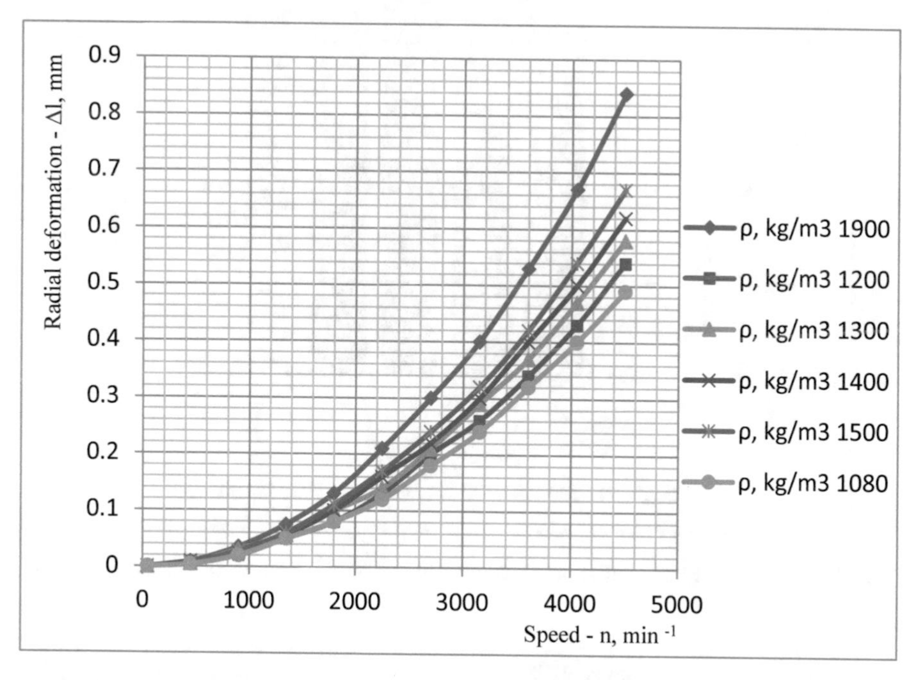

Fig. 7. Change of radial deformation Δl depending on the rotation speed of the coupling for polymers with different densities.

The results of the simulation tests for the dependence of the critical rotational speed on the density of the polymer are graphically presented in Fig. 7.

The critical rotational speed of the polymers with different density, corresponding to a radial deformation Δ1 = 0,5 mm, is shown in Fig. 8. The elastic element with a material density ρ = 1080 kg/m^3 and modulus of elasticity E = 6 MPa is not radially deformed above the permissible Δl = 0,5 mm. It can be clearly seen that the critical rotation speed reduces about 23% when the material density of the elastic element increases from 1200 to 1900 kg/m^3. The radial deformation of the elastic element is shown on Fig. 9.

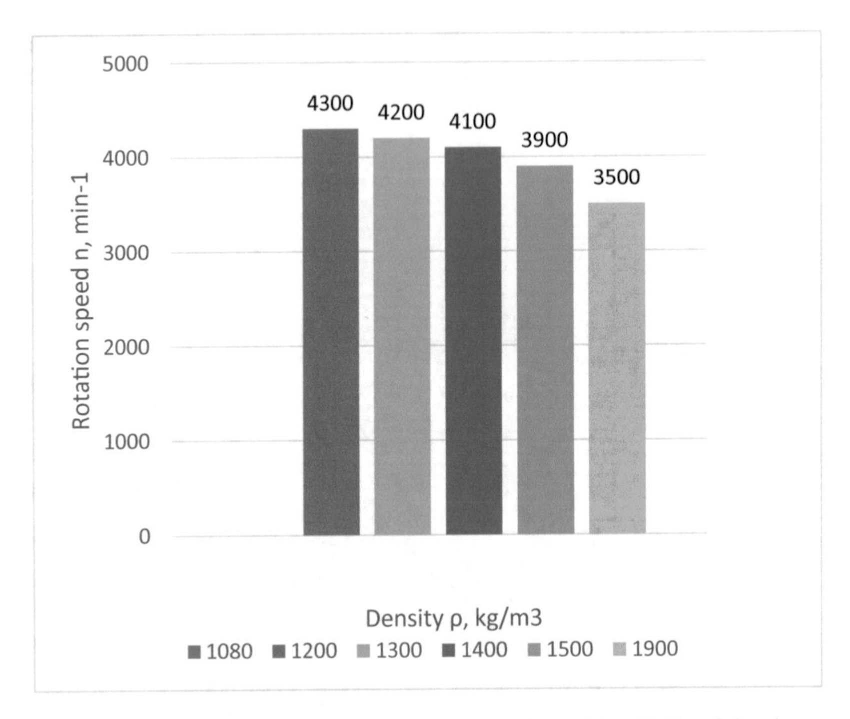

Fig. 8. Critical speed of rotation in constant modulus of elasticity 6 MPa of the elastomer.

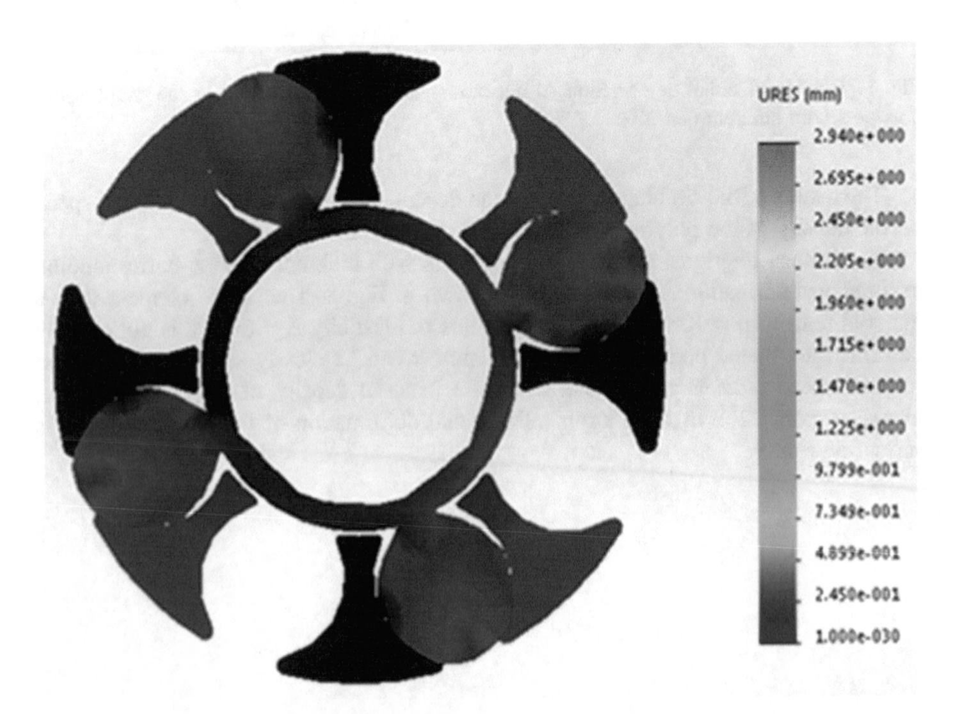

Fig. 9. Radial deformation of the elastic element at maximum load.

4 Conclusions

Present paper deals with FEM analysis of flexible coupling with rubber elastic element. Radial deformation of the loaded cylinders is determined by FEA method using CAE software. Critical speeds are defined, resulting in pinching of rubber elastic element that depends on properties of the material used. Critical rotational speed increases with increase elastic modulus of the elastic element of 4 to 8 MPa and decreases with increased density of the material from 1080 to 1900 kg/m^3. Results show, that achieving maximum rotational speed for investigated flexible coupling without plastic deformation requires elastomer with a modulus of elasticity of $E \geq 6$ MPa and density $\rho \leq 1100$ kg/m^3.

Results of these tests can be used for correction of the geometric parameters of the elastic element or development of more suitable design of metal fingers to ensure the required coupling load.

References

1. Walker, J.: Elastomer Engineering Guide (2017). https://www.jameswalker.biz/de/pdf_-docs/148-elastomer-engineering-guide
2. Axinte, A., Bejan, L., Ţăranu, N., Ciobanu, P.: Modern approaches on the optimization of composite structures. Buletinul institutului politehnic din Iaşi (2013). http://www.bipcons.ce.tuiasi.ro/Archive/424.pdf
3. Micro Max Trade (2019). http://www.micromaxtrade.com/prod.php?lng=bg&cat=8-&subcat=2
4. Elastomer Materials (2003). https://www.allsealsinc.com/03_Elastomers-Materials.-pdf
5. Bauman, J.T.: Fatigue, Stress and Strain of Rubber Components – Guide for Design Engineer. Hanser, Munich (2008)
6. Fediuc, D.O., Budescu, M., Fediuc, V., Venghiac,VM.: Compression modulus of elastomers. Buletinul institutului politehnic din Iaşi (2013). http://www.bipcons.ce.tuiasi.ro/Archive/369.pdf
7. Georgiev, et al.: Programen product za avtomatizirano planirane, provejdane I obarabotvane na rezultatite sled dvu-, tree- I chetri factorni planirani rotatabelni eksperimenti. EXPLAN, сп. МТТ изд.ТО на НТС Varna, pp. 73–79 (2008). ISSN1312-0859. (in Bulgarian)
8. Ivanova, E.: Izsledvane vliqnieto na vida na deformacioniq process varhu eksploatacionite harakteristiki na elastièn syedinitel s nemetalen element, Disertaciq Ph.D. TU-Varna (2016). (in Bulgarian)
9. Ivanova, E., Vasilev, T., Hristov, H.: Vliqnie na fiziko - mexanichnite svojstva na elastichniq element varhu deformacionioniq process na saedinitel SEGE. Mashinostroene I Mashinoznanie, b. 1, pp. 105–108 (2018). ISSN1312-8612. (in Bulgarian)
10. Ivanova, E., Vasilev, T., Hristov, H.: Study the influence of rotation speed on deformation process for flexible coupler with rubber elastic element. Machines, Technologies, Materials, Issue 7, pp. 336–339. Scientific Technical Union of Mechanical Engineering Industry 4.0 Year XI, Sofia (2017). ISSN WEB 1313-0226 (print), ISSN1314-507X (online)
11. BDS 16420 – 86: Saediniteli Elastichni s Gumen element. Osnovni parametric I texnièski izizskvaniq. (in Bulgarian)
12. Lefterov, L., Baltadjiev, A., Atanasov, C.: Saediniteli, Tehnika, Sofia (1986). (in Bulgarian)

13. Ivanova, E., Tenev, S., Vasilev, T.: The theoretical model for determinig critical rotation speed flexible coupling type SEGE. Sci. Bull. Nav. Acad. **XXI**, 326–329 (2017). ISSN: 2392-8956, ISSN-L: 1454-864X
14. Kallinikidou, E.: Nonlinear Analysis of Plastic and Rubber Components with SolidWorks Simulation. Documentation for Simulation Products Dassault Systems, p. 33 (2010)
15. Boulanger, et al.: Finite amplitude waves in Mooney–Rivlin and Hadamard materials. In: Hayes, M.A., Soccomandi, G. (eds.) Topics in Finite Elasticity (2001)
16. Altidis, et al: Analyzing hyperelastic materials with some practical considerations, Midwest ANSYS Users Group, 18 May 2005
17. Becker, W.: Material 70 EPDM 291, Freudenberg FST – MCE Material Center EUROPE. https://www.fst.com/-/media/files/materialdatasheet/70%20epdm%20291-en.ashx

Identifying Influence of Supplier Benefits on Collaboration Between Partners

Jasmina Dlacic[1(✉)], Toni Petrinic[2], and Borut Milfelner[3]

[1] Faculty of Economic and Business, University of Rijeka, Rijeka, Croatia
jasmina.dlacic@efri.hr
[2] Domeni Ltd., Matulji, Croatia
[3] Faculty of Economics and Business, University of Maribor, Maribor, Slovenia

Abstract. Collaboration between company and its suppliers is important in providing value to customers, especially in the age when companies are dominantly trying to focus on lowering costs to be more effective. But not just cost-effectiveness is a driving principle for all companies. Some companies are oriented towards establishing relationships with suppliers in order to provide value to customers. Hence, purpose of this paper is to explore relationships between companies and their key suppliers and to identify what influences their collaboration activities. Primary research was conducted on 182 Croatian companies, applying scales identified in previous research. Factor analysis and multivariate linear regression analysis were applied in analyzing research results. Results indicate that there are several benefits that companies perceive key suppliers are providing to them. Market and scout function, quality function, social support function, rescue function and innovation development function are some of them. These functions (supplier benefits), influence company future collaboration with key suppliers and consequently influences co-creation. Research results identified that quality, social support and innovation development function influence company's future collaboration activities. Based on these results implications for marketing managers are offered and ideas for co-creation provided.

Keywords: B2B marketing · Supplier benefits · Future collaboration · Co-creation · Croatia

1 Introduction

Collaboration between business partners is something that is considered as a building block for developing business in future. But, not all relationships between different business partners are equally important and it is not just that any collaboration will consequently provide value to both sides. Providing superior value to customers, in relationship marketing context we could call them partners, is important as it establishes competitive advantage for them in the long run [1]. Relationship marketing perspective suggests the establishing, developing and maintaining successful relational exchanges [2] between all partners. Hence, it is important to build relationships and collaboration in order to provide value for both sides.

N. Mitrovic et al. (Eds.): CNNTech 2019, LNNS 90, pp. 275–287, 2020.
https://doi.org/10.1007/978-3-030-30853-7_16

In order to consider company's offer as valuable, especially on business-to-business market, it is important that it is co-created together with partners. In the core of co-creation process is the notion of service-dominant logic (S-D logic) that service is the base and the beginning of value creation process [3]. Hence, in value creation process, or better to say co-creation, business partners are engaged throughout the process. Still, it is not extensively researched how business partners decide to enter this co-creation process and how they perceive benefits they receive from each other. As the focus of our research are relationships with suppliers, the main purpose of our paper is to explore relationships between companies and their key suppliers and to identify what influences their collaboration activities and consequently motivates them to engage in the value co-creation process.

Such understanding of relationships between business partners in collaboration process is important in order to understand benefits partners perceive they are gaining. Business partners are deciding to participate in value co-creation process for various motives [4]. The motives or benefits they perceive will shape their decision-making. Therefore, if benefits are perceived to be high, it is more likely that business partners will start the collaboration process and experience co-creation process. Following this, the paper aims to answer what benefits are perceived as important and how they influence business partners' decision to continue collaborating in the long run.

Paper is structured as follows. After Introduction, a Theoretical framework is explained and research model established. This is followed by Empirical research with research results. Paper concludes with Discussion and Conclusion section.

2 Theoretical Framework

Companies nowadays focus on long-term orientation between business partners. In the heart of long-term orientation is relationship marketing and value co-creation. Establishing long-term relationships with consumers and business partners is the basic of relationship marketing. Relationship marketing includes establishing, maintaining and enhancing relationships with customers and other partners with a profitable outcome [5, p 138]. Long-term relationships are the essence of relationship marketing [6]. Value co-creation process builds on that notion, as it assumes that all parties involved into the value creation process are adding to company offering. Hence, company is treating them as partners [7]. Therefore, value is created through collaboration between the company and its partners [8]. Building a relationship embraces collaboration as its constituting element.

Collaboration builds on establishing long-term relationships between business partners that are based on mutual dependence and trust [9]. Business partners consequently extend trust from one collaboration to all relationships they have between each other. Trust, together with commitment and satisfaction adds to defining relationship quality [10]. When these elements are evaluated as important and present in a specific relationship it is considered that relationship quality is present. So, through building on trust, together with satisfaction related to specific products and services and with adding commitment to continue collaboration, partners will establish long-term relationships. Relationship quality influences business channel performance and relational benefits [10] through which the customer loyalty is enhanced [11]. Therefore, business partners are willing to continue to stay in the specific business relationship as it adds to their profitability on the long run.

Also, Chowdhury, Gruber and Zolkievski [12] argue on plethora of positive value co-creation outcomes divided into direct and indirect ones. Direct ones include financial value, brand growth, consistency and quality, time efficiency and cost efficiency. On the other hand, indirect outcomes include, client/supplier learning, enhanced reputation, innovation, awareness about competitor tactics and access to other network actors. Consequently, value co-creation has positive outcomes on company profitability (e.g. [13]).

Investing long-term relationships and relationship quality develops positive attributes embedded within them [14] and adds to process of value co-creation. Through collaboration, business partners are developing strong relationships and are focusing on long-term relationships with a valuable outcome [15, 16] and develop willingness to collaborate in future [17]. Value in process of co-creation is deriving from interaction between business partners and companies. Collaboration, cooperation and communication are the key element in this process [18].

Previous literature [19] suggests that selection of companies' suppliers is based on benefits they provide to the company. This influences collaboration and co-creation activities among partners. Companies are deciding to collaborate in the long run as perceived benefits are adding to business partners' willingness to participate in co-creation process [4]. Benefits from these processes are multiple. Walter et al. [19] acknowledges the following ones: cost reduction, volume, quality, safeguard, market, scout, social support and innovation development. Business partners contribute to lowering costs for supplied products/services and they provide good value for money (cost-reduction benefit). They also provide long-term delivery promises and assure that they will deliver what was ordered (volume benefit); they provide functional, reliable and constant quality products (quality benefit); assure that company is not dependent on other suppliers, offer flexible handling of supply agreements and give the opportunity to change the delivered quantity (safeguard benefit). Suppliers also offer information on new buyers, new suppliers if they are not possible to deliver ordered product or connect company with third parties like governmental bodies, local government or different agencies (market benefit); they provide information on procurement markets, competitors or new technological advancements on market (scout benefit). They also appreciate company employees, provide opportunities to collaborate and motivate employees towards better performance (social support benefit). Additionally, they may provide ideas for new products/services, help in developing new products/services and help in establishing new production or service delivery processes (innovation development processes).

Perceived benefits are adding to business partners' willingness to participate in co-creation process and are related to relationship quality [19]. This adds to importance of determining future collaboration possibilities, as they are based on relationship quality. A high quality customer–supplier relationship motivates business partners to establish long-term relationships [20]. As can be seen, future is in the collaborative co-creation of value by mutual commitment of resources from all included parties in a specific relationship [21].

3 Empirical Research

3.1 Research Design

Proposed research framework was tested on Croatian companies operating on business-to-business market. Initial database for research was obtained from Bureau Van Dijk's Amadeus database. Companies were selected based on their dominant market (business to business). They were selected if they were still active on Croatian market, and if they had an e-mail address. As survey was conducted using Limesurvey online questionnaire platform, it was necessary for companies to have the active e-mail address. Data collection was conducted in 2014 and 181 questionnaires were collected. This represents a response rate of 6%. Despite the fact that company database was new and up to date, it has to be noted that a significant number of e-mail addresses was found nonexistent and that the large number of e-mail addresses was general (e.g. info@-companyname or companyname@companyname). This influenced the final response rate, since no real person was contacted in the first place.

The questionnaire was designed using several pre-established scales related to the benefits that company receive through the interaction with producers from suppliers' perspective, namely: market benefit, scout benefit, quality benefit, social support benefit, rescue benefit and innovation development benefit [19]. For determining the future collaboration with suppliers a scale from [17] was applied. All items used were rated on a 7-point Likert scale ranging from "strongly disagree" to "strongly agree".

In data collection process respondents were asked to focus on suppliers that are the companies key suppliers, as smaller ones probably were not focusing on developing long-term relationships. Since the majority of work is done with the key suppliers it is more likely they will be more eager to collaborate and engage in co-creation process with them.

Analysis was performed using the SPSS 21 software package.

3.2 Research Sample

The collected sample was analyzed and the profile of 182 respondents is presented in Table 1.

Table 1. Research sample characteristics

Characteristic	Percentage
Industry	
Services	48.4%
Production	28%
Combination of services and production	23.6%
Company size	
Micro company	7.5%
Small company	42.2%

(*continued*)

Table 1. (*continued*)

Characteristic	Percentage
Medium company	37.9%
Large company	12.4%
Capital structure	
Private capital	73.62%
Public capital	26.38%
Domestic capital	68.68%
Foreign capital	31.32%

Source: Research results

From previous table it is evident that most of our respondents are from services (48.4%), with 23.6% of them having combination of services and production. The largest number of companies in the sample is from small companies (42.2%), followed by medium sized companies (37.9%). Capital structure is as following: 73.62% of the companies in the sample were in the private capital ownership and 26.38% were publically owned. 68.68% were owned by domestic capital and 31.32% by foreign owners.

We also analyzed some additional characteristics in our sample in order to better understand our respondents. The youngest company among our respondents has been in business for 2 years, while the oldest has been in business for 140 years. On average, companies in our sample were 26.81 years old. In the last five years, companies have on average introduced 100.85 new products or services to the market. New product and services account for 32.07% of total company sales. New markets that company entered in the last five years, account for 23.28% of total company sales.

3.3 Research Results

We continue our analysis with analyzing single scales that were used. In Table 2 average values and standard deviation are presented.

Table 2. Descriptive characteristics of scales used in the research

Constructs	Mean	SD
Market benefit	*3.44*	*1.323*
Connects us with new possible buyers for our products/services (MB1)	3.63	1.567
Connects us with new possible distributors for products/services that he can't provide (MB2)	3.56	1.754
Connects us with relevant third parties (state institutions, governmental bodies…) (MB3)	3.12	1.648
Scout benefit	*3.66*	*1.648*
Provides us with information on procurement markets (SB1)	3.80	1.701

(*continued*)

Table 2. (*continued*)

Constructs	Mean	SD
Provides us with information on competitors (SB2)	3.34	1.637
Provides us with information on relevant third parties (state institutions, governmental bodies…) (SB3)	3.25	1.570
Provides us with information on technological advancements on our market (SB4)	4.27	1.685
Volume benefit	*5.56*	*1.312*
Provides us with long-delivery promises for offered products/services (VB1)	5.63	1.177
Provides us with promise that every order will be processed and delivered (VB2)	5.49	1.086
Quality benefit	*5.53*	*1.104*
They deliver products/services that are functioning properly (QB1)	5.60	1.066
They deliver products/services that are reliable (QB2)	5.53	1.183
They deliver products/services with constant quality (QB3)	5.51	1.150
They deliver ordered products/services within agreed conditions (QB4)	5.46	1.017
Social support benefit	*4.68*	*1.498*
They appreciate our employees (SSB1)	5.25	1.335
They encourage communication and exchange of information between our and their employees (SSB2)	4.68	1.530
They encourage our employees to collaborate (SSB3)	4.78	1.444
Cooperation with them gives our employees sense of employment security (SSB4)	4.01	1.682
Rescue benefit	*4.95*	*1.289*
They lower our dependence on other suppliers (RB1)	5.04	1.274
They give us opportunity to variate ordered quantity (RB2)	5.31	1.079
They give us opportunity to have flexible handling with other suppliers (RB3)	4.51	1.515
Innovation development benefit	*3.95*	*1.688*
Provides us with ideas for new products/services (IDB1)	4.23	1.686
Provides us with ideas for developing new products/services (IDB2)	3.98	1.685
Provides us with ideas for advancing manufacturing/services processes (IDB3)	3.53	1.777
Provides us with ideas for technological know-how (IDB4)	4.07	1.605
Future collaboration	*5.66*	*0.968*
We are willing to include our suppliers in future projects (FC1)	5.60	1.007
We are willing to collaborate with our suppliers in realizing future projects (FC2)	5.62	0.971
We plan to collaborate with our suppliers on future projects if they occur (FC3)	5.76	0.926

Source: Research results

As can be observed from Table 2 companies rate the importance for following benefits in ascending importance: volume ($\bar{x} = 5.56$), quality ($\bar{x} = 5.53$), rescue ($\bar{x} = 4.95$), social support ($\bar{x} = 4.68$), innovation development ($\bar{x} = 3.95$), scout ($\bar{x} = 3.66$) and market ($\bar{x} = 3.44$). Hence, quality of delivered products/services and what suppliers promise they will deliver on time is regarded as the most important benefits they receive from their suppliers. Contrary to that, the scout benefit, that is providing information of other parties on market like competitors or procurement, together with market benefits, that include connecting company with new possible buyers and distributors, are considered to have the least importance for companies. Possibility of future collaboration is evaluated as high with average value ($\bar{x} = 5.66$) on scale from 1 to 7, meaning that companies are willing to collaborate with their key suppliers in the future.

To test our research question we continued with the testing the appropriateness of the used scales for further analysis. Exploratory factor analysis (EFA) together with testing the reliability of used constructs was performed. Then multivariate regression analysis was used in order to test the influence of benefits on future collaboration with suppliers. EFA was conducted using Principal axis factoring with oblimin rotation and Kaiser Normalization. This method was used as theoretical background for all research constructs is present [22] and since it was assumed that benefits could correlate [23].

After preforming EFA and analyzing results, one item ("Cooperation with them gives our employees sense of employment security") was excluded from further analysis. This was done as it loaded into two factors. Results of Kaiser-Meyer-Olkin (KMO) measure of sampling adequacy and Bartlett's test of sphericity indicate: KMO = 0.880 and $\chi2 = 2613.198$ (df = 253, $p < 0.05$). As these results are in accordance with suggested threshold value for KMO as 0.7 and with Bartlett's test that is statistically significant [23] we continued with the analysis. EFA returned five factor solution (Table 3) explaining 69.250% of variance in the results.

Table 3. Factor loadings

Items	Factor				
	Market & Scout benefit	Volume & Quality benefit	Social support benefit	Rescue benefit	Innovation development benefit
VB1		0.568			
VB2		0.709			
QB1		0.743			
QB2		0.909			
QB3		0.783			
QB4		0.701			
RB1				0.436	
RB2				0.461	
RB3				0.579	

(*continued*)

Table 3. (*continued*)

Items	Factor				
	Market & Scout benefit	Volume & Quality benefit	Social support benefit	Rescue benefit	Innovation development benefit
MB1	0.607				
MB2	0.642				
MB3	0.912				
SB1	0.708				
SB2	0.727				
SB3	0.848				
SB4					0.442
IDB1					0.508
IDB2					0.774
IDB3					0.828
IDB4					0.569
SSB1			0.689		
SSB2			0.845		
SSB3			0.806		

Note: Loadings less than 0.3 were not presented. Rotation converged in 9 iterations. VB = Volume benefit, QB = Quality benefit, RB = Rescue benefit, MB = Market benefit, SB = Scout benefit, IDB = Innovation development benefit, SSB = Social support benefit
Source: Research results

From previous table it can be noted that some benefits from previous literature have joined together. This is probably due to market specifics. A previous research [19] was conducted on developed market (Germany) and this research on transitional market (Croatia). It can mean that market specifics influence the reactions of companies on the market or that our respondents find difficult to distinguish between these two benefits. Also, some items (like "Provides us with information on technological advancements on our market") have loaded into different factor (in Innovation development benefit instead of previously identified Scout benefit). This is probably due to the fact that it focuses on providing information on technological innovation, and this perceptive is dominant in factor Innovation development benefit.

Constructs reliability (Cronbach alpha) was checked and results as well as the reliability of the future collaboration factor are presented in Table 4. From the Table 4 it can be noted that majority of scales have reliability above the suggested value of 0.7 [24]. Cronbach alphas are in the range of 0.665 to 0.941. Cronbach alpha of Rescue benefit is lower than suggested rule of 0.7, so this scale is not reliable and won't be used in further analysis. Other scales are reliable and can be used for further analysis. In subsequent analysis, scales were composed as an average index of items that constitute the factor.

Table 4. Constructs characteristics

Factor	Characteristics					
	Cronbach alpha	% of explained variance	Eigenvalue	Number of items	Scale mean	Items mean
Market & Scout benefit	0.897	37.111%	8.536	6	20.70	3.450
Volume & Quality benefit	0.903	13.367%	3.074	6	33.21	5.536
Social support benefit	0.857	7.581%	1.744	3	14.71	4.903
Rescue benefit	0.665	6.320%	1.454	3	14.86	4.954
Innovation development benefit	0.854	4.871%	1.120	5	20.08	4.016
Future collaboration	0.941	N/A	N/A	3	16.97	5.657

Source: Research results

Based on conducted analysis, it can be concluded that benefits can be divided into following groups (when observing benefits from suppliers): market and scout benefit, volume and quality benefits, social support benefit and innovation development benefit.

To further test the research question related to identifying what benefits are perceived as important and influence business partners' decision to continue collaborating it in the long run, multivariate regression analysis was conducted. Future collaboration was used as the dependent variable. The Enter method of the selection of variables in regression analysis was used. Results are presented in Table 5.

Table 5. Multivariate regression results

	B	Beta	T-value
Constant	2.613 (0.358)		7.307
Market & Scout benefit	−0.017 (0.054)	−0.025	−0.318
Volume & Quality benefits	0.349 (0.070)	0.348**	4.951
Social support benefit	0.128 (0.056)	0.177**	2.259
Innovation development benefit	0.137 (0.058)	0.200**	2.349
R^2	0.321		
R^2 (adj)	0.306		
F-value	20.910**		

Note: N = 182; standard error is in parenthesis; ** $p < 0,05$
Source: Research results

Table 5 shows that perceived benefits of market and scout benefit, volume and quality benefits, social support benefit and innovation development benefit explain 30.6% of variance in the research results that is variance in future collaboration. This influence is statistically significant, but still remains to further explore other benefits that might influence future collaboration activities. We can conclude that future collaboration is influenced by volume and quality benefits (β = 0.348), innovation development benefit (β = 0.200) and social support benefit (β = 0.177). Market and scout function does not influence future collaboration activities, as its influence is not statistically significant.

In order to exclude the possibility of multicollinearity, we also tested the relationships between independent variables in our model. Results are presented in Table 6.

Table 6. Collinearity analysis

	Tolerance	VIF
Market & Scout benefit	0.616	1.624
Volume & Quality benefit	0.776	1.289
Social support benefit	0.627	1.597
Innovation development benefit	0.528	1.893

Note: N = 182; standard error is in parenthesis;
** p < 0,05
Source: Research results

Results indicate that collinearity is not a problem, as VIF values and Tolerance are below acceptable levels [22]. Highest VIF, is 1.893, hence not substantially larger than 1 and well below 10. Lowest tolerance is 0.528 that is also acceptable. The Durbin-Watson test is 2.148, indicating that residuals are not correlated.

Based on our analysis we can conclude that original formula:

$$Y_i = \beta_0 + \beta_i X_i + \varepsilon_i$$

Can be modified into:

$$\text{Future collaboration} = 2.613 + 0.348 * \text{Volume \& Quality benefits} + 0.177 * \text{Social support benefit} + 0.200 * \text{Innovation development benefit} + \varepsilon_i$$

4 Discussion and Conclusion

This paper contributes to recognition of different benefits' influence on future collaboration between company and its suppliers. Research results indicate several benefits that influence company decision to continue collaboration with specific supplier. These benefits are identified as market and scout benefit, volume and quality benefits, social support benefit and innovation development benefit. Although, previous research [19] indicated more different benefits, research results show that some benefits, depending of market structure, can be clustered together. Hence, in transitional markets benefits that companies consider in relation with suppliers are related to: (1) providing information from market like indicating new possible buyers or providing information on competitors and other institutions on the market – market and scout benefit; (2) delivering products/services that are of high quality, reliable and functioning properly, the promise that products/services will be delivered as ordered - volume and quality benefit; (3) encouraging collaboration between employees of company and its suppliers as well as communication and exchange of ideas between company and its suppliers - social support benefit; (4) providing ideas for new products or advancing processes of manufacturing or service delivery – innovation development benefit.

Paper also contributes to identifying that some benefits from supplier-company relationship boost future collaboration while others, like market and scout benefit have no statistically significant influence on this future collaboration activity. Volume and quality function, among researched benefits, has the greatest influence on developing future collaboration. Therefore, companies that want to continue collaboration with current suppliers have to carefully scrutinize and compare their suppliers according to volume and quality benefits.

Value co-creation process is getting more and more attention on business-to-business markets. It provides benefits not only for customers but also for companies [4]. At the beginning of this process is a notion that companies want to collaborate in the long run. Hence, with identifying benefits that enhance future collaboration, companies also create possibilities for value co-creation process to occur.

Some managerial implications can be drawn based on this research. Firstly, it is important that marketing managers distinguish different benefits that are perceived in relationship between company and its suppliers. To focus just on benefits like volume and quality benefits is not enough, as also other benefits like social support and innovation development benefit do contribute to the process. Therefore, encouraging collaboration in joint projects, fostering communication and exchange of ideas with business partners, as well as starting innovation processes together with suppliers will contribute to better recognition of business partners that are willing and possible to collaborate with. Secondly, future collaboration depends on on-time delivery of promised reliable and properly functioning products and services. Alone this is however not enough as also communication and exchange of ideas in joint projects are contributing to future collaboration intention.

This research also has some limitations like sample size, which is relevant for Croatia and reflects the structure of companies in this country. Additionally it would be interesting to analyze companies and their relationships with its suppliers related to different industries and to explore the results between industries. Further it would be interesting to explore why market and scout benefit, doesn't have influence on future collaboration. Maybe it is considered by companies as hygienic factor and therefore not evaluated consequently.

Acknowledgement. This paper has been supported by the University of Rijeka under the project number Uniri-drustv-18-235-1399.

References

1. Ravald, A., Grönroos, C.: The value concept and relationship marketing. Eur. J. Mark. **30**(2), 19–30 (1996)
2. Morgan, R.M., Hunt, S.D.: The commitment-trust theory of relationship marketing. J. Mark. **58**(3), 20–39 (1994)
3. Chathoth, P., Altinay, L., Harrington, R.J., Okumus, F., Chan, E.S.: Co-production versus co-creation: a process based continuum in the hotel service context. Int. J. Hosp. Manag. **32**, 11–20 (2013)
4. Saarijärvi, H., Kannan, P.K., Kuusela, H.: Value co-creation: theoretical approaches and practical implications. Eur. Bus. Rev. **25**(1), 6–19 (2013)
5. Grönroos, C.: Service Management and Marketing: Managing the Moments of Truth in Service Competition. Jossey-Bass, San Francisco (1990)
6. Reinartz, W., Kumar, V.: The mismanagement of customer loyalty. Harv. Bus. Rev. **80**(7), 86–94 (2002)
7. Prahalad, C.K., Ramaswamy, V.: Co-opting customer competence. Harv. Bus. Rev. **78**(1), 79–90 (2000)
8. O'Cass, A., Ngo, V.L.: Examining the firm's value creation process: a managerial perspective of the firm's value offering strategy and performance. Br. J. Manag. **22**(4), 646–671 (2011)
9. Ganesan, S.: Determinants of long-term orientation in buyer-seller relationships. J. Mark. **58** (2), 1–19 (1994)
10. Athanasopoulou, P.: Relationship quality: a critical literature review and research agenda. Eur. J. Mark. **43**(5/6), 583–610 (2009)
11. Vesel, P., Žabkar, V.: Relationship quality evaluation in retailers' relationships with consumers. Eur. J. Mark. **44**(9/10), 1334–1365 (2010)
12. Chowdhury, I.N., Gruber, T., Zolkiewski, J.: Every cloud has a silver lining—exploring the dark side of value co-creation in B2B service networks. Ind. Mark. Manag. **55**, 97–109 (2016)
13. Payne, A., Storbacka, K., Frow, P.: Managing the co-creation of value. J. Acad. Mark. Sci. **36**(1), 83–96 (2008)
14. Lahiri, S., Kedia, B.L.: Determining quality of business-to-business relationships: a study of Indian IT-enabled service providers. Eur. Manag. J. **29**, 11–24 (2011)
15. Hagedoorn, J.: Understanding the cross-level embeddedness of inter-firm partnership formation. Acad. Manag. Rev. **31**, 670–680 (2006)
16. Lado, A.A., Dant, R.R., Tekleab, A.G.: Trust opportunism paradox, relationalism, and performance in interfirm relationships: Evidence from the retail industry. Strat. Manag. J. **29**, 401–423 (2008)
17. Wagner, S.M., Eggert, A., Lindemann, E.: Creating and appropriating value in collaborative relationships. J. Bus. Res. **63**(8), 840–848 (2010)
18. Sheth, J.N., Sisodia, R.S., Sharma, A.: The antecedents and consequences of customer centric marketing. J. Acad. Mark. Sci. **28**(1), 55–66 (2000)
19. Walter, A., Muller, T.A., Helfert, G., Ritter, T.: Functions of industrial supplier relationships and their impact on relationship quality. Ind. Mark. Manag. **32**, 159–169 (2003)
20. Palmatier, R.W., Dant, R.P., Grewal, D., Evans, K.R.: Factors influencing the effectiveness of relationship marketing: a meta-analysis. J. Mark. **70**(4), 136–153 (2006)
21. Sheth, J.: Revitalizing relationship marketing. J. Serv. Mark. **31**(1), 6–10 (2017)

22. Field, A.: Discovering Statistics Using SPSS, 3rd edn. Sage publications, London (2009)
23. Hair Jr., J.F., Black, W.C., Babin, B.J., Anderson, R.E.: Multivariate Data Analysis, 7th edn. Pearson, Upper Saddle River (2010)
24. Nunnally, J.C.: Psychometric Theory. McGraw-Hill, New York (1967)

An Application of VLES Turbulent Flow Simulation Methodology to Flow over Smooth Hills

Nikola Mirkov(✉)

Institute of Nuclear Sciences - Vinča, University of Belgrade,
Mike Petrovića Alasa 12-14, Belgrade, Serbia
nmirkov@vin.bg.ac.rs

Abstract. This study aims at examining predictive capabilities of a specific global (non-zonal) hybrid RANS/LES turbulent flow simulation strategy in the case of flow over smooth hills. The rationale behind recent popularity of such an approach are constraints on computational resources when one is faced with practical LES for realistic engineering flows. In present approach, which originates from the seminal work of Speziale, simulation is performed on coarser computational meshes then required by LES constraints, where the large part of turbulent kinetic energy is unresolved. A challenge of accurate prediction of wall-bounded flows is approached in hybrid models by blending two approaches (RANS and LES) into a single global model, whereby switching is applied seamlessly based on considerations of turbulent flow and grid scales. The RANS part of the hybrid model is supposed to be active in the vicinity of the wall, and the swift transition to LES model should be performed away from the wall, on a desired distance defined by calibration of switching parameters. In our study we perform transient VLES simulation on a model of smooth hill at laboratory scale, resembling well documented wind tunnel experiment, and study in detail statistical flow properties by comparisons with the experimental results. Special attention is given to hybrid model switching behavior in the said case. The study gives very useful insight into simulation of flows of the industrial and environmental types.

Keywords: Turbulent flows · Numerical simulation · Turbulence model validation · Hybrid RANS/LES · Very Large Eddy Simulation

1 Introduction

Although there is a substantial evidence on progress of Large-Eddy Simulations (LES) approach in turbulence modeling during past five decades, it had limited impact on simulation of industrial type of flows, mostly because of high computational cost. Instead of passively waiting for increase in computer power, current efforts in turbulence modeling is to find a way to make LES useful in industrial types of CFD simulations. It is, in part, accomplished by hybridization techniques, which can roughly be divided into two broad groups of zonal and seamless methods. In the current paper

N. Mitrovic et al. (Eds.): CNNTech 2019, LNNS 90, pp. 288–304, 2020.
https://doi.org/10.1007/978-3-030-30853-7_17

we are interested in the later approach. The mechanisms are different in respective models, but the final goal is the same - follow different routes to a decreasing turbulent eddy-viscosity as the artifact of Reynolds averaging procedure.

As Speziale notes [1], simplicity of Smagorinsky SGS model guarantees that it will be useful for sole purpose of dissipation of turbulence kinetic energy at smallest (dissipative) scales, just before molecular viscosity effects take over. To render this type of model effective, numerical meshes need to be as fine as resolution of finest scales at particular Reynolds number requires. This becomes increasingly difficult taking into account the scaling of computational resolution and computational time with *Re* number [2, 3]. In many cases, this requirement is too strict for practical computations. Idea is then to decrease the need for very fine numerical mesh, and to compensate this by using more elaborate sub-grid models, those that would "contain more physics" as turbulence modelers often point out. By saying this it is suggested that there are partial differential equations to describe transport processes of Reynolds stress tensor components at best (c.f. RSM modeling), or scalar quantities such as turbulence kinetic energy, and dissipation rate enough for calculating single velocity and time scale of turbulence (cf. eddy viscosity models - EVM) and perform single point closure.

Experience in technology of RANS turbulence modeling comes to rescue claiming that models used in this modeling practice contain enough physical simulacrum needed for LES simulation on coarse meshes, where mesh resolution doesn't reach dissipative scales. This (using elaborate RANS models in place of SGS models and performing LES simulation on under resolving meshes) is sometimes denoted as Very Large Eddy Simulation (VLES) although the use of the acronym is sometimes misplaced and confusing, as noted by some authors [4].

Another approach which is useful, if there is a scale separation in Fourier-space representation of turbulence signal at a particular location in flow determined by a fixed coordinate system is Unsteady Reynolds Averaged Simulation (URANS). Scale separation implies that certain low wave-number mode has substantially bigger amplitude (energy in L_2 norm). In physical space this is seen as large scale instability, mental picture of a wake behind a cylinder is recognized here as most often recalled in relation to this notion. Even if there is no "spectral gap" between overly excited mode and dissipative range, and if we conduct RANS simulation with time stepping we can claim that we are performing URANS simulation [5]. If we want greater resolution we will make the time step smaller. Some authors claim there is an equivalence between VLES and URANS procedure. This is specially backed up by the fact that filtered Navier-Stokes equations used in LES have the same form as those of URANS.

Among seamless RANS/LES hybridization approaches detached eddy simulation (DES) of Spalart et al. [6] seems to have had greatest success and has been implemented in major commercial CFD packages. Schiestel et al. [7, 8] have created and promoted a special continuous approach for hybrid RANS/LES modeling that allows seamless coupling between two respective regions, known as partially integrated transport modeling (PITM). Considerations of numerical stability, mathematical simplicity, as well as simplicity of implementation, have lead Hanjalić and Kenjereš to modify it and put it into a form of so-called "seamless alpha" model [9]. Computer program implementation amounts to changing about of dozen lines of $k - \epsilon$ code, and running simulation in time accurate mode.

The originally proposed VLES method of Speziale evolved through proposals such as limited-numerical-scales (LNS) of Batten et al. [10] and later by Han and Krajnović [11]. In [11] authors propose a VLES model similar to one proposed by Speziale [1] with the original resolution control function that resembles the ratio of the unresolved turbulent kinetic energy to the total kinetic energy. This model seemed interesting for present study. Contrary to the original paper, in the present study VLES model was combined not with the standard $k - \epsilon$ model, but with realizable $k - \epsilon$ model of Shih et al. [15] and with $k - \omega$ Shear Stress Transport model of Menter [12, 13].

The study of hybrid LES/RANS models in wall bounded flows with smooth curvature has been part of recent European turbulence modeling projects, and notable results are reported in [14]. Different seamless hybrid methods were tested there in the case of smooth periodic hill. In contrast to that study here we consider a flow configuration studied in wind tunnel with the aim to replicate true atmospheric conditions at certain scale.

The study of flow over hilly terrain in general has importance for wide range of applications. Among those are the wind energy application, pollutant dispersion, safety of structures, wind impact to agriculture and forestry, and aviation safety. Most of the experimental studies are focused on the effect of hill presence on mean flow characteristics and turbulence. The experimental studies on laboratory scale have advantage of well defined geometry and inflow boundary conditions, besides the resolution, when compared to atmospheric scale flow studies, such as Askervein hill and Bolund hill, which makes them interesting for turbulence model validation.

2 Mathematical Model and Numerical Implementation

The VLES approach amounts to introducing the resolution control function which aims at damping the Reynolds stress tensor as a way of modelling sub-grid scale stresses $(\tau_{ij})_{sgs} = F_r(\tau_{ij})_{RANS}$. The predictive capability of such turbulence modeling approach depends on choice of particular RANS model and on resolution function.

In [11], authors propose a resolution control function, proposed on claims that it roughly resembles the ratio of the unresolved turbulent kinetic energy to the total kinetic energy. The formulation includes three important length scales, the cut-off (grid) length scale, L_c, integral scale L_i and Kolmogorov scale L_k, and is given as follows,

$$Fr = min\left(1, \left[\frac{1 - exp(-\beta L_c/L_k)}{1 - exp(-\beta L_i/L_k)}\right]^n\right), \tag{1}$$

where $n = 2$ and $\beta = 2.0 \times 10^{-3}$. Length scales are defined in following way,

$$L_c = 0.61\left(\Delta_x \Delta_y \Delta_z\right)^{1/3}, L_i = k^{\frac{3}{2}}/\epsilon, L_k = \nu^{\frac{3}{4}}/\epsilon^{\frac{1}{4}}. \tag{2}$$

Governing equations for turbulent flow described by conservation for mass and momentum in Reynolds averaged form are,

$$\frac{\partial U_j}{\partial x_j} = 0, \tag{3}$$

$$\frac{\partial U_i}{\partial t} + U_j \frac{\partial U_i}{\partial x_j} = -\frac{1}{\rho}\frac{\partial P}{\partial x_i} + \frac{\partial}{\partial x_j}\left[\nu \left(\frac{\partial U_i}{\partial x_j} + \frac{\partial U_j}{\partial x_i} \right) - \overline{u_i u_j} \right], \tag{4}$$

where Reynolds stresses are calculated from relation incorporating Boussinesq hypothesis

$$\overline{u_i u_j} = \frac{2}{3} k \delta_{ij} - \nu_t \left(\frac{\partial U_i}{\partial x_j} + \frac{\partial U_j}{\partial x_i} \right). \tag{5}$$

Calculation of eddy viscosity requires solution of additional transport equations. In the case of $k - \epsilon$ model, eddy viscosity is computed using relation between turbulent kinetic energy and turbulent kinetic energy dissipation rate, each obtained from its own transport equation as

$$\nu_t = FrC_\mu \frac{k^2}{\epsilon}. \tag{6}$$

Here turbulent eddy-viscosity is modulated by VLES resolution control function F_r.

The so called “realizable” $k - \epsilon$ turbulence model is an implementation of the model presented in [15]. In this model, realizability is achieved by functional variability of the C_μ coefficient in the formulation of eddy viscosity. This is experimentally motivated. While it has a value of $C_\mu = 0.09$ in inertial sublayer of boundary layer flow, in homogeneous shear flow it is around 0.05. Therefore some way of changing the C_μ coefficient depending on the flow situation is desirable.

The turbulence model realizability considers constraints any model should meet. These are following inequalities regarding Reynolds stress tensor components:

$$\overline{u_\alpha^2} > 0, (\alpha = 1, 2, 3),$$
$$\frac{\overline{u_\alpha u_\beta}^2}{\overline{u_\alpha^2 u_\beta^2}} \leq 1, (\alpha = 1, 2, 3;\ \beta = 1, 2, 3).$$

Second condition is known as Schwarz’s inequality. What is needed is positivity of normal stresses, and that Reynolds stress tensor components obey Schwartz’s inequality. Based on these realizability constraints various authors, including W. Reynolds, as well as Shih, Zhu and Lumley, proposed following form of an equation for C_μ:

$$C_\mu = \frac{1}{A_0 + A_s U^{(*)} \frac{k}{\epsilon}}. \tag{7}$$

Shih, Zhu, Lumley give following formulation for the constituents of the previous equation

$$U^{(*)} = \sqrt{S_{ij}S_{ij} + \tilde{\Omega}_{ij}\tilde{\Omega}_{ij}}, \tag{8}$$

where in current implementation

$$\tilde{\Omega}_{ij} = \overline{\Omega_{ij}} - \epsilon_{ijk}\omega_k, \tag{9}$$

where $\overline{\Omega_{ij}}$ is the mean rate-of-rotation tensor viewed in a rotating reference frame with the angular velocity ω_k. The $-2\epsilon_{ijk}\omega_k$ term, appearing in the equation for $\tilde{\Omega}_{ij}$ is not included in present implementation, following recommendation from [20].

Other model parameters are,

$$A_s = \sqrt{6}cos\phi, \phi = \frac{1}{3}arccos\left(\sqrt{6}W\right), W = \frac{S_{ij}S_{jk}S_{ki}}{\tilde{S}^3}, \tilde{S} = \sqrt{S_{ij}S_{ij}}. \tag{10}$$

For stable implementation of the model, the argument of the *arccos* function should be in interval [− 1, 1], therefore, the suggested implementation of ϕ is in a following way,

$$\phi = \frac{1}{3}arccos\left(max\left(-1, min\left(\sqrt{6}W, 1\right)\right)\right). \tag{11}$$

In the original reference, and in current implementation, the coefficient A_0 has value $A_0 = 4.0$. This difference in [20] where model uses calibrated value of $A_0 = 4.04$.

The transport equation for turbulence kinetic energy k remains the same as in standard $k - \epsilon$ model, while there are some changes with equation for dissipation. The transport equations of Realizable $k - \epsilon$ model implemented in present code have following form,

$$\frac{\partial}{\partial t}(\rho k) + \frac{\partial}{\partial x_j}(\rho k u_j) = \frac{\partial}{\partial x_j}\left[\left(\mu + \frac{\mu_t}{\sigma_k}\right)\frac{\partial k}{\partial x_j}\right] + P_k - \rho\epsilon, \tag{12}$$

$$\frac{\partial}{\partial t}(\rho\epsilon) + \frac{\partial}{\partial x_j}(\rho\epsilon u_j) = \frac{\partial}{\partial x_j}\left[\left(\mu + \frac{\mu_t}{\sigma_\epsilon}\right)\frac{\partial \epsilon}{\partial x_j}\right] + \rho C_1 S\epsilon - \rho C_2\frac{\epsilon^2}{k + \sqrt{\nu\epsilon}}, \tag{13}$$

where

$$C_1 = max\left[0.43, \frac{\eta}{\eta + 5}\right], \eta = S\frac{k}{\epsilon}, S = \sqrt{2S_{ij}S_{ij}}. \tag{14}$$

Production of turbulence kinetic energy due to velocity gradients P_k, is expressed as,

$$P_k = -\overline{u_i u_j}\frac{\partial U_i}{\partial x_j}, \tag{15}$$

where Reynolds stress tensor is found using Eq. (5).

The production of turbulence kinetic due to buoyancy P_b, is left out in current implementation, since we deal with situation replicating conditions of neutral atmospheric stability.

The constants of the model are,

$$C_{1\epsilon} = 1.44, C_2 = 1.9, \sigma_k = 1.0, \sigma_\epsilon = 1.2. \tag{16}$$

In another tested variant, the $k-\omega$ SST model is used as eddy-viscosity model in VLES approach. Here we briefly describe its implemented form.

The transport equations for the Shear Stress Transport (SST) $k-\omega$ model [12, 13] have similar form as those of the standard $k-\omega$ model. Here it is shown in the common form using dynamic viscosity - μ. The turbulence kinetic energy - k, and specific dissipation rate - ω, are obtained from the following equations,

$$\frac{\partial}{\partial t}(\rho k) + \frac{\partial}{\partial x_i}(\rho k u_i) = P_k - \beta^* \rho \omega k + \frac{\partial}{\partial x_j}\left[(\mu + \sigma_k \mu_t)\frac{\partial k}{\partial x_j}\right], \tag{17}$$

$$\frac{\partial}{\partial t}(\rho \omega) + \frac{\partial}{\partial x_i}(\rho \omega u_i) = \frac{\gamma}{\nu_t} P_k - \beta \rho \omega^2 + \frac{\partial}{\partial x_j}\left[(\mu + \mu_t \sigma_\omega)\frac{\partial \omega}{\partial x_j}\right] + D_\omega, \tag{18}$$

where P_k represents production term for turbulence kinetic energy, calculated exactly, as previously described in the case of Realizable $k-\epsilon$ model, and D_ω is the cross diffusion term is defined as,

$$D_\omega = 2(1-F_1)\frac{\rho \sigma_{\omega 2}}{\omega}\frac{\partial k}{\partial x_j}\frac{\partial \omega}{\partial x_j}. \tag{19}$$

There are two sets of constants, denoted by subscripts 1 for inner, and 2 for outer, respectively. Blending is achieved using blending function

$$C = F_1 C_1 + (1-F_1) C_2, \tag{20}$$

where C represents any of the model coefficients. The blending function F_1 is given by

$$F_1 = tanh\left(\left(min\left(max\left(\frac{k^{1/2}}{\beta^* \omega d}, \frac{500\nu}{d^2 \omega}\right), \frac{4\rho \sigma_{\omega 2} k}{CD_{k\omega} d^2}\right)\right)^4\right), \tag{21}$$

with

$$CD_{k\omega} = max\left(2\rho \sigma_{\omega 2}\left(\frac{1}{\omega}\right)\left(\frac{\partial k}{\partial x_i}\right)\left(\frac{\partial \omega}{\partial x_i}\right), 10^{-10}\right). \tag{22}$$

and d is the distance to the nearest wall. The coefficient values are, $\beta^* = 0.09$, $\gamma_1 = 5/9$, $\gamma_2 = 0.44$, $\beta_1 = 0.075$, $\beta_2 = 0.0828$, $\sigma_{k1} = 0.85$, $\sigma_{k2} = 1.0$, $\sigma_{\omega 1} = 0.5$, $\sigma_{\omega 2} = 0.856$.

Turbulent eddy viscosity, including VLES modulation, is defined as

$$\mu_t = F_r \frac{\rho a_1 k}{max[a_1 \omega, SF_2]}, \tag{23}$$

where S is scalar invariant of the strain rate tensor $S = \sqrt{2S_{ij}S_{ij}}$, $a_1 = 0.31$, and F_2 is second blending function, defined as,

$$F_2 = tanh\left(max\left(\frac{k^{1/2}}{\beta^* \omega d}, \frac{500\nu}{d^2 \omega} \right)^2 \right). \tag{24}$$

It is recommended to use limited form of production P_k, defined as

$$\widetilde{P_k} = min(P_k, 10\beta^* \rho \omega k). \tag{25}$$

2.1 Wall Boundary Conditions

In our work we use following implementation of automatic wall treatment boundary condition which provide accurate values in cases of varying grid accuracy near walls.

The non-dimensional variables we use are,

$$y^+ = \frac{y u_\tau}{\nu}, U^+ = \frac{u}{u_\tau}, \omega^+ = \frac{\omega \nu}{u_\tau^2}. \tag{26}$$

In the first inner cells ω is set to

$$\omega_w = \frac{\rho (u_\tau)^2}{\mu} \omega^+. \tag{27}$$

The task is to find proper formulation for friction velocity, u_τ, and non-dimensional specific dissipation - ω^+. We start by defining proper form of ω^+ equation in viscous sub-region and in log-region, and then its blended form. In the viscous sub-region we define ω as a function of non-dimensional distance from wall $\omega = f(y^+)$ in a following way,

$$\omega_{vis} = \frac{6\nu}{0.075 y^2} = \frac{6\nu u_\tau^2}{0.075 \nu^2 (y^+)^2} = \frac{6 u_\tau^2}{0.075 \nu (y^+)^2}. \tag{28}$$

Taking the Law of the wall into account, the expression for ω can be developed by supplying definition for friction velocity in viscous sub-region,

$$u_\tau = \frac{U_1}{y^+} \Rightarrow \omega = \frac{6 U_1^2}{0.075 \nu (y^+)^4}. \tag{29}$$

The subscript 1, denotes the value of the first near wall cell.

In the log-region ω can be expressed as,

$$\omega_{log} = \frac{u_\tau}{\sqrt{\beta^*}\kappa y} = \frac{u_\tau}{0.3\kappa}\frac{u_\tau}{\nu y^+} = \frac{u_\tau^2}{0.3\kappa\nu y^+}. \quad (31)$$

Same as in the case of viscous sub-region, we can further develop the expression for ω by taking the Law of the wall into account supplying definition for friction velocity in viscous log-region.

Blending is used to compute proper value of ω in first cell near wall using following expression,

$$\omega_1(y^+) = \sqrt{\omega_{vis}^2(y^+) + \omega_{log}^2(y^+)}. \quad (32)$$

A Note on Approach of Han and Krajnović. In their paper [11], authors use commercial code ANSYS Fluent with active *Enhanced Wall Function* option. We find that the ability to activate Low-Re model in the vicinity of the wall at some levels of resolution, which is intrinsic to that approach [20], is crucial for cases with separation from smooth walls. Authors use standard High-Re $k - \epsilon$ model as underlying RANS model in VLES, and had used meshes with higher wall resolution than in typical High-Re model use-case. We conclude that the full description of their approach would be of a three components model, with significant contribution of Wolfstein's one-equation model in the overall region of RANS model activity, which is near wall. This is important consideration since it was noted previously, that for simulation of flows over smooth steep hills with flow separation, standard wall function treatment used for High-Re turbulence models based on equation for dissipation rate of turbulence kinetic energy, ϵ, didn't give satisfactory results, even when so-called scalable wall functions [20] were used. This seems to be improved with said *Enhanced Wall Function*, which allows activation of one-equation model near wall.

3 Numerical Algorithm

For the present study we used open-source code *freeCappuccino* [16–18], a fully unstructured, parallelized code implementing Finite Volume Method, based on collocated variable arrangement [19]. Time dependent terms are discretized by Backward Differentiation Formula of the second order (BDF2). Cell centered gradients are computed using efficient implementation of least-square gradient reconstruction algorithm. Diffusion terms are treated in such a way to properly take into account the mesh non-orthogonality (and skewness) by explicit correction terms, as described in [16]. Convection terms are discretized using TVD schemes, in flux limiter form. Most notably the MUSCL scheme was used for RANS simulations. For VLES runs, the central differentiation scheme (CDS) was used in whole domain, since it is most natural choice for LES. This choice is motivated by the fact that the LES region dominates the simulation domain away from the near wall region, where RANS becomes activated, and since the switching of the convection discretization approach between different turbulent model regions was found unpractical (yet it should be investigated). However, due to lower grid resolution than in practical LES, some minimal amount of numerical dissipation is added to keep the simulation stable by blending CDS scheme with only 2% of upwind differentiation scheme trough γ coefficient which controls the deferred correction (see [16, 19]).

Pressure-velocity coupling was handled by segregated, SIMPLE algorithm. In VLES we perform nonstationary simulations, which is also handled using SIMPLE because of stability issues of noniterative time-advancement approaches, since the chosen timestep size produced maximum Co numbers around $Co_{max} \approx 2$. In average 15–20 SIMPLE iterations were used per timestep, which were enough for residuals to drop between two and three levels of magnitude before moving to next time-level. In VLES runs, the under-relaxation factors were set to $\alpha_u = 0.7$ for velocity components, and same for turbulence scalars, while $\alpha_p = 0.3$ was used for pressure. The pressure correction equation may be assembled and solved multiple times to alleviate accuracy and stability issues of mesh skewness. Most often two passes were enough.

Linear equations are solved using BiCGStab linear solver with SIP preconditioner for velocity components and scalars, while CG(SIP), a SIP preconditioned Conjugate gradient method was used for symmetric linear system resulting from pressure-correction Poisson equation discretization. The use of SIP preconditioner was enabled by a structured grid in the case of steep, smooth cosine-squared hill of Ishihara et al. which produced seven diagonal matrix. More details on SIP used as a preconditioner for Krylov subspace linear equation solvers may be found in [16].

Post processing of time dependent simulation demands division of output files to one with averaged fields, which contain ensemble average fields resulting from all time steps and files with instantaneous field values. Statistical averaging is performed at the end of each time step.

4 Results

The VLES model is examined on a problem of turbulent flow over a smooth hill at laboratory scale, experimentally studied in [21]. The setup considered turbulent flow over a hill with circular base, a cosine-squared cross section and a maximum slope of about 32. The shape of the hill is defined by

$$z_s(x,y) = hcos^2\left(\pi\left(x^2+y^2\right)^{1/2}/2L\right),$$

where $h = 40\,\text{mm}$ and $L = 100\,\text{mm}$. The Reynolds number based on free stream velocity and boundary-layer thickness was $Re = U_\infty \delta / \nu = 1.4 \times 10^5$. The simulated boundary layer in the experiment represented atmospheric boundary layer in the scale 1:1000 [21]. Inflow boundary condition for streamwise velocity is defined using power law in the form $U/U_{ref} = \left(z/z_{ref}\right)^n$, where exponent $n = 0.135$, reference height $z_{ref} = 200$, $U_{ref} = 5.5$ m/s. The original reference [21] mentions that the logarithmic profile should be used in surface layer of the boundary layer, where a parameter is

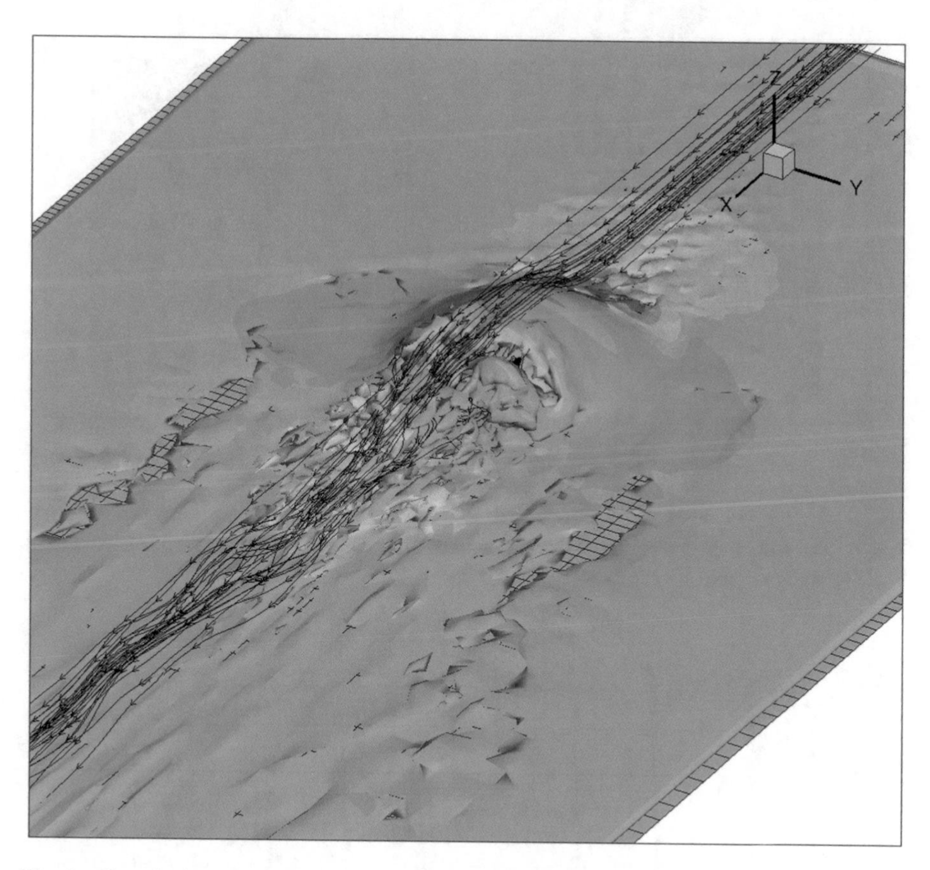

Fig. 1. Complexity of the flow in the separation region behind the hill. Iso-surfaces of Q-criteria, colored by pressure, and streamlines. Snapshot of flow simulated with VLES using $k - \omega$ SST model.

present which emulates presence of low grass in real atmospheric setting. We decided to use power law profile described previously since it was shown in the original paper that it approximates measured mean velocity profile to a very good approximation. The wall in our simulation is considered as aerodynamically smooth. The rough wall boundary condition influence is left for future study.

Flow domain consists of $120 \times 61 \times 48$ (351360) cells in streamwise, spanwise and wall-normal direction respectively. Computational domain extends approximately four hill heights to the front of the hill. The mesh is structured, produces by algebraic extrusion from bottom surface where geometry is defined to approximately ten hill heights, in wall normal direction, to allow proper definition of free-slip boundary conditions on top of the domain. The non-dimensional distance of first cells near bottom wall is $y^+ \approx 5$ i.e. first cells are in the edge of viscous sublayer. This is an unusual circumstance that is only enabled by the use of automatic wall treatment. It is found that higher resolution near wall is needed than is usual for High-Re model use cases, to

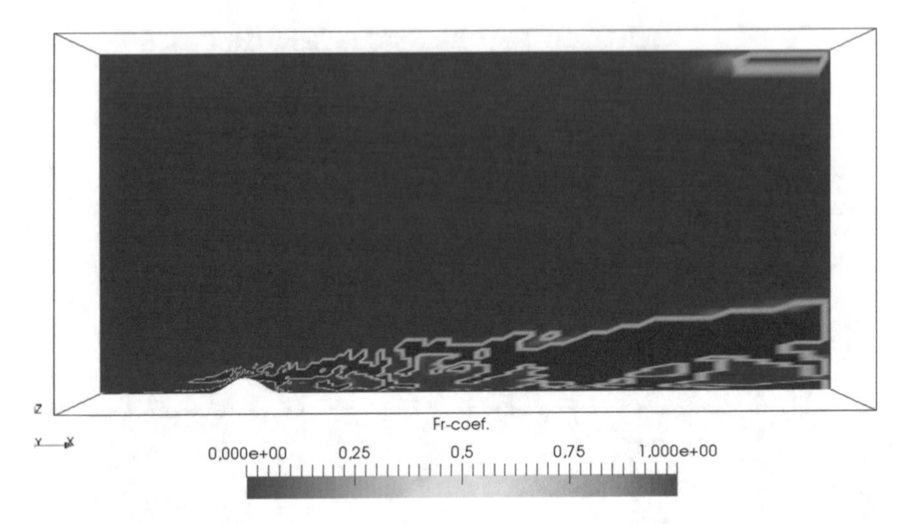

Fig. 2. The instantaneous values of resolution control function F_r in center plane cross-section. VLES with Realizable $k - \epsilon$ model.

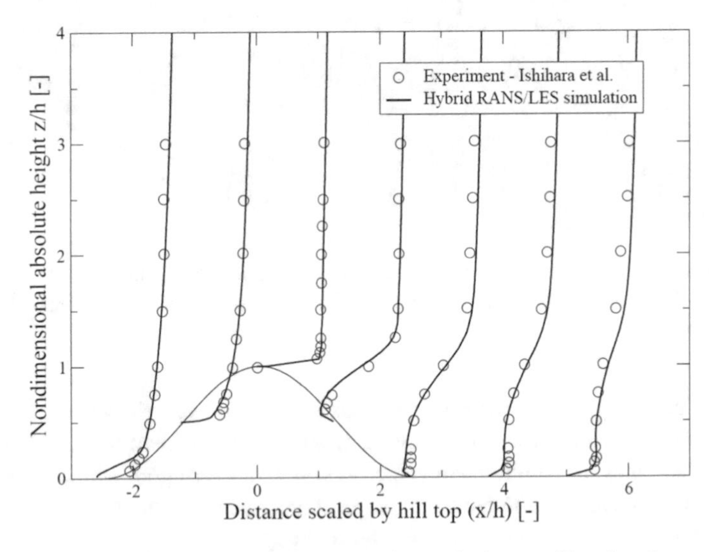

Fig. 3. Development of time-averaged streamwise velocity profiles in the center-plane, compared to the experiment of Isihara et al. [21]. VLES using $k - \omega$ SST model.

produce flow separation from smooth hill, yet since Low-Re model was not used (in the case of $k - \epsilon$ model), there was not incentive to fully resolve viscous sublayer.

The complexity of the analyzed flow situation is illustrated by Fig. 1. where effects of unsteadiness, flow separation and reattachment is present. Behind the hill strong instability is observed, shown here by isosurfaces of Q-criteria and streamlines. In the experiment boundary layer separated at the hill crest, a recirculation region is formed behind the hill, and reattachment point is near to the hill base on the lee ward side.

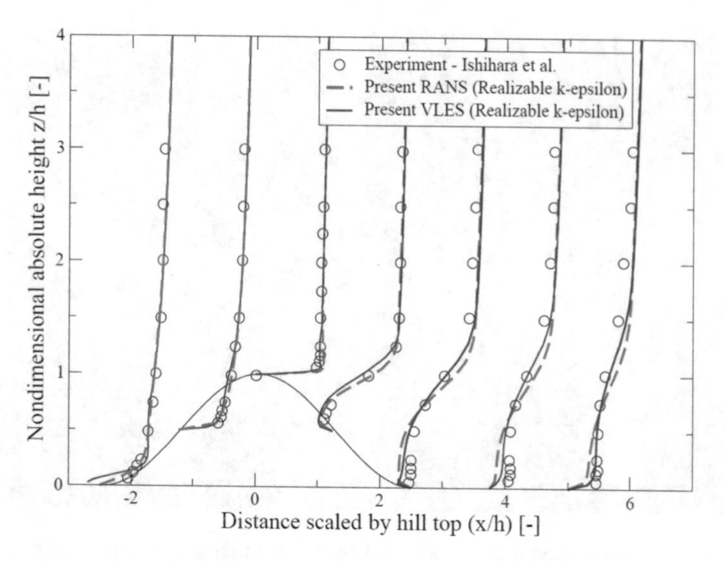

Fig. 4. Development of time-averaged streamwise velocity profiles in the center-plane, compared to the experiment of Isihara et al. [21]. VLES with Realizable $k - \epsilon$ model and RANS results.

The ability to capture flow unsteadiness in the wake behind a hill is a property of the VLES method used here, which is not possible with a standard RANS approach. Being able to resolve in time accurate manner the fluctuations behind the hill is important in many applications. The performance of the model in this region is determined by the resolution function F_r which modulates turbulent eddy viscosity. The values of resolution control function vary spatially and temporally. The instantaneous values of F_r in the center plane cross-section is shown in Fig. 2 for the case of VLES with Realizable $k - \epsilon$ model. Based on this, we can give a qualitative description of spatial behavior of resolution control function. It is visible that in the undisturbed boundary layer, in front of the hill, F_r has value of one in first couple of cell layers next to wall. This means that model shows URANS behavior in the vicinity of wall, as expected. In our simulations F_r has sharp transition to value of zero at certain distance away from the wall and model switches to ILES behavior. Around and behind the hill the situation is more complex. The URANS and ILES zones are both present with some regularity. Right next to wall, behind the hill, the URANS is active in a region of thickness comparable to the one in front of the hill. On the lee side of the hill, in the separation zone, and in part of the flow separation region, the ILES region is present. URANS zones are also present in significant area found on the interface of growing wake region behind the hill and undisturbed flow that passed above the hill. The switching zone between two models is again very thin.

Fig. 5. The instantaneous streamwise velocity field in the center-plane cross section.

In the original experiment, three velocity components were measured on the center plane of the hill on various positions. We will focus only on streamwise velocity component as most significant one. For comparison, both the data obtained from the simulation and the data from [21] are represented in Figs. 3 and 4 for VLES models based on $k-\omega$ SST model and Realizable $k-\epsilon$ model, respectively. Figure 4 also includes RANS results with the same model for comparison. In these figures the mean velocity components are normalized by the mean velocity obtained at the hill height in the undisturbed boundary layer $(U_h = 4.7m/s)$.

All models are capable of capturing most distinguishing features of the flow in various level of precision. The first one is increased flow velocity on the hill top. In applications such as wind energy, estimation of flow speed-up throughout the flow field and especially taking place on the hill top is of greatest importance in problems of wind-farm siting. Both VLES simulations give almost perfect matching with measurements. The RANS simulation produces slight over-predicted value at hill top compared to VLES approach with the same model. In figure showing resolution control function it is visible that strong modulation on turbulence eddy viscosity takes place on very top of the hill and we see that this has favorable effect on results.

An interesting and unexpected disadvantage of VLES predictive capabilities turns to be the slight deceleration at the upwind hill foot. Both VLES simulations give large over-prediction of deceleration in mean values. The RANS simulation gave close prediction and eliminated the doubts that inaccuracy in inlet profile might be the cause. It is clearly a modeling error and the reason is still investigated. One of the potential reasons may be the switching between zones in VLES that happens on approximately the same distance from wall. The over-predicted deceleration is maintained in mid-slope position in front of the hill in the case of the VLES with $k-\omega$ SST model, while the VLES with Realizable $k-\epsilon$ model recovers and shows excellent matching.

The slight kink in this profile may have the switching between models as cause.

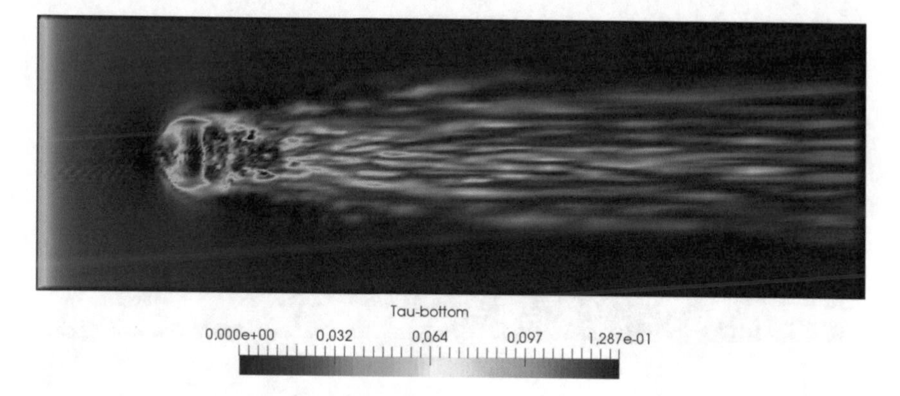

Fig. 6. The instantaneous values of wall shear stress τ_w.

The instantaneous velocity field in center plane is shown in Fig. 5. The figure illustrates the capability of the modelling approach to reproduce nonstationary flow structures important in various applications.

An advantage of VLES approach is ability to show temporal as well as spatial development of wall friction in the wake region behind the hill. This was not covered in cited measurements and is presented here only for illustration purposes in Fig. 6.

Another important flow feature is flow separation behind the hill. The VLES approach gave excellent predictions of this zone in several ways with only minor flaws. The VLES approach using $k - \omega$ SST model gave best results in present study. The profile at mid-slope position of VLES with Realizable $k - \epsilon$ model produced slight under-prediction in mean velocity in zone of the wake, compared to measurements and to RANS simulation, which showed very good matching at this position. Situation changes with vertical mean velocity profile that originates from the hill foot on lee side. The growth in thickness of the wake zone is much better predicted in this one and all later vertical profiles with VLES than in RANS simulation with the same model. Another quantitative feature is recovery in near wall velocity after the shallow recirculation bubble that ends in hill foot in the experiments. We see that both VLES approaches give improved results when comparing the last two profiles to the left in near wall region. Both VLES models slightly over-predict the length of the bubble, although the VLES approach using $k - \omega$ SST model gives better estimate, according to spatial development of mean velocity profile.

Time averaged contours of velocity magnitude, streamwise velocity and turbulence kinetic energy are shown in Fig. 7. The middle picture shows negative iso-values of streamwise velocity by dashed line. In between is the zero streamwise velocity iso-contour depicting the size and shape of recirculation bubble, slightly over-predicted in present case.

However it was found useful in this regard to have some small and controlled numerical instabilities and not to damp all out by numerical diffusion. This was achieved through proper choice of convection term discretisation. The CDS scheme was blended with only a small percent of upwind scheme (2% of UDS). Small nonphysical instabilities are shown in Fig. 5 above the hill top, which were found more favorable in present simulation than overly diffusive effect, also nonphysical of other schemes.

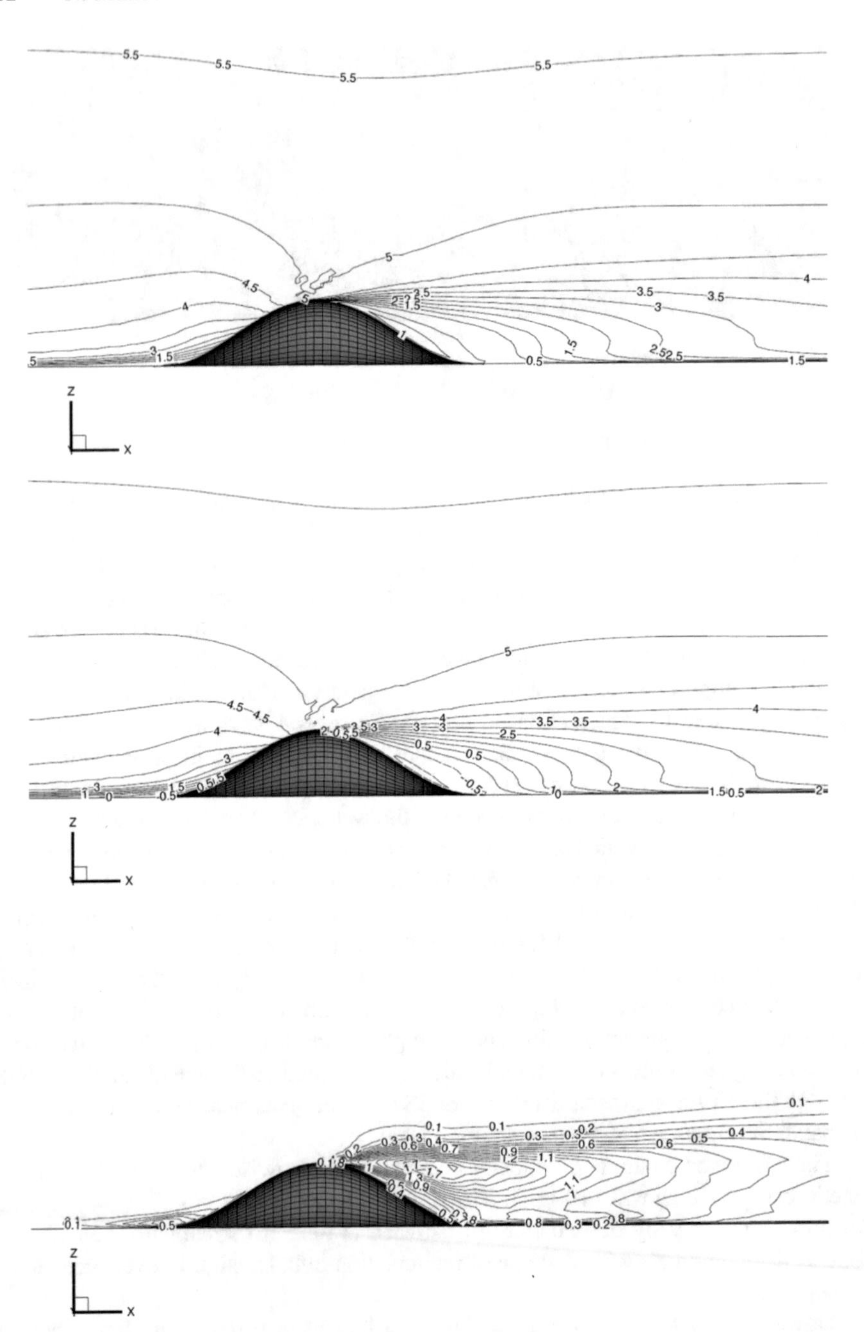

Fig. 7. The time-averaged contours of: Velocity magnitude (top), Streamvise velocity (negative values shown by dashed lines, center), and Turbulent kinetic energy (bottom), center-plane cross section. VLES using $k - \omega$ SST model.

5 Conclusions

The study showed performance of VLES simulation methodology when applied to turbulent flows over smooth steep hills. The study used existing VLES resolution control function and applied it with different RANS models than proposed in the original method. Combination of resolution control function of Han and Krajnović [11] with Realizable $k - \epsilon$ model and $k - \omega$ SST model might be original contribution of the present study. We are careful with this assertion because this is very active area of development and combinatorial possibilities are huge. Almost any eddy-viscosity model can be incorporated in present VLES approach and give unique model. We have seen that in the case of smooth hill used for comparative assessment the VLES approach indeed gave improved predictive performance regarding some important flow features such as speed-up at hill top, wake region thickness growth and flow velocity recovery after the recirculation bubble. Moderate improvement is achieved with prediction of recirculation zone length. Further model improvements are possible and more detailed investigation of model behavior is left for future studies. We believe that in practical environmental flow cases in atmospheric micro-scale, proposed hybrid flow simulation methodology may give better results than RANS models and should be given a consideration by CFD practitioners.

Acknowledgements. Support of the Ministry of Education, Science and Technological Development of Republic of Serbia, trough project TR-33036 is greatly acknowledged.

References

1. Speziale, C.G.: Turbulence modeling for time-dependent RANS and VLES: a review. AIAA J. **36**, 173–184 (1998)
2. Chaouat, B.: The state of the art of hybrid RANS/LES modeling for the simulation of turbulent flows. Flow Turbul. Combust. **99**, 279–327 (2017)
3. Sagaut, P., Deck, S., Teraccol, M.: Multiscale and Multiresolution Approaches in Turbulence, 2nd edn. Imperial College Press, London (2013)
4. Spalart, P.R.: Strategies for turbulence modelling and simulations. Int. J. Heat Fluid Flow **21**, 252–263 (2000)
5. Menter, F.R., Egorov, Y.: The scale-adaptive simulation method for unsteady turbulent flow predictions. Part 1: theory and model description. Flow Turbul. Combust. **85**, 113–138 (2010)
6. Spalart, P.R., Jou, W., Strelets, M., Allmaras, S.R.: Comments on the feasibility of LES for wings, and on a hybrid RANS/LES approach. In: Liu, C., Liu, Z. (eds.) Advances in DNS/LES, pp. 137–147. Greyden Press, Columbus (1997)
7. Schiestel, R., Dejoan, A.: Towards a new partially integrated transport model for coarse grid and unsteady turbulent flow simulations. Theor. Comput. Fluid Dyn. **18**, 443–468 (2005)
8. Chaouat, B., Schiestel, R.: Progress in subgrid-scale transport modelling for continuous hybrid non-zonal RANS/LES simulations. Int. J. Heat Fluid Flow **30**, 602–616 (2009)
9. Hanjalić, K., Kenjereš, S.: Some developments in turbulence modeling for wind and environmental engineering. J. Wind Eng. Ind. Aerodyn. **96**, 1537–1570 (2008)

10. Batten, P., Goldberg, U., Chakravarthy, S.: Interfacing statistical turbulence closures with large-Eddy simulation. AIAA J. **42**, 485–492 (2004)
11. Han, X., Krajnović, S.: An efficient very large eddy simulation model for simulation of turbulent flow. Int. J. Numer. Methods Fluids **71**, 1341–1360 (2012)
12. Menter, F.R.: Two-equation eddy-viscosity turbulence models for engineering applications. AIAA J. **32**(8), 1598–1605 (1994)
13. Menter, F.R., Kuntz, M., Langtry, R.: Ten years of industrial experience with the SST turbulence model. In: Hanjalić, K., Nagano, Y., Tummers, M. (eds.) Turbulence, Heat and Mass Transfer 4, pp. 625–632. Begell House, Inc. (2003)
14. Jakirlić S., Chang, C.-Y., Kadavelil, G., Kniesner, B., Maduta, R., Šarić, S., Basara, B.: Critical evaluation of some popular hybrid LES/RANS methods by reference to a flow separation at a curved wall. In: 6th AIAA Theoretical Fluid Mechanics Conference, AIAA 2011-3473 (2011)
15. Shih, T.H., Liou, W.W., Shabbir, A., Yang, Z., Zhu, J.: A new k-ε eddy viscosity model for high reynolds number turbulent flows. Comput. Fluids **24**(3), 227–238 (1995)
16. Mirkov, N., Rašuo, B., Kenjereš, S.: On the improved finite volume procedure for simulation of turbulent flows over real complex terrains. J. Comput. Phys. **287**, 18–45 (2015)
17. Mirkov, N., Stevanović, Ž.: New non-orthogonality treatment for atmospheric boundary layer flow simulation above highly non-uniform terrains. Therm. Sci. **20**(Suppl. 1), 223–233 (2016)
18. Mirkov, N., Vidanović, N., Kastratović, G.: freeCappuccino - an open source software library for computational continuum mechanics. In: Mitrovic, N., Milosevic, M., Mladenovic, G. (eds.) Experimental and Numerical Investigations in Materials Science and Engineering, CNNTech 2018. Lecture Notes in Networks and Systems, vol. 54. Springer, Cham (2019). http://www.github.com/nikola-m/freeCappuccino/
19. Ferziger, J.H., Perić, M.: Computational Methods for Fluid Dynamics, 2nd edn. Springer, Cham (1999)
20. ANSYS Fluent Theory Guide, Release 15.0, ANSYS, Inc. Southpointe 275 Technology Drive, Canonsburg, PA 15317
21. Ishihara, T., Hibi, K., Oikawa, S.: A wind tunnel study of turbulent flow over a three dimensional steep hill. J. Wind Eng. Ind. Aerodyn. **83**, 95–107 (1999)

Experimental and Numerical Investigation of the T-Stub Elements with Four Bolts in a Row Until Bolt Fracture

Đorđe Jovanović[1](✉), Nenad Mitrović[2], Zlatko Marković[3], Dragiša Vilotić[1], and Boris Kosić[2]

[1] Faculty of Technical Sciences, University of Novi Sad, 21000 Novi Sad, Serbia
djordje.jovanovic@uns.ac.rs
[2] Faculty of Mechanical Engineering, University of Belgrade, 11000 Belgrade, Serbia
[3] Faculty of Civil Engineering, University of Belgrade, 11000 Belgrade, Serbia

Abstract. For the past several decades, a codified design of steel connections in civil-engineering has been based on the component approach. For a very common end-plate connection, a tension component, named T-stub, usually dictates the connections' behavior. This T-stub element is greatly investigated in the configuration with two bolts in a row, but the configuration with four bolts in a row is usually neglected, both in the studies and codes. This paper presents an experimental investigation of T-stub elements and important aspects of their numerical modeling. Special attention is dedicated to the material testing and modeling, since all of the tests were performed until bolt fracture. Uniaxial tests of steel specimens were performed using extensometers, strain gauges, and Aramis system, while the bolt material is additionally tested by microscopic examination and hardness testing. In order to obtain satisfactory calibration of numerical models developed in Abaqus, knowing material parameters including damage initiation and propagation is crucial. Several iterative numerical-experimental procedures for obtaining the true stress-strain curves are outlined and compared, along with the well-known Bridgman method. The advantages of using Aramis system in calibrating numerical model, for both material and assembly are demonstrated. In the end, comparisons of numerical and experimental behavior curves are presented and satisfactory results are obtained.

Keywords: T-stub element · Experimental investigation · True stress-strain curve · Abaqus · Aramis system · Digital Image Correlation method

1 Introduction

Steel structures in civil engineering usually comprise beam-to-column connections, and more than often this connection is designed as a bolted end-plate connection. This type of connection is recognized as effective in every aspect, and is used as long as bolts. It consists of an end-plate welded to the end of the beam and connected to column flange

N. Mitrovic et al. (Eds.): CNNTech 2019, LNNS 90, pp. 305–322, 2020.
https://doi.org/10.1007/978-3-030-30853-7_18

(or web) by bolts arranged in several rows. A modern approach in the design of this connection type is a component method, which is based on the mechanical model of the connection. Every relevant part of the connection is represented by an equivalent spring, and the springs are interconnected in the same relation as the components in the actual connection. Design codes provide analytical expressions for both the stiffness and ultimate strength of every component, in order to evaluate connection's stiffness and resistance.

The component method was firstly proposed by Zoetmeijer in 1972 [1], who focused on one of the tensioned components – T-stub element. The basic idea was to calculate ultimate resistance of end plate with bolts (or column flange with bolts) without employing the plate theory, which would resulted in too complicated formulas for everyday use. Therefore, Zoetmeijer observed what is called the equivalent T-stub element, which is a hypothetical element consisting of a plate lying on the rigid foundation connected with it by bolts, and a perpendicular web in tension connected to the middle of the plate. This is the same problem which was used by Douty and McGuire [2] to express the location and magnitude of the prying forces. Zoetmeijer defined three modes of failure: (1) plate failure by formation of four plastic hinges, (2) bolt failure and (3) simultaneous failure (of both plate and bolts). Additionally, in order to calculate the resistance of the component in an actual connection, the length of the equivalent T-stub must be equal to the effective length, which is obtained by selecting a corresponding yield line pattern of the actual plate. Prying forces were not explicitly incorporated in the formula, but some of the recommendations about the position of prying forces' resultant are taken from the work of Douty and McGuire.

Significant advances in the field were made by Jaspart [3], who explored the stiffness of a T-stub component, and participated in the development of the Eurocode EN 1993-1-8 [4]. On the American continent, very similar results were obtained by Dranger [5] and Mann and Morris [6]. The main difference introduced in the Dranger's solution is an application of the limit analysis (kinematic theorem) which will be adopted later as an elegant technique to implicitly include prying forces. The T-stub element is extensively investigated afterwards, both experimentally and numerically, and initial formulas are slightly adjusted and proven satisfactory. The only problem is that Eurocode does not acknowledge the design of T-stub elements with four bolts in a row, even though they have been used broadly from the first half of the twentieth century. American code [7] also adopted the component method, but similarly disregards connections with four bolts in a row.

In this paper, a part of experimental research of T-stub elements with four bolts in a row is presented. The complete investigation comprised 36 specimens, with 9 different configurations. Results for six of them are presented here. One of the main goals of the on-going research program is to perform a comprehensive parametric numerical analysis. Thus, a sophisticated numerical model needs to be developed. Moreover, the material representation, arose from material testing, presents a key element in capturing the true behavior all up to the point of fracture. Different measurement methods, as well as various techniques of material modeling are presented, discussed and tested herein. This led to the precise calibration of a numerical model and satisfactory numerical representation of tested T-stub elements.

2 Material Testing and FEM Calibration

2.1 Uniaxial Tensile Testing

The most commonly used test for steel material examination is a tensile test with the extensometer used for measuring displacements. Hence, engineering stress-strain relations will always show an elastic range, strain hardening range, and a strain softening range. Someplace in the material will be a weak zone, which will not be able to carry the further load. In this zone, the ductile fracture will start to develop. This will cause the weaker area to strain further than the rest of the model, even if the load is not increased, which is known as the "necking" of the section. In general, the strain softening rendered on the engineering stress-strain curve is associated with the necking range of the test. Due to the non-uniform strain distributions existent at the neck at large strain values, it has long been acknowledged that the changes in the geometry of the specimen must be considered, if the material response up to failure is needed to be identified [8, 9]. Calculation of the true stress-strain relation from the experimental data, intended to be used in a numerical investigation based on the Finite Element Method - FEM (which is inherently a discrete method), is investigated in this section. Firstly, a comprehensive review of the available methods on this subject is presented. For large uniform deformations, "true" stress and strain measures are defined by Eq. (1) which is valid only up to necking (n) point:

$$\sigma_t = \sigma_e \cdot (1 + \varepsilon_e); \quad \varepsilon_t = \ln(1 + \varepsilon_e) \tag{1}$$

where index e denotes engineering and t the true values of stress and strain.

The necking problem has troubled the researchers for a long time. Considere formulated his criterion back in 1885 [10]. There are different approaches to the problem (e.g. deformation strengthening theory, plastic instability, FEM), depending on the field of interest. Nevertheless, judging by recent developments in the field [11], the problem is still present.

Basically, among these approaches, there are three groups of procedures for the determination of true stress-strain curves from tensile tests: analytical, iterative, and most recently, quasi-direct. The most prominent analytical method was developed by Bridgman [9]. He assumed that the necking principal stress plane is circular and the circumferential strain is equal to the radial strain. These assumptions lead to the reduction factor k, which enlarges the average (nominal) stress σ_a, as given in Eq. (2).

$$\sigma_e = k \cdot (\sigma_a)_{av} \quad k = \frac{1}{\left(1 + \frac{2R}{a}\right) \cdot \ln\left(1 + \frac{a}{2R}\right)} \tag{2}$$

where σ_e is effective stress, R is radius of curvature of neck and a is radius of smallest cross section.

Some researchers found that Bridgman's formula does not comply with their results (justly, with numerical results). The problem with the formula is that stress triaxiality is related to material properties, which the formula does not take into account, as well as the assumption of the constant equivalent strain on the main stress surface, which is not true [11]. Chen [12] replaced the shape of the outer surface to hyperbola, and Xie et al. [13] modified material properties. Recently, a comprehensive study of the issue, with the aim to correct the approximations of the Bridgman method has been presented by Dong et al. [11].

The second group of procedures is introduced by the work of Zhang and Li [14], who proposed that the parameters for a true stress-strain relation should be determined by performing iterative FE simulations with an experimental load-extension (P-Δ) curve as a target. Since the method treats the stress-strain curve during necking as unknown, it is computationally intensive and time-consuming [8]. Ling [15] identified the upper and lower bound of the curve in the region after necking, as the linear extension and power-law fit, respectively. He proposed the weighted-average method of these two functions, which was adopted by Arasaratnam et al. [8]. Wang et al. [16] proposed a similar iterative algorithm but with some modifications. Namely, they developed a program and implemented their solution scheme, which iterates at the same strain increment until convergence in P-Δ response is achieved, and then steps to the next strain increment. Also, their weighted function assumes that the real curve (in the domain of the current strain increment) lies between the horizontal line and the tangent line at the previous strain increment.

In [17], Pavlovic et al. investigated the behavior of composite steel-concrete sections where bolts are used as shear connectors. Since a detailed FEM analysis is performed in order to capture complex phenomenon of bolt and concrete failure, the authors proposed an iterative approximate algorithm for obtaining the true stress-strain curve and to define a ductile damage law. Even though the solution technique is indeterminate and requires numerical iterations as it is the case in the already mentioned procedures [8, 14], this approach is very interesting. It consists of measuring the length of the necking zone (l^{loc}), which is possible even on the destroyed specimen, and definition of the trial stress-strain relationship on the basis of Eqs. (3–4):

$$l_i = \begin{cases} l^0, & i<n \\ l^0 + (l^{loc} - l^0)[(\Delta l_i - \Delta l_n)/(\Delta l_r - \Delta l_n)]^{\alpha_L}, & i>n \end{cases} \quad (3)$$

$$\varepsilon_i^{nom} = \begin{cases} \Delta l_i / l_i, & i<n \\ \varepsilon_{i-1}^{nom} + (\Delta l_i - \Delta l_{i-1})/l_i, & i>n \end{cases}; \quad \overline{\sigma_i} = \begin{cases} \sigma_i, & i<n \\ \sigma_n^{nom}(1+\varepsilon_i^{nom}), & i>n \end{cases} \quad (4)$$

where index n denotes quantity at point of necking formation, while r denotes fracture and l^0 initial gauge length.

Strain localization is governed by the power law through the localization rate factor α_L, which is iteratively obtained. This concept is appealing since it is easy to obtain the required input data and to calculate the trial curve. The weaknesses of the method regard the lack of the theoretic foundations of some assumptions, such as the damage initiation criterion or power law of stress localization.

It is often easier to make tensile specimens with a rectangular cross-section, as it is the case for plated steel elements. However, the necking of flat sample has been found to be much more troublesome than one of the bar [16]. The distribution of strain inside the neck is highly non-uniform and the strain gradients rise with the increasing of the aspect ratio. Several studies concentrated on this issue [16, 18, 19]. Since Bridgeman's correction for the rectangular cross section did not show satisfactory predictions, most of the studies appeared with the development of the Digital Image Correlation (DIC) measuring systems.

Even though the mathematical expressions for exact behavior throughout necking are not yet unequivocally established, the problem of obtaining the true stress-strain curve is considerably alleviated with the introduction of measuring systems based on the DIC. This technique allows the strain identification on the specimen's surface. There were numerous studies that provided such measurement data for a variety of materials [11, 18–22]. The material properties are usually used for further numerical investigations. Therefore, this opens a spectrum of additional issues. One of them, when using DIC for strain measurements, is the strain reference length. Ehlers investigated this issue in [19], and pointed out that the required strain data for numerical simulation are dependent of finite element length. Thus, the finite element length has to be equal to the strain reference length, which depends on the facet size used in DIC.

Another problem arises with extrapolation of the fracture strain obtained in the tensile test to the different stress states. According to the present knowledge of the materials behavior, stress triaxiality can affect the rupture mainly in two ways [23]. The first way is simply by changing the portion of the deviatoric part of the stress, hence the influence on the plastic deformation is apparent. The second mechanism affects a void growth inside the material. Triaxiality dictates the enlargement of the micro-voids until void coalescence and anticipated ductile rupture occurs. This dual nature of the failure by triaxiality makes it difficult to represent the relationship between stress triaxiality and local ductility reduction by a unique curve [23]. The triaxiality is defined as:

$$\theta = \frac{\sigma_m}{\sigma_e} = \frac{(\sigma_{11} + \sigma_{22} + \sigma_{33})/3}{\sqrt{\frac{(\sigma_{11}-\sigma_{22})^2 + (\sigma_{22}-\sigma_{33})^2 + (\sigma_{33}-\sigma_{11})^2}{2}}} \tag{5}$$

i.e. the ratio of mean σ_m to effective (von-Mises) stress, σ_e.

At the specimen periphery, triaxiality equals to 1/3, and it grows by approaching the specimen axis. If the Bridgeman's assumptions are adopted, the triaxiality at the center of the specimen inside the necking zone can be calculated as $\theta = 1/3 + ln(a/2R)$.

There are two possibilities to overcome this problem. The first is to perform tests on different types of specimens, in order to experimentally acquire the fracture strain for different stress triaxiality. This is usually performed by upsetting differently shaped specimens, such as disks, cylinders or rings, as demonstrated in [24, 25]. This method can effectively replace the uniaxial tensile test with the collar test for obtaining fracture criterion for the similar value of average triaxiality ratio, as it is shown in [24]. The second approach is needed if the uniaxial tensile test is the only one available. It entails some predictions about the relation between failure strain and triaxiality. One such prediction, which is used often, and also adopted in this study, is proposed by Rice and Tracey [26], and is given by Eq. (6):

$$\overline{\varepsilon}_f^{pl} = \alpha \cdot \exp(-\beta \cdot \theta) \tag{6}$$

where α and β are material constants, and ε_f^{pl} is the equivalent plastic strain at fracture. Equivalent plastic strain is usually expressed through the corresponding strain rate as:

$$\overline{\varepsilon}_{pl} = \int_0^t \dot{\overline{\varepsilon}}_{pl} d\tau = \int_0^t \sqrt{\frac{2}{3} \dot{\varepsilon}'^{pl}_{ij} \dot{\varepsilon}'^{pl}_{ij}} \, d\tau \tag{7}$$

This problem can be extended further (see for example [27]). Since the stress triaxiality alone cannot determine the complete state of stress at a material point, one would need to involve another parameter, for which purpose Lode parameter is commonly adopted. For example, in the case of $\theta = 1$, the value of Lode parameter (L) defines whether the stress state is axisymmetric tension ($L = -1$), plane strain ($L = 0$) or axisymmetric compression ($L = 1$). However, the number of tested specimens is in this case further increased. Since this study is primarily interested in bolt fracture, and the bolt will be primarily tensioned, it will be shown that relevant data can be extracted from just the tensile test.

2.2 Experimental Testing of Materials

The main focus of this research is to investigate the behavior of T-stub element with four bolts in a row in the context of ultimate resistance, stiffness evaluation, prying forces and failure mode. There are many failure criteria defined, and the choice is made in regard to the problem investigated. Since T-stub elements are part of the end-plate connection in civil steel structures, the failure criterion defined by relevant technical provisions (e.g. Eurocode) would be a reasonable choice. Eurocode EN 1993-1-5 defines failure as the 5% equivalent dilatation. The reason for this threshold is adopted small deformation throughout this design standard that does not apply in the case of larger strains. In this study, a failure defined as the physical fracture of a specimen, not limited to the small deformation theory, is investigated. Therefore, the specimens are tested until this criterion, and in every case, bolts were the element that experienced fracture, due to much higher ductility of steel used for plated elements. Since one of the goals is to provide a precise numerical simulation of the aforementioned tests, material properties (especially regarding failure), must be examined in detail.

All of the materials are tested in the uniaxial tensile test. The steel material used for plated elements is tested in two directions, with two or three specimens per material. Bolts are tested more rigorously, since they will ultimately dictate the failure of the whole T-stub element. The specimens used for material testing are dimensioned according to the relevant standard [28]. Since some of the steel plates have somewhat considerable thickness (19 mm), it was really important to use the appropriate cutting technology. The influence of some of the modern cutting technologies on the length of heat affected zone is investigated in [29]. The specimens used for material testing taken from rolled profiles used as T-stub elements, as well as for plates used for welded T-stub elements, are cut by a water-jet, in order to limit the thermal effect on material structure. The same procedure was used for all specimens, but for conciseness only specimens taken from bolts will be discussed further. Bolt specimens are produced on the lathe machine. Additionally, bolts are tested as a whole, according to standard [30]. None of the aforementioned tests had suggested any bolt defects. Both material and the whole bolts proved satisfactory strength. Before the complete testing, one trial specimen of the T-stub element was tested, and unsatisfactory bolt fracture due to thread stripping was observed (see Fig. 1). In order to clarify the reasons, both hardening test using Buehler Wilson Tucon 1102 machine and examination of microstructure were performed. The results of the latter are presented in Fig. 1. It can be seen that decarbonisation took place in the outer layer, approximately 100 μm in thickness.

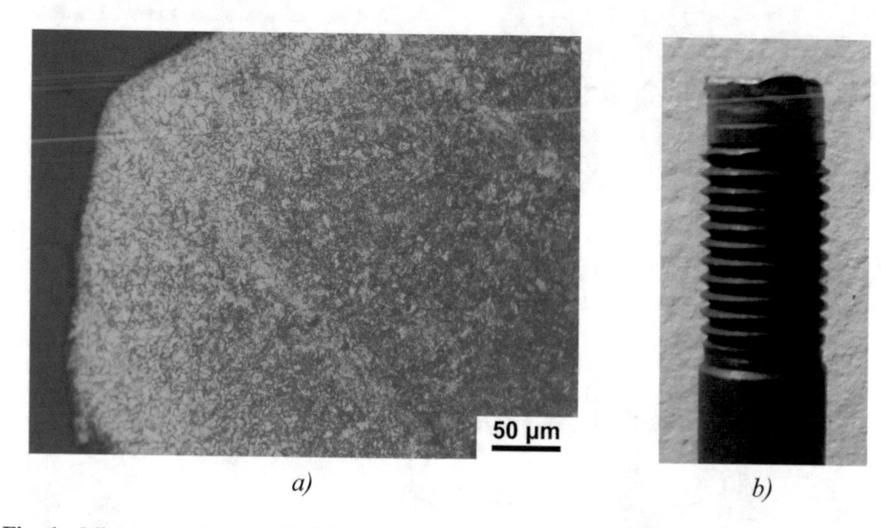

Fig. 1. Microscopic image of bolt's microstructure with decarbonization (a) and bolt failed by thread stripping (b)

The four tensile specimens of bolt material were tested with different strain measurement devices. The displacement-controlled experiments are carried out with a tensile test machine SHIMADZU AGS-X with 300 kN load cell capacity, at Belgrade University, Faculty of Civil Engineering. Strain rate is kept below 0.008 s^{-1} as specified in [28]. The force and stroke were recorded in 10 Hz frequency with multichannel acquisition device HBM MGCPlus. Additionally, the strain is measured by three methods: strain gauges, an extensometer and the DIC–based system - ARAMIS (GOM, Germany). Two-dimensional Aramis 2M system was used for the in-plane measurement. The optical measuring system is positioned on a tripod independent of the tensile machine. A stochastic pattern was sprayed onto the specimen's surface by applying a matt white-base layer followed by a layer of black dots (see Figs. 2 and 3). A 2-megapixel camera automatically recorded the images of the specimen with a delay time between the consecutive images of 1 s.

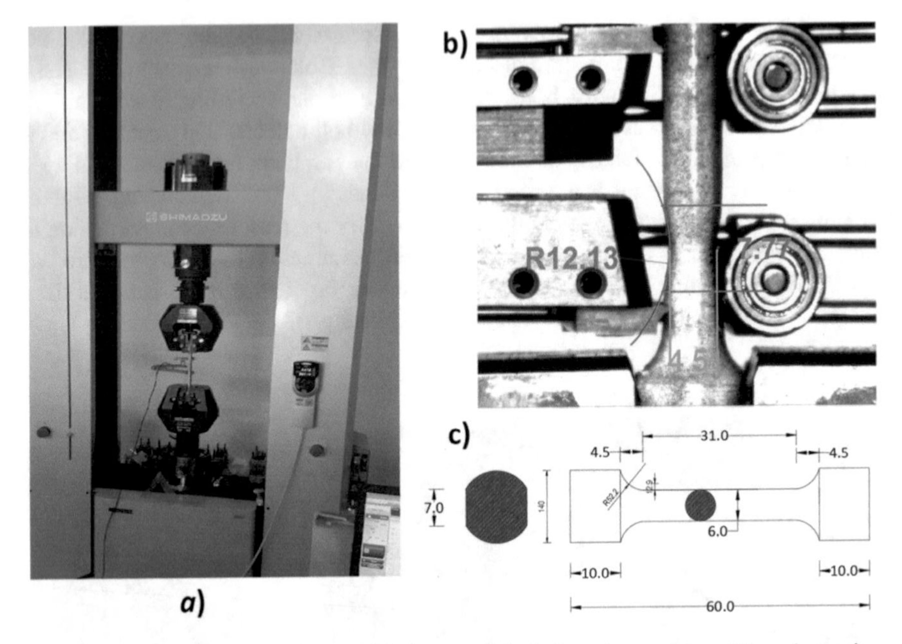

Fig. 2. Tensile test (a), specimen of bolt material during the necking (b) and specimen dimensions (c)

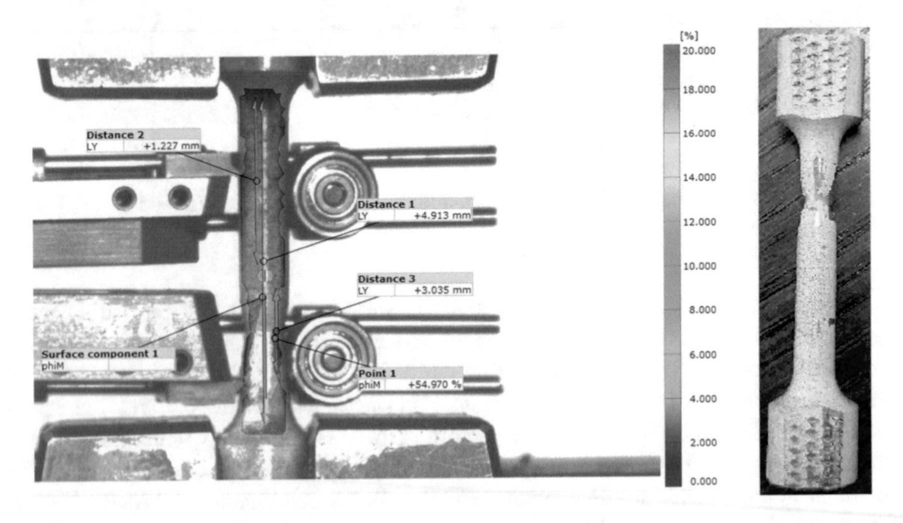

Fig. 3. Aramis results for one specimen of bolt material with the specimen after tensile failure

Stroke data obtained from the AGS-X tensile machine was not usable, due to the initial slippage between the specimen and grips. The results from strain gauges (SG), extensometer (Ext) and DIC are presented in Fig. 4. As it can be noticed, strain gauges are satisfactory in elastic range, but they cannot record the strain until failure. Further, equivalent strain calculated by Aramis was unattainable in the necking zone due to color layer depletion due to large straining (see Fig. 3). The Aramis system was also used to provide strain measurements analogous to extensometer, with the one gauge length outside of the necking zone (L_0 = 22.19 mm) and the other encompassing it (L_0 = 40.58 mm).

2.3 FEM Simulation of Tensile Test

For the purpose of the material model calibration, numerical simulation of a tensile test is performed using the Abaqus explicit code. A quasi-static nonlinear (geometric and material) analysis was performed with a dynamic explicit solver. The choice of explicit analysis is made because it does not demonstrate convergence issues often encountered with implicit static solver [17]. Due to symmetry, only one quarter of the test specimen was modeled. The load is applied by controlling displacement and a set of specific boundary conditions are imposed to satisfy the symmetry conditions. Because of the fracture strain dependency on the mesh size, the finite element dimension are chosen as 1 mm and 2 mm in length (2 models), as would be used in models of T-stub elements. Eight-nodded quadrilateral elements (C3D8R) with reduced integration have been employed in all simulations. Proportional mass scaling is used with factor 20.000, keeping the kinetic energy of the system less than 0.1% of internal energy during the course of the entire analysis. All the analysis regarding the tensile experiment and tuning of the material properties is done using double precision. No strain rate effect is considered.

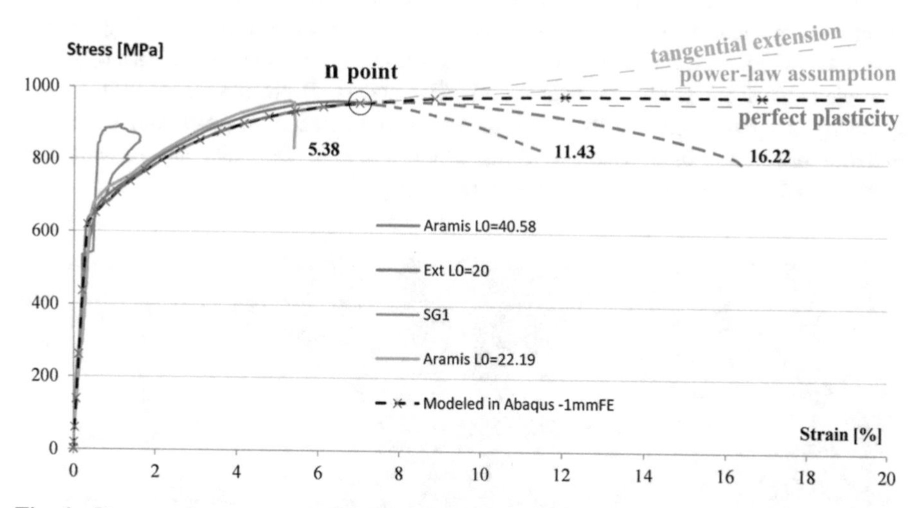

Fig. 4. Stress-strain curves calculated by Eq. (1) from different measurements and the true stress-strain relation adopted for FEM analysis

The nonlinear strain hardening of each specimen is characterized by the von-Mises criterion. To relieve the convergence difficulty caused by the large disparity of the derived work-hardening rate, which is encountered just before and after necking, the tabular representation of the stress-strain curve is defined with no fewer than 30 data points, to provide a stable change of the work-hardening rate.

The choice of the damage model is usually narrowed down to a ductile damage model (DDM), which is the damage model selected in this study as well. This model is based on the work of Hillerborg [31]. Still, there are two ways of employing any damage criterion in this kind of analysis. The first one is to set the damage initiation criterion to the onset of necking derived by the aforementioned procedures (Considere criterion was used here), as well as to define damage propagation through plastic dilatation up to the point of fracture, where the damage factor becomes equal to 1.0. This is done in [17]. The second one is to obtain true stress-strain curve which will simulate the behavior all up to the point of fracture, and to apply instant damage for that equivalent strain. The second procedure is easier to implement, since the first one is indeterminate and has many possible solutions. The only pitfall of this procedure is that the neck formation is dependent of the mesh size. The considered mesh with elements' size of 1 mm and 2 mm were fine enough to simulate neck formation. Ductile damage initiation criterion was defined at fracture equivalent plastic strain, and related to stress triaxiality by employing Eq. (6) with $\beta = 1.5$.

3 T-Stubs Test

3.1 Experimental Set up

Full-scale T-stub elements were tested in three different configurations, with two specimens of each. All of the bolts that were used in this study were M16 of grade 8.8. The plate thickness and bolt position on the T-stub elements' plate were varied. The T-stub elements were produced from rolled HEA and HEB profiles, in order to obtain geometry as perfect as possible, and also to avoid residual stresses and change in material properties in case of welding. Dimensions of elements are presented in Table 1, with symbol explanation presented in Fig. 5.

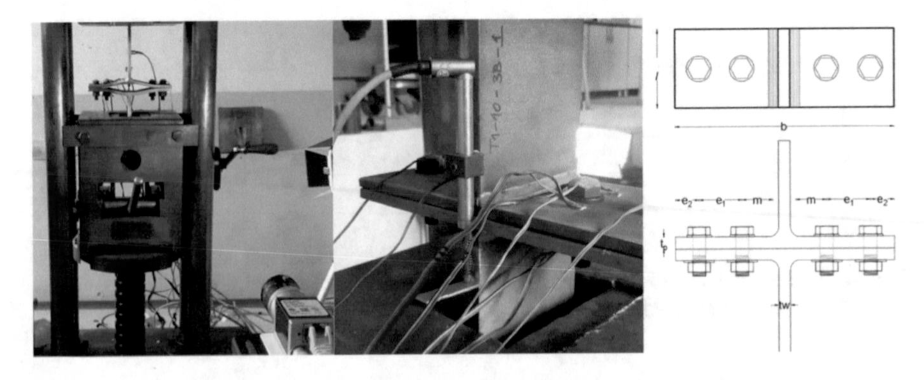

Fig. 5. Experimental set-up for testing of a T-stub element

The bolts were not pre-tensioned, as pre-tensioning does not affect ultimate resistance of the T-stub element, only the stiffness. The specimens were tested on the adapted Schaffhauzen hydraulic tension machine of the capacity of 1000 kN. Every test included two T-stub elements arranged in symmetric position and connected with four bolts. The experiment is performed with double T-stub assemblage in order to achieve the symmetry condition as much as possible. A similar set-up was used by [32], as opposed to one used in [33], where thick plates were used to imitate a rigid surface. The deformations are measured with one extensometer connected to the web-plate joints of each specimen. Again, Aramis system was employed for both strain and displacement measurement. The camera was oriented to the side of the assembly in order to obtain strain distribution through the plate's thickness, since the theory of T-stub elements treats them as linear elements. Also, strain gauges were used on the upper side of the plate to measure the strain in the plate at the inner bolt line. The force is applied by the speed of 1 kN/s in elastic range, and was automatically decreased during the plastic deforming. Also, KFG type strain gauges produced by Kyowa were embedded into a hole, 2 mm diameter, bored from the top head of the bolt. The same data acquisition (MGC Plus) was used. The strain gauges in bolts, along with load cell in the tensile machine, enabled recoding of prying forces in these assemblages throughout the entire test.

Table 1. Specimen dimensions

Specimen designation	t_{plate}	m	e_2	e_1	l	b	t_{web}	d_0
T-14-1-A1	13.9	59.6	26.3	39.6	140.0	298.2	8.7	17.5
T-14-1-A2	13.9	59.4	26.2	40.0	139.8	298.2	8.7	17.9
T-14-3-A1	13.9	29.2	26.2	70.1	140.0	298.2	8.8	17.9
T-14-3-A2	14.0	29.3	26.3	69.9	140.0	298.2	8.7	17.8
T-19-1-A1	18.5	58.2	28.4	39.6	139.9	302.4	11.5	17.8
T-19-1-A2	18.5	58.1	28.2	40.0	140.0	302.5	11.5	17.9

All dimensions are in mm

3.2 Finite Element Analysis

The numerical model of one half of the T-stub elements is developed in accordance with procedures used for tensile specimens, explained already in Sect. 2.3. For another half of the T-stub, the symmetry condition was used in the numerical model. Still, another symmetry condition could not be used to represent the second T-stub element in the assembly. For this purpose, rigid surface was adopted. The bolts were modelled as one half of the whole bolt assembly. It was found redundant to model the bolt with threads, as it is used in [17]. The upper part consisted of the bolt head and washer making one part with the bolts shank. The symmetry boundary condition was applied at the bottom surface of this half-bolt assembly, while the plate was in contact with the rigid surface, so it could lift from it. The bolts were also in contact with the upper surface of the plate. General contact interaction procedure, provided in Abaqus inside the explicit solver, was used with normal behavior defined as hard formulation and the tangential behavior as frictional with friction coefficient of 0.3. The mesh size is kept the same as for the tensile test model. Some of the material properties that are used in the model are presented in Table 2, while the multi-linear hardening rule is omitted, since more than 50 points are defined for each material. The numerical model of one specimen after the bolt failure is presented in Fig. 6.

Table 2. Mechanical properties of the materials used in numerical simulation

Part name	E [GPa]	f_y [Mpa]	ε_n [%]	σ_u [MPa]
T-14-1 & T-14-3 (A1 & A2)	210	313.6	18.58	519.8
T-19-1 (A1 & A2)	210	308.4	18.36	541.1
Bolts	209	611.0	6.89	957.4

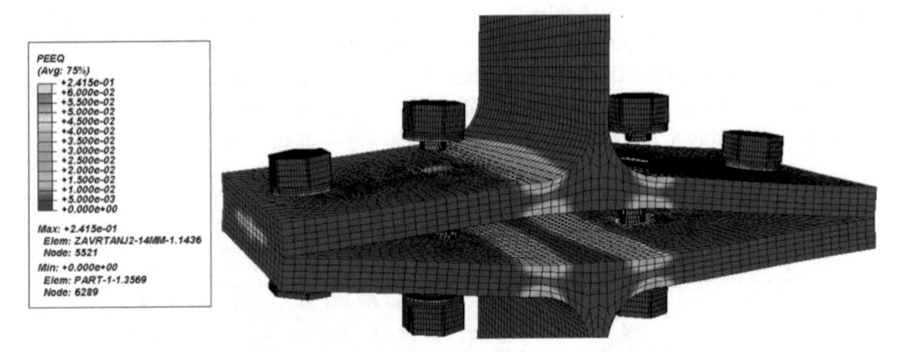

Fig. 6. Numerical model of specimen T-14-1A – moments after the bolt failure

4 Results and Discussion

4.1 Material Testing Results

The extensometer provided results that are affected by the problems discussed in Sect. 2. However, in order to evaluate the presented methods, these results were used in the numerical modeling. Aramis results of strain, extracted from displacement of two points, are also presented in Fig. 4, while the equivalent strain calculated by the Aramis is shown on Fig. 3. The Aramis' results for distance 2, suggest that necking strain is 5.38%, even though higher necking strain was obtained after numerical calibration.

The results of numerical simulations are presented in Fig. 7. Bridgman's formula, Eq. (2), predicted 77% increase of stress at the axis of specimen, while numerical simulations showed increase of 27% for 2 mm mesh and 50% for 1 mm mesh. The true stress-strain curve shown on Fig. 4 was found between power-law prediction and perfect plasticity curve, hence assumptions made in [8, 15] were not valid in this case.

Hardness test performed on bolts (along with microscopic examination) showed average values of 263 HV in surface layer, as oppose to 363 HV at the inner material. This was observed in one batch of bolts, most probably because of the unprotected atmosphere during the thermal treatment of bolts. This batch was replaced, but this problem is addressed here because rigorous following the procedure prescribed in ISO 898-1 [30] does not guarantee that this defect will be noticed. Therefore, hardness testing of the bolt surface, or preferably a pre-test is more than desired before any large-scale testing of assemblies with bolts.

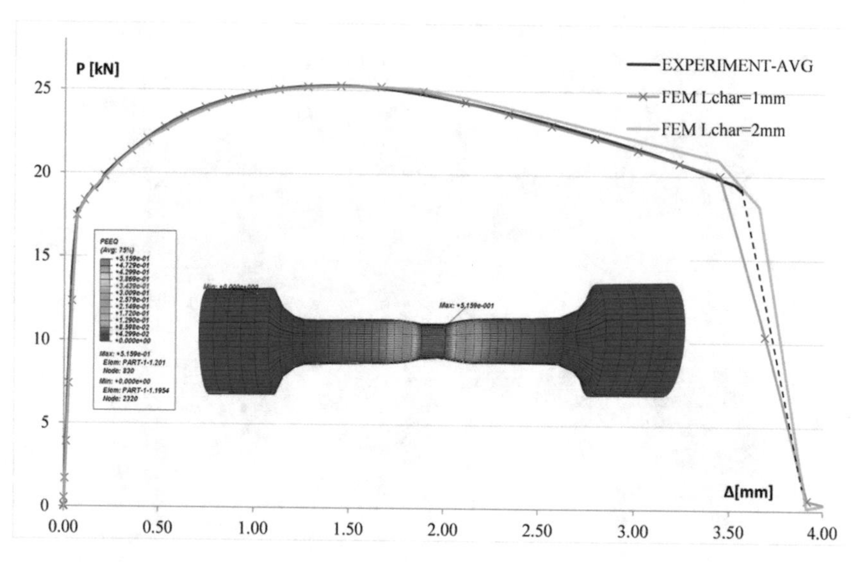

Fig. 7. Force-elongation (P-Δ) for bolt's specimens simulations and averaged test data

4.2 T-Stub Elements Results

The results of experimental tests are briefly presented through Figs. 8, 9, 10, which display equivalent von-Mises strains is the T-stub elements' plates just before the bolts' failure. Even though specimen T-14-1A sustained considerable plastic deformation (>6% at the maximum load), the bolts dictated the failure of each specimen. If the plate thickness is the only parameter varied, as in the case of T-14-1A and T-19-1A, 36% of thickness increase lead to 24% resistance increase, 37.5% stiffness increase and twice smaller ductility. Reduction of distance between inner bolts (m – see Table 1) for specimens T-14-1A and T-14-3A increased both the stiffness (77%) and strength (44%) with the expense of ductility reduction of 75%.

The comparison of numerical and experimental results of the force-deformation relationship is depicted in Fig. 11. Some discrepancy can be explained by the non-uniformity of some parts of the mesh, and also by imperfections of the tested specimens. Initial stiffness [kN/mm] of each specimen is also indicated in the Fig. 11. It is noticeable that reduction of the dimension *m* has a greater effect than the plate thickness increase.

The failure modes of the tested specimens are depicted in Fig. 12. The tests were performed by applying the load until the failure of both of the inner two bolts. This explains asymmetric deformation of failed specimens presented in Fig. 12. Severe bending of the bolts can be noticed here. Eurocode does not incorporate the bolt bending in the design expressions, but American standard does. It can be also noticed from the figures that position of the plastic hinges in the plate of the T-stub elements corresponds well with the Eurocode predictions of 0.8r from the web, where r is the roll radius.

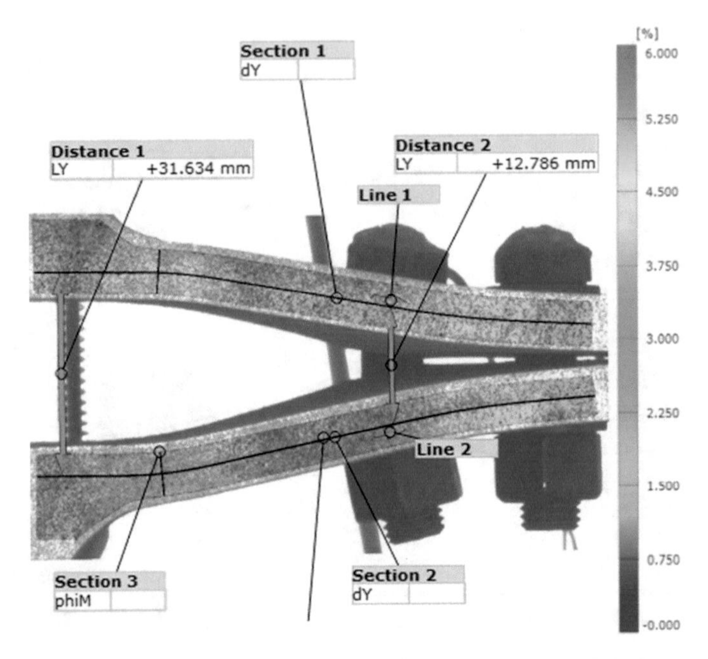

Fig. 8. Equivalent strain of tested T-stub specimen T-14-1-A (1&2)

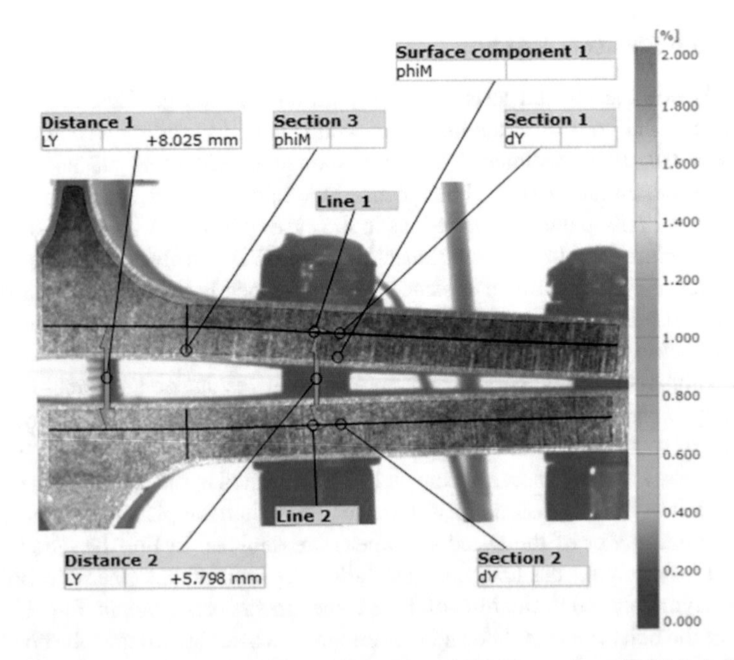

Fig. 9. Equivalent strain and failure modes of tested T-stub specimen T-14-3-A (1&2)

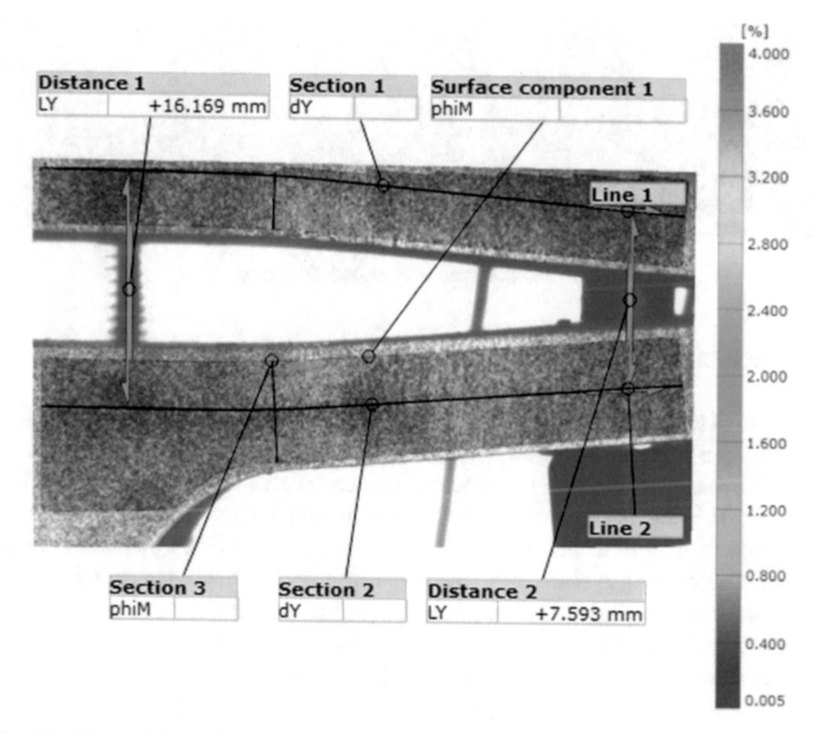

Fig. 10. Equivalent strain and failure modes of tested T-stub specimen T-19-1-A (1&2)

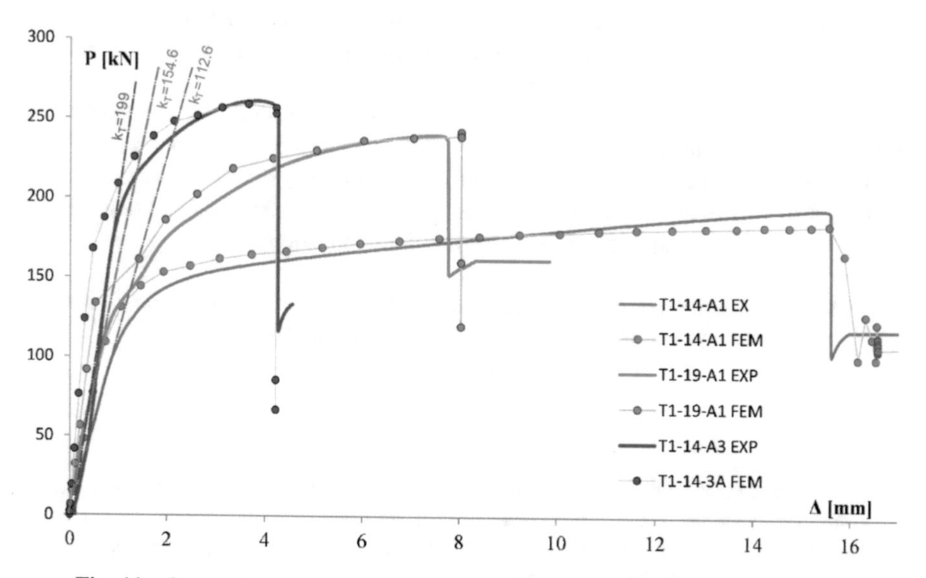

Fig. 11. Comparison of numerical and experimental P-Δ curves for all specimens

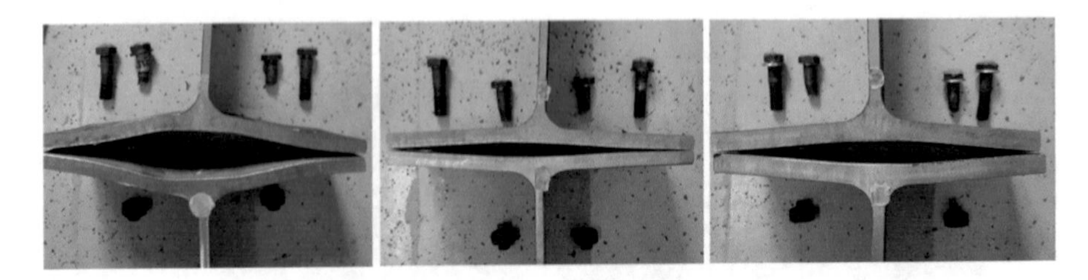

Fig. 12. Failure modes of tested T-stub specimen

5 Conclusion

There is an evident need for further investigation of T-stub element with four bolts in a row, since the relevant formulas for their design are not adopted in Eurocode. This problem appears elementary, but actually it incorporates many traditionally difficult problems, such are prying forces, bolt bending and highly plastic deforming with several sorts of nonlinearities. Since it is encountered in every-day practice, simple and expressible solutions are desired. It was practically impossible to measure many of the relevant information during the test, only a few decades ago.

All of this incited the on-going study, with the aim to contribute to the knowledge on the behavior of T-stub elements, from both experimental and numerical aspect. Since full scale experiments are an expensive and time-consuming endeavor, numerical simulations are a plausible substitute as an investigation technique. This paper is concentrated on both of these methods. Experimental investigation of six T-stub specimens is presented, as well as different methods for steel material testing. Also, numerical modeling guidance is presented in detail, and confirmed by results, both for tensile specimens and T-stub assemblages.

Future work will concentrate on confirmation of an existing or derivation of a new expression for the ultimate resistance of a T-stub element with four bolts in a row, as well as the selection of the most appropriate model of prying forces for these elements. Furthermore, T-stub elements with four bolts proved envious ductility and redundancy, which are both rather advantageous properties for connections.

Acknowledgment. The research of the first author was supported by the Serbian Ministry of Education, Science and Technological Development, Grant No. 36043.

References

1. Zoetemeijer, P.: A design method for the tension side of statically loaded, bolted beam-to-column connections. Heron Delft Univ. **20**(1), 1–59 (1974)
2. Douty, R.T., McGuire, W.: High strength bolted connections with applications to plastic design. University of Missouri, Columbia (1965)
3. Jaspart, J.P.: Etude de la semi-rigidite des noeuds poutre-colonne et son influence sur la resistance et la stabilite des ossatures en acier. Université de Liège, Belgium (1991)
4. Eurocode 3: Design of steel structures - Part 1–8: Design of joints. CEN, Brussels (2005)

5. Dranger, T.S.: Yield line analysis of bolted hanging connections. Eng. J. **14**(3), 92–97 (1977)
6. Mann, A.P., Morris, L.J.: Limit design of extended end-plate connections. J. Struct. Div. **105**(3), 511–526 (1979)
7. Specification for Structural Steel Buildings. AISC, USA (2010)
8. Arasaratnam, P., Sivakumaran, K.S., Tait, M.J.: True stress-true strain models for structural steel elements. ISRN Civ. Eng. **2011**, 1–11 (2011)
9. Bridgman, P.: The stress distribution at the neck of a tension specimen. Trans. Am. Soc. Metals **32**, 553–574 (1944)
10. Considère, M.: Annales des Ponts et Chaussées **9**, 574–775 (1885)
11. Dong, S., Xian, A., Lian, Z., Mohamed, H.S., Ren, H.: Necking phenomenon based on the Aramis system. Proc. Inst. Mech. Eng. Part C J. Mech. Eng. Sci. **233**(11), 3904–3916 (2018)
12. Chen, C.: Study on Metal Fracture. Metallurgical, Industry Press of China, Beijing (1978)
13. Xie, F., Zhang, T., Chen, J.E., Liu, T.-G.: Updating of the stress triaxiality by finite element analysis. J. Vib. Shock **32**, 8–14 (2012)
14. Zhang, L., Li, Z.H.: Numerical analysis of the stress-strain curve and fracture initiation for ductile material. Eng. Fract. Mech. **49**(2), 235–241 (1994)
15. Ling, Y.: Uniaxial true stress-strain after necking. AMP J. Technol. **5**, 37–48 (1996)
16. Wang, Y., Xu, S., Ren, S., Wang, H.: An experimental-numerical combined method to determine the true constitutive relation of tensile specimens after necking. Adv. Mater. Sci. Eng. **2016**, 1–12 (2016)
17. Pavlović, M., Marković, Z., Veljković, M., Buđevac, D.: Bolted shear connectors vs. headed studs behaviour in push-out tests. J. Constr. Steel Res. **88**, 134–149 (2013)
18. Scheider, I., Brocks, W., Cornec, A.: Procedure for the determination of true stress–strain curves from tensile tests with rectangular cross section specimens. J. Eng. Mater. Technol. **126**(1), 70–76 (2004)
19. Ehlers, S., Varsta, P.: Strain and stress relation for non-linear finite element simulations. Thin Walled Struct. **47**(11), 1203–1217 (2009)
20. Hoffmann, H., Vogl, C.: Determination of true stress-strain-curves and normal anisotropy in tensile tests with optical strain measurement. CIRP Ann. **52**(1), 217–220 (2003)
21. Milosevic, M., Milosevic, N., Sedmak, S., Tatic, U., Mitrovic, N., Hloch, S., Jovicic, R.: Digital image correlation in analysis of stiffness in local zones of welded joints. Tech. Gaz. **23**(1), 19–24 (2016)
22. Milosevic, M., Mitrovic, N., Jovicic, R., Sedmak, A., Maneski, T., Petrovic, A., Aburuga, T.: Measurement of local tensile properties of welded joint using Digital Image Correlation method. Chemicke Listy **106**, 485–488 (2012)
23. Maresca, G., Milella, P.P., Pino, G.: A critical review of triaxiality based failure criteria. XIII Convegno Nazionale IGF, vol. 13, Cassino, Italy (1997)
24. Alexandrov, S., Vilotic, D., Konjovic, Z., Vilotic, M.: An improved experimental method for determining the workability diagram. Exp. Mech. **53**(4), 699–711 (2012)
25. Vilotic, D., Chikanova, N., Alexandrov, S.: Disk upsetting between spherical dies and its application to the determination of forming limit curves. J. Strain Anal. **34**(1), 17–22 (1999)
26. Rice, J.R., Tracey, D.M.: On the ductile enlargement of voids in triaxial stress fields. J. Mech. Phys. Solids **17**(3), 201–217 (1969)
27. Kiran, R., Khandelwal, K.: A triaxiality and Lode parameter dependent ductile fracture criterion. Eng. Fract. Mech. **128**, 121–138 (2014)
28. EN ISO 6892-1:2016: Metallic materials—Tensile testing, Part 1: Method of test at room temperature. CEN, Brussels (2016)

29. Harničárová, M., Zajac, J., Stoić, A.: Comparison of different material cutting technologies in terms of their impact on the cutting quality of structural steel. Tech. Gaz. **17**(3), 371–376 (2010)
30. ISO 898-1: Mechanical properties of fasteners made of carbon steel and alloy steel. ISO, Geneva, Switzerland (2009)
31. Hillerborg, A., Modeer, M., Peterson, P.E.: Analysis of crack formation and crack growth in concrete by means of fracture mechanics and finite elements. Cem. Concr. Res. **6**, 773–782 (1976)
32. Pisarek, Z., Kozłowski, A.: End-plate steel joint with four bolts in the row. In: Gizejowski, M., Kozlowski, A., Sleczka, L., Ziólko, J. (eds.) Progress in Steel, Composite and Aluminium Structures, vol. 1, pp. 257–266. Taylor & Francis Group, London (2006)
33. Massimo, L., Gianvittorio, R., Aldina, S., da Silva Luis, S.: Experimental analysis and mechanical modeling of T-stubs with four bolts per row. J. Constr. Steel Res. **101**, 158–174 (2014)

Experimental Investigations on Bound Frequency of Axial Ball Bearings for Fixing the Ball Screws

Vladislav Krstic[1], Dragan Milcic[2(✉)], and Miodrag Milcic[2]

[1] "Ljubex International" d.o.o, Čede Mijatovića 3, 11120 Belgrade, Serbia
[2] Faculty of Mechanical Engineering in Niš, University of Niš, Aleksandra Medvedeva 14, 18000 Niš, Serbia
dragan.milcic@masfak.ni.ac.rs

Abstract. In contemporary machine tools as well as processing centres, threaded spindles are getting increasingly used. On this occasion, the higher speed of work and high accuracy of guidance are expected. Regarding this, the issue of fixing is very delicate resulting in axial angular contact ball bearings, which regarding their structure and performances, successfully meet the specified requirements. At greater work rates, there is heat load of these bearings. Due to that reason, in the designing phase, it is necessary to specify whether the temperature load shall be significant. If this is the case, the attention shall be paid to its thermal stability in the designing phase. In this paper, the experimental testing of thermal bound frequency of bearing type ZKLF, manufacturer "INA" - "Schaeffler Technologies" GmbH & Co. KG will be presented.

Keywords: Axial ball bearing · Experimental testing · Thermal stability · Revolution frequency

1 Introduction

Main demand of the modern market is wide palette of different products. It means that the producers have to offer wide palette of products with an appropriate level of quality with additional requirement on productivity. In order to insure quality of the product, it is necessary to be provided as much as possible precise guidance of the tools as well as workpieces.

Regarding to these demands the most frequent solutions for linear guidance are based on threaded spindles. Threaded spindles have to fulfill following demands: higher speed of work and high accuracy of guidance. Under these operating conditions, i.e. requirements, higher thermal load is expected.

It has to be noted that the main load of the threaded spindle will be realized through axial force. Significant axial force will be induced by screw nut. At higher number of revolutions friction in pair screw nut- threaded spindle as well as in bearing assembly will be significantly increased.

Friction level will defined thermal load of the system. The best way to analyze thermal load is to define temperature which will appear under operating conditions.

N. Mitrovic et al. (Eds.): CNNTech 2019, LNNS 90, pp. 323–339, 2020.
https://doi.org/10.1007/978-3-030-30853-7_19

Temperature field is useful when some another additional thermal sources or thermal losses are expected.

Thermal load/overload leads to elastic i.e. plastic deformations which will influence on accuracy of the guidance.

With this background it would be very useful to know in design phase reference speed i.e. number of revolutions of the bearing which will be applied in assembly. As it already known standard procedure for determination of the reference speed is presented in DIN 732 part 1 and 2 [1, 2]. It has to be noted that DIN732 part 1 and 2 does not possess needed parameters for determination of the reference speed for the axial ball bearings for type ZKLF, ZKLFA, DKLF, DKLFA, ZKLN, etc. The reason for this is specific contact angle of the mentioned bearing which is 60° as well as existence of the screw connection, precise nut for preload of the bearing and housing.

At start research previous work in this area was analyzed.

Thermal process in system screw nut- threaded spindle is presented in [3]. This paper shows the correlation between thermal load and accuracy of the guidance.

An aim of [4] is to find out a way for description of correlation between structure of bearing assembly and load which is induced by temperature load. For solving of this task combination of different models was used with different degrees of freedom. At the end, new designed system gave required correlation

In order to predict thermal behavior of one numerically controlled machine in [5] is presented a model which describes observed system regarding thermal load. It allows different combinations of the base thermal parameters such as geometry, different characteristics of materials etc. Difference between results and real measurement was technically acceptable.

Due to determinate deformation of balls and thermal load in [6] a three-dimensional measurement was used. The result gave a proposal for selection of the bearing.

The base of research in [7] is model of thermo-mechanical coupling. Via this model thermal load was analyzed at higher number of revolutions.

Axial error compensation regarding heat generation by guidance via spindle is presented in [8].

During rotation there is appropriate displacement of the inner ring. This displacement appears as a result of the existence of the centrifugal forces. Correlation between dynamic characteristics and centrifugal forces is presented in [9].

Start point by research of the reference speed by angular contact ball bearings is presented in [10]. The main approach to determination of the reference speed is shown in this paper. Focus is on the reference conditions as well as main thermal basics which will be used in analyze of the system "spindle- bearing- screw connections- housing- precise nut for pre-stress of the bearing".

Extended research of the reference speed of the angular contact ball bearing is presented in [11]. Real bearing assembly was observed and it has given some conclusions regarding possibility of determination of the reference speed.

In order to get thermal load of the real bearing assembly of the threaded spindle, it was necessary to do FEA analysis [12]. FEA analysis was done in ABAQUS 6.9-3. Complete thermal numerical calculations was performed regarding all reference conditions. As the result temperature field of real system was obtained.

In [13] general approach to thermal analysis of the axial angular contact ball bearing, type ZKLF is presented. All important and main aspects are shown. The paper is extended research of thermal load of real system with threaded spindle. Conclusion has given main recommendation which helps in pre-selection of the bearing type already in design phase.

In [16] was analyzed spindle-bearing. For that purpose thermo-mechanical coupling analysis was used. The analysis was focused on centrifugal force, preload, gyroscopic moment as well as condition of lubrication. The analysis was carried out thorough two stages, experimental test and simulation. It was concluded that significant influence parameter by operating of spindle bearing is preload. Because of this cooling has very important role to provide balance with generated heat which is result of the friction in the bearing.

In [17] is presented analysis of heat generation in bearing. The models which were used in this paper take into account different items such as axial load, velocity, sliding between rolling elements and cage, as well as between cage and outer ring. Special focus of analysis is on bearings with shaft of a gyro motor.

Yanfei et al. had presented correlation between thermal load under different external conditions such as constant preload, pure radial load [18]. Balamurugan and Bhagyanathan are presented in [19] correlation between number of revolutions and speed of temperature changing. For that purpose steady state thermal-stress simulations were used. Liu and Bian analyzed the effect of load and rotation speed on bearing temperature rise by ANSYS [20, 21]. Chen consideration of thermal contact resistance, influence of rotation speed and load on temperature field of angular contact ball bearing is studied [22]. Method for assessment of the optimal preload of the bearing was presented in [23].

At design phase of driven spindle it is good to estimate expected thermal behavior. For this purpose in [24] was presented a model which takes into account generated heat in the bearing as well as the heat influence from the bearing to other elements of observed system. Zivkovic et al. have presented temperature field of the machine tool spindle using 3D FE thermal model [25]. In order to get temperature field of a bearing ball new analytical model is presented in [26]. In [27] was shown a calculation method which uses the values i.e. results of the multi point temperature measurement. Measurement was carried out using fiber Bragg grating. Chao et al. have defined thermal model which includes friction at the elliptical contact and gyroscopic moment [28].

As it already known, kinematics of the bearing is very important for proper operation. That is why kinematics of the bearing must not be disturbed. Correlation between heat generation and kinematics of the bearing was presented in [29]. Van-The Than and Jin H. Huanghave presented an algorithm based on a quasi-static model and finite different method which will be used for prediction of nonlinear thermal characteristics of a spindle bearing [30].

Functioning of tool machines is not possible without threaded spindles. The basic requirements for tool machines are primarily high accuracy of guidance. The threaded spindles of tool machines – main and auxiliary, are manufactured with high level of accuracy, and special attention is paid to fixing the threaded spindles regarding the rigidity, oscillation damping and thermal deformations.

Fixing of threaded spindles is done by special bearings requiring high level of performances, which they successfully meet.

It is well known that fixing of ball screws is one of the most demanding and challenging tasks, because in these cases the high level of performance accuracy is required, whereas there are increased loads in radial and axial direction.

A long engineering experience resulted in real solutions for this type of bearings. Regarding the fact that there is a high number of revolutions (especially with drill machines, grinding machines, etc.), with the high axial forces, the rolled members are most frequently angular contact ball bearings convenient for high number of revolutions. Since there are also axial loads, the reception, distribution and compensation issues of axial loads had to be solved. It was solved by angular contact ball bearing with angular ball contact with rolling paths, as well as bias load in the mere bearing. Bias load is realized by two-piece internal ring planning the certain distance between the parts of rings acting as damper of axial forces occurring in the bearing. In the Fig. 1 there are two bearing structures for ball screws.

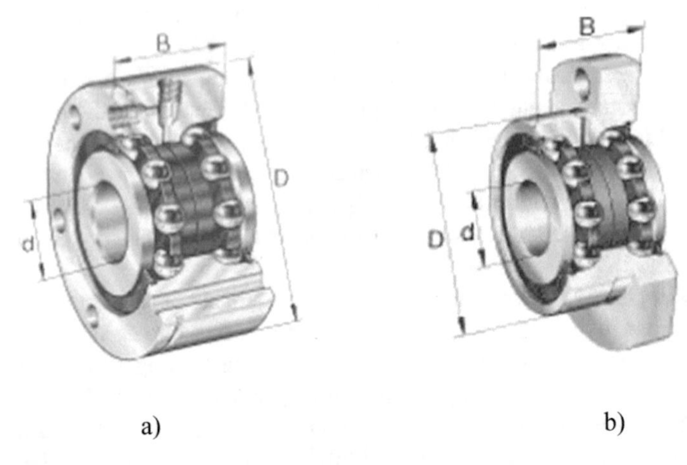

Fig. 1. Bearings for threaded spindles: (a) Bearing of series ZKLF…-2RS, ZKLF…-2Z; (b) Bearing of series ZKLFA…-2RS, ZKLFA…2Z [14] *d*- internal diameter of bearing; *D*- outside bearing diameter; *B*- total bearing width

In some cases it is possible to couple two bearings so that there is a pair of coupled bearings providing higher bearing capacity as well as higher rigidity (Fig. 2).

Principally, the bearings for fixing of ball screws are specified as bearings of high precision. It refers mainly to the manufacture quality of all its integral parts.

Depending on the grade of bearing measurements, the angular contact ball bearings for threaded spindles can receive both great radial and axial forces which make them more specific related to other bearing types. Internal ring, crest with balls and outside ring are placed on each other making one adjusted unit. Using such structure, it is possible to precisely axially load these bearings.

The protection against impurities and moisture is made by rubber sealing rings. In case of higher speeds, sealing by metal rings is recommended.

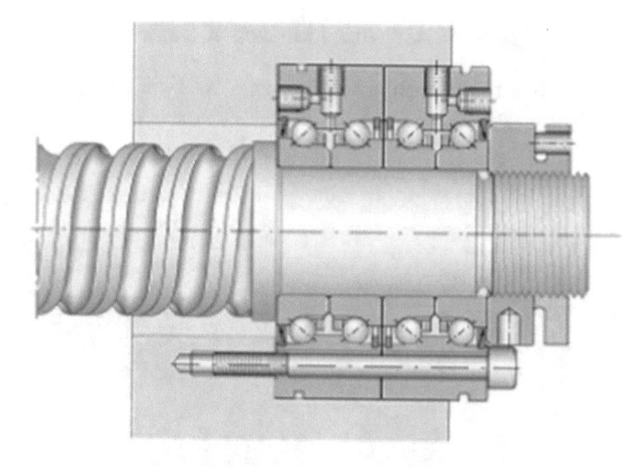

Fig. 2. Coupled ZKLF…-2RS-2AP bearings [14]

These bearings are manufactured with and without the strengthening openings (implying the bolt connection) on the outside ring. The bearings with strengthening opening on the outside ring can be strengthened to machine using the bolts at the place planned for fixing. The strengthening openings are clean industrial solution because they decrease the expenses of lid and machine opening processing.

With machines used in timber industry or means of transportation with ball screws, bearings of less precision are used, thus there are bearings with higher tolerance values.

2 Axial Ball Bearings with Angular Contact for Ball Screws

The requirements for fixing of ball screws are laid so that the bearings with their structure cannot use them at their optimum. That is why "INA" and "FAG" proposed a great palette of angular contact ball bearings which by its structure can meet the requirements regarding bearing capacity, rigidity, slight friction at highly dynamically loaded plant of ball screws. This range of products successfully meets all industrial requirements set by fixing of threaded spindles.

Axial ball bearings are manufactured as single line, two lines or three lines. They include the outside ring of high thickness, ball crest and single line or two line internal rings.

In its manufacture program, INA provides the axial ball bearings of series ZKLN, ZKLF, ZKLFA and DKLFA, for fixing of threaded spindles with two line internal ring.

Series in the manufacture program of "INA" have, with many number of types, openings for direct strengthening to the machine where ball screws fixing is done. The rings with these bearings are made using the precise nuts for adjustment of bias load in the mere bearing.

It is important to specify that these bearings due to pressure angle (grazing- contact angle) of 60° besides the radial one can also receive the high axial load [14].

2.1 Axial Ball Bearings with Angular Contact of Series "ZKLF"

Axial ball bearing with angular contact of series ZKLF is directly connected to the machine using the bolts (Fig. 3).

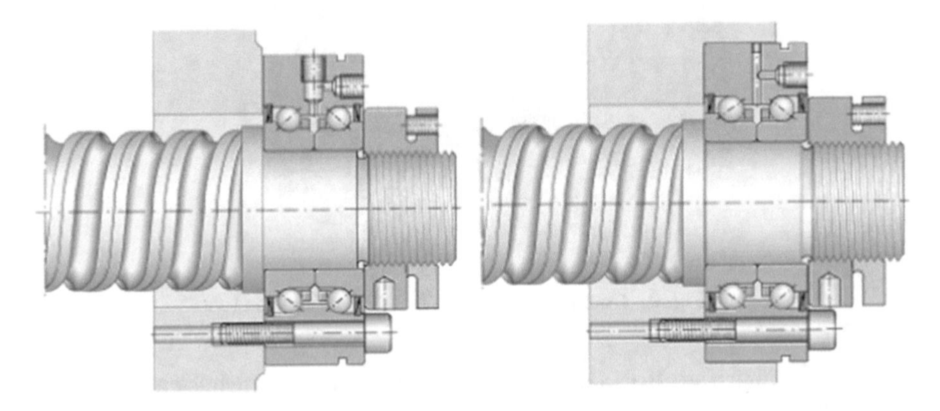

Fig. 3. Outside bearing ring is fixed to flat machine front and bias loaded by precise nut

Axial ball bearings with angular contact of series ZKLF…-2RS-PE match the bearing series ZKLF but they differ in tolerance and quality. Their precision grade is P5 in comparison to DIN 620 and therefore the diameter tolerance values are wider. The label "PE" directly specifies that the quality if of lower precision grade [14]. They are applied when the precision of positioning can be lower. Then also the requirements for other structural members are lower regarding precision.

Axial ball bearings with angular contact of series ZKLF…- 2RS and ZKLF…-2Z are also available in greater dimensions. It is clear that higher cross sections imply greater bearing capacity.

The bearings of series ZKLF-…2RS-2AP, include two coupled bearings ZKLF…-2RS, (Fig. 2). The coupled bearings at outside surface of the external ring in this case are labeled with arrow, assisting in defining the exact arrangement of coupled bearings. With regular assembly and arrangement of coupled bearings, the sealants should be directed outwards (Fig. 2). During the package mounting, both bearings must have their axes completely mated with the threaded spindle axis. As it is shown in Fig. 2, the coupled bearings are strengthened by bolt to machine case, where bias load is adjusted by precise nut.

The bearings with suffix "2RS" are mutually sealed by rubber sealants, convenient for mounting at the positions with fine dust or liquid, whereas the bearings with suffix "2Z" are mutually sealed with metal sealants which are convenient in cases of high number of revolutions.

All bearings are factory lubricated with grease based on lithium soaps according to GA28, but they have the possibility of additional lubrication made by the lubrication opening located on the external ring. In most cases, the factory lubrication is sufficient for entire working life of bearing [14].

3 Experimental Testing

Complete experimental research for doctoral thesis [13], regarding research of construction- tribological parameters of bearing type ZKLF was carried out at company Schaeffler in Herzogenaurach – Germany. The goal of experimental testing of axial ball bearings “ZKLF2575-2Z” and “ZKLF50115-2Z” is specifying the following parameters:

- Revolution frequency
- External ring temperature
- Surrounding temperature
- Friction force, i.e. friction moment
- Preload axial forces.

3.1 Defining the Testing Protocol

The testing protocol stipulates that four units of bearings of both types shall undertake the specified metering. The sample selection shall be done using the method of “random sample”.

Regarding the fact that it is the specific type of bearing, it is necessary to perform certain preparations of sample in order to do the regular mounting and setting the sample at the test bench.

The test bench comprises the shaft simulating the ball screw and by which from the engine the required revolution frequency is obtained on the internal ring of bearing, whereas the external ring remains free and it does not receive any load – it remains stationary during entire testing time.

During metering, also the surrounding temperature shall be measured, i.e. room temperature, which should be 20 °C.

Testing every bearing sample shall be done in three phases as follows:

- Testing with sealing rings lubricated with factory grease,
- Testing without sealing rings lubricated with factory grease,
- Testing without sealing rings which was previously thoroughly cleaned from factory grease and lightly lubricated.

The idea is to do comparison of results in order to obtain the average values after every performed metering series of every sample.

Every sample after setting on the testing desk shaft must be prestressed by certain axial force of prestress stipulated by Schaeffler amounting:

- for bearing ZKLF2575-2Z $F_x = 1945$ N
- for bearing ZKLF50115-2Z $F_x = 3148$ N

Prestress is done by precise nut for prestress of subject bearings also manufactured by Schaeffler (AM25 and AM50). The specified prestress must remain at the constant level prior to, during and after the completed testing. Regarding the fact that there is no case to which the sample would be additionally strengthened by bolted connection, due to maintenance of constant prestress, the set of plate springs is inserted into the entire compound functioning as additional strengthening for the sample. The test bench with the sample is shown in Fig. 4.

Fig. 4. View of test bench with the sample: 1- Precise nut for prestress; 2- Set of plate springs; 3- Bearing being tested; 4- Temperature sensor (lubrication opening) for temperature metering of internal ring; 5- Temperature sensor - temperature metering of external ring; 6- Friction force metering sensor [13]

3.2 Testing Program

The testing comprises:

1. Prestress metering for cold bearing,
2. Grease distribution control within the bearing,
3. Loads of tested sample with various levels of revolution frequency,
4. Prestress metering for warm bearing.

In order to obtain the precise results, it is necessary to make the equal distribution of grease within the bearing (factory filling). For both types of bearing, the lubrication L192 was used with the following quantities:

- for bearing ZKLF2575-2Z 1,9 g
- for bearing ZKLF50115-2Z 4,7 g

In order to distribute the grease equally, at the beginning of every metering, the operation of both types of sample is foreseen with two levels of revolution frequency, 500 rpm and 1000 rpm during 15 min. Afterwards, it is foreseen that the lubricant is equally distributed within the bearing.

When the lubricant it distributed at the bearing, then every sample is loaded until the planned maximum revolution frequency (8000 i.e. 6000 rpm) for five minutes continuously.

The following step is loading every sample with various levels of revolution frequency. The scope of revolution frequency is:

- for bearing ZKLF2575-2Z 0–8000 rpm,
- for bearing ZKLF50115-2Z 0–6000 rpm.

The revolution frequency level is 1000th part of maximum revolution frequency. The load duration of every revolution frequency level is 30 min. After the last load level, the prestress value is necessary to be metered again. In the following phase, from all samples, the sealing rings were removed and the load procedure is repeated based on grades and afterwards the prestress value was metered again.

In the last step, there are two bearing samples ZKLF2575-2Z completely cleaned – degreased, lubricated VG68 and repeated testing for the same load diagram. Prestress metering was necessary at the end of load.

The results of the experimental testing, which show the friction moment from bearing operating speed, in rpm, are given in the Figs. 5 and 6.

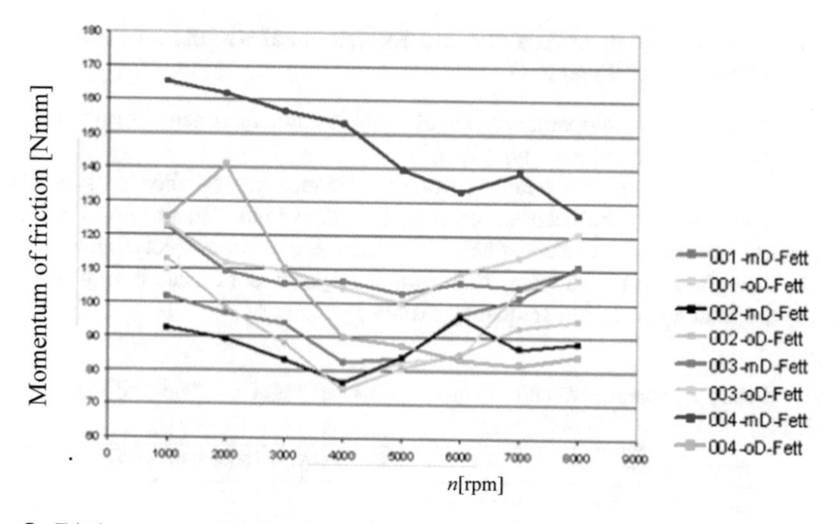

Fig. 5. Friction moment diagrams for ZKLF2575-2Z (mD- Fett – with sealings and fat, oD- Fett – without sealings and with fat)

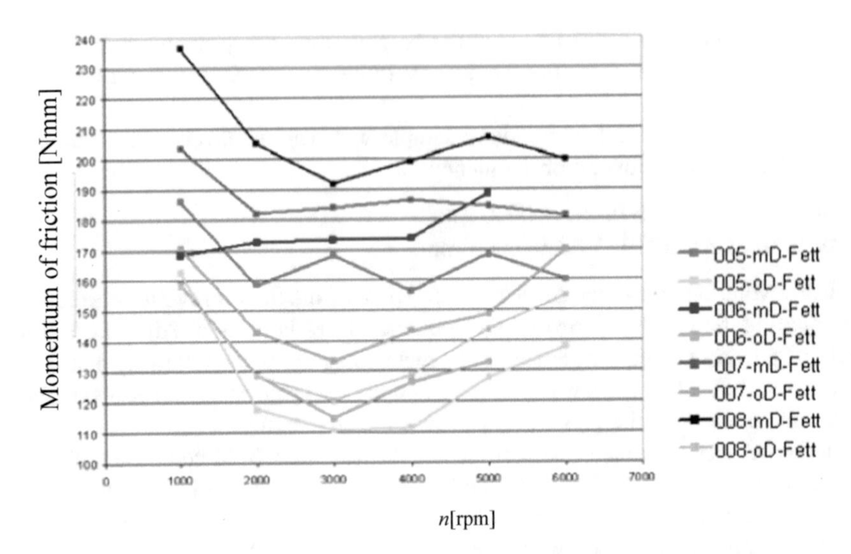

Fig. 6. Friction moment diagrams for ZKLF50115-2Z (mD- Fett – with sealings and fat, oD-Fett – without sealings and with fat)

3.3 Comparative View of Analytic and Experimental Results for Axial Ball Bearings of Type ZKLF

This chapter shows the comparative view of analytic and experimental results for axial angular contact ball bearings of type ZKLF.

For instant defining the bound revolution frequency, so-called fast operation characteristic $n{\cdot}d_m$, can be used which is at the same time also the basic guideline (internal recommendation of Schaeffler). This characteristic for subject bearings clearly provides analytical result showing that with the specified bearing it is possible to increase the bound revolution frequency (Table 1).

Table 1. Values of bound revolution frequency obtained based on simplified fast operation formula [11]

Bearing types		ZKLF2575-2Z	ZKLF50115-2Z
ZKLF-2Z	Based on formula (1)	11527	6435
	Catalogue [14]	4700	3000
ZKLF..-2RS	Vmax = 12 m/s	5582	3140
	Catalogue [14]	2600	1500

Regarding the fact that the subject of paper is testing the bound revolution frequency in relation to thermal stability, in the further testing the analytic analysis of temperature load was done resulting from friction in the bearing. The following iteration was experimental testing and identification of effective parameters.

Due to easier comparison in Figs. 7 and 8, there are comparative diagrams obtained by analytical procedure in the program “BEARINX” [15] and experiment. Bearinx is patented program of Schaeffler, leading bearing producer in the world, located in Germany. This software is also available online and offers many calculations in area of bearings and bearing arrangements. It consists of different modules which provide different calculations. One module is “Caligula” which provides calculation of thermal load of the bearings, which was used in this paper.

$$n_g = \frac{n \cdot d_m}{d_m} \tag{1}$$

For $d_m < d_G$ =>

$$n \cdot d_m = 750000\left(\frac{d_m}{d_G}\right)^{0,2}\left(1 + \frac{2D_w \cdot \cos\alpha}{d_m}\right)^{-0,8} \cdot f_s \cdot f_M \tag{2}$$

For $d_m \geq d_G$ =>

$$n \cdot d_m = 750000\left(\frac{d_m}{d_G}\right)^{-0,08}\left(1 + \frac{2D_w \cdot \cos\alpha}{d_m}\right)^{-0,8} \cdot f_s \cdot f_M \tag{3}$$

where:

d_m - Average diameter of bearing in mm
D_w - Ball diameter, i.e. rolling item in mm
f_S - Lubrication factor $f_S = 0{,}75$
f_M - Cage type factor (1,3 – for brass; 1 for steel cage)
α - Contact angle of rolling items
d_g - Bearing form factor.

For bearings which are rubber sealed from both sides (2RS) the maximum revolution frequency is obtained from the condition that the maximum sliding speed at contact sealants is v = 12 m/s (internal recommendation of Schaeffler).

Figures 7 and 8 show comparative diagram of friction moment. Shown diagrams are selected by samples which had the least deviation by experimental testing, and they are sample 003 in Fig. 5 and sample 007 in Fig. 6. Since Bearinx gave dependency friction moment- number of revolutions it had to be shown same dependency which was obtained by experimental test in order to make correct comparison.

In order to get more precisely description of the bearing condition for samples 003 and 007, in Tables 2 and 3 are shown the measured temperatures respectively.

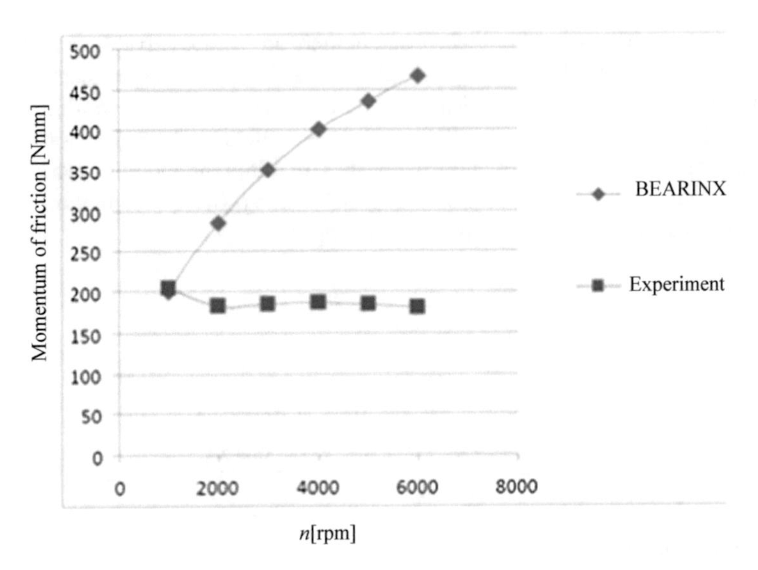

Fig. 7. Comparative diagram of friction moment for bearing ZKLF2575-2Z

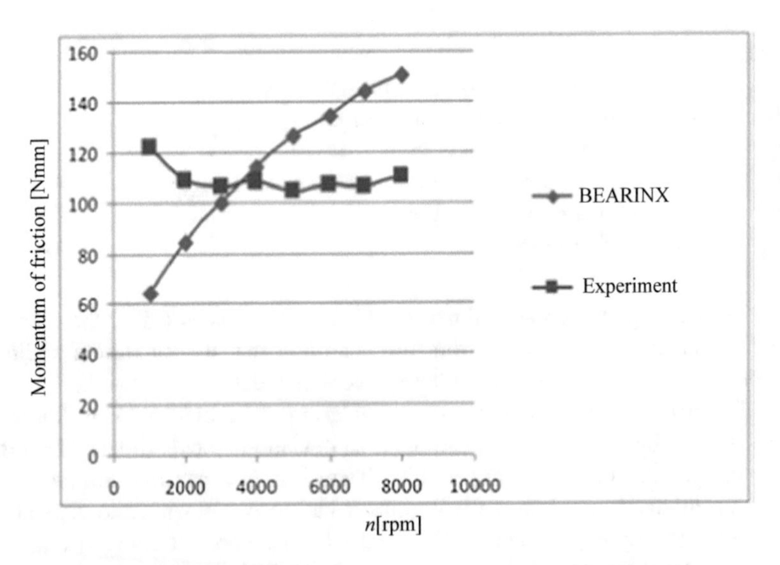

Fig. 8. Comparative diagram of friction moment for bearing ZKLF50115-2Z

Table 2. Measured temperatures for sample 003 (bearing ZKLF2575-2Z)

Number of revolutions in rpm	Friction moment in Nmm	Temperature of the outer ring in °C	Temperature of the inner ring in °C
1006	122,3	31,6	33,6
2002	109,4	36,9	40,6
3002	105,5	39,7	44,8
4000	106,5	42,5	49
5008	102,8	47,4	55,3
6007	106,0	50	59,4
7004	104,5	52,9	63,7
8001	110,6	55,5	68,1

Table 3. Measured temperatures for sample 007 (bearing ZKLF50115-2Z)

Number of revolutions in rpm	Friction moment in Nmm	Temperature of the outer ring in °C	Temperature of the inner ring in °C
1004	203,5	29,8	31,5
2002	182,1	32,8	35,5
3001	184	35,6	39,7
3999	186,5	38,9	44,4
5010	184,4	41,3	48,2
6007	181,2	44,2	52,1

The above mentioned temperatures were measured by temperature sensors which were located on outer ring of the bearing as well as through the lubrication hole to the inner ring. As it shown in Tables 2 and 3 the samples did not reach reference temperature of 70 °C and because of this thermal balance could not be used for determination of the thermal limit number of revolutions.

It must be noted that measured realized axial forces of prestress were 1935 N before and 1929 N after testing by sample 003 and 3171 N before and 3292 N after testing by sample 007. The difference between the axial forces by prestressing is technically acceptable in sense of identification of the influence parameters. Correlation between actual and variable loads is presented by diagram of friction moment Figs. 7 and 8.

Experimental test was used only for identification of the influence parameters on thermal stability of the bearing. It is important to note that complete experimental test as well as complete analysis in this paper are valid under above mentioned conditions and will be used as start point for validation of future tests.

The reasons for occurred discrepancy are the following ones:

- Technical possibilities of the test bench are such that it was not possible to simulate the axial load
- Factors of lubrication and loads in program "BEARINX" are not completely defined for this type of bearing.

As it is known, the bearings of type ZKLF are axial ball bearings with angular contact for fixing of threaded spindles. Their main characteristic is possibility of transfer of great axial forces, i.e. loads with high guide precision. During experimental testing, the test bench was used whose technical characteristics could not have provided the load in axial direction. The only axial load which was present during the experimental testing is an axial force induced by precise nut for prestress and set of plate springs. The specified axial force in the bearing cannot be treated as real axial load because in practice the axial loads are significantly higher.

Due to the specified reason, the only conclusion to be made is that the bearing loads are only the consequence of revolution, thus certain discrepancies are found compared to analytical expectation.

In the analytical procedure performed in the program "BEARINX" [15], the standard procedure for specification of thermal frequency of revolutions is used which is shown in code DIN 732 part 1 and 2 [1, 2]. All coefficients necessary for calculation are automatically taken from the specified code. It should be specified that the code DIN 732 does not consider the effect of machine case to which the bearing is connected using the bolted connection nor the effect of mere bolted connection.

Regarding the fact that the aim of paper is testing the bound thermal revolution frequency for bearings of series ZKLF, as well as significant parameter identification, the occurred discrepancy can be considered technically acceptable.

The second part of the task refers to the parameter identification affecting the bound thermal revolution frequency. It was noted that there are certain discrepancies between analytic and experimental results.

The obtained experimental results imply the main significant effects: grease distribution in bearing, bearing acceleration as well as effect of centrifugal force to cage and rolling items (balls).

For specifying additional parameters, the thorough analysis is required which should include a great number of experiments in order to define the grease quantity for lubrication, coefficient resulting from the type of bearing and lubrication for referential conditions, relation coefficient for number of revolutions, coefficient which shall take into consideration the mixed friction in the bearing at the acceleration as well as coefficient which shall define the effect of centrifugal force to the cage and rolling items.

4 Conclusion

The actual task was research of bound thermal revolution frequency of axial angular contact ball bearings type ZKLF. As already known, the existing code DIN 732 part 1 and 2 defines calculation of the reference number i.e. bound thermal revolution frequency. Main problem by DIN 732 part 1 and 2 is that the bearing type such as ZKLF, ZKLFA, DKLF, DKLFA are not included in it. Because of this it is not possible to get precise calculation for these types. The reason is additional influence parameters which are not defined for the mentioned bearings. The background for this is specific construction, i.e. specific angular contact of 60°, additional precision nut for pre-stressing of the bearing as well as screw connection for additional fixing of bearing on machine housing.

In order to get some information about thermal load and stability at design level, it is necessary to calculate bound thermal revolution frequency, i.e. to define thermal load of the bearing assembly at high revolution frequency. Because of this, it was tried to get reference number of revolutions by experimental test.

As result of the experimental testing it was obtained dependence spectrum which is presented in this paper. The resulting spectrum of dependence show a result deviation. Because of this, two samples with the least result deviation were selected for further analysis. It has to be noted that measured temperatures of outer rings for the selected samples were under 70 °C, which further means that was not possible to use thermal balance for bound thermal revolution frequency determination. Due to this fact, obtained dependence spectrum was used for identification of influence parameters.

The comparison of analytical and experimental results lead to the conclusion that the significant parameters for the observed case of mounting are: grease distribution effect in bearing, bearing acceleration effect as well as centrifugal force effect on cage and rolling elements. Next conclusion is that thermal load of the bearing assembly realized by axial angular contact ball bearings depends on many parameters such as, material and length of the threaded spindle, geometry and material of the machine housing on which the bearing will be additional fixed, geometry of bearings, i.e. on the reference surface through which will the heat i.e. thermal load will be transferred.

That is why general recommendation to make FEA for each separate installation situation which will precise describe temperature field of the observed bearing assembly. It should be emphasized that the kinematics of the bearing has to be taken into account especially in case where the bearing does not reach bound thermal revolution frequency but endangers it's kinematics.

Acknowledgement. This paper presents the results of the research conducted within the project "Research and development of new generation machine systems in the function of the technological development of Serbia" funded by the Faculty of Mechanical Engineering, University of Niš, Serbia.

References

1. Deutsches Institut für Normung: Thermische Bezugsdrehzahl- Teil 1 (1994)
2. Deutsches Institut für Normung: Thermische Bezugsdrehzahl- Teil 2 (1994)
3. Mahmmod, A.M.: The effect of the heat generated by friction in the ball- screw-nut system on the precision of high speed machine. J. Al-Taqani **24**(6), 112–123 (2011)
4. Zahedi, A., Movahhedy, M.R.: Thermo-mechanical modeling of high speed spindles. Sci. Iran. **19**(2), 282–293 (2012)

5. Xiaolei, D., Jianzhong, F., Yuwen, Z.: A predictive model for temperature rise of spindle–bearing integrated system. J. Manuf. Sci. Eng. **137**(2), 1–10 (2015)
6. Yoshida, T., Tozaki, Y., Omokawa, H., Hamanaka, K.: Tribological technology. Three-dimensional ball motion in angular contact ball bearing for high-speed machine tool spindle. J-GLOBAL **38**(6), 304–307 (2001)
7. Xiao, S., Guo, J., Zhang, B.: Research on the motorized spindle's thermal properties based on thermo-mechanical coupling analysis. In: Technology and Innovation Conference, pp. 1479–1483. ITIC, Hangzhou (2006)
8. Yang, L., Wanhua, Z.: Axial thermal error compensation method for the spindle of a precision horizontal machining. In: International conference Mechatronics and Automation, pp. 2319–2323. ICMA (2012)
9. Wang, B., Mei, X., Hu, C., Wu, Z.: Effect of inner ring centrifugal displacement on the dynamic characteristics of high-speed angular contact ball bearing. In: International Conference Mechatronics and Automation, pp. 951–956. ICMA (2010)
10. Krstić, V., Miltenović, A., Banić, M., Miltenović, Đ.: Grenzdrehzahlermittlung an axialschrägkugellager für Gewindetriebe. In: Proceedings of the 7th International Conference "Research and Development of Mechanical Elements and Systems", IRMES, Zlatibor-Republic of Serbia (2011)
11. Krstić, V., Miltenović, A., Banić, M.: Thermal speed limit of axial roller bearings used in support of screw- nut transmissions. Balk. J. Mech. Transm. **1**(2), 39–44 (2011). ISSN 2069-5497
12. Krstić, V., Milčić, D., Milčić, M.: A thermal analysis of the threaded spindle bearing assembly in numerically controlled machine tools. Facta Univ. **16**(2), 261–272 (2018). ISSN 0354-2025
13. Krstić, V.: Research of construction- tribological parameters of ball bearings with angular contact type ZKLF in terms of optimal basic function. Doctoral dissertation. University in Niš (2018)
14. Schaeffler Gruppe Industrie: Lager für Gewindetriebe, Katalog TPI 123D-D: Schaeffler Gruppe (2009)
15. Schaeffler Technologies GmbH&Co. KG: Bearinx program package: Schaeffler Technologies GmbH&Co. KG
16. Wu, L., Tan, Q.: Thermal characteristic analysis and experimental study of a spindle-bearing system, entropy. Open Access J. **18**(7), 271 (2016)
17. Wang, Y.S., Liu, Z., Zhu, H.F.: Heat generation of bearing. Key Eng. Mater. **480–481**, 962–967 (2011)
18. Zhang, Y., hu Li, X., Hong, J., Yan, K., Li, S.: Uneven heat generation and thermal performance of spindle bearings. In: Tribology International, vol. 126, pp. 324–335 (2018)
19. Balamurugan, N., Bhagyanathan, C.: Analysis and investigation on thermal behaviours of ball bearing in high speed spindle. Int. J. Innov. Sci. Mod. Eng. (IJISME) **2**(4), 16–20 (2014)
20. Liu, X.W., Wang, W., Wang, Q.L.: Analysis of temperature field of high speed angular contact ball bearing based on ANSYS. Comb. Mach. Tool Autom. Mach. Technol. **3**, 13–15 (2015)
21. Bian, W., Wang, Z., Yuan, J., Xu, W.: Thermo-mechanical analysis of angular contact ball bearing. J. Mech. Sci. Technol. **30**(1), 297–306 (2016)
22. Chen, G., Wang, L., Gu, L., Zheng, D.-Z.: Heating analysis of the high speed ball bearing. J. Aerosp. Power **22**(1), 163–168 (2007)
23. Dong, Y., Zhou, Z., Liu, M.: Bearing preload optimization for machine tool spindle by the influencing multiple parameters on the bearing performance. Adv. Mech. Eng. **9**(2), 1–9 (2017)

24. Brecher, C., Shneor, Y., Neus, S., Bakarinow, K., Fey, M.: Thermal behavior of externally driven spindle: experimental study and modelling. Engineering **7**, 73–92 (2015)
25. Živković, A., Zeljković, M., Mladjenović, C., Tabaković, S., Milojević, Z., Hadžistević, M.: A study of thermal behavior of the machine tool spindle. Thermal Science (2018). ISSN 0354-9836
26. Baïri, A., Alilat, N., Bauzin, J.G., Laraqi, N.: Three-dimensional stationary thermal behavior of a bearing ball. Int. J. Therm. Sci. **43**(6), 561–568 (2004)
27. Zhou, X., Zhang, H., Hao, X., Liao, X., Han, Q.: Investigation on thermal behavior and temperature distribution of bearing inner and outer rings. Tribol. Int. **130**, 289–298 (2019)
28. Jin, C., Wu, B., Hu, Y.: Heat generation modeling of ball bearing based on internal load distribution. Tribol. Int. **45**(1), 8–15 (2012)
29. Rabréau, C., Kekula, J., Ritou, M., Sulitka, M., Shim, J., LeLoch, S., Furet, B.: Influence of bearing kinematics hypotheses on ball bearing heat generation. Proc. CIRP **77**, 622–625 (2018)
30. Than, V.-T., Huang, J.H.: Nonlinear thermal effects on high-speed spindle bearings subjected to preload. Tribol. Int. **96**, 361–372 (2016)

Geometry Optimization of Flight Simulator Mechanism Using Genetic Algorithm

Milos Petrasinovic(✉), Aleksandar Grbovic, and Danilo Petrasinovic

Faculty of Mechanical Engineering, University of Belgrade, Belgrade, Serbia
petrasinovicmp@gmail.com

Abstract. Flight simulators are motion platforms used to train pilots in various flight regimes. Theoretically, they need to simulate all flight cases and all forces acting upon the pilot during flight. In order to successfully simulate real flight, a moving part of the simulator needs to have six degrees of freedom. The pilot's body is moved and oriented in the space according to the video shown to him. Among many existing designs, parallel mechanisms based on the Stewart platform are most frequently used. In this paper, geometry optimization of the Stewart platform with rotary actuators (6-RUS) is done with a genetic algorithm. For the sake of optimization, it is necessary to define a minimum number of parameters that fully define mechanism with all constraints. The purpose of geometry optimization is to find a mechanism with a workspace that is suitable for simulating flight.

Keywords: Flight simulator · Rotary Stewart platform · Geometry optimization · Genetic algorithm

1 Introduction

Today, flight simulators play an irreplaceable role in pilot training and certification. The first flight simulators were developed well before World War II, but their full potential was discovered when there was not enough time to train a large number of pilots on real planes. With these machines, pilots could learn how to fly under conditions in which flight by visual reference isn't safe anymore. In such situations, the pilot needs to know how to control the aircraft, relying only on information collected from the instruments and its own vestibular system.

Over time, flight simulators have become complex mechatronic systems, but their main purpose, to enable pilots to feel what they see with their eyes, remains the same. In order to achieve realistic simulation of all flight cases and all forces acting upon the pilot during the flight, a moving part of the simulator needs to have six degrees of freedom. The chosen mechanism with six degrees of freedom has to have a sufficiently large workspace to simulate airplane motion, which is why the determination of motion boundaries is very important.

A free rigid body has six degrees of freedom, it can perform six independent motions, three translations and three rotations. A kinematic chain is a series of bodies or segments connected together by joints. A kinematic chain can be simple or complex, and both can be divided into open and closed [1]. One segment of a simple kinematic

N. Mitrovic et al. (Eds.): CNNTech 2019, LNNS 90, pp. 340–358, 2020.
https://doi.org/10.1007/978-3-030-30853-7_20

chain can't be connected to more than two other segments, while one segment of a complex kinematic chain can be connected to more than two other segments. Open kinematic chains have segments that are connected to only one other segment, while all closed chain segments are connected to at least two other segments, as shown in Fig. 1.

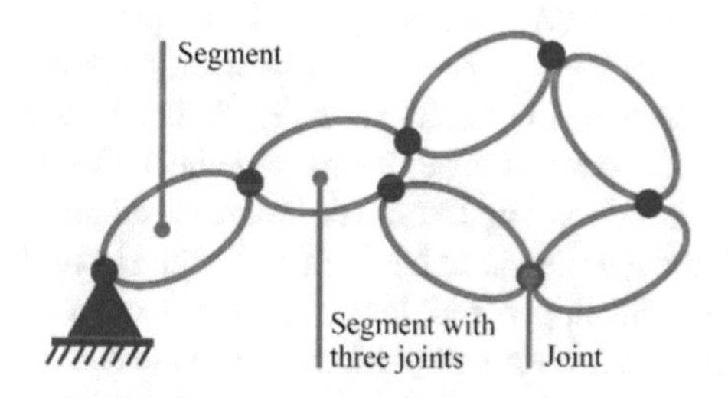

Fig. 1. Closed kinematic chain.

Among many existing designs, parallel mechanisms based on the Stewart platform, defined in [2], are most frequently used for flight simulators. From this first work until today this is a popular research topic in robotics with many published research papers about inverse and forward kinematics problem [3, 4], dynamics [3, 5, 6], singularities and workspace estimation [3, 7–9].

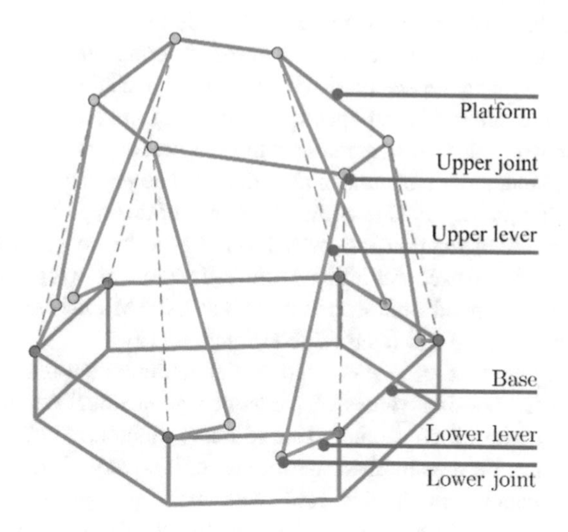

Fig. 2. Stewart platform with rotary actuators.

Even after so many years of research, there is still the opportunity to find a new or better way of practical use of this mechanism. Beside flight simulators, this mechanism is currently used, for instance, within telescope and satellite dish positioning system, for machining tools, as part of medical instruments, and for vibration damping.

In aerospace engineering, it is necessary to take into account vibrations and noise that can affect the physical condition of the pilot and his passengers. Sources of vibrations and passive damping are shown in [10], instead of a passive damping component that connects a seat and the rest of the aircraft, mechanism based on Stewart platform can be used for active isolation and damping of vibrations.

There are many different types of geometrical interpretations of the Stewart platform, but the main objective of this paper is geometry optimization of the Stewart platform with rotary actuators [11], as shown in Fig. 2. The Stewart platform with rotary actuators has six pairs of upper and lower levers (together called legs), and movement of the platform is achieved by rotating the lower lever around an axis going through one end of a lever. This is the revolute joint between base and a lower lever that is controlled by an electric motor. The upper and lower lever are connected with universal or spherical joint, and the upper lever and platform are connected by spherical joints. This is the reason why this type is also called 6-RUS, in accordance with types of joints.

In order to be able to control a position and orientation of the platform, the part of the simulator that connects base and platform (leg) needs to have an actuator. Some of the other geometrical types use six levers with variable length (linear or prismatic actuators) or six pairs of two levers of which one lever is fixed but has variable length. The most important design aspects of the mechanism besides the type of actuators are spatial configuration (locations of connections) and the type of connections (joints) [11].

A design solution based on rotary actuators, in this case, electric motors with proper gearboxes, is adopted due to its simplicity in terms of simulation, production, and maintenance. If the servo motor with servo drive is not cost-effective, an induction motor or asynchronous motor can be controlled using variable-frequency drive (VFD) and both products are standardized and widely available. The current angle of the shaft can be measured with an encoder or potentiometer.

The platform with this type of actuator is the simplest one to scale down and test on a smaller model, kinematics and control algorithm can be physically tested even before investing in the first prototype. One of the drawbacks of this type is that due to working condition, fatigue life estimation for lower levers has to be done. Same experimental and numerical methods used for fatigue life estimation of wing spar [12] can be applied. Using the extended finite element method (XFEM), it is possible to numerically obtained crack paths and fatigue life estimation [13].

After selecting the type of actuator and spatial configuration remaining part of the simulator design is choosing values of predefined geometrical parameters that fully define mechanism with all constraints. Due to many different combinations and their interrelated influence on the simulator performance, it is very difficult to find optimal values for these parameters. In this case, with many parameters often some optimization algorithms must be employed to be able to get sufficiently good results in the desired time. For problems like this, the optimization algorithms are often numerical and are not providing the solution in the closed form. These algorithms are traditionally iterative, constrained minimization problems can be solved with minimization of the objective function by successive approaching within a sequence of calculations [14]. There are already good efforts in employing the genetic algorithm [15, 16] for optimization of this kind of problems, it is suitable because there is a large number of local minima and it is difficult to find global one.

The goal of optimization is defined by the implementation of objective (fitness) function. There are many performance indicators that can be taken into account while calculating objective functions such as the maximum feasible orientation of the platform, the volume of workspace, needed joint movements, dexterity, singularities, and physical interference between segments.

2 Geometric Parameters

In order to analyze whether the shown type of mechanism can be successfully applied in practice for the flight simulator, first of all, the geometry must be defined. The geometry should be defined with a minimum number of parameters that fully define mechanism with all constraints. Changing the values of these parameters within the optimization process should provide better characteristics of the flight simulator.

In kinematics, joint movements are referred to as generalized or internal coordinates and external coordinates are the ones that define the position and orientation of segments in reference to a global reference frame. Changing internal coordinates changes external coordinates, i.e. position and orientation of the platform. The process of determining external coordinates for a given set of internal coordinates (joint movements) is referred to as forward kinematics. Determining internal coordinates (joint movements) for a given set of external coordinates (position and orientation) is referred to as inverse kinematics.

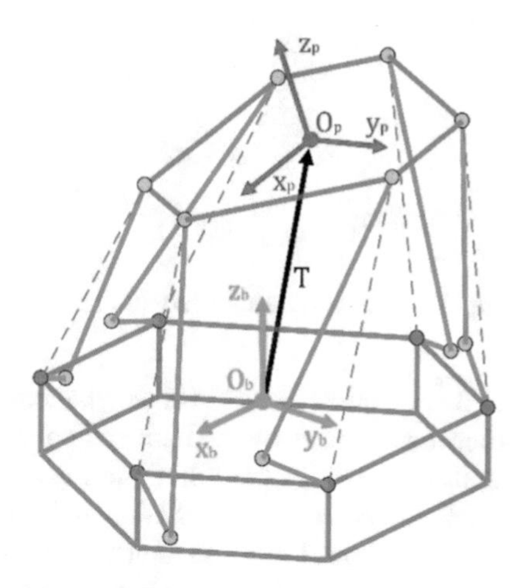

Fig. 3. Global and platform's reference frame.

Calculating the required rotations of rotary actuators and lower levers for a given position and orientation of the platform requires solving the inverse kinematics. The most intuitive way of solving this type of problem is often referred to as a geometric method. As shown in Fig. 3, the first step is to define two reference frames. A global reference frame $(O_b x_b y_b z_b)$ fixed to the center of the base and a moving platform reference frame $\left(O_p x_p y_p z_p\right)$ fixed to the center of the platform.

Control values are defined by vector $\boldsymbol{q} = [x, y, z, \psi, \theta, \varphi]^T$, and they represent external coordinates of the moving platform's local reference frame in reference to the global frame fixed to the nonmoving base of the platform. In other words translation of platform's frame in reference to the global frame is defined by $\boldsymbol{T} = [x, y, z]^T$ while orientation is defined using Euler angles (ψ, θ, φ), respectively, yaw, pitch, and roll about z, y, and x global axes.

In this case, the external coordinates are known, but if it is necessary to determine the external coordinates based on the known lower levers orientation, it would require solving forward kinematics. The forward kinematics of this system is very complex and a lot of effort has been invested in solving it.

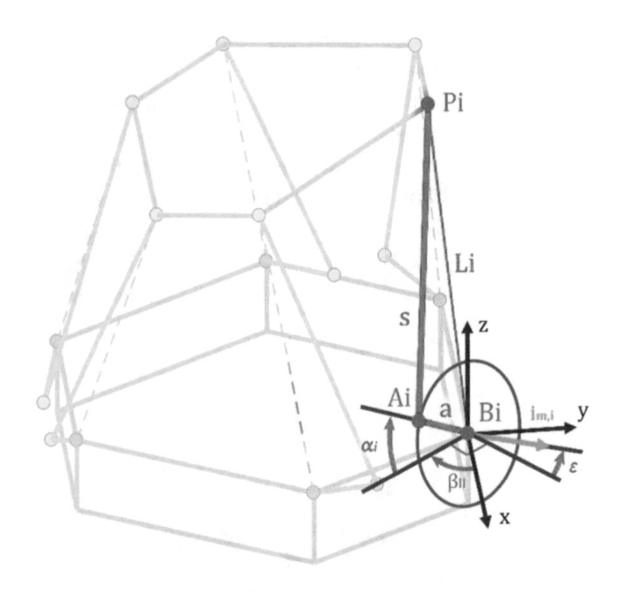

Fig. 4. Geometric parameters of one leg.

For a given position and orientation (control values $\boldsymbol{q}$), the needed length of one leg or intensity of vector $\boldsymbol{L}_i$ (between point B_i on the base and point P_i on the platform) has to be determined, this length is shown for the i-th leg in Fig. 4. Due to six legs of this mechanism $i = 1, 2, \ldots, 6$. Point B_i is the intersection point of the actuator axis of rotation and the lower lever. Lower lever has constant length and it is equal to parameter a, while the constant length of the upper lever is equal to parameter s. Point P_i is a point on the platform around which the spherical joint rotates, the center of the sphere. For the introduced reference frames, the following relation determines the vector of the i-th leg.

$$\boldsymbol{L}_i = \boldsymbol{T} + \boldsymbol{R}\boldsymbol{p}_i - \boldsymbol{b}_i \tag{1}$$

Vector $\boldsymbol{T}$ is already mentioned, while $\boldsymbol{R}$ represents the rotation matrix whose elements are functions of the three Euler angles that determine the orientation of the moving platform's reference frame in reference to the global frame. In order to ease writing, a condensed notation of the trigonometric functions sine and the cosine, "s" and "c", respectively are used to define rotation matrix $\boldsymbol{R}$ as follows:

$$\boldsymbol{R} = \begin{bmatrix} c\psi \cdot c\theta & c\psi \cdot s\theta \cdot s\varphi - s\psi \cdot c\varphi & s\psi \cdot s\varphi + c\psi \cdot s\theta \cdot c\varphi \\ s\psi \cdot c\theta & c\psi \cdot c\varphi + s\psi \cdot s\theta \cdot s\varphi & s\psi \cdot s\theta \cdot c\varphi - c\psi \cdot s\varphi \\ -s\theta & c\theta \cdot s\varphi & c\theta \cdot c\varphi \end{bmatrix} \quad (2)$$

Vector $\boldsymbol{p}_i = [x_{Pi}, y_{Pi}, z_{Pi}]^T$ is the position vector of a point P_i in the reference frame of the platform. Vector $\boldsymbol{b}_i = [x_{Bi}, y_{Bi}, z_{Bi}]^T$ is the position vector of a point B_i in the global reference frame. The necessary length of the leg can now be computed by the Euclidean norm of the vector $\boldsymbol{L}_i$ as $l_i = \|\boldsymbol{L}_i\|_2$.

After calculating the vector $\boldsymbol{L}_i$, the next step is to find the angle of rotation $\boldsymbol{\alpha}_i$ for the rotary actuator. This angle defines rotation around the axis of the actuator, while this axis is defined by unite vector $\boldsymbol{i}_{mi}$. The purpose of finding the angle $\boldsymbol{\alpha}_i$ is to define the position of point A_i which provides the correct position of upper and lower levers for the defined length of the leg. Angle $\boldsymbol{\alpha}_i$ is measured from the initial vector $\boldsymbol{a}_i$, which lies in the horizontal plane. Vector $\boldsymbol{a}_i = [x_{Ai}, y_{Ai}, z_{Ai}]^T$ is the position vector of a point A_i in the global reference frame.

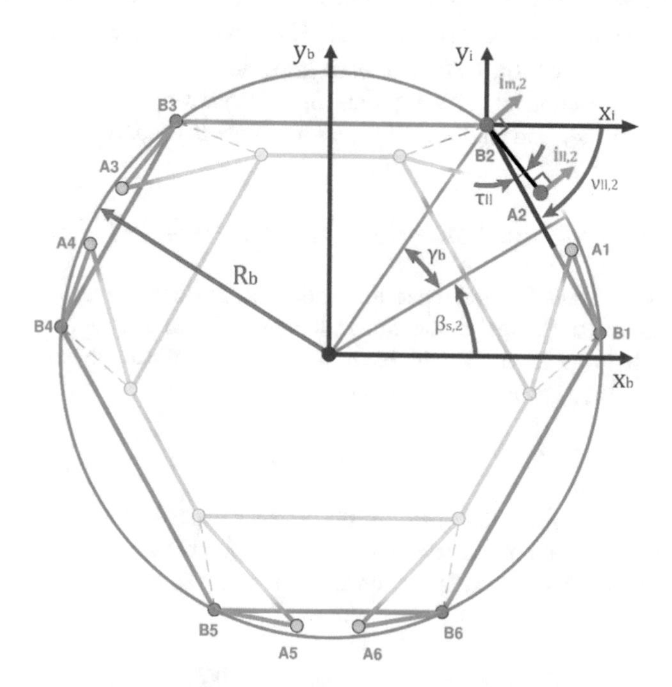

Fig. 5. Geometric parameters of the mechanism's base.

The actuator's axis of rotation and the circle representing all possible positions of point A_i are shown in Fig. 4. In reference to the horizontal plane in which the base lies, the axis of rotation is tilted by ε angle. In order to fully define upper and lower levers, first coordinates of point B_i and P_i needs to be determined in corresponding reference frames. The intuitive way to do this is to use polar coordinates, noticing that points B_i and P_i lie on constant diameter circles and that there are 3 axes of symmetry between these points (point B_1 can be mapped to point B_2 while the axis of symmetry is at a certain angle to the x - axis of global frame).

As shown in Fig. 5, the axis of symmetry, between two B_i points, is defined by angle $\beta_{s,i}$, this angle has a constant value for all considered geometries. It is not a geometrical parameter for optimization but it is necessary for determining coordinates of point B_i and P_i with other parameters. The value of this angle for all legs is defined with vector $\boldsymbol{\beta}_s = [30°, 30°, 150°, 150°, -90°, -90°]^T$.

Because point B_i can be located on both sides of the axis while the absolute value of angle γ_b remains the same, an additional vector that defines angle signs is defined. Vector $\boldsymbol{\sigma} = [-1, 1, -1, 1, -1]^T$ also has constant value and its purpose is to define on which side in reference to the axis of symmetry are points B_i and P_i, and to define different rotation directions of even and odd actuators.

It is necessary to define the referent orientation for the lower levers in the horizontal plane of the base. This orientation is defined with angle $\boldsymbol{v}_{ll,i}$ measured from x - axis of the global reference frame. The value of this angle is defined with vector $\boldsymbol{v}_{ll} = [120°, -30°, -120°, 30°, 0°, 180°]^T$ for each lower lever.

Finally, the position of all six points B_i is defined by two variable geometric parameters, radius R_b, and angle γ_b. The constant angle $\beta_{s,i}$ and the value σ_i are used as follows:

$$x_{B,i} = R_b \cos(\delta_{b,i}), y_{B,i} = R_b \sin(\delta_{b,i}), z_{B,i} = 0, \delta_{b,i} = \beta_{s,i} + \gamma_b\, \sigma_i \tag{3}$$

The angle between the reference orientation of the lower lever in the horizontal plane of the base and the real orientation in this plane is the parameter τ_{ll}, which is the same for all six lower levers. Again, based on $\beta_{s,i}$ and σ_i, the angle between the lower lever and x - axis is given by the following equation:

$$\beta_{ll,i} = \beta_{s,i} + \tau_{ll}\, \sigma_i \tag{4}$$

The projections of the unit vector of the actuator's axis of rotation and axis of the joint between lower and upper lever (which is explained later) in the horizontal plane are normal to the lower lever. Based on that, the unit vector of the actuator is given below:

$$\boldsymbol{i}_{m,i} = \begin{bmatrix} \cos(\beta_{ll,i} + 90°\, \sigma_i)\cos(\varepsilon) & \sin(\beta_{ll,i} + 90°\sigma_i)\cos(\varepsilon) & \sin(\varepsilon) \end{bmatrix}^T \tag{5}$$

Coordinates of the vector $\boldsymbol{a}_i$ can be expressed as follows:

$$x_{A,i} = a\left(\cos(\alpha_i)\cos\left(\beta_{II,i}\right) + \sin(\alpha_i)\sin(\varepsilon)\sin\left(\beta_{II,i}\right)\sigma\right) + x_{B,i} \tag{6}$$

$$y_{A,i} = a\left(\cos(\alpha_i)\sin\left(\beta_{II,i}\right) - \sin(\alpha_i)\sin(\varepsilon)\cos\left(\beta_{II,i}\right)\sigma\right) + y_{B,i} \tag{7}$$

$$z_{A,i} = a\,\sin(\alpha_i)\cos(\varepsilon) + z_{B,i} \tag{8}$$

In the previous equations, only angle α_i is unknown, in order to find this angle, it is first necessary to define the geometric parameters of the platform.

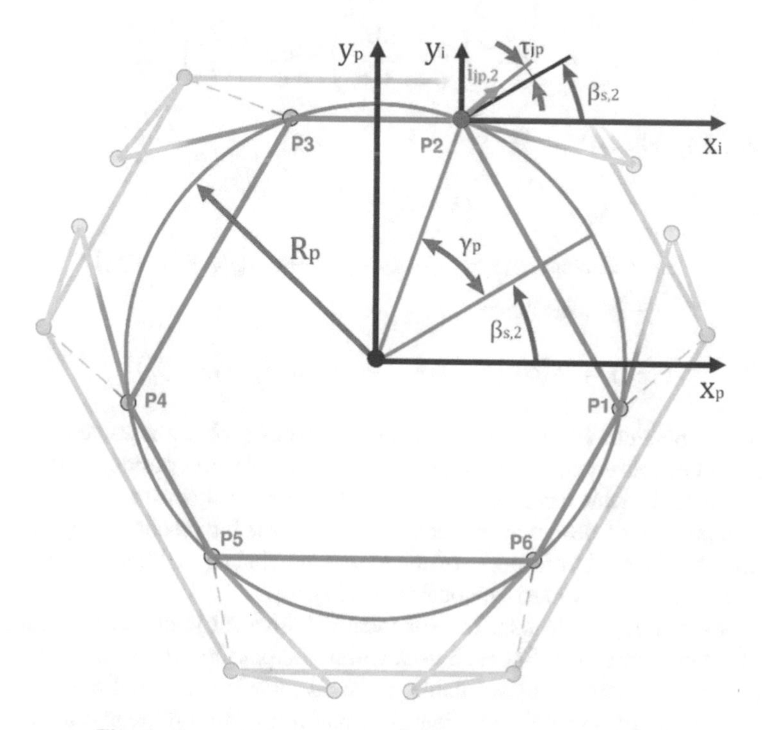

Fig. 6. Geometric parameters of the mechanism's platform.

The position of all six points P_i in reference to the platform's reference frame (coordinates of vector $\boldsymbol{p}_i$) is just like points B_i, beside $\beta_{s,i}$ and σ_i, defined by two variable geometric parameters, radius R_p, and angle γ_p.

$$x_{P,i} = R_p\cos\left(\delta_{p,i}\right), y_{P,i} = R_p\sin\left(\delta_{p,i}\right), z_{P,i} = 0, \delta_{p,i} = \beta_{s,i} + \gamma_{p,i}\sigma_i \tag{9}$$

Using translation vector $\boldsymbol{T}$ and rotation matrix $\boldsymbol{R}$ it is possible to obtain a position vector of a point P_i in reference to the global reference frame.

Besides parameters for the definition of points P_i position, in Fig. 6 is shown a projection of unit vector of the upper's joint axis in the plane defined by x_p and y_p. The orientation of this vector in this plane (angle between vector projection and x_p axis) is defined as follows:

$$\beta_{jp,i} = \beta_{s,i} + \tau_{jp,i}\sigma_i \tag{10}$$

Using now known vectors $\boldsymbol{p}_i$ and $\boldsymbol{b}_i$, and known lengths a, s, and L_i, angle α_i can be obtained from coordinates of vector $\boldsymbol{a}_i$. After solving a system of equations and using trigonometric identities, the following equations are obtained:

$$\alpha_i = \sin^{-1}\left(\frac{\operatorname{sgn}(E_i)D_i}{\sqrt{E_i^2 + F_i^2}}\right) - \tan^{-1}\left(\frac{F_i}{E_i}\right) \tag{11}$$

where D_i, E_i, and F_i are:

$$D_i = L_i^2 + a^2 - s^2 \tag{12}$$

$$\begin{aligned} E_i = {} & 2a\,\sigma\sin(\varepsilon)\left(\sin\left(\beta_{ll,i}\right)(x_P - x_B) - \cos\left(\beta_{ll,i}\right)(y_P - y_B)\right) \\ & + 2a\cos(\varepsilon)(z_P - z_B) \end{aligned} \tag{13}$$

$$F_i = 2a\left(\cos\left(\beta_{ll,i}\right)(x_P - x_B) + \sin\left(\beta_{ll,i}\right)(y_P - y_B)\right) \tag{14}$$

With the obtained Eq. (11), the inverse kinematics problem is solved. If there is a real solution of this equation for all angles α_i then the desired position and orientation of the platform is achievable, if there are no other physical constraints.

In the process of designing a mechanism like this, for practical usage as a flight simulator, the physical constraints of the kinematic chains, such as lever interference and limitations of joints must be considered [17].

Whenever some mechanism has spherical or universal joints, their angular limits should be considered [16]. For each joint within the platform range of work has to be defined. The most commonly available spherical joint is called rod end bearing, the nominal position of levers for this joint is when levers are orthogonal to each other. Based on this, if the rotation between the ends of the upper lever is possible, then constraints of the upper and lower joints can be defined as follows:

$$\lambda_{p,i} = \left|\cos^{-1}\left(\frac{\overrightarrow{P_iA_i} \cdot \boldsymbol{i}_{jp,i}}{s}\right) - 90^\circ\right| < \lambda_{max} \tag{15}$$

$$\lambda_{b,i} = \left|\cos^{-1}\left(\frac{\overrightarrow{P_iA_i} \cdot \boldsymbol{i}_{ll,i}}{s}\right) - 90^\circ\right| < \lambda_{max} \tag{16}$$

The maximal cone angle of this type of joints is λ_{max}, and it needs to be prior determined for a specific joint. The unit vectors $\boldsymbol{i}_{jp,i}$ and $\boldsymbol{i}_{ll,i}$, used in the previous

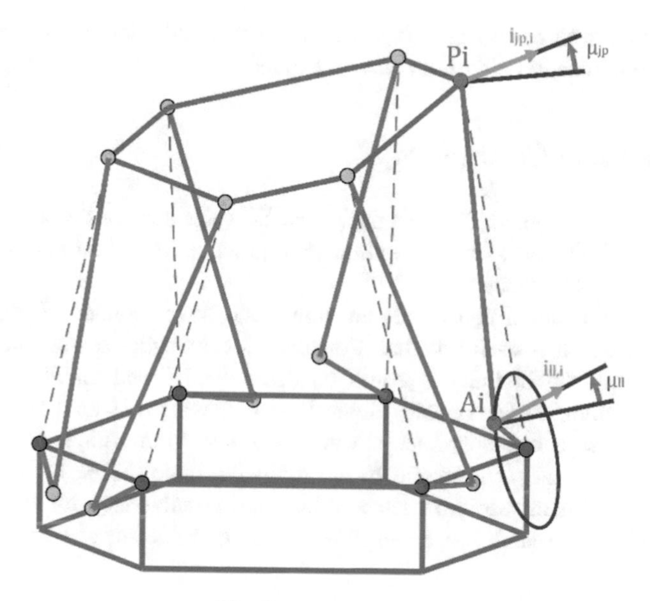

Fig. 7. Axes of joints.

equations, define axes of joints. These vectors are shown in Fig. 7, their orientation in planes of the platform and the base are already discussed, but two more geometric parameters are introduced for total orientation. Vectors are tilted in reference to the corresponding planes by angles μ_{jp} and μ_{ll}. The same principle is already used for the actuators, and the following equations are obtained:

$$\boldsymbol{i}_{jp,i} = \left[\cos\left(\beta_{jp,i}\right)\cos\left(\mu_{jp}\right) \quad \sin\left(\beta_{jp,i}\right)\cos\left(\mu_{jp}\right) \quad \sin\left(\mu_{jp}\right) \right] \tag{17}$$

$$\boldsymbol{i}_{ll,i} = \begin{bmatrix} \cos\left(\beta_{ll,i} + 90^\circ\sigma_i\right)\cos(\varepsilon + \mu_{ll}) \\ \sin\left(\beta_{ll,i} + 90^\circ\sigma_i\right)\cos(\varepsilon + \mu_{ll}) \\ \sin(\varepsilon + \mu_{ll}) \end{bmatrix} \tag{18}$$

In order to optimize geometry, eleven geometric parameters that determine the geometry of the mechanism for flight simulator are defined. They unambiguously describe platform geometry. It should be noted that the geometry could be defined in other ways but since no restrictions are imposed, besides symmetry between joints, the same coordinates can always be obtained.

For the initial orientation of all lower levers can be selected case when they are in plane with the base, and the initial position of the platform in that case is:

$$z_0 = \sqrt{s^2 - \left(x_{P,1} - x_{A,1}\right)^2 - \left(y_{P,1} - y_{A,1}\right)^2} + z_{A,1} \tag{19}$$

After defining all necessary geometric parameters and adopting all the necessary vectors, the optimization algorithm can be applied.

3 Optimization Constraints

Engineering design problem, like this one, often has some constraints like, for example, manufacturing limitations. For the optimization process, these limitations must be, in some way, taken into account.

The first optimization constraints are upper and lower bounds of proposed geometric parameters. It is easier to find a solution for a smaller search space. For this specific problem, the boundaries could be determined based on allowable overall dimensions of flight simulator and dimensions of payload. Even when bounds are known, often there is a standard set of value to choose from. This type of problem is known as a mixed integer problem. In the following table (Table 1) are shown predefined ranges of geometric parameters values and possible step for values between bounds. In this way, search space and standard set for values of geometric parameters are defined.

Table 1. Value range of geometric parameters for optimization.

Parameter	Value range	Step
a	100 mm to 400 mm	10 mm
s	600 mm to 1200 mm	10 mm
R_p	500 mm to 800 mm	10 mm
R_b	600 mm to 1500 mm	10 mm
γ_p	5° to 55°	1°
γ_b	5° to 55°	1°
ε	−90° to 90°	1°
τ_{ll}	−180° to 180°	1°
τ_{jp}	−90° to 90°	1°
μ_{ll}	−90° to 90°	1°
μ_{jp}	−90° to 90°	1°

The next step is the definition of nonlinear constraints, one constraint of this type is given with Eqs. (15) and (16). If these inequalities are satisfied, then the desired position and orientation of the platform are achievable in terms of joints motion In addition, there must be no interference between the legs of the mechanism. This type of constraints for lower levers can be defined as follows:

$$\min\left(\left\|\overrightarrow{D_iD_j}\right\|_2\right) > r_{min}, i \neq j \tag{20}$$

where D_i and D_j are any points on closed line segments (from B_i to A_i and from B_j to A_j respectively) of two lower levers that are part of different legs. This ensures that minimum distance (expressed with Euclidean norm) between any two lower levers is always larger than predefined value r_{min}, and this can be adapted for any parts of a mechanism.

In order to find a necessary torque of actuators for payload on the platform, the Jacobian matrix for this type of mechanism needs to be obtained. The Jacobian matrix maps the change of internal coordinates to change of external coordinates with respect to the time [18]. Based on the principle of virtual work, the Jacobian matrix, as a relationship between joint velocities and the platform velocities, is also a direct relation between necessary torque of actuators and applied load to the platform [19]. For each leg of the mechanism, the loop closure equation is given by:

$$\overrightarrow{O_bB_i} + \overrightarrow{B_iA_i} + \overrightarrow{A_iP_i} = \overrightarrow{O_bO_p} + \overrightarrow{O_pP_i} \tag{21}$$

By differentiating Eq. (21) with respect to the time, the following equation is obtained:

$$\boldsymbol{\omega}_{\alpha,i} \times \overrightarrow{B_iA_i} + \boldsymbol{\omega}_{ul,i} \times \overrightarrow{A_iP_i} = \boldsymbol{v}_p + \boldsymbol{\omega}_p \times \overrightarrow{O_pP_i} \tag{22}$$

where $\boldsymbol{v}_p$ is platform linear velocity, $\boldsymbol{\omega}_{\alpha,i}$, $\boldsymbol{\omega}_{l.i}$, and $\boldsymbol{\omega}_p$ are angular velocities of the actuator (and lower lever), upper lever, and platform, respectively. By introducing the unit vector of the upper lever as $\overrightarrow{A_iP_i} = s\boldsymbol{\lambda}_{ul,i}$, multiplying both sides of the equality with it, and using characteristics of the cross product, the equation becomes:

$$\boldsymbol{\lambda}_{ul,i}\left(\boldsymbol{\omega}_{\alpha,i} \times \overrightarrow{B_iA_i}\right) = \boldsymbol{\lambda}_{ul,i} \cdot \boldsymbol{v}_p + \left(\overrightarrow{O_pP_i} \times \boldsymbol{\lambda}_{ul,i}\right)\boldsymbol{\omega}_p \tag{23}$$

By introducing the unit vector of rotary actuator $\boldsymbol{i}_{m,i}$ and his angular velocity $\dot{\alpha}_i$, the following equation is obtained:

$$\boldsymbol{\lambda}_{ul,i}\left(\boldsymbol{i}_{m,i} \times \overrightarrow{B_iA_i}\right)\dot{\alpha}_i = \boldsymbol{\lambda}_{ul,i} \cdot \boldsymbol{v}_p + \left(\overrightarrow{O_pP_i} \times \boldsymbol{\lambda}_{ul,i}\right)\boldsymbol{\omega}_p \tag{24}$$

Based on the previous equation, the Jacobian matrix of external coordinates $\boldsymbol{J}_q$ and internal coordinates $\boldsymbol{J}_\alpha$ can be defined as follows:

$$\boldsymbol{J}_q\dot{\boldsymbol{q}} = \boldsymbol{J}_\alpha\dot{\boldsymbol{\alpha}} \tag{25}$$

$$\boldsymbol{J}_q = \begin{bmatrix} \boldsymbol{\lambda}_{ul,1} & \overrightarrow{O_pP_1} \times \boldsymbol{\lambda}_{ul,1} \\ \vdots & \vdots \\ \boldsymbol{\lambda}_{ul,6} & \overrightarrow{O_pP_6} \times \boldsymbol{\lambda}_{ul,6} \end{bmatrix}, \boldsymbol{J}_\alpha = \mathrm{diag}\left(\boldsymbol{\lambda}_{ul,i}\left(\boldsymbol{i}_{m,i} \times \overrightarrow{B_iA_i}\right)\right) \tag{26}$$

$$\dot{\boldsymbol{q}} = \begin{bmatrix} v_{px} & v_{py} & v_{pz} & \omega_{px} & \omega_{py} & \omega_{pz} \end{bmatrix}^T, \dot{\boldsymbol{\alpha}} = \begin{bmatrix} \dot{\alpha}_1 & \cdots & \dot{\alpha}_6 \end{bmatrix}^T \tag{27}$$

Finally, the Jacobian matrix $\boldsymbol{J}$ can be expressed with the following equation:

$$\boldsymbol{J} = \boldsymbol{J}_{\alpha}^{-1}\boldsymbol{J}_{q} \tag{28}$$

As earlier noted, with the inverse of matrix $\boldsymbol{J}$ and if load acting upon the platform $\boldsymbol{F}$ is known, necessary torque of actuators $\boldsymbol{T}$ can be found with:

$$\boldsymbol{T} = \left(\boldsymbol{J}^{T}\right)^{-1}\boldsymbol{F} \tag{29}$$

In case of static equilibrium, if $\boldsymbol{C_{in}}$ is center of inertia of payload, and if just weight of payload ($m\boldsymbol{g}$) is acting upon the platform, then vector $\boldsymbol{F}$ can be written as:

$$\boldsymbol{F} = \left[m\boldsymbol{g} \quad \boldsymbol{C_{in}} \times m\boldsymbol{g}\right]^{T} \tag{30}$$

Nonlinear constraint based on necessary torque of actuators is given with the following inequality:

$$|T_i| < T_{max} \tag{31}$$

where T_{max} is maximum allowable torque.

4 Geometry Optimization with Genetic Algorithm

The Genetic Algorithms (GA) represents a methodology for solving optimization problems based on the concept of natural selection [14]. Problem is solved for a different set of inputs and for each set is given some score which is then used as a selection criterion. This set of inputs is called the population of one generation. Before solving the problem again, the set of inputs is changed using elite selection, cross-over, and mutation rules. The score which is given to individual solution is in fact value of an objective function for a given set of inputs. The idea is to provide an algorithm that evolves toward a global optimal solution.

This particular algorithm is selected because of a large number of inputs (in this case, eleven geometric parameters), and highly nonlinear and discontinuous nature of the problem, with many local minima.

If all defined constraints are satisfied for a set of variable values (values of geometric parameters), then for this set, the corresponding score in the objective function must be calculated. Researchers have proposed various characteristics for the goal of optimization, some of them are based on the properties of the Jacobian matrix and the properties of the workspace. In this case, the implementation of objective (fitness) function needs to ensure that the final solution is optimal for the flight simulator. It should be noted that the implementation of the objective function should be tailored to a specific application. From many performance indicators that can be taken into account while calculating objective functions, dexterity and the maximum feasible Euler angles have been selected as the most valuable for flight simulation.

In order to calculate Local Dexterity Index (LDI) η as the reciprocal value of the Jacobian matrix condition number [16], a homogeneous Jacobian matrix $\boldsymbol{J}_h$ should first be obtained. As shown in [20], the Jacobian matrix of external coordinates $\boldsymbol{J}_q$ can be separated into two matrices that have different physical dimensions. Then some sort of normalization can be applied to obtain a homogeneous matrix. The first matrix, obtained as the first three columns, is related to forces while the last three columns are related to torques.

$$\boldsymbol{J}_q = [\boldsymbol{J}_{qF} \quad \boldsymbol{J}_{qT}] \tag{32}$$

Method of normalization based on the isotropy of the system [20] proposes the characteristic length L_c, in order to divide $\boldsymbol{J}_{qT}$ and obtain necessary dimension. Homogeneous Jacobian matrix is then defined as follows:

$$\boldsymbol{J}_h = \boldsymbol{J}_{\alpha}^{-1}\left[\boldsymbol{J}_{qF} \quad \tfrac{1}{L_c}\boldsymbol{J}_{qT}\right], L_c = \sqrt{\frac{\mathrm{trace}\left(\boldsymbol{J}_{qT}^T\boldsymbol{J}_{qT}\right)}{\mathrm{trace}\left(\boldsymbol{J}_{qF}^T\boldsymbol{J}_{qF}\right)}} \tag{33}$$

For single position and orientation of platform, Local Dexterity Index can be calculated from the following equation:

$$\eta = \frac{1}{\mathrm{cond}(\boldsymbol{J}_h)} \tag{34}$$

Second and third performance indicators are based on the maximum absolute value of feasible independent Euler angles pitch and roll about y and x global axes, for the given position. To determine the value of the performance indicators, 27 points in space have been selected. These points are obtained as a mesh grid constructed with the following three vectors (dimensions are in mm) (Fig. 8):

$$\boldsymbol{x}_{ws} = [-150, 0, 150]^T, \boldsymbol{y}_{ws} = [-150, 0, 150]^T, \boldsymbol{z}_{ws} = z_0[-75, 0, 75]^T \tag{35}$$

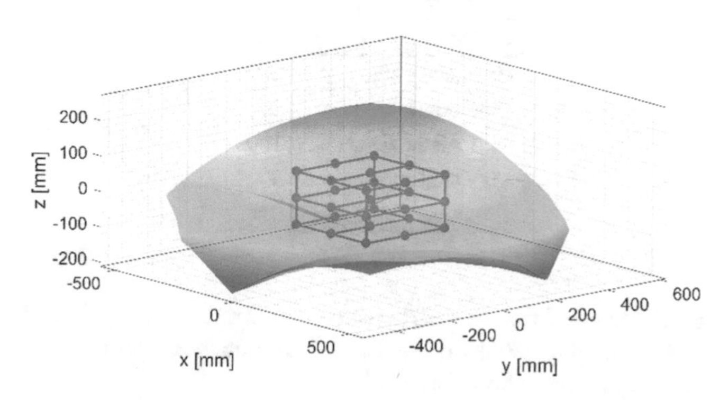

Fig. 8. Points for evaluation of the performance indicators.

Values of η_j, $\theta_{j,max}$, and $\varphi_{j,max}$, where $j = 1, 2, \ldots, 27$, are combined within objective function using a weighted sum approach. In addition, the boundaries for angles are introduced to obtain values between 0 and 1, while the value of η_j is already in this interval. For each point, the largest achievable angles, with all constraints, are determined, a search was done within defined boundaries for angles.

$$y_{of} = 100 - \sum\nolimits_{k=1}^{3} \frac{100}{27} \sum\nolimits_{j=1}^{27} \omega_k I_{k,j}, \sum\nolimits_{k=1}^{3} \omega_k = 1 \tag{36}$$

After introducing performance indicators, the previous equation for calculation of penalty value (score) within the objective function becomes:

$$y_{of} = 100 - \frac{100}{27} \sum\nolimits_{j=1}^{27} \omega_1 \eta_j + \omega_2 \frac{\theta_{j,max} - \theta_{j,min}}{2\theta_{max}} + \omega_3 \frac{\varphi_{j,max} - \varphi_{j,min}}{2\varphi_{max}} \tag{37}$$

In the following table are shown variables, which are necessary for the optimization, and their values.

Table 2. Necessary optimization variables and their values.

Variable	Value
Vector of sum weights	$\boldsymbol{\omega} = [0.4, 0.3, 0.3]^T$
Boundaries for angles	$\theta_{max} = 35°, \varphi_{max} = 35°, \lambda_{max} = 35°$
Max allowed torque	$T_{max} = 200\,\mathrm{Nm}$
Allowed lever distance	$r_{min} = 100\,\mathrm{mm}$
Payload mass and center of inertia	$m = 250\,\mathrm{kg}, \boldsymbol{C_{in}} = [0, 0, 500]\mathrm{mm}$

Implementation of the Genetic Algorithm from the Global Optimization Toolbox within MATLAB is used, with settings shown in the following table (Table 3).

Table 3. Settings for the Genetic Algorithm.

Option	Value
Population size	200
Elite count	10
Max generations	1000
Max stall generations	300
Crossover fraction	0.8

The obtained values of the geometric parameters for the final optimization solution are given in the following table (Table 4). This is obtained optimal solution for the values that are adopted and shown in Table 2.

Table 4. Values of geometric parameters for the final solution.

Parameter	Value	Parameter	Value
a	230 mm	ε	−39°
s	1110 mm	τ_{ll}	−122°
R_p	520 mm	τ_{jp}	−12°
R_b	840 mm	μ_{ll}	−18°
γ_p	7°	μ_{jp}	−1°
γ_b	35°		

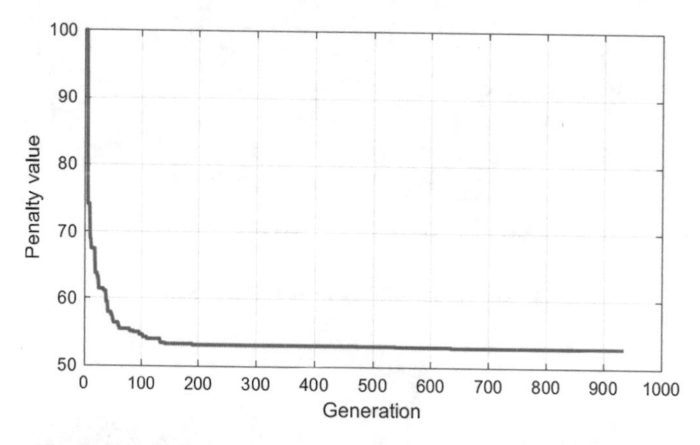

Fig. 9. Change of penalty value (score) in reference to generation.

Termination of optimization occurred because the max stall generation stopping criterion was reached. As shown in Fig. 9, optimization stopped after 935 generations with the best penalty value of 52.97. Optimization reached a penalty value of 55 after the first 100 generations (Fig. 10).

The workspace of the mechanism is embedded in a six-dimensional space that cannot be graphically represented. Because of that in the following figures (Fig. 11) is shown workspace for constant orientation, in this case when all Euler angles are zero.

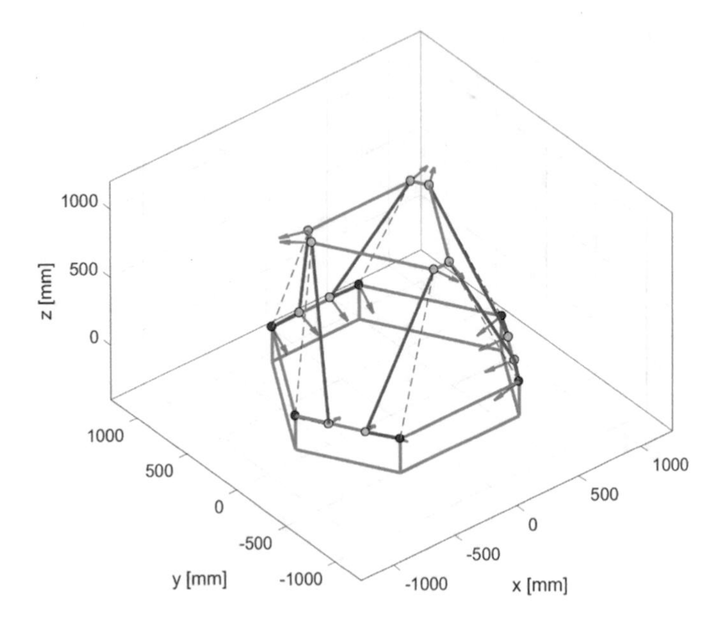

Fig. 10. The geometry of the final solution.

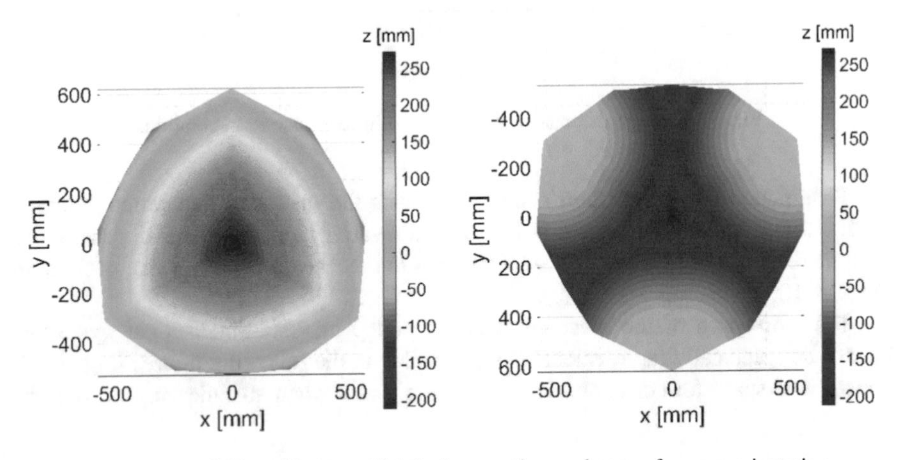

Fig. 11. Top (left) and bottom (right) view on the workspace for zero orientation.

5 Conclusions

In this paper, an algorithm for obtaining optimal geometry for the mechanism of the flight simulator based on the Stewart platform with rotary actuators is shown. There are algorithms that can be used in cases when the simulator cannot reach a certain position, due to saturation of some degrees of freedom, to provide the feeling as if the real motion has been achieved (washout algorithms). Besides that, good workspace and dynamic response are still a critical part of the design. Eleven geometric parameters are proposed, they can fully define the geometry of simulator with all constraints. Optimal values of geometric parameters were obtained using the genetic algorithm with the appropriate objective function that can be defined to suit many different design criteria and requirements. The considered optimization process is time-efficient and can take advantage of parallel computing.

The optimization approach shown in this paper is not specific to this mechanism and can be easily adapted to other types of parallel mechanisms mentioned earlier.

References

1. Čović, V., Lazarević, M.: Mehanika robota, 1st edn. Mašinski fakultet Univerziteta u Beogradu, Beograd (2008). (in Serbian)
2. Stewart, D.: A platform with six degrees of freedom. Proc. Inst. Mech. Eng. **180**(1), 371–386 (1965)
3. Fichter, E.: A Stewart platform-based manipulator: general theory and practical construction. Int. J. Robot. Res. **5**(2), 157–182 (1986)
4. Gao, X., Lei, D., Liao, Q., Zhang, G.: Generalized Stewart-Gough platforms and their direct kinematics. IEEE Trans. Rob. **21**(2), 141–151 (2005)
5. Lee, J., Geng, Z.: A dynamic model of a flexible Stewart platform. Comput. Struct. **48**(3), 367–374 (1993)
6. Dasgupta, B., Mruthyunjaya, T.: Closed-form dynamic equations of the general Stewart platform through the Newton-Euler approach. Mech. Mach. Theory **33**(7), 993–1012 (1998)
7. Jiang, Q., Gosselin, C.: The maximal singularity-free workspace of the Gough-Stewart platform for a given orientation. J. Mech. Des. **130**(1), 112304-112304-8 (2008)
8. Jiang, Q.: Maximal singularity-free total orientation workspace of the Gough-Stewart platform. J. Mech. Robot.-Trans. ASME **1**(3), 034501-034501-4 (2008)
9. Ji, Z.: Analysis of design parameters in platform manipulators. J. Mech. Des. **118**(4), 526–531 (1996)
10. Ilić, Z., Rašuo, B., Jovanović, M., Jovičić, S., Tomić, Lj., Janković, M., Petrašinović, D.: The efficiency of passive vibration damping on the pilot seat of piston propeller aircraft. Measurement **95**, 21–32 (2017)
11. Szufnarowski, F.: Stewart platform with fixed rotary actuators: a low cost design study. In: Advances in Medical Robotics, chap. 4, 1st edn. Uniwersytet Rzeszowski, Rzeszow (2013)
12. Petrašinović, D., Rašuo, B., Petrašinović, N.: Extended finite element method (XFEM) applied to aircraft duralumin spar fatigue life estimation. Tech. Gaz. **19**(3), 557–562 (2012)
13. Grbović, A., Kastratović, G., Sedmak, A., Balać, I., Popović, M.: Fatigue crack paths in light aircraft wing spars. Int. J. Fatigue **123**, 96–104 (2019)
14. Buljak, V.: Inverse Analyses with Model Reduction, 1st edn. Springer, Heidelberg (2012)
15. Su, Y., Duan, B., Zheng, C.: Genetic design of kinematically optimal fine tuning Stewart platform for large spherical radio telescope. Mechatronics **11**, 821–835 (2001)
16. Joumah, A., Albitar, C.: Design optimization of 6-RUS parallel manipulator using hybrid algorithm. Int. J. Inf. Technol. Comput. Sci. **2**, 83–95 (2018)
17. Hua, C., Weishan, C., Junkao, L.: Optimal design of stewart platform safety mechanism. Chin. J. Aeronaut. **20**, 370–377 (2007)
18. Stoughton, R., Arai, T.: A modified stewart platform manipulator with improved dexterity. IEEE Trans. Robot. Autom. **9**(2), 166–173 (1993)

19. Xie, Z., Li, G., Liu, G., Zhao, J.: Optimal design of a Stewart platform using the global transmission index under determinate constraint of workspace. Adv. Mech. Eng. **9**(10), 1–14 (2017)
20. Fassi, I., Legnani, G., Tosi, D.: Geometrical conditions for the design of partial or full isotropic hexapods. J. Robotic Syst. **22**(10), 507–518 (2005)

A Method for Collision Avoidance in 4π External Beam Radiation Therapy

Ivan M. Buzurovic[1(✉)], Slavisa Salinic[2], Dragutin Lj. Debeljkovic[3], and Robert A. Cormack[1]

[1] Harvard Medical School, Harvard University, Boston, MA 02115, USA
ibuzurovic@bwh.harvard.edu
[2] Faculty of Mechanical and Civil Engineering, University of Kragujevac, 36000 Kraljevo, Serbia
[3] Faculty of Civil Aviation, Megatrend University, 11000 Belgrade, Serbia

Abstract. In this study, a method for collision avoidance (CA) in external beam radiation therapy (EBRT) is proposed. The method encompasses the analysis of all positions of the moving components of the beam delivery system, such as the treatment table and gantry, including patient specific information obtained from the computed tomography (CT) images. This method eliminates the need for time-consuming dry runs prior to the actual treatments. This method includes a rigorous computer simulation and CA check prior to each treatment. With this treatment simulation, it is possible to quantify and graphically represent all positions and corresponding trajectories of all points of the moving parts during treatment delivery. The development of the workflow includes several steps: (a) derivation of combined dynamic equation of motion of the EBRT delivery systems, (b) developing the simulation model capable of drawing the motion trajectories of the specific points, (c) developing the interface between the model and the treatment plan parameters, such as couch and gantry parameters for each field. The patient CT images were registered to the treatment couch, so the patient dimensions were included into the simulation. The treatment field parameters were structured in an XML file that was used as an input into the dynamic equations. The trajectories of the moving components were plotted on the same graph using the dynamic equations. If the trajectories intersect, it was the signal that collision exists. This CA method is effective in the simulation of the treatment delivery.

Keywords: Dynamic equations · Radiation therapy · Collision avoidance

1 Introduction

Currently, the medical physics community experiences a rapid growth and development of the technology used for various types of radiation treatments. External beam radiation therapy (EBRT) is a part of that process, [1, 2]. With undivided opinion, the development and implementation of advanced treatment techniques such as intensity modulated radiation therapy (IMRT) [3], or volumetric modulated arc therapy (VMAT) [4], are already a standard of treatment for many malignancies. These techniques initially included dynamic motion of the multi-leaf collimator (MLC) leaves during the

N. Mitrovic et al. (Eds.): CNNTech 2019, LNNS 90, pp. 359–374, 2020.
https://doi.org/10.1007/978-3-030-30853-7_21

treatment delivery, as in IMRT. Further development included the real-time motion of the collimator and gantry of the linear accelerator without treatment interruptions, as in the VMAT technology. Furthermore, the various motion tracking and compensation techniques require dynamical motion of the all machine components involved in the radiation treatments such as dynamically moving MLC, couch motion compensation, and real-time imaging of the patients during treatments [5–19], etc. Dynamic model approach is becoming more utilized in other radiation therapy modalities [20].

Therefore, there are strong indications that it is a matter of time when 4π treatments delivery technique will be introduced. This technology currently exists in pieces and the proper established links will result in dynamic dose delivery that involves the real-time motion of all moving components such as machine gantry and collimators together with treatment couch movement without treatment interaction, i.e. the beam would need to be turned off during the delivery of single fraction. This is partially achieved in the IMRT and VMAT; however, the treatment couch maintains its position requiring the radiation therapist to make the adequate set-up for each treatment beam. In the current clinical workflows, couch rotation between the dose deliveries for each radiation beam is done manually while the beam is turned off. For that purpose, a radiation therapist needs to go into the treatment room and to adjust the couch position manually between the deliveries of the beams. Therefore, that process required additional time resulting in an increase of the total treatment time in IMRT techniques or in arc-type plans in stereotactic treatments. For arc VMT treatments, the treatment couch is steady. If the treatment couch can be moved in real-time during the radiation delivery, more favorable dosimetry is likely to be expected. This concept is still work in progress for both vendors and researchers.

Furthermore, the quality assurance procedure for EBRT requires that the collision should be checked prior treatment, requiring off-line delivery of the treatment plan by a therapist or physicist so that a possible collision should be detected between the equipment and patients. There are some known incident reports describing the event related to this issue. Based on the current clinical practice and based on the future directions and tendencies in EBRT, it is required to develop a procedure for collision avoidance in EBRT that can be implemented in a straightforward manner without requirements for additional manpower or extensive off-line repetitions of the treatments.

In this article, we proposed a method for collision avoidance in 4π EBRT. The proposed methodology can check the position of all moving components of the beam delivery system, including the treatment table and gantry with specific information related to the patient dimensions on the treatment couch without time-consuming repetition of the treatments. This was achieved by a rigorous computer simulation of all moving components including specifics, such as positions of the couch and gantry during delivery, position of the patients, and imaging equipment. The proposed methodology can be applied to other clinical procedures, such as tracking and motion compensation, quality assurance (QA) program, and other treatments including the QA of the isocenter and other components of the linear accelerator. In addition, the proposed method can be used for the future development of 4π treatment delivery in which the couch motion can be performed in real-time without treatment interactions.

2 Materials and Method

The problem of collision avoidance, in general, can be formulated in the following way: is it possible to know all positions and corresponding trajectories of all points of the moving parts of the linear accelerator and treatment couch during delivery of the radiation dose in the absolute (non-movable) coordinate system? If the answer to this question is yes, then the problem of collision could be solved by checking if the motion trajectories would intersect or have identical coordinates in the absolute coordinate system.

The development of this procedure required accurate derivation of the equation of motion for the system. We presented the equation of motion for one of the most used designs of the equipment, so the procedure can be implemented for other equipment in a similar manner. The development of the workflow for implementation of the rapid collision avoidance includes several steps: (a) derivation of the accurate equation of motion for the EBRT delivery systems (linear accelerators and tables), (b) developing the simulation model capable of drawing the motion trajectories of the specific points, (c) developing the interface between the model and the treatment plan parameters, such as couch and gantry parameters for each field in the treatment plans.

2.1 Equations of Motion for a Patient Positioning System

A mechanism shown in Fig. 1(a) contains of the turntable 1 of mass m_1, the beam 2 of mass m_2 and length L connected to the turntable at the point O_1 by a revolute joint, the motor 3 with ball screws, the beam 4 of mass m_4 and the length L connected to the beam 2 by a revolute joint, and the treatment couch with patient 5. The coordinate frame $O\xi_1\eta_1\zeta_1$ is fixed to the turntable in the manner shown in Fig. 1(a) while the inertial coordinate frame $Oxyz$ is positioned in the manner shown in Fig. 1(b). In addition, the local coordinate frames $C_2\xi_2\eta_2\zeta_2$, $C_4\xi_4\eta_4\zeta_4$, and $C_5\xi_5\eta_5\zeta_5$, are fixed to bodies 2, 4, and 5, respectively.

The couch with the patient performs the translation relative to the couch support fixed to the end B of beam 4. At point B, the coordinate frame $B\xi\eta\zeta$ is attached to the couch support (Fig. 1(a)). The transformation matrix $\mathbf{R} \in R^{3\times3}$ from the coordinate frame $O\xi_1\eta_1\zeta_1$ to the inertial frame $Oxyz$ has the form (Fig. 1(b)):

$$\mathbf{R} = \begin{bmatrix} \cos\theta & -\sin\theta & 0 \\ \sin\theta & \cos\theta & 0 \\ 0 & 0 & 1 \end{bmatrix}. \tag{1}$$

The motor governs the ball screws that are responsible for the vertical motion of the couch with the patient in a manner that the end of beam 4 moves along the vertical line, passing through point O_1 (length b is a constant quantity).

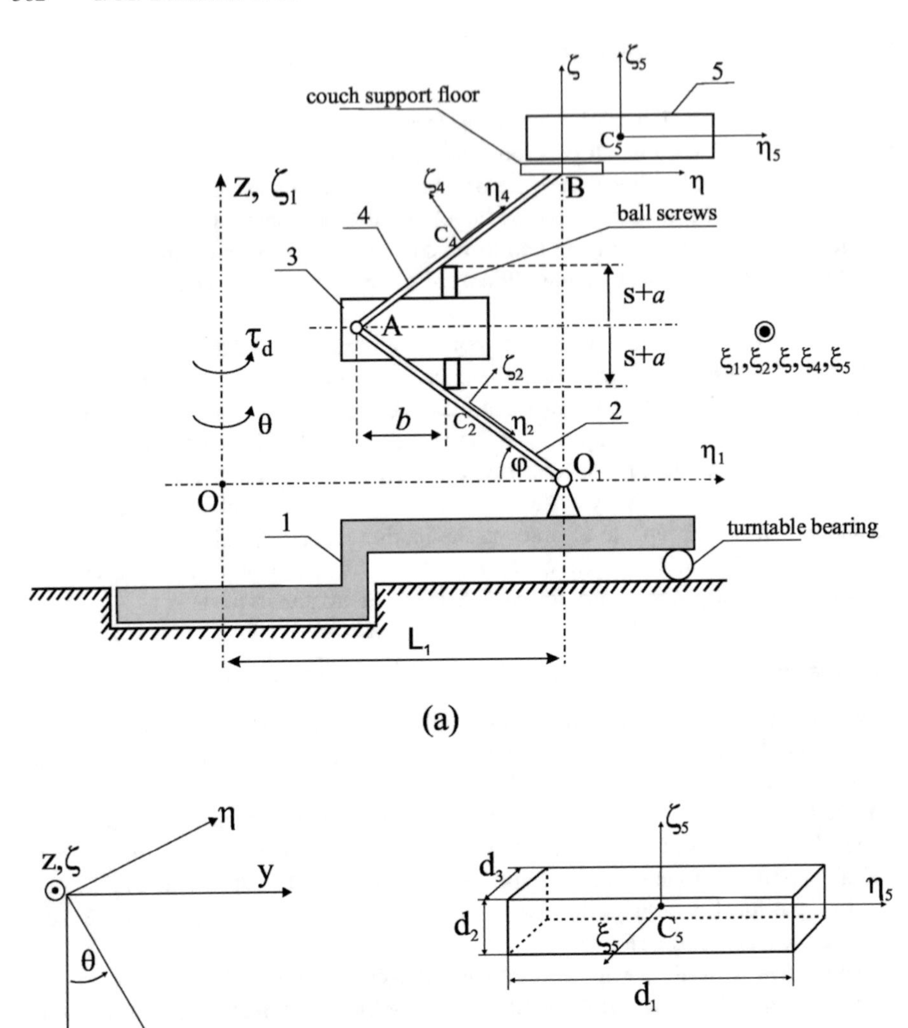

Fig. 1. (a) The lifting mechanism that is entirely above the ground level; (b) position of the frame $O\xi_1\eta_1\zeta_1$ relative to the frame $Oxyz$; (c) mechanical model of the treatment couch with patient coordinate system.

Displacement s generated by the ball screws (Fig. 1(a)) with a single thread is determined as:

$$s = nh \tag{2}$$

where n is the number of revolutions for each ball screw, and h is the lead of the threads. According to this, the rotation angle of the ball screws is given as follows:

$$\psi = 2\pi n = \frac{2\pi s}{h}. \tag{3}$$

Note that geometric parameter a introduced in Fig. 1(a) represents a constant quantity. In the sequel, all the vectors and matrices, whose elements are given in $C_i\xi_i\eta_i\zeta_i$ local frame, will be denoted by the corresponding superscript (i). The position of the system with respect to the inertial frame is determined by the generalized coordinates $\mathbf{q} = [\theta,\ \varphi,\ \xi,\ \eta]^T$ where θ represents the angle of rotation of the turntable about the z axis, φ is the relative rotation of beam 2 with respect to the turntable, and ξ and η are the relative Cartesian coordinates of the center of mass, C_5, of the couch with the patient in the frame $B\xi\eta\zeta$. For $\xi = \eta = 0$ the point C_5 is on the direction O_1B. The kinetic energy of the turntable is:

$$T_1 = \frac{1}{2} J_1 \dot{\theta}^2 \tag{4}$$

where J_1 is the mass moment of inertia of the turntable about the z axis. The kinetic energy of the beam 2 is determined as:

$$T_2 = \frac{1}{2} m_2 \dot{\mathbf{r}}_{C_2}^T \dot{\mathbf{r}}_{C_2} + \frac{1}{2} \boldsymbol{\omega}_2^{(2)T} \mathbf{J}_{C_2}^{(2)} \boldsymbol{\omega}_2^{(2)} \tag{5}$$

where $\mathbf{r}_{C_2}$ is the position vector of the center of mass C_2 with respect to the inertial frame given by:

$$\mathbf{r}_{C_2} = [-L_1 \sin\theta,\ L_1 \cos\theta,\ 0]^T + \mathbf{R}\left[0,\ \ -\frac{L}{2}\cos\varphi,\ \ \frac{L}{2}\sin\varphi\right]^T \tag{6}$$

whose time derivative is denoted by $\dot{\mathbf{r}}_{C_2}$, $\omega_2^{(2)} = \left[-\dot{\varphi},\ \ -\dot{\theta}\sin\varphi,\ \ \dot{\theta}\cos\varphi\right]^T$ is the angular velocity of beam 2, and $\mathbf{J}_{C_2}^{(2)} = diag(m_2L^2/12,\ 0,\ m_2L^2/12)$ is the inertial tensor referred to the center of mass of beam 2. It is assumed that the mass of the motor 3 is concentrated at the point A. In accordance to this assumption, the kinetic energy of the motor is:

$$T_3 = \frac{1}{2} m_3 \dot{\mathbf{r}}_A^T \dot{\mathbf{r}}_A \tag{7}$$

where $\mathbf{r}_A$ is the position vector of point A with respect to the inertial frame given by:

$$\mathbf{r}_A = [-L_1 \sin\theta,\ L_1 \cos\theta,\ 0]^T + \mathbf{R}[0,\ \ -L\cos\varphi,\ \ L\sin\varphi]^T \tag{8}$$

whose time derivative is denoted by $\dot{\mathbf{r}}_A$. The beam 4 has the kinetic energy determined by the following expression:

$$T_4 = \frac{1}{2} m_4 \dot{\mathbf{r}}_{C_4}^T \dot{\mathbf{r}}_{C_4} + \frac{1}{2} \boldsymbol{\omega}_4^{(4)T} \mathbf{J}_{C_4}^{(4)} \boldsymbol{\omega}_4^{(4)} \tag{9}$$

where $\mathbf{r}_{C_4}$ is the position vector of the center of mass C_4 with respect to the inertial frame given by:

$$\mathbf{r}_{C_4} = [-L_1 \sin\theta,\ L_1 \cos\theta,\ 0]^T + \mathbf{R}\left[0,\ -\frac{L}{2}\cos\varphi,\ \frac{3L}{2}\sin\varphi\right]^T \tag{10}$$

whose time derivative is denoted by $\dot{\mathbf{r}}_{C_4}$, $\omega_4^{(4)} = \left[\dot{\varphi},\ \dot{\theta}\sin\varphi,\ \dot{\theta}\cos\varphi\right]^T$ is the angular velocity of beam 4, and $\mathbf{J}_{C_4}^{(4)} = diag(m_4 L^2/12,\ 0,\ m_4 L^2/12)$ is the inertial tensor referred to the center of mass of beam 4. Finally, the kinetic energy of the treatment couch with the patient is given by:

$$T_5 = \frac{1}{2} m_5 \dot{\mathbf{r}}_{C_5}^T \dot{\mathbf{r}}_{C_5} + \frac{1}{2} \boldsymbol{\omega}_5^{(5)T} \mathbf{J}_{C_5}^{(5)} \boldsymbol{\omega}_5^{(5)} \tag{11}$$

where $\mathbf{r}_{C_5}$ is the position vector of the center of mass C_5 with respect to the inertial frame given by:

$$\mathbf{r}_{C_5} = \mathbf{R}[\xi,\ \ \eta + L_1,\ \ d_2/2 + 2L\sin\varphi]^T \tag{12}$$

whose time derivative is denoted by $\dot{\mathbf{r}}_{C_5}$, $\omega_5^{(5)} = \left[0,\ 0,\ \dot{\theta}\right]^T$ is the angular velocity of the treatment couch with the patient, and $\mathbf{J}_{C_5}^{(5)} = diag(m_5(d_1^2 + d_2^2)/12,\ \ m_5(d_2^2 + d_3^2)/12,$ $m_4(d_1^2 + d_3^2)/12)$ is the inertial tensor referred to the center of mass of the couch with patient. Here, the treatment couch with patient are modeled as a uniform rectangular parallelepiped of edges d_1, d_2, and d_3 and mass m_5 (Fig. 1(c)). For $\xi = \eta = 0$ the point C_5 is on the axis ζ. Consequently, the kinetic energy of the system is:

$$T = \sum_{i=1}^{5} T_i = \frac{1}{2}\dot{\mathbf{q}}^T \mathbf{M}\mathbf{q} \tag{13}$$

where $\mathbf{M}(\mathbf{q}) \in R^{4\times4}$ is the mass matrix whose elements are:

$$M_{11} = \frac{1}{12}\left(\begin{array}{l} 12J_1 + 2m_2L^2 + 12m_2L_1^2 + 6m_3L^2 \\ + 12m_3L_1^2 + 2m_4L^2 + 12m_4L_1^2 + m_5d_1^2 + m_5d_3^2 + 12m_5L_1^2 \\ + 24L_1m_5\eta + 12m_5\eta^2 + 12m_5\xi^2 - 12LL_1(m_2 + 2m_3 + m_4)\cos\varphi \\ + 2L^2(m_2 + 3m_3 + m_4)\cos 2\varphi \end{array}\right), \tag{14}$$

$$M_{22} = \frac{1}{12}\left(\begin{array}{l} 4m_2L^2 + 12m_3L^2 + 16m_4L^2 \\ + 24m_5L^2 + 12(m_4 + 2m_5)L^2\cos 2\varphi \end{array}\right), \tag{15}$$

$$M_{33} = M_{44} = m_5 \tag{16}$$

$$M_{13} = M_{31} = -m_5\eta - m_5L_1 \quad (17)$$

$$M_{14} = M_{41} = m_5\xi \quad (18)$$

$$\begin{array}{c} M_{12} = M_{21} = 0, \quad M_{23} = M_{32} = 0, \\ M_{24} = M_{42} = 0, \quad M_{34} = M_{43} = 0 \end{array} \quad (19)$$

The potential energy of the mechanism considered is

$$\Pi = \left(\frac{m_2}{2} + m_3 + \frac{3m_4}{2} + 2m_5\right) gL\sin\varphi. \quad (20)$$

The total virtual work of the control forces and torques are given as:

$$\delta A = \tau_d\delta\theta + F_\xi\delta\xi + F_\eta\delta\eta + 2\tau_M\frac{2\pi}{h}\delta s \quad (21)$$

where τ_d is the driving torque for the motion of the turntable, F_ξ and F_η are the internal control forces responsible for translational motion of the couch with the patient along the directions parallel to axes ξ and η, respectively, τ_M is the internal driving torque of the motor for revolutions of ball screws. From Eq. (21), the generalized control forces vector is obtained as follows:

$$\mathbf{Q}^{(c)} = \left[\tau_d,\ \frac{4\pi\tau_M b}{h\cos^2\varphi}\delta\varphi,\ F_\xi,\ F_\eta\right]^T \quad (22)$$

where the relation between the displacement s and the angle φ:

$$\tan\varphi = \frac{s+a}{b} \Rightarrow \delta s = \frac{b}{\cos^2\varphi}\delta\varphi \quad (23)$$

is taken into account. Note that a and b are constant geometric parameters of the mechanism (Fig. 1(a)). Using Lagrange's equations [21, 22]:

$$\frac{d}{dt}\left(\frac{\partial T}{\partial\dot{\mathbf{q}}}\right)^T - \left(\frac{\partial T}{\partial\mathbf{q}}\right)^T = \mathbf{Q}^{(g)} + \mathbf{Q}^{(c)}, \quad (24)$$

the differential equations of motion of the mechanism are obtained, which can be written in the following matrix form:

$$\mathbf{M}(\mathbf{q})\ddot{\mathbf{q}} + \boldsymbol{\Gamma}(\mathbf{q},\dot{\mathbf{q}}) = \mathbf{Q}^{(g)} + \mathbf{Q}^{(c)} \quad (25)$$

where

$$\begin{aligned}\mathbf{Q}^{(g)} &= -\left(\partial \Pi / \partial \mathbf{q}\right)^T \\ &= [0, \; -(m_2/2 + m_3 + 3m_4/2 + 2m_5) g L \cos\varphi, \; 0, \; 0]^T,\end{aligned} \tag{26}$$

and $\boldsymbol{\Gamma}(\mathbf{q}, \dot{\mathbf{q}}) = [\Gamma_1, \ldots, \Gamma_4]^T \in R^{4\times 1}$ is a column matrix whose elements are given by:

$$\Gamma_i = \sum_{j=1}^{4} \sum_{k=1}^{4} [jk, \, i] \dot{q}_j \dot{q}_k, \quad i = 1, \ldots, 4, \tag{27}$$

where:

$$[jk, \, i] = \frac{1}{2} \left(\frac{\partial M_{ij}}{\partial q_k} + \frac{\partial M_{ik}}{\partial q_j} - \frac{\partial M_{jk}}{\partial q_i} \right), \qquad i, j, k = 1, \ldots, 4 \tag{28}$$

are Christoffel's symbols of the first kind [21], or in its developed form:

$$\begin{aligned}\Gamma_1 &= 2m_5 \eta \dot{\theta} \dot{\eta} + 2m_5 L_1 \dot{\theta} \dot{\eta} + 2m_5 \xi \dot{\theta} \dot{\xi} \\ &+ \frac{1}{12} \dot{\theta} \dot{\varphi} \begin{bmatrix} 12 L L_1 (m_2 + 2m_3 + m_4) \sin\varphi \\ -4L^2 (m_2 + 3m_3 + m_4) \sin 2\varphi \end{bmatrix},\end{aligned} \tag{29}$$

$$\begin{aligned}\Gamma_2 &= -(2m_5 + m_4) L^2 \dot{\varphi}^2 \sin 2\varphi \\ &+ \frac{1}{24} \dot{\theta}^2 \begin{bmatrix} -12 L L_1 (m_2 + 2m_3 + m_4) \sin\varphi \\ +4L^2 (m_2 + 3m_3 + m_4) \sin 2\varphi \end{bmatrix},\end{aligned} \tag{30}$$

$$\Gamma_3 = -2m_5 \dot{\theta} \dot{\eta} - m_5 \xi \dot{\theta}^2, \tag{31}$$

$$\Gamma_4 = 2m_5 \dot{\theta} \dot{\xi} - m_5 \eta \dot{\theta}^2. \tag{32}$$

In the following part, the dynamic equation of motion of the linear accelerator in conjunction with the treatment table are presented. The combination of the motion trajectories from both the treatment table and the gantry of linear accelerator will reveal the possible collisions. Patient specifics, such as exact position of the treatment table and size, were extracted from the previously obtained CT images during the simulation phase.

2.2 Equations of Motion for Collision Detection and Avoidance

A simplified model of the gantry and treatment couch is shown in Fig. 2. The gantry rotates around fixed axis y_1 situated in the coordinate system, and it is parallel to y. In Fig. 2, the notations τ_G and θ_G represent the gantry driving torque and the gantry angle of rotation about axis y_1, respectively. Also, with P_G and P_{IC} are denoted, respectively, the characteristic point on the gantry treatment head and the isocenter, where

$\overline{P_G P_{IC}} = const. = h_1$. The point P_{IC} belongs to the axis z. The inertial coordinate frame $Oxyz$ shown in Fig. 2 is defined in the same manner as in Sects. 2.1. Note that the point P_G is moving along the circular trajectory with radius h_1 which belongs to the coordinate plane Oxz.

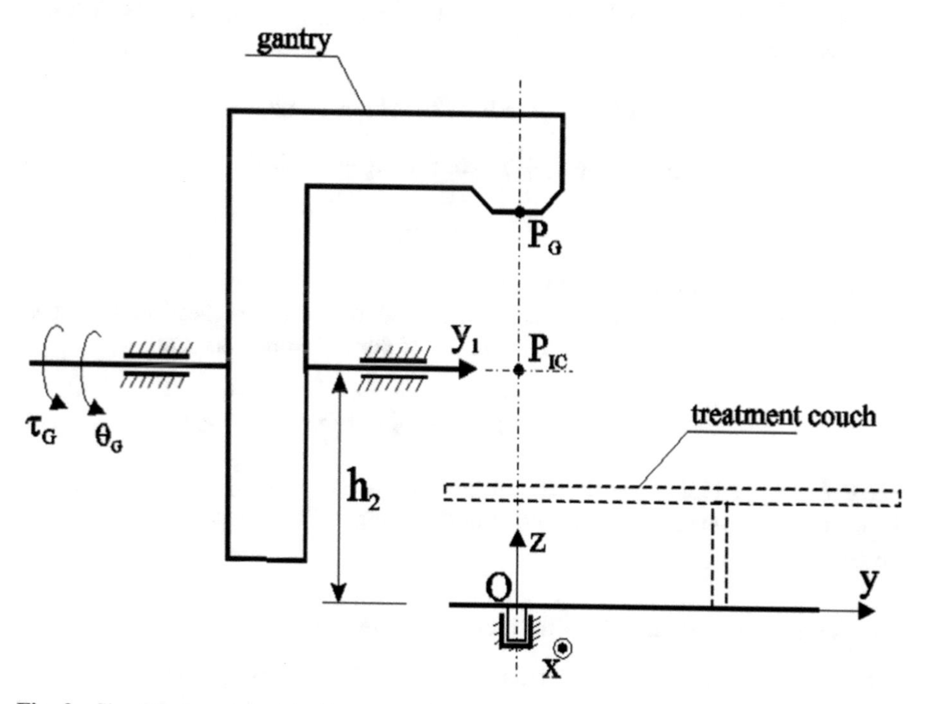

Fig. 2. Simplified model of the linear accelerator with a treatment head (gantry) and treatment couch

The coordinates of point P_G with respect to the inertial frame are:

$$x_{P_G} = h_1 \sin\theta_G, \quad y_{P_G} = 0, \quad z_{P_G} = h_2 + h_1 \cos\theta_G \tag{33}$$

The differential equation of the gantry is:

$$J_G \ddot{\theta}_G = \tau_G \tag{34}$$

where J_G is the mass moment of inertia of the gantry about the axis y_1. The position vector of arbitrary patient point P is given by:

$$\mathbf{r}_P = \mathbf{r}_{C_5} + \mathbf{r}^* \tag{35}$$

where vector $\mathbf{r}_{C_5}$ is determined in Eq. (12) and the vector $\mathbf{r}^* = \overrightarrow{C_5P}$ is defined as:

$$\mathbf{r}^* = \mathbf{R}[\xi^* + \xi,\ \eta^* + \eta,\ \zeta^* + d_2/2]^T \tag{36}$$

where $\xi^* = \text{const.}$, $\eta^* = \text{const.}$, and $\zeta^* = \text{const.}$, are the coordinates of point P with respect to the frame $C_5\xi_5\eta_5\zeta_5$. Based on Eqs. (33), (34), and (36), the coordinates of the point P are:

$$x_P = (2\xi + \xi^*)\cos\theta - (2\eta + L_1 + \eta^*)\sin\theta\ , \tag{37}$$

$$y_P = (2\eta + L_1 + \eta^*)\cos\theta + (2\xi + \xi^*)\sin\theta\ , \tag{38}$$

$$z_P = 2L\sin\varphi + d_2 + \zeta^* \tag{39}$$

for the mechanism described in Sect. 2.1.

A possible collision between the patient and the treatment head of the linear accelerator can be determined by the analysis of the following condition:

$$\overline{PP_G} = \sqrt{(x_P - x_{P_G})^2 + (y_P - y_{P_G})^2 + (z_P - z_{P_G})^2} \geq \ell \tag{40}$$

where ℓ is an arbitrary prescribed safety constant (separation). Correlation of the dynamics of the gantry and the dynamics of the lifting mechanism are achieved through the quantity $\overline{PP_G}$.

2.3 Treatment Plan and Patient-Specific Parameters

A treatment plan consists of the treatment fields. The following relevant geometry parameters has been defined for each treatment field during the dosimetry treatment planning process: gantry position (angle), and couch position (height, rotation, and tabletop position). Collimator position is defined as well, however, it is not relevant for the CA checks. All these parameters were obtained from the final treatment plan. The CA is checked for the motion of the gantry and couch during the transition from one to the other treatment field and for the motion of the gantry during VMAT treatment delivery. For this purpose, simulation is performed using Eqs. (25) and (37–39). Graphical representation is obtained using Eq. (40).

Patient-specific parameters such as position on the treatment table and separations are obtained from the patient simulation (CT images), as in Fig. 3.

These parameters were considered during the CA checks and trajectories of motion analysis.

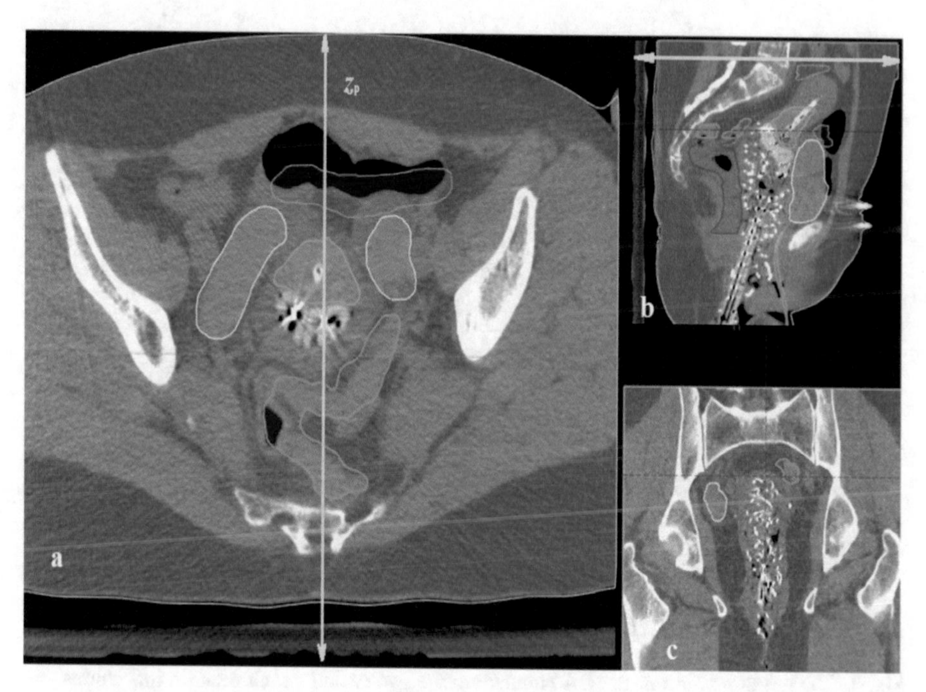

Fig. 3. Patient-specific parameters include position of the patient on the treatment table; they are used to calculate coordinates of point $P(x_p, y_p, z_p)$ in absolute coordinate system. (a) axial, (b) sagittal, and (c) coronal view of a representative CT image set.

3 Results

The CT images of the patients were registered to the treatment couch so the patient dimensions were included into the simulation. The treatment field parameters (gantry and couch positions) were structured in the xml file that was used as an input into the dynamic equations. The trajectories of the moving components were plotted on the same graph using the dynamic equations of motion. If the trajectories intersect, it is a signal that a collision exists; otherwise, the treatment field did not cause a collision.

For computer simulation, we have considered mass of the tabletop with payload m = 200 kg, mass of the moving rods $m_1 = m_2 = 50$ kg, mass of the motor with the holder M = 50 kg, length of the rod $L = 80$ cm, a = b = 3 cm, and the lead of the thread $h = 0.01$ m. The sampling frequency is $v = 5$ Hz, and total time of simulation vary depending on the treatment plans.

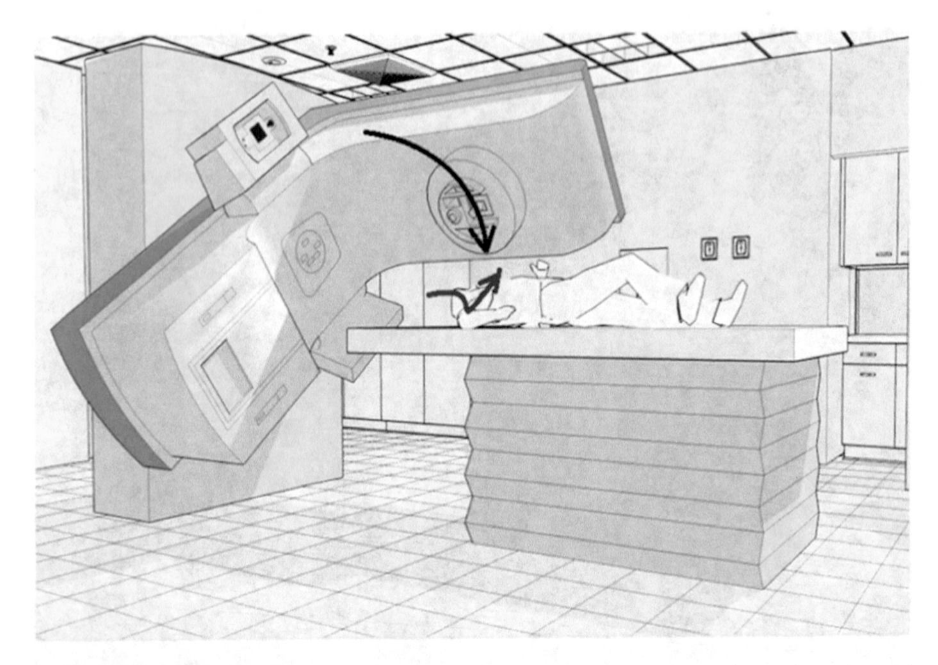

Fig. 4. 3D representation of the CA simulation; the gantry and couch rotation trajectories with the patient in the supine position. Patient specific positions and separations are included into the simulation using generalized CT coordinates.

3D representation of simulation is presented in Fig. 4. This type of simulation can be suitable for clinical implementation of the proposed methodology. The end user does not need to run simulation using the outlined equations. All treatment plan parameters can be imported in the CA check module as it was done in this work - through an xml file. The computer simulation results for a representative case and robotic couch as in Fig. 1 have been presented in the Figs. 5 and 6 below.

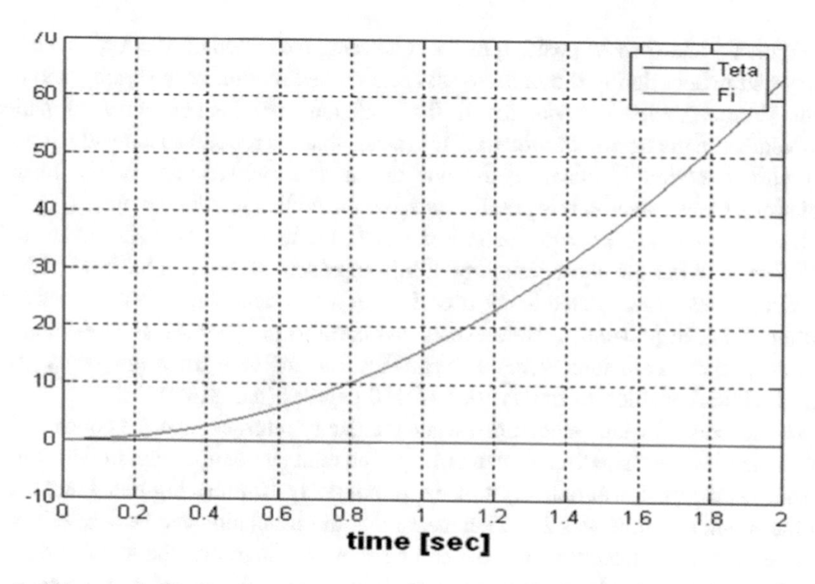

Fig. 5. The equations of motion for the table and the gantry (represented by the generalized coordinates) are simulated using the treatment field parameters as an input. Collision was detected in this case.

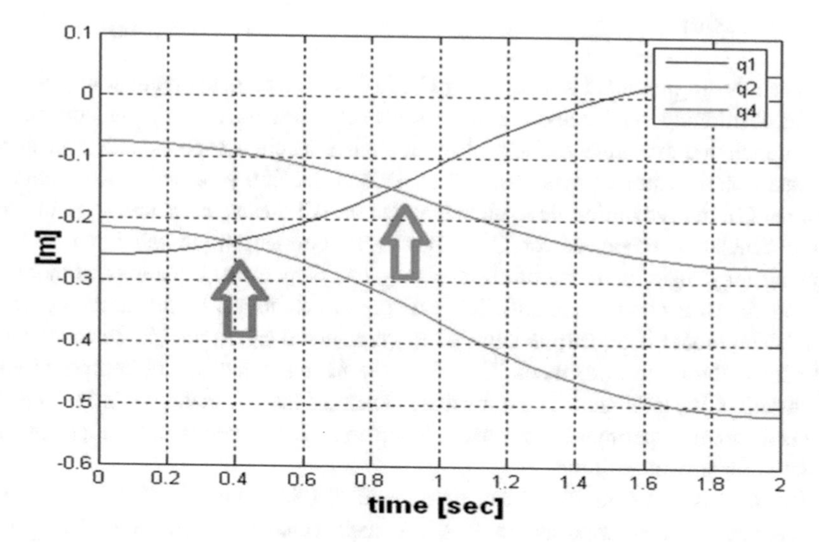

Fig. 6. The equations of motion for the table and the gantry (represented by the generalized coordinates) are simulated using the treatment field parameters as an input. Collision was detected in this case.

Figures 5 and 6 show position of the representative generalized coordinates of the gantry and patient during the motion tasks. Combined motion of the gantry and couch results in possibility to visualize if the collision will happen prior to patient's involvement in the treatment process. This means that the robotic couch and gantry will start virtual motion identical as the one during the real delivery of the treatment (radiation). Consequently, it is possible to see if the collision will take place (Fig. 5), or not (Fig. 6). In a case when collision is detected, the treatment plan should be revised and adjusted. After the commissioning of this approach, all test can be done offline, so the valuable resources related to the use of linear accelerator should not be blocked. In addition, this can prevent possible events and harm to the patients. Figure 5 indicates that two possible collisions were detected. That was the case when the gantry rotated from the initial position to the position of 180 degrees (red arrow)

The second collision was noticed when the gantry returned, and the couch position was changed at the same time (green arrow). The total simulation time in this case was the same as the treatment delivery time (approximately 10 min). Figures 5 and 6 show that the simulation time was 2 s. This means that the simulation can be accelerated and the time to detect if the collision exist can be shorten. Therefore, the total time on this task can be saved, unlike in traditional approach where the complete treatment plan must be delivered prior to the real treatment. In addition, the simulation allows to identify a treatment field and all positions in which collision would occur.

4 Discussion

In this study we presented a method for CA in EBRT. The mathematical model of the patient positioning system and the linear accelerator was developed. The equations of motion included dynamics of the system, and other specifics such as patient positions, separation and treatment plan parameters. Patient specific parameters were obtained from the CT images during the patient simulation. All these parameters were included in the simulation model of for CA. They were summarized in the input xml file, allowing for simple inclusion of all relevant parameters to the equations of motion.

The rigorous computer simulation was performed to verify the accuracy of the model. The model itself can be adjusted to guarantee a safety margin between gantry and patient during the treatment. Therefore, the patient safety is not compromised in any instant of the treatment. The immediate benefit of the CA method is that it replaces time-consuming experiments prior to the treatments. Consequently, all equipment stays available for the longer time.

The proposed methodology can be applied to other clinical procedures such as tracking and motion compensation, [4–9]. In such scenario, the position of the patient can be calculated based on the given equations, and the treatment can be interrupted if the predefined motion limits were reached. The CA method us suitable for the improvement of the general QA program in radiation treatments [23], including the QA of the isocenter of the linear accelerators and other moving components such as imaging system [24]. In this approach, equation of motion can be utilized to confirm position of the mechanical isocenter of the linear accelerator more rigorously.

Such results can be compared against the radiation isocenter, so the possible deviation can be tested more frequently without additional resources.

In addition, the proposed method can be used for the future development of 4π treatment delivery in which the couch motion can be performed in real-time without treatment interruptions. This method would decrease treatment time and potentially improve the patients' dosimetry. Implementation of the proposed technique can potentially improve real-time tracking of the tumor-volume to deliver highly conformal precise radiation dose at almost 100% duty cycle while minimizing irradiation to health tissues and sparing critical organs. This, in turn, will potentially improve the quality of patient treatment by lowering the toxicity level and increasing survival.

5 Conclusion

The proposed collision avoidance method was proved to be effective in the simulation of the treatment delivery. The proper implementation of this system can potentially improve the QA program and it can increase the efficacy and efficiency in the clinical setup. This method can be implemented as a standard QA procedure since it eliminates need for occupying the physical resources. Adequately developed 3D visualization can increase usability.

References

1. Dawson, L.A., Sharpe, M.B.: Image-guided radiotherapy: rationale, benefits, and limitations. Lancet Oncol. **7**(10), 848–858 (2006)
2. Lo, S.S., Fakiris, A.J., Chang, E.L., Mayr, N.A., Wang, J.Z., Papiez, L., Teh, B.S., McGarry, R.C., Cardenes, H.R., Timmerman, R.D.: Stereotactic body radiation therapy: a novel treatment modality. Nature reviews Clinical oncology **7**(1), 44 (2010)
3. Webb, S.: Intensity-Modulated Radiation Therapy. CRC Press, London (2015)
4. Palma, D., Vollans, E., James, K., Nakano, S., Moiseenko, V., Shaffer, R., McKenzie, M., Morris, J., Otto, K.: Volumetric modulated arc therapy for delivery of prostate radiotherapy: comparison with intensity-modulated radiotherapy and three-dimensional conformal radiotherapy. Int. J. Radiat. Oncol.* Biol.* Phys. **72**(4), 996–1001 (2008)
5. Barnes, E.A., Murray, B.R., Robinson, D.M., Underwood, L.J., Hanson, J., Roa, W.H.: Dosimetric evaluation of lung tumor immobilization using breath hold at deep inspiration. Int. J. Radiat.* Oncol.* Biol. Phys. **50**(4), 1091–1098 (2001)
6. Buzurovic, I., Huang, K., Yu, Y., Podder, T.K.: A robotic approach to 4D real-time tumor tracking for radiotherapy. Phys. Med. Biol. **56**(5), 1299 (2011)
7. Buzurovic, I., Yu, Y., Werner-Wasik, M., Biswas, T., Anne, P.R., Dicker, A.P., Podder, T. K.: Implementation and experimental results of 4D tumor tracking using robotic couch. Med. Phys. **39**(11), 6957–6967 (2012)
8. Buzurovic, I., Podder, T. K., Huang, K., Yu, Y.: Tumor motion prediction and tracking in adaptive radiotherapy. In: 2010 IEEE International Conference on BioInformatics and BioEngineering, pp. 273–278 (2010)
9. Chen, Q.S., Weinhous, M.S., Deibel, F.C., Ciezki, J.P., Macklis, R.M.: Fluoroscopic study of tumor motion due to breathing: facilitating precise radiation therapy for lung cancer patients. Med. Phys. **28**(9), 1850–1856 (2001)

10. Cho, B., Poulsen, P.R., Keall, P.J.: Real-time tumor tracking using sequential kV imaging combined with respiratory monitoring: a general framework applicable to commonly used IGRT systems. Phys. Med. Biol. **55**(12), 3299 (2010)
11. D'Souza, W.D., McAvoy, T.J.: An analysis of the treatment couch and control system dynamics for respiration-induced motion compensation. Med. Phys. **33**(12), 4701–4709 (2006)
12. Erridge, S.C., Seppenwoolde, Y., Muller, S.H., van Herk, M., De Jaeger, K., Belderbos, J.S., Boersma, L.J., Lebesque, J.V.: Portal imaging to assess set-up errors, tumor motion and tumor shrinkage during conformal radiotherapy of non-small cell lung cancer. Radiother. Oncol. **66**(1), 75–85 (2003)
13. Shirato, H., Suzuki, K., Sharp, G.C., Fujita, K., Onimaru, R., Fujino, M., Kato, N., Osaka, Y., Kinoshita, R., Taguchi, H., Onodera, S.: Speed and amplitude of lung tumor motion precisely detected in four-dimensional setup and in real-time tumor-tracking radiotherapy. Int. J. Radiat. Oncol.* Biol.* Phys. **64**(4), 1229–1236 (2006)
14. Huang, K., Buzurovic, I., Yu, Y., Podder, T.K.: A comparative study of a novel AE-nLMS filter and two traditional filters in predicting respiration induced motion of the tumor. In: 2010 IEEE International Conference on BioInformatics and BioEngineering, pp. 281–282 (2010)
15. Buzurovic, I., Huang, K., Podder, T.K., Yu, Y.: Comparison between acceleration-enhanced adaptive filters and neural network filters for respiratory motion prediction. In: 11th Symposium on Neural Network Applications in Electrical Engineering, pp. 181–184 (2012)
16. Kamino, Y., Takayama, K., Kokubo, M., Narita, Y., Hirai, E., Kawawda, N., Mizowaki, T., Nagata, Y., Nishidai, T., Hiraoka, M.: Development of a four-dimensional image-guided radiotherapy system with a gimbaled X-ray head. Int. J. Radiat. Oncol.* Biol.* Phys. **66**(1), 271–278 (2006)
17. Ozhasoglu, C., Murphy, M.J.: Issues in respiratory motion compensation during external-beam radiotherapy. Int. J. Radiat. Oncol.* Biol.* Phys. **52**(5), 1389–1399 (2002)
18. Podder, T.K., Buzurovic, I., Galvin, J.M., Yu, Y.: Dynamics-based decentralized control of robotic couch and multi-leaf collimators for tracking tumor motion. In: 2008 IEEE International Conference on Robotics and Automation, pp. 2496–2502 (2008)
19. Podder, T.K., Buzurovic, I., Hu, Y., Galvin, J.M., Yu, Y.: Partial transmission high-speed continuous tracking multi-leaf collimator for 4D adaptive radiation therapy. In: 2007 IEEE 7th International Symposium on BioInformatics and BioEngineering, pp. 1108–1112 (2007)
20. Buzurovic, I.M., Salinic, S., Orio, P.F., Nguyen, P.L., Cormack, R.A.: A novel approach to an automated needle insertion in brachytherapy procedures. Med. Biol. Eng. Compu. **56**(2), 273–287 (2017)
21. Greenwood, D.T.: Classical dynamics. Courier Corporation, New York (1997)
22. Salinic, S., Nikolić, A.: A new pseudo-rigid-body model approach for modeling the quasi-static response of planar flexure-hinge mechanisms. Mech. Mach. Theory **124**, 150–161 (2018)
23. Klein, E.E., Hanley, J., Bayouth, J., Yin, F.F., Simon, W., Dresser, S., Serago, C., Aguirre, F., Ma, L., Arjomandy, B., Liu, C.: Task Group 142 report: quality assurance of medical accelerators a. Med. Phys. **36**(9), 4197–4212 (2009)
24. Sun, B., Goddu, S.M., Yaddanapudi, S., Noel, C., Li, H., Cai, B., Kavanaugh, J., Mutic, S.: Daily QA of linear accelerators using only EPID and OBI. Med. Phys. **42**(10), 5584–5594 (2015)

Dynamic Simulation of Dual Mass Flywheel

Jozef Bucha[1(✉)], Jan Danko[1], Tomas Milesich[1], Radivoje Mitrovic[2], and Zarko Miskovic[2]

[1] Faculty of Mechanical Engineering, Institute of Transport Technology and Engineering Design, Slovak University of Technology in Bratislava, 812 31 Bratislava, Slovakia
jozef.bucha@stuba.sk

[2] Faculty of Mechanical Engineering, University of Belgrade, Kraljice Marije 16, 11120 Belgrade 35, Serbia

Abstract. New powertrain design poses new challenges in terms of driveline vibration or ultimately in terms of NVH (Noise Vibration Harshness). Current state of the technology for vibration damping in car powertrains is the so-called Dual Mass Flywheel (DMF). The paper deals with creation of MBD model of double mass flywheel based on CAD model of dual mass flywheel produced by ZF. Parameters of virtual model are tuned and compared with experimental measures of dual mass flywheel. Second part of paper deals with dynamic simulation of DMF. Model used for dynamic simulation consists of two bodies and torsion spring with stiffness computed from simulation of full DMF model.

Keywords: Dual mass flywheel · Multi body dynamic · MSC/Adams · NVH

1 Introduction

Global trend in automotive technology such as electrification of car powertrain leads to higher diversification of powertrain designs i.e. its topology. New powertrain design poses new challenges in terms of driveline vibration or ultimately in terms of NVH (Noise Vibration Harshness). To come to terms with NVH issues could be a fairly difficult process, however a methodical approach including mathematical description of powertrain topology, using multi body simulations and real tests allow not only to understand the nature of the problem but also to develop applicable solutions. Experimental approach to the investigation of described processes is presented in numerous papers, such as [1, 2].

Current state of the technology for vibration damping in car powertrains is the called Dual Mass Flywheel (DMF). The dual mass flywheel allows not only the damping of vibrations which stem from the combustion engine itself (combustion of the fuel as well as its crankshaft mechanics) but also offers improvements in other operation conditions.

N. Mitrovic et al. (Eds.): CNNTech 2019, LNNS 90, pp. 375–392, 2020.
https://doi.org/10.1007/978-3-030-30853-7_22

2 Structural Model of Powertrain Components

The increasing vehicle agility, constantly increasing from car generation to car generation engine torques and requirements of NVH has made the dynamic behavior of such vehicles more critical, and the problems of vehicle comfort, safety, stability, and control, more complex and difficult [3]. Basically, the objective is to design a vehicle such that it satisfies the requirements of the personnel or material transported, as well as vehicles' reliability and efficiency [4]. The goal is to obtain the optimum solution to this problem in the most economical way. Analysis of the vehicle and a determination of the validity of analysis are required.

In practice, however simulation of specific powertrains is fundamental although most designs are evolved by testing [5]. A global trend of powertrain electrification leads to a fundamental change of topology different needs in damping and decoupling systems, and so it seems require a reappraisal of the problem of analyzing a vehicle and thus predicting its behavior. It is possible to analyze the vehicle as a mechanical system of many degrees of freedom in order to describe all motions of the vehicle components, such as wheels, gearboxes, engine, flywheel, clutch, differential, hub and propeller shaft.

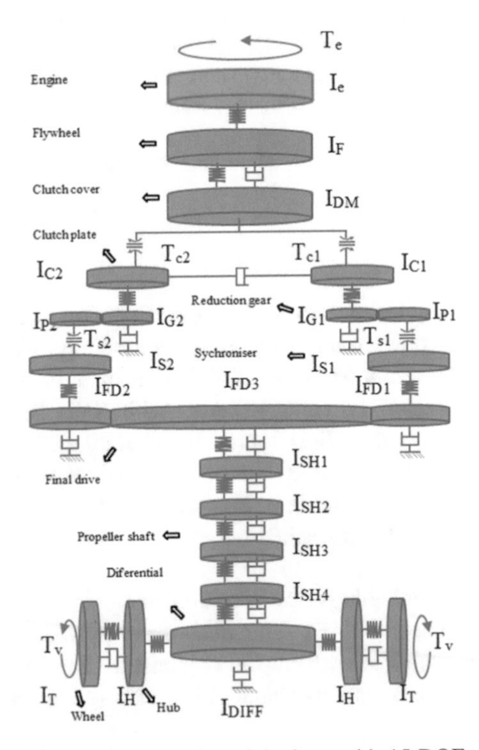

Fig. 1. Dynamical model of car with 15 DOF

Figure 1 shows dynamic model of car; this system can be taken to represent an automobile and its suspension for small vertical motions if one assumes that the engine and the body and its contents are rigidly connected to the frame. The tire is assumed to contact the ground at a point directly below the wheel. The equations of motion for this system may be obtained from Lagrange's Eq. (1) for non-conservative system,

$$\frac{d}{dt}\left(\frac{\partial E_k}{\partial \dot{q}_j}\right) - \frac{\partial E_k}{\partial q_j} + \frac{\partial E_p}{\partial q_j} + \frac{\partial E_{dis}}{\partial \dot{q}_j} = Q_{n,j} \tag{1}$$

where

E_k - kinetical energy of system (function of general coordinate velocity and time),
E_p - potentials energy of system (function of general coordinate positions),
E_{dis} - dissipative (Rayleigh) function represents damping energy in coordinate,
$Q_{n,j}$ - general forces corresponding to exciting forces in coordinate (other power force are in E_p and E_{dis})

Full procedure of equations determination and derivation for this model can be found in [3].

$$I_E \varepsilon_E - K_E(\varphi_F - \varphi_E) = T_E \tag{2}$$

$$I_F \varepsilon_F - K_E(\varphi_F - \varphi_E) - K_F(\varphi_{DM} - \varphi_F) - C_F(\omega_{DM} - \omega_F) = 0 \tag{3}$$

$$I_{DM} \varepsilon_{DM} + K_F(\varphi_{DM} - \varphi_F) + C_F(\omega_{DM} - \omega_F) = -(T_{C1} + T_{C2}) \tag{4}$$

$$I_{C1} \varepsilon_{C1} - K_{C1}(\varphi_{G1} - \varphi_{C1}) - C_{WC}(\omega_{C2} - \omega_{C1}) = T_{C1} \tag{5}$$

$$I_{C2} \varepsilon_{C2} - K_{C2}(\varphi_{G2} - \varphi_{C2}) + C_{WC}(\omega_{C2} - \omega_{C1}) = T_{C2} \tag{6}$$

$$\left[\frac{I_{P1}}{\gamma_{G1}^2} + I_{G1}\right] \varepsilon_{G1} + K_{C1}(\varphi_{G1} - \varphi_{C1}) - C_{D1}\omega_{G1} = -\frac{T_{S1}}{\gamma_{G1}} \tag{7}$$

$$\left[\frac{I_{P2}}{\gamma_{G2}^2} + I_{G2}\right] \varepsilon_{G2} + K_{C2}(\varphi_{G2} - \varphi_{C2}) - C_{D2}\omega_{G2} = -\frac{T_{S2}}{\gamma_{G2}} \tag{8}$$

$$I_{S1} \varepsilon_{S1} - K_{S1}(\gamma_{FD1}\varphi_{FD3} - \varphi_{S1}) = T_{S1} \tag{9}$$

$$I_{S2} \varepsilon_{S2} - K_{S2}(\gamma_{FD2}\varphi_{FD3} - \varphi_{S2}) = T_{S2} \tag{10}$$

$$\begin{aligned} &(I_{SH1}\varepsilon_{SH1} + I_{SH2}\varepsilon_{SH2} + I_{SH3}\varepsilon_{SH3} + I_{SH4}\varepsilon_{SH4}) + K_{SH}(\phi_{SH1} - \phi_{FD3}) - K_{SH1}(\phi_{SH2} - \phi_{SH1}) \\ &+ K_{SH2}(\phi_{SH3} - \phi_{SH2}) - K_{SH3}(\phi_{SH4} - \phi_{SH3}) + K_{SH4}(\phi_{DIFF} - \phi_{SH4}) \\ &+ C_{SH}(\omega_{SH1} - \omega_{FD3}) - C_{SH1}(\omega_{SH2} - \omega_{SH1}) + C_{SH2}(\omega_{SH3} - \omega_{SH2}) \\ &- C_{SH3}(\omega_{SH4} - \omega_{SH3}) + C_{SH4}(\omega_{DIFF} - \omega_{SH4}) = 0 \end{aligned} \tag{11}$$

$$\begin{aligned} &(\gamma_{FD1}^2 I_{FD1} + \gamma_{FD2}^2 I_{FD2} + I_{FD3})\varepsilon_{FD3} + \gamma_{FD1}K_{S1}(\gamma_{FD1}\varphi_{FD1} - \varphi_{S1}) \\ &+ \gamma_{FD2}K_{S2}(\gamma_{FD2}\varphi_{FD2} - \varphi_{S2}) - K_{SH}(\varphi_{SH1} - \varphi_{FD3}) \\ &- (\gamma_{FD1}C_{FD1} + \gamma_{FD2}C_{FD2})\omega_{FD3} - C_{SH}(\omega_{SH1} - \omega_{FD3}) = 0 \end{aligned} \tag{12}$$

$$
\begin{aligned}
I_{DIFF}\varepsilon_{DIFF} - K_{SH4}(\varphi_{DIFF} - \varphi_{SH4}) - 2K_{AX}(\varphi_H - \varphi_{DIFF}) \\
+ C_{SH4}(\omega_{DIFF} - \omega_{SH4}) - C_{DIFF}\omega_{DIFF} = 0)
\end{aligned} \tag{13}
$$

$$
2I_H\varepsilon_H + 2K_{AX}(\varphi_H - \varphi_{DIFF}) - 2K_H(\varphi_T - \varphi_H) - 2C_H(\omega_T - \omega_H) = 0 \tag{14}
$$

$$
2I_T\varepsilon_T + 2K_H(\varphi_T - \varphi_H) + 2C_H(\omega_T - \omega_H) = -T_W \tag{15}
$$

3 Flywheel

As far back as the 1930s, it was established that torsional vibration (and thus transmission noise), can be reduced effectively by lowering the torsional rigidity between the engine and the transmission. This knowledge led to the development of the torsional vibration damper, which is integrated into either the clutch disc or the flywheel [6, 7].

Since ignition pressures will continue to rise and operating and idle speeds will drop, the demand for refinement will increase, and unwanted noise will be heard with greater clarity due to general lowering of the interior noise level. Torsional vibration must therefore be damped with increasing efficiency. To optimize the torsional vibration pattern, the oscillation forms and natural frequencies for the entire power train must be known. Calculation methods, measurements in the vehicle and subjective assessments of vibration and noise represent a comprehensive set of tools for selective optimization [8, 9].

The trends in powertrain technology such as engine downsizing, switching off the specific engine cylinders require better and better torsional damping. The current solutions for torsional damping are the following:

1. Torsional damper in clutch disc.
2. Dual Mass Flywheel.
3. Dual Mass Flywheel with pendulum systems.

In the Figs. 2, 3 and 4 are shown some variants of flywheel that can be used to reduce frequency of oscillation. The first variant uses the series of springs to reduce oscillation defined with maximal torque of engine, the second variant better reduces the oscillation that use the inner damper and third variant reduces the maximum of oscillation, because it uses the principles of pendulum systems and serial springs together.

Fig. 2. Dual mass flywheel with serial spring

Fig. 3. Dual mass flywheel with inner damper

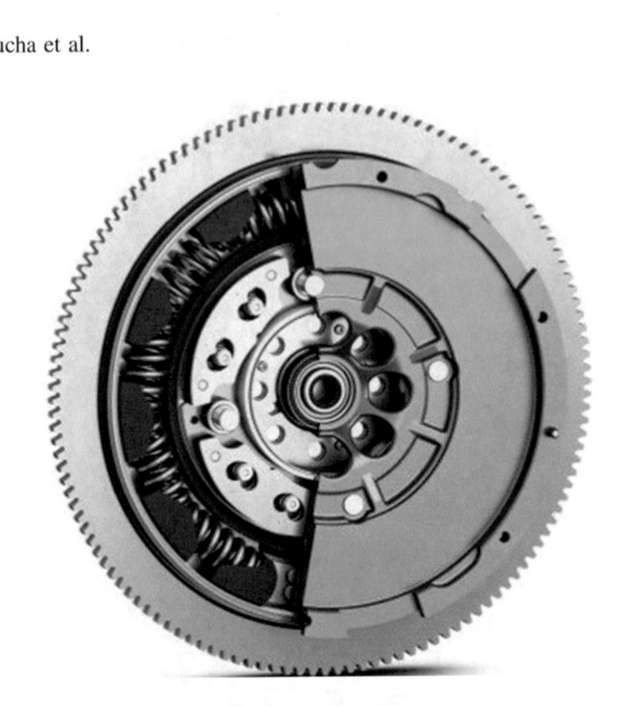

Fig. 4. Dual mass flywheel with pendulum system

The MBD model of dual mass flywheel with serial springs is presented. Parts used in primary side of DMF are shown in Fig. 5.

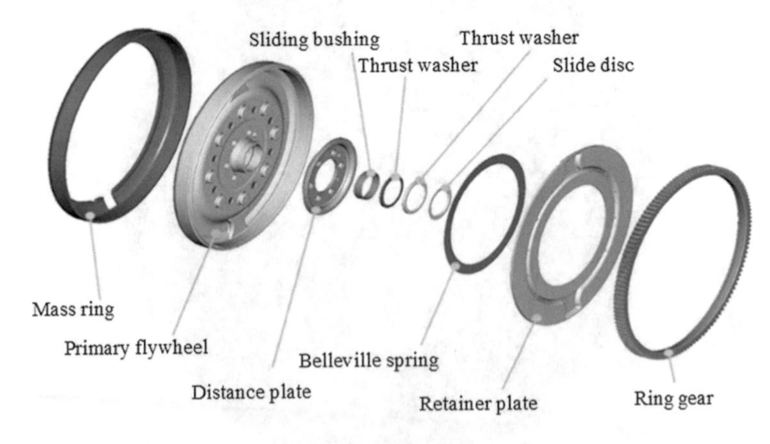

Fig. 5. Construction parts of primary side of flywheel

Parts used in secondary side of flywheel are shown in Fig. 6.

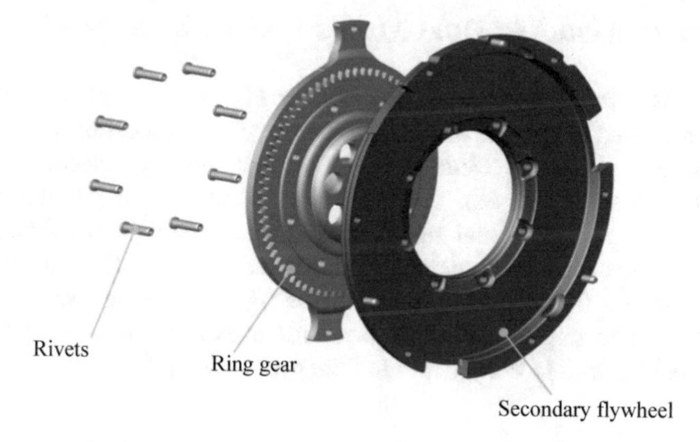

Fig. 6. Construction parts of secondary side of flywheel

In Fig. 7 is depicted pressure spring set used in dual mass flywheel. Two spring sets are presented in DMF.

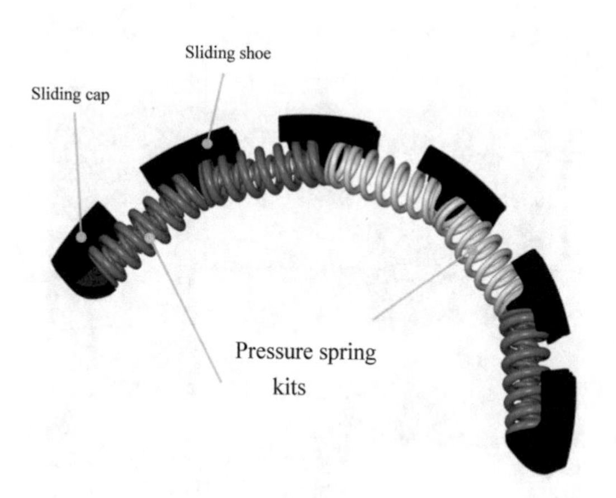

Fig. 7. Pressure spring set

One common solution is to create progressive torsional damper characteristics and so to ensure the low stiffness i.e. low natural frequency by low torque and enable the higher stiffness by high torque in order to be able to transmit the maximum engine torque. DMF presented in paper consist of three stages stiffness characteristic.

- The first stage is for low torque vibrations.
- The second stage is for high torque vibrations.
- The third stage ensure the stop torque by very high amplitudes (e.g. by load change).

4 Multibody Model of Dual Mass Flywheel in Adams/View

Process of MBD model workflow is shown in Fig. 8. CAD model of dual mass flywheel was created in Creo 2. All necessary parts were then exported as..stp files. In Adams/View 3 sub models were created (Primary flywheel, Secondary flywheel, Spring Caps and sliding shoes). For every part in all sub models density of used material was defined. Mass and moment of inertia was then computed for every part. After that, two other sub-models were created: Primary flywheel with merged parts and secondary flywheel with merged parts. For computing mass and inertia of merged model the aggregate mass function in Adams/View was used. Both flywheels then were modeled as one body with multiple geometries [10].

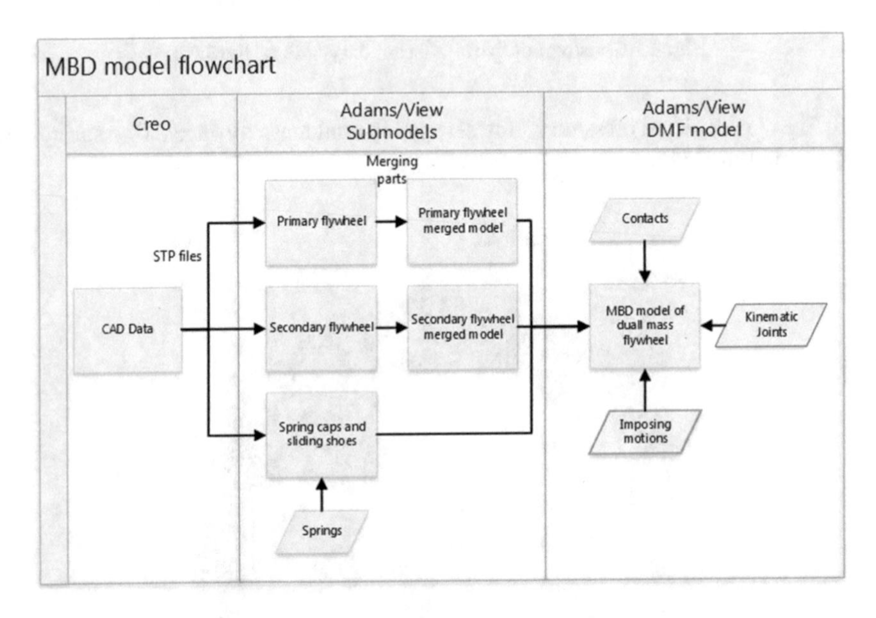

Fig. 8. Workflow of dual mass MBD model creation

Figures 7 and 8 shows merged of models of primary flywheel and secondary flywheel with computed Centre of Gravity and orientation of inertia.

Fig. 9. Primary flywheel sub model with merged parts.

Fig. 10. Secondary flywheel sub model with merged parts

Spring caps and sliding shoes were created in another sub-model (Fig. 11). Each spring in dual mass flywheel was created with 3 Adams force elements (1 SForce – used for modeling longitudinal spring properties and 2 Bushings – used for modeling lateral spring properties). Longitudinal stiffness of the springs was defined according spring drawings, Lateral stiffness of the spring is calculating during simulation by Adams according Eq. (16)

$$k_p = \frac{2.6k}{1 + 0.77\left(\frac{H}{D}\right)^2}\left(1 - \frac{P}{U_1 H_0 k}\right) \tag{16}$$

where

k - longitudinal stiffness,
H - length of compressed spring,
D - diameter of spring,
P - longitudinal spring load,
U_1 - spring aspect ratio
H_0 - free length of spring

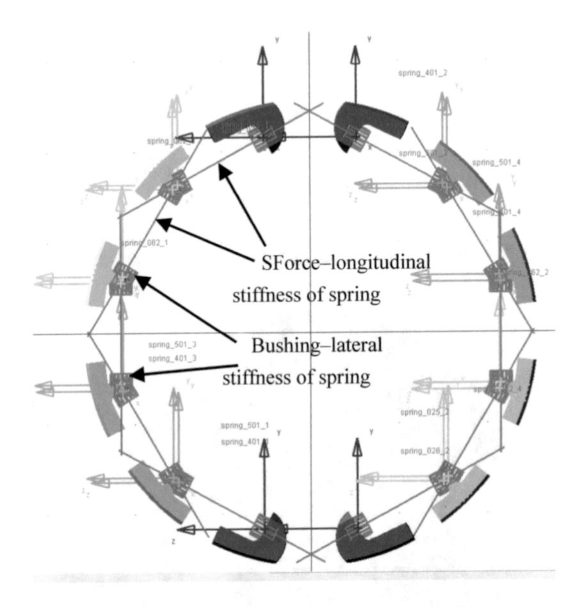

Fig. 11. Spring caps and sliding shoes sub-model. Markers show attachment points of spring and bushing elements

All three sub-models (Primary flywheel with merged parts (Fig. 9), Secondary flywheel with merged parts (Fig. 10) and Spring caps and sliding shoes sub-model (Fig. 11)) were assembled together, thus creating complete MBD model of Dual Mass Flywheel (Fig. 12). In this model, needed kinematic joints, and all geometric contacts were defined. All contacts were defined as geometric contacts (CAD geometries were used). Contacts are defined between spring caps and sliding shoes, between ring gear (part of secondary flywheel) and spring caps, between primary flywheel, retainer platter (part of primary flywheel) and spring caps.

For connection of primary side of flywheel and secondary side of flywheel, bearing element from Adams/Machinery was used. Properties of bearing are shown in Table 1. Type of bearing is Deep Groove Ball Bearing – such as one investigated in [11–13].

Table 1. Properties of bearing

Ball pitch diameter	34.5 mm
Inner raceway radius	2.34 mm
Outer raceway radius	2.34 mm
Number of balls	13
Ball diameter	4.5 mm
Diametral clearance	1.6E-2 mm

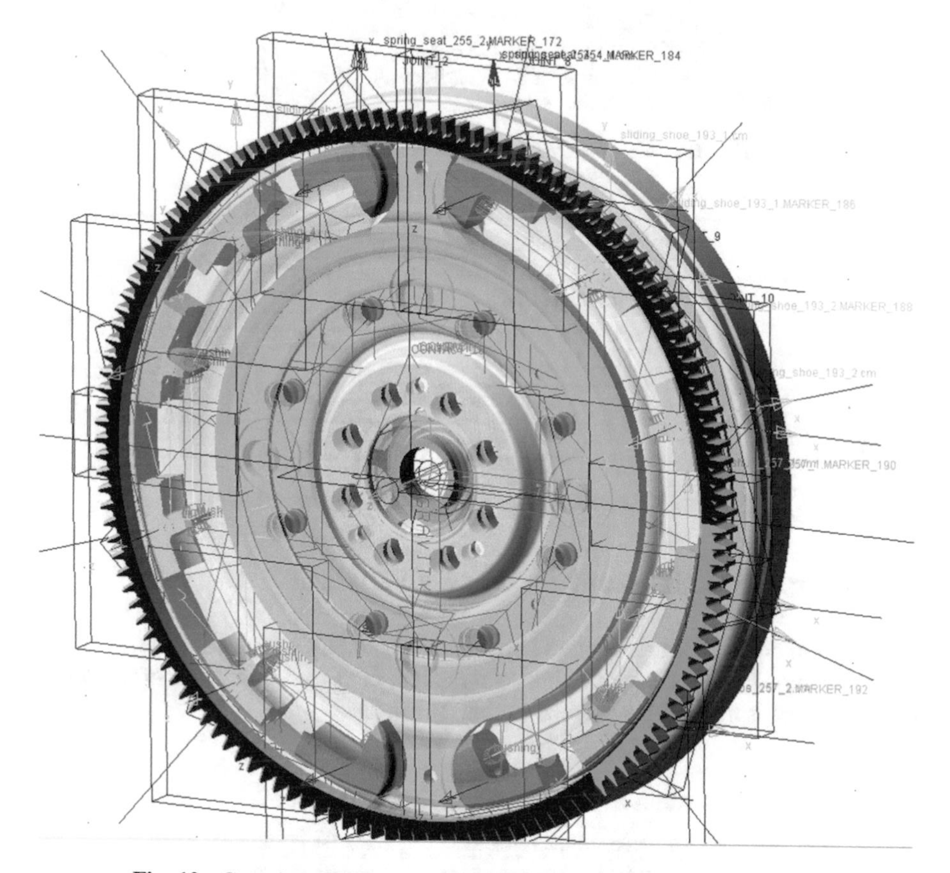

Fig. 12. Complete MBD model of Dual Mass Flywheel in Adams/View

5 Comparison of MBD Model with Experimental Data

Created MBD model was compared with measured results. Figure 13 shows measured stiffness of Dual Mass Flywheel; Fig. 14 shows computed stiffness of MBD model. Difference between measured and simulated MBD model is 5 Nm for maximum torsion angle 58°.

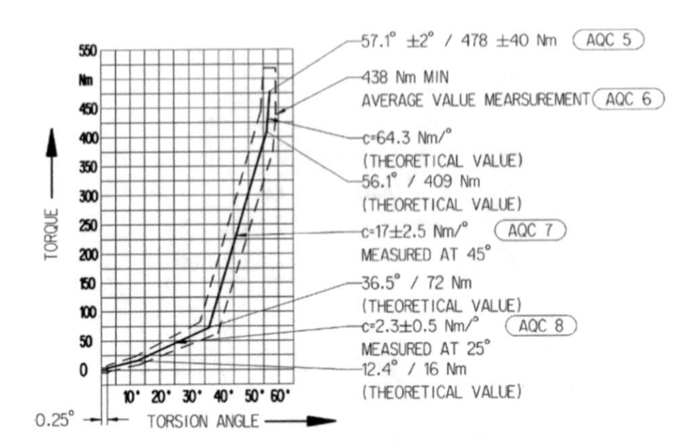

Fig. 13. Measured value of Dual Mass Flywheel stiffness

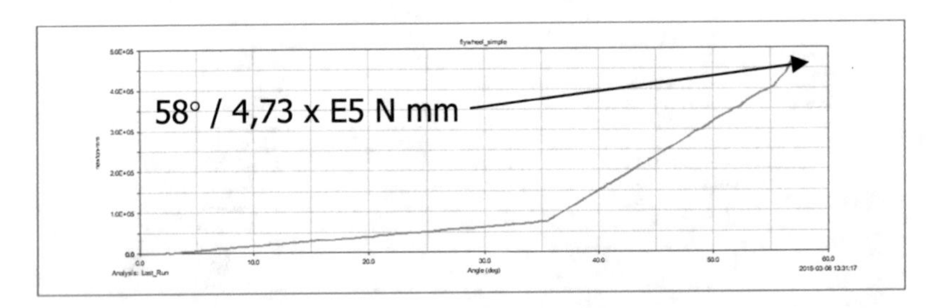

Fig. 14. Computed value of Dual Mass Flywheel stiffness

6 Dynamic Simulation of Dual Mass Flywheel

In Fig. 15 id depicted block schema of dynamic simulation model. In the primary side of double mass flywheel is applied imposed motion, in the secondary side of the double mass flywheel is applied load torque (Table 2).

Table 2. Basic parameters of dynamic model

Primary side, Moment of Inertia	9.5E+04 kg mm^2
Secondary side Moment of Inertia	3.8E+04 kg mm^2
Torsional stiffness	Spline, Fig. 14
Torsional damping	16 Nm ms/°

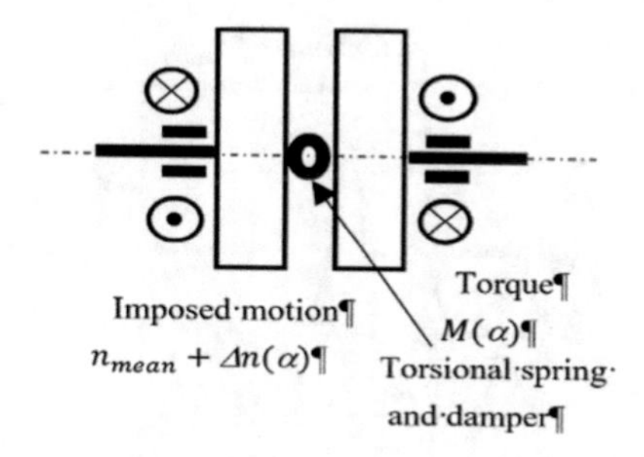

Fig. 15. Block diagram of dynamic simulation model

In Fig. 16 is depicted change of angular velocity for various types of engines for mean speed 3500 RPM in dependence on crankshaft angular displacement.

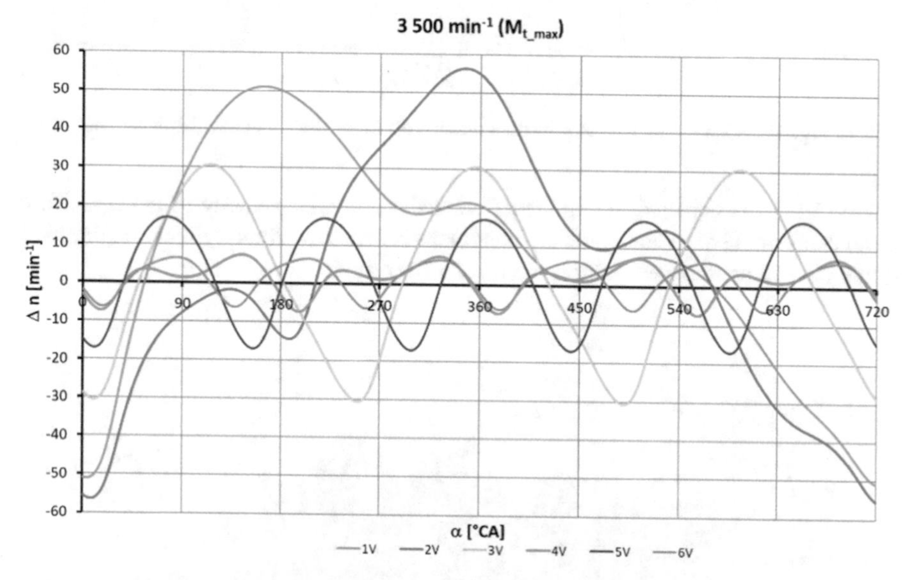

Fig. 16. Change of angular velocity in dependence on crankshaft angular displacement

In Fig. 17 is depicted change of moment for various types of engines for mean speed 3500 RPM in dependence on crankshaft angular displacement.

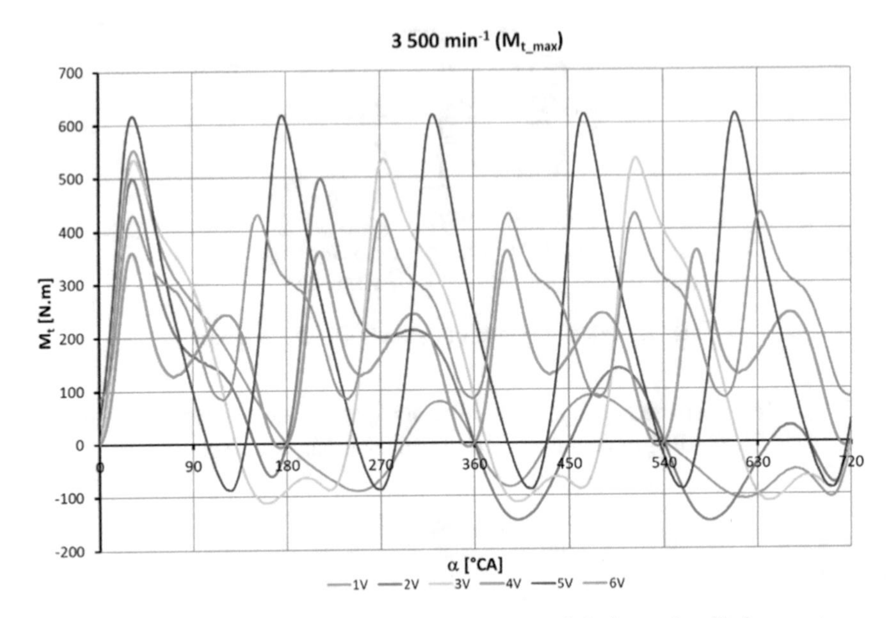

Fig. 17. Change of moment in dependence on crankshaft angular displacement

In Fig. 18 is depicted 3D spline of crankshaft angular velocity in dependence of mean crankshaft speed and crankshaft angular displacement for 4-cylinder engine. This spline is used as imposed motion input for MBD model

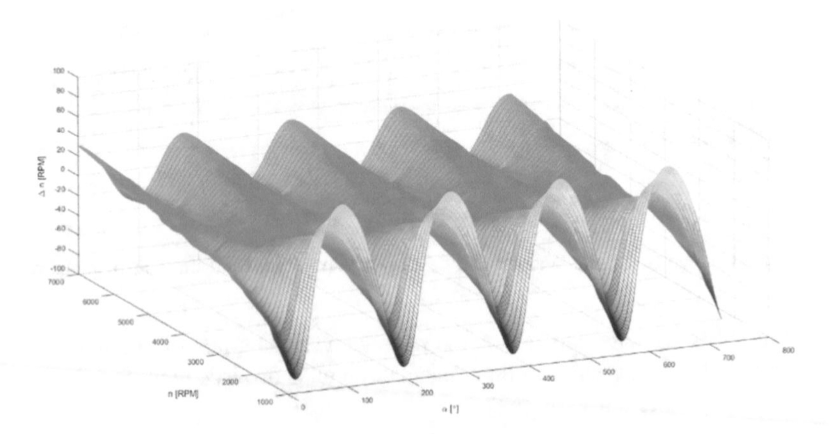

Fig. 18. 3D spline of Dual Mass Flywheel input motion characteristic

Figure 19 shows angular velocity of primary side of DMF, Fig. 20 shows angular velocity of secondary side of DMF for mean speed 3500 RPM.

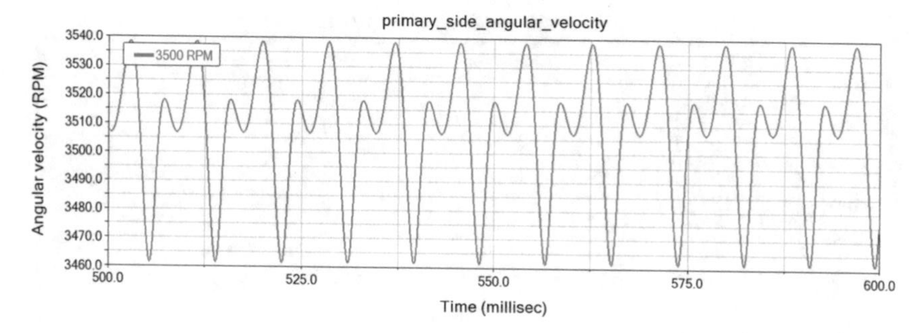

Fig. 19. Angular velocity of primary side of DMF – Mean angular velocity 3500 RPM

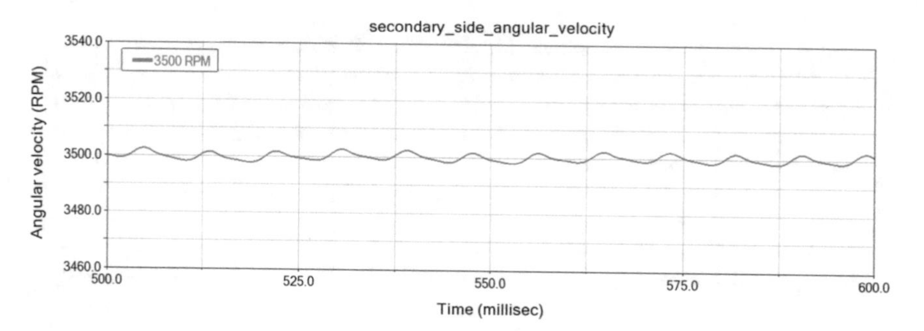

Fig. 20. Angular velocity of secondary side of DMF – Mean angular velocity 3500 RPM

Figure 21 shows torsional deformation of torsional spring for mean speed 3500 RPM.

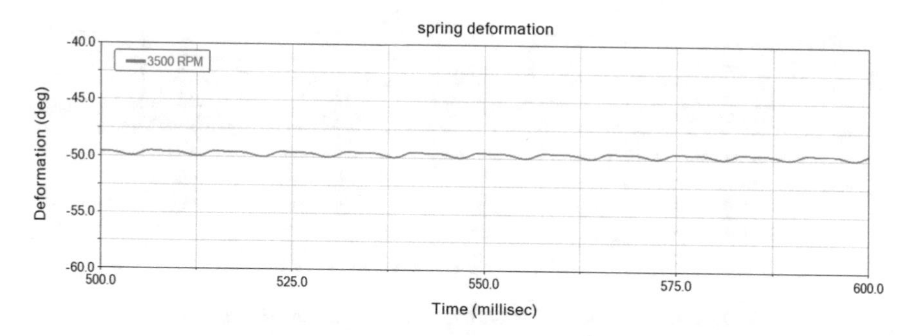

Fig. 21. Deformation of torsion spring – Mean angular velocity 6000 RPM

Figure 22 shows angular velocity of primary side of DMF, Fig. 23 shows angular velocity of secondary side of DMF for mean speed 6000 RPM.

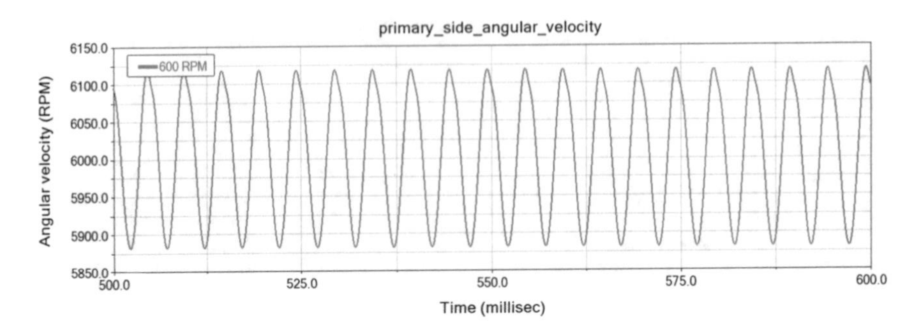

Fig. 22. Angular velocity of primary side of DMF – Mean angular velocity 6000 RPM

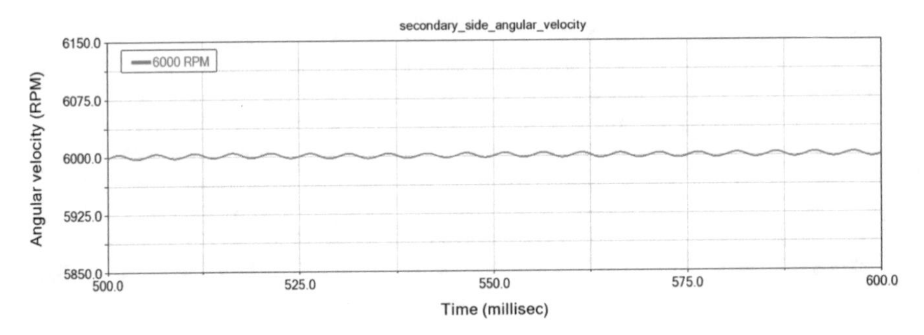

Fig. 23. Angular velocity of secondary side of DMF – Mean angular velocity 6000 RPM

Figure 24 shows torsional deformation of torsional spring for mean speed 6000 RPM.

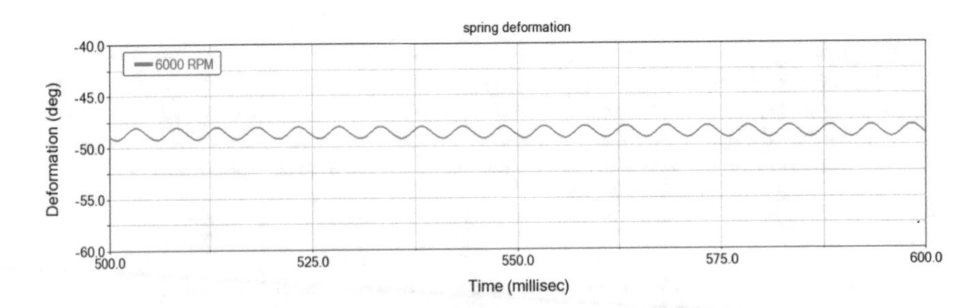

Fig. 24. Deformation of torsion spring – Mean angular velocity 6000 RPM

In Figs. 25 and 26 are depicted comparison of primary and secondary side of DMF for Mean speed 3500 RPM (Fig. 25) and 6000 RPM (Fig. 26).

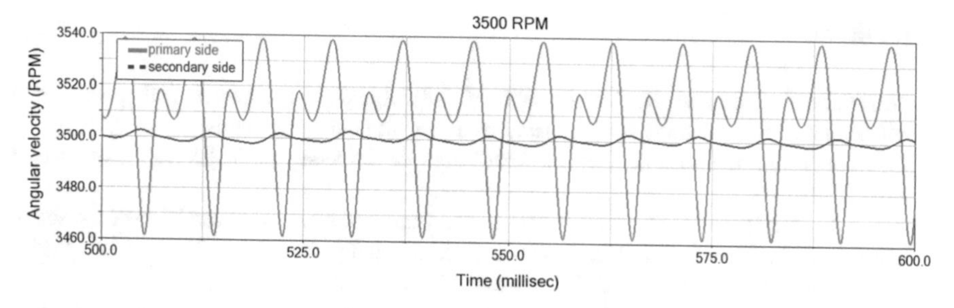

Fig. 25. Comparison of angular velocity – Mean angular velocity 3500 RPM

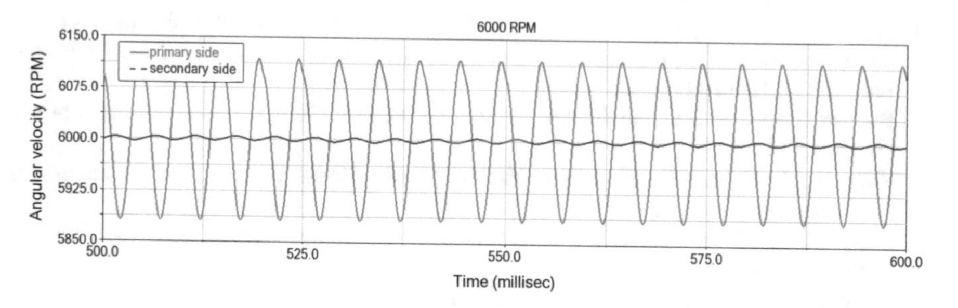

Fig. 26. Comparison of angular velocity – Mean angular velocity 6000 RPM

In Table 3 are summarized simulation results.

Table 3. Comparison of simulation results

Mean speed	3500 RPM	
Primary side of DMF	Peak angular velocity	76.9688 RPM
Secondary side of DMF	Peak angular velocity	4.8063 RPM
Mean speed	6000 RPM	
Primary side of DMF	Peak angular velocity	236.3914 RPM
Secondary side of DMF	Peak angular velocity	7.6651 RPM

7 Conclusion

Using MBD modeling can help engineers and designers to create precise models of vehicle components such as Dual Mass Flywheel. Main advantage is that is not necessary to create models described by equations like presented in second chapter. MBD model in Adams/View is closer to traditional CAD model.

References

1. Mišković, Ž., Mitrović, R., Maksimović, V., Milivojević, A.: Analysis and prediction of vibrations of ball bearings contaminated by open pit coal mine debris particles, Technical Gazzete (Tehnicki Vjesnik). Josip Juraj Strossmayer University of Osijek **24**(6), 1941–1950 (2017)
2. Tasic, M., Mitrovic, R., Popovic, P., Tasic, M.: Influence of running conditions on resonant oscillations in fresh-air ventilator blades used in thermal power plants. Thermal Sci. Institut za nuklearne nauke "Vinča" **13**(1), 139–146 (2009)
3. Test based Optimization of NVH for hybrid and electric vehicles. https://community.plm.automation.siemens.com/siemensplm/attachments/siemensplm/Simcenter_event_tkb/109/6/26_Vortrag.pdf. Accessed 22 May 2019
4. Popovic, P., Ivanovic, G., Mitrovic, R., Subic, A.: Design for reliability of a vehicle transmission system. Proc. Inst. Mech. Eng. Part D: J. Autom. Eng. Prof. Eng. Publishing Ltd. **226**(2), 194–209 (2012)
5. Happian-Smith, J.: An introduction to Modern Vehicle Design. Butterworth-Heinmann, Oxford (2001)
6. Kister, I.: Smart Damping in car powertrains, 1st edn. Slovak University of Technology in Bratislava, Bratislava (2018)
7. Drexl, H.-J.: Motor Vehicle Clutches: Function and Design. Verlag Moderne Industie, Landsberg (1999)
8. Steinhauser, J.: Methods of Computer Modeling and Simulation of Drive Systems, 1st edn. Slovak University of Technology in Bratislava, Trnava (2017)
9. Danko, J., Milesich, T., Bucha, J.: Nonlinear model of the passenger car seat suspension system. J. Mech. Eng. **67**(1), 23–28 (2017)
10. Bucha J., Kister I.: Multi Body model of double mass flywheel. In: 10th International Symposium KOD 2018 Machine and Industrial Design in Mechanical Engineering Proceedings, University of Novy Sad, Balatonfüred (2018)
11. Lazovic, T., Mitrovic, R., Ristivojevic, M.: Influence of internal radial clearance on the ball bearing service life. J. Balkan Tribol. Assoc. Sci. Bul. Commun. **16**(1), 1–8 (2010)
12. Miškovic, Ž., Mitrović, R., Stamenić, Z.: Analysis of grease contamination influence on the internal radial clearance of ball bearings by thermographic inspection. Thermal Sci. Institut za nuklearne nauke "Vinča" **20**(1), 255–265 (2016)
13. Grujičić, R., Tomović, R., Mitrović, R., Jovanović, J., Atanasovska, I.: The analysis of impact of intesity of contact load and angular shaft speed on the heat generation within radial ball bearing. Thermal Sci. Institut za nuklearne nauke "Vinča" **20**(5), 1765–1776 (2016)

Determination of Dynamic Properties of Rubber-Metal Motor Mount of Electric Powertrain

Ján Danko[1(✉)], Jozef Bucha[1], Tomás Milesich[1], Radivoje Mitrovic[2], and Zarko Miskovic[2]

[1] Faculty of Mechanical Engineering, Institute of Transport Technology and Engineering Design, Slovak University of Technology in Bratislava, Nam. Slobody 17, 812 31 Bratislava, Slovakia
jan.danko@stuba.sk

[2] Faculty of Mechanical Engineering, University of Belgrade, Kraljice Marije 16, 11120 Belgrade 35, Serbia

Abstract. Increasing the volume of the production of vehicles with electric or hybrid powertrain also leads to specific activities for research and development in this segment. One of the current areas that present the paper deals with the idea of reducing structure-borne vibration from electric powertrain to the vehicle interior. Electric powertrain and mainly electric motor are a source of high-frequency vibration up to 3000 Hz. On the other side, the internal combustion engine is a source of significantly lower frequencies. For this reason, a different approach is needed when designing electric powertrain mounts. To reduce the vibration transfer to the structures of the vehicle, rubber-metal motor mounts are used. Rubber-metal motor mount must have specific dynamic properties such as the dynamic stiffness, the loss angle and hysteresis properties. Hysteresis properties are important to define the damping properties of the rubber-metal motor mount. The Bouc-Wen model was used for modeling of hysteresis properties. The results from simulations are compared with experimental data.

Keywords: Rubber-metal motor mount · Electric vehicle · Powertrain · Hysteresis properties

1 NVH Challenges in Electric Vehicles

Currently, the automotive industry is under pressure from both legislation and consumers to provide cleaner vehicles. To meet this goal, it is necessary to reduce fuel consumption and emissions. Also, the vehicle reliability is permanently in the focus of scientific research worldwide [1]. One of the ways to achieve more eco-friendly vehicles is by electrification of the vehicle powertrain. As a result, vehicles using electric powertrains in combination with, or instead of, internal combustion engines (ICE) are becoming increasingly common. This leads to the use of new concepts those properties are different from vehicles with combustion engine powertrain. One of the most used methodologies to monitor the performance and characteristics of different mechanical systems and components in operating conditions are vibration diagnostics [2, 3] and

N. Mitrovic et al. (Eds.): CNNTech 2019, LNNS 90, pp. 393–409, 2020.
https://doi.org/10.1007/978-3-030-30853-7_23

thermographic inspection [4]. This paper will focus predominantly at the NVH (noise-vibration-harshness). The electrification of vehicles powertrain causes a radical change in the area of NVH as new powertrains create new requirements for NVH. The term NVH is usually used to refer to the study of acoustic noise and mechanical vibration in vehicles and their subjective feelings by humans (see Fig. 1) [5].

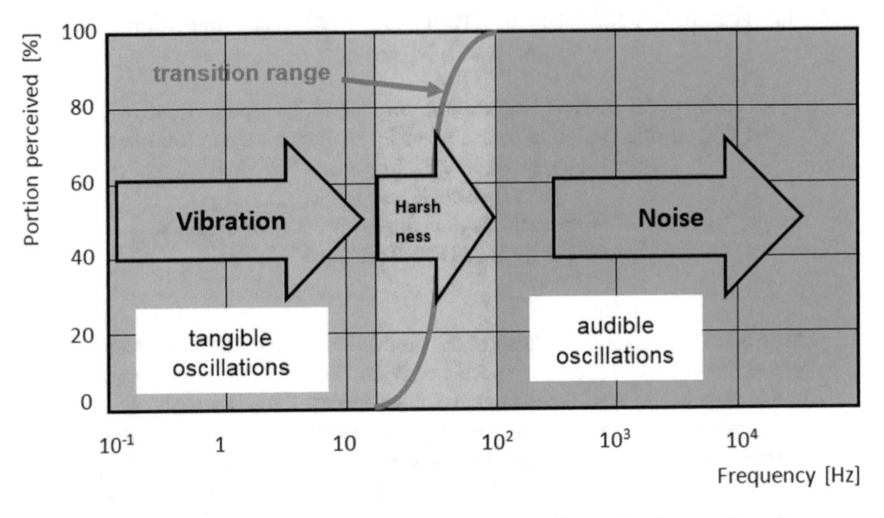

Fig. 1. Frequency ranges associated with the terms noise, vibration, and harshness.

Years of experience and skills in NVH design to reduce oscillations generated by combustion engine, cannot be simply used to reduce oscillations of the electrified vehicle powertrains [6]. Although electric vehicles (EV) are markedly quieter than vehicle with combustion engine. The noise contains high-frequency tonal components with base frequencies that are appeared as unpleasant. The use of electronics for electric powertrain, like as the pulse width modulation (PWM) control, creates a specific sound modulation footprint. The usual wideband noise sources like tire and wind noise are emitting from transmission systems and various other auxiliaries, are no longer hidden by the dominant noise of combustion engine [7]. By removing the internal combustion engine, these disturbing sounds become plainly audible and annoying. In conjunction with the electric drive, new areas are emerging that need to be solved in NVH solutions. Thus, there are new possibilities to achieve much lower levels of interior noise in the vehicle that was previously possible with the use of an internal combustion engine (see Fig. 2). Electric vehicle NVH must deal with [8]:

- Electric powertrain noise and vibration,
- Wind noise,
- Tire noise,
- Auxiliary systems noise and vibration,
- Other noise and vibration

1.1 Electric Powertrain Noise and Vibration

Typical sources of vibration and vibration in the electrical powertrain are:

- Electric motor(s),
- Direct-drive, single-speed gearbox,
- Power electronics unit,
- Battery pack cooling system.

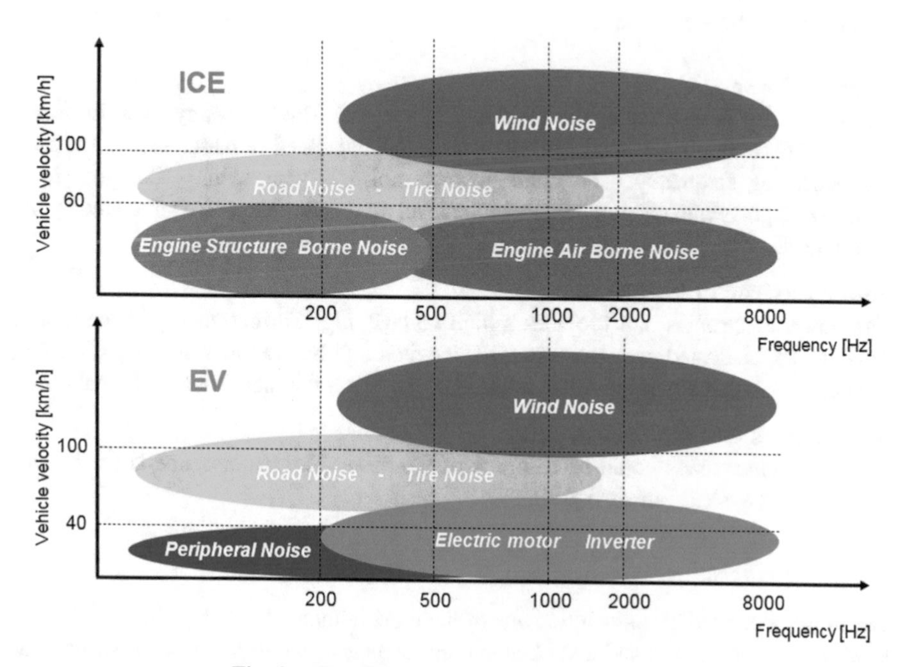

Fig. 2. ICE - EV Frequency ranges of interest.

Electric Motor Noise and Vibration

An important source of vibration and noise in electric motors are electromagnetic pulses. The different types of electric motors are used in electric vehicles:

- asynchronous motor (ASM);
- electrically excited synchronous motor (EESM);
- permanent magnet synchronous motor (PMSM);
- switched reluctance motor (SRM).

Depending on the electric motor design, the electromagnetic pulses creates torque pulses (cogging), and can be very strong. Torque pulses are emitted as noise directly from the electric motor housing and they are also transmitted structurally to the vehicle body structure through the motor mounts [8, 9]. Electromagnetic forces from electromagnetic pulses are lower than the combustion forces and reciprocating mass forces of a combustion engine, and markedly, they are at a much higher frequency (up to

8000 Hz). These electromagnetic forces produce tonal noises (harmonics of rotor speed), also called electric motor whistling. The source could be also the electric motor cooling system.

The sources of vibration and noise of the electric motor ultimately can be:

- cogging torques,
- torque irregularities due to the electrical power supply,
- rotor bearings,
- modes of the motor housing,
- modes of the motor shaft.

Gearbox Noise
Gearbox noise is also tonal like electric motor noise. The frequency of noise is the product of the number of teeth and the gear rotational speed in hertz. This is the base tooth-meshing frequency. The main gearbox noise is gear whine. The goal is to minimize high-frequency gear whine on the level no greater than noise generated by the motor itself.

Power Electronics Unit Noise
The power electronics unit provides a high voltage, high current energy to the motor. This is a sophisticated computer-controlled switch that has variable or fixed switching frequencies almost always in range 5–20 kHz and this is the main source of tonal noise.

Battery Pack Cooling System Noise
Also, the battery pack could be a source of the noise. The main source is the battery pack cooling system (air or liquid).

1.2 Wind Noise

The noise generated by the wind is one of the most influential (see Fig. 3). The normal level of wind noise will suddenly become unacceptable due to the absence of internal engine noise, especially up to medium speed (less than 80 km/h). The main source of wind noise is air flowing around the vehicle, air turbulences on exterior parts like body parts, underbody parts, mirrors, etc.

1.3 Tire Noise

The tire noise also is the most significant due to the absence of internal combustion engine noise (see Fig. 3). The tire noise is emitting from the contact of the rotating wheels and the road and it is divided into an airborne and a structure-borne contribution. Air-borne noise is radiated from the tire and enters the interior. Structure-borne road noise is transferred through the chassis elements to the interior.

1.4 Auxiliary Systems Noise and Vibration

Any electric vehicle must still have an air conditioning and heating system, properly functioning brakes, ABS/ ESP systems, and other various mechanical systems.

Without the engine to drive these systems, a range of electrically driven ancillary devices will be needed to provide this functionality. The noise and vibration generated by these ancillary devices are by themselves perhaps relatively minor but in total could become significant.

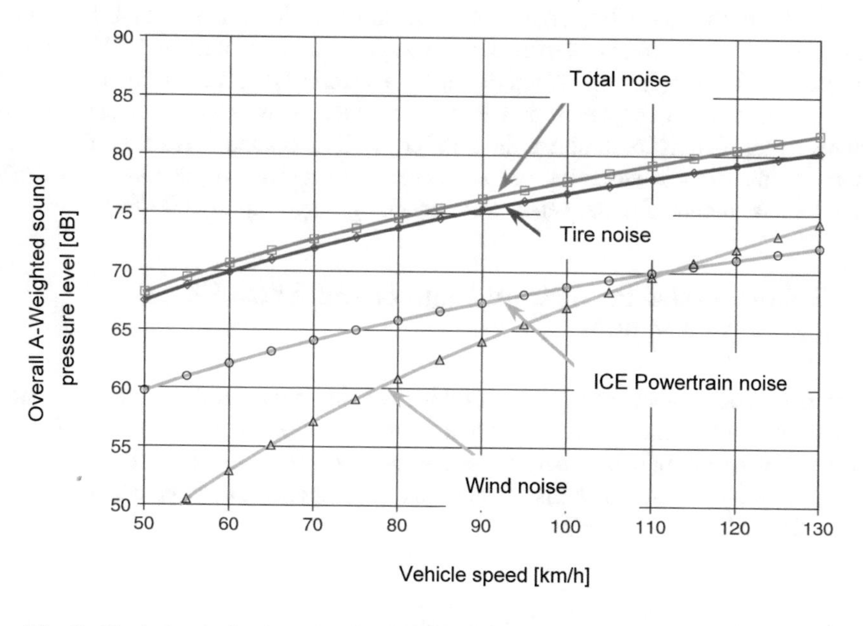

Fig. 3. Typical contribution of ICE vehicle noise sources as a function of speed [10]

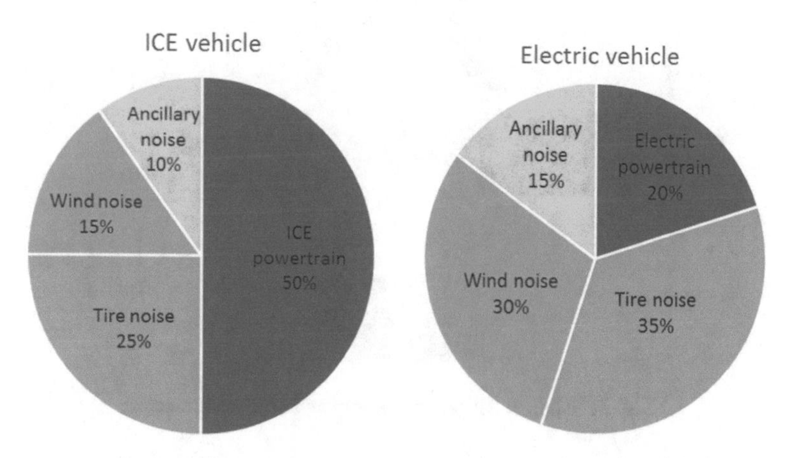

Fig. 4. Distribution of emphasis on various NVH areas for ICE and electric vehicle

1.5 Summarization of NVH Challenges in Electric Vehicles

As it is shown (see Fig. 4), in vehicles with internal combustion engines the main source of vibration and noise is powertrain – engine, transmission.

In electric vehicles, the main source of vibration and noise is tire noise due to the absence of a combustion engine and due to the lower noise of electric powertrain. But the electric powertrain vibration and noise are also significant due to high frequencies [11, 12]. As mentioned above there are two ways for transfer noise and vibration from the electric motor to the vehicle interior, the one is airborne and the second is structure borne. Due to high frequency and lower amplitude, it is possible to use common rubber-metal mounts to eliminate the transfer of vibration and noise from the electric motor to the vehicle structure and passengers. However, it is necessary to design rubber-metal mounts for frequency wide operating range (up to 2000 Hz).

2 Rubber-Metal Powertrain Mounts and Their Effect on NVH of Vehicle

A rubber-metal component consists of rubber bushing permanently vulcanized usually to a metal subcomponent. This permanent connection allows the transfer of oscillations from metal to the rubber, where these are damped and eliminated. The figure (see Fig. 5) shows the main functions of rubber-metal components and interactions between each other [5].

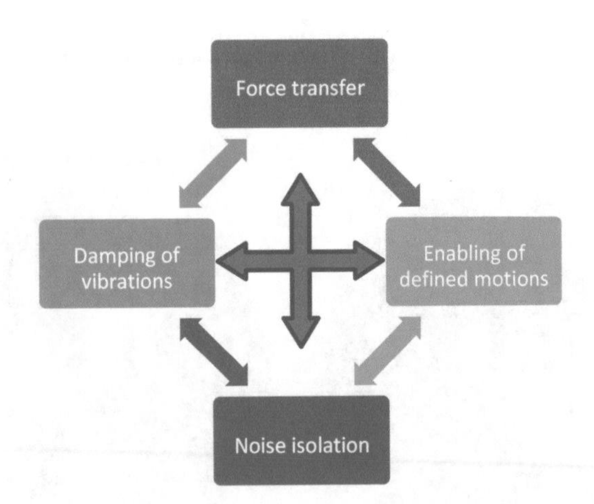

Fig. 5. Main functions of the rubber-metal component

To perform these four functions, rubber-metal components are subjected to several different requirements, many of which conflict with one another. These requirements are [5]:

- Force Transfer
  - hard bushing, minimal compression,
  - minimal damping.
- Ride Comfort
  - the soft bushing in longitudinal direction,
  - the hard bushing in lateral direction,
  - low torsional stiffness.
- Noise Isolation
  - soft bushing,
  - minimal damping.
- Vibration Damping
  - maximum damping.

For mounting purposes, a vehicle's engine and transmission are considered as a single unit - powertrain. The powertrain is mounted to the vehicle using engine mounts, transmission mounts, and torque supports. The mounting components are based on vulcanized rubber-metal connections.

The mounts used at the engine and transmission attachment points serve to counteract the static loads of the powertrain and limit the maximum displacements caused by load shifts or high torques. The engine's low-frequency oscillations can be considerably reduced by specifying frequency-selective dampers. The engine and transmission mount also effectively limit the transfer of engine and transmission excitations to the vehicle's chassis and body, which reduces structure-borne noise. This allows vibration and noise levels in the vehicle interior to be configured and tuned to increase passengers comfort.

Engine and transmission mounts are required to provide maximum damping for low frequencies and large excitation amplitudes while minimizing structure-borne noise for high frequencies and small amplitudes.

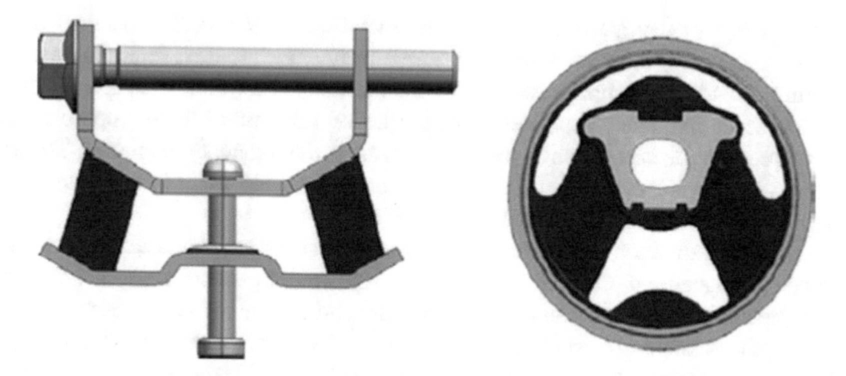

Fig. 6. The contour of complex mounts - wedge mount

The contours of the rubber mounts are changed from standard parts (such as block or circular shapes) to complex mounts.

The stiffness of the wedge mount (see Fig. 6) in three axes can be modified by changing the dimension and angle orientation of the rubber elements. Bush-type rubber mounts (see Fig. 7) used in a powertrain are used for controlling motion and vibration of a powertrain with compact space [13].

Also, the electric powertrain needs to be mounted in the vehicle with rubber-metal mounts. But there are little bit different conditions as listed above. Amplitudes are lower but the frequency range is wider (up to 2000 Hz) compared with ICE powertrain.

Fig. 7. Bush type rubber-metal mount

Electric vehicle powertrains produce excitations in mid and high-frequency range and create an important structure-borne disturbance that transferred through the vehicle structural parts into the interior and makes unwanted noise and vibration [14].

2.1 Dynamic Properties of Rubber-Metal Mounts

The high-frequency properties of the rubber-metal components are dependent on the materials properties, and on the geometry and assembly technology. Important material properties include the damping factor, the modulus of elasticity, and the shear modulus, all of which are frequency dependent. The frequency response of these properties leads to an increase in the component's overall stiffness as the frequency increases. To reach a high level of noise attenuation, one of the goals of development is to minimize the dynamic hardening of the bushings and mounts [5].

Except for geometry, the rubber component properties are characterized also with its **force-displacement curve** ($F_{A,}$ s_A). The rubber component force-displacement curves depend on used material (Shore hardness) and the mount's geometry (cut-outs, bump stops). A range of curves can be generated. The force-displacement curves of a rubber-metal component feature some degree of **hysteresis**. The amount of hysteresis is determined mainly by the damping properties of the rubber and the type and shape of the rubber-metal mount.

With the changing of the rubber compound composition, the viscoelastic characteristics of elastomers can be adjusted to fulfill given requirements. In the automotive industry, the most commonly used indicator of damping functions is the **loss angle,** δ (see Fig. 8). Loss angle is one of the important parameters when designing powertrain mounts.

The second important parameter is **dynamic stiffness,** c_{dyn} (see Fig. 8). The viscoelastic properties of rubbers result in a dynamic stiffness that is proportional to the excitation velocity.

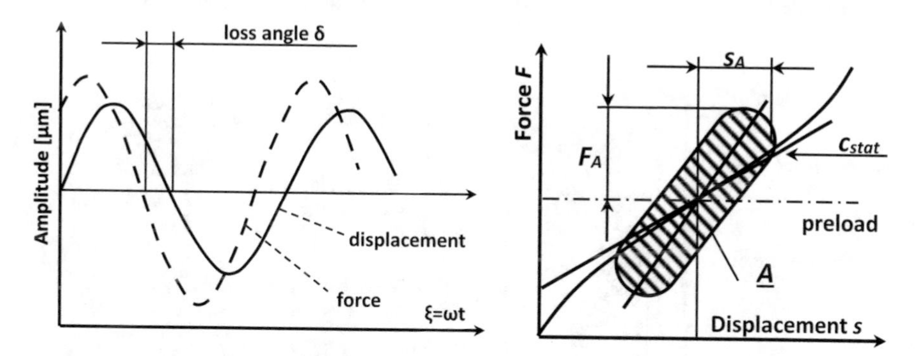

Fig. 8. Loss angle and dynamic stiffness as dynamic properties of rubber-metal mount

Dynamic stiffness and loss angle are possible to compute from the following equations:

$$c_{dyn} = \frac{F_A}{s_A} \tag{1}$$

$$\delta = arcsin \frac{A}{F_A . s_A . \pi} \tag{2}$$

F_A – force [N],
s_A – displacement [mm],
A – area of force-displacement hysteresis loop [mm^2].

3 Experimental Measurement of Dynamic Properties of Rubber-Metal Motor Mount

To determine the dynamic properties, it is necessary to excite the electric motor rubber mount with a sinusoidal signal with a constant amplitude of acceleration in the specific frequency range. Force-displacement curves are obtained from the logged force and the displacement. Displacement is determined from logged acceleration. Force-displacement curves shown the hysteresis loop. The area of the hysteresis loop shows damping energy from which it can be determined a loss angle. Dynamic stiffness is determined by dividing the force and displacement for each frequency separately.

3.1 Experimental Measurement Test Rig Configuration

For identifying dynamic properties at high frequencies, it is appropriate to use an electromagnetic shaker for excitation of the rubber-metal mounts because the hydraulic pulsators are frequency limited [15, 16]. Due to the wide frequency measurement range, the measurement may be affected by resonances e.g. test rig frame or test jig. Therefore, this must be considered, modal analyses of the test jig must be carried out and also the test rig frame itself must be designed outside the resonances.

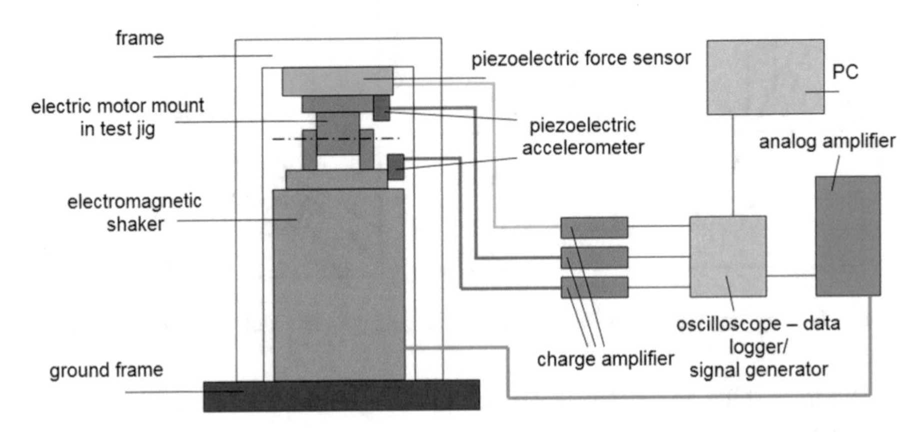

Fig. 9. Schematic view of test rig for determination of dynamic properties of rubber-metal mounts

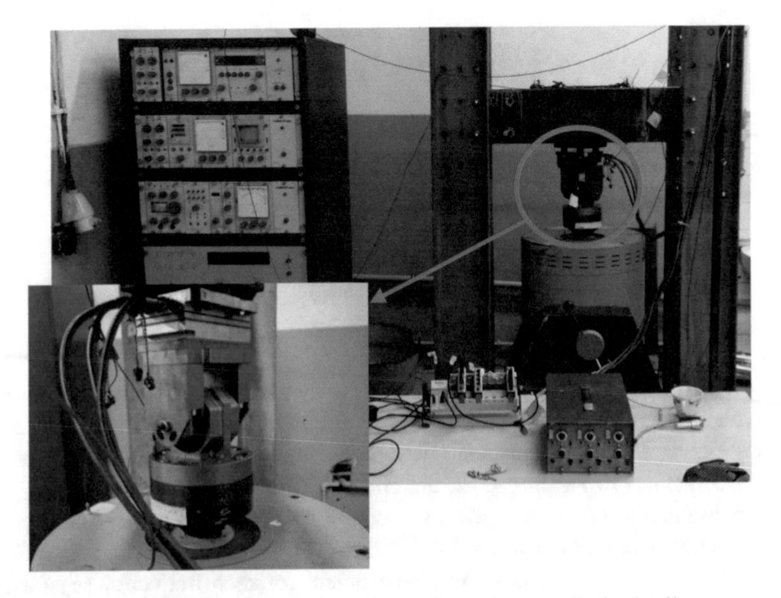

Fig. 10. Measurement test rig configuration, test jig in detail

The electromagnetic shaker test rig is shown (see Figs. 9 and 10), which shows the complex measurement chain also with data logging. Due to the need to measure at frequencies up to 2000 Hz, piezoelectric acceleration sensors (Bruel & Kjaer) and force sensor (Kistler) with charge amplifiers were selected. The oscilloscope (Pico 5000) was selected to transfer the data to the PC, which includes a signal generator for the electromagnetic shaker.

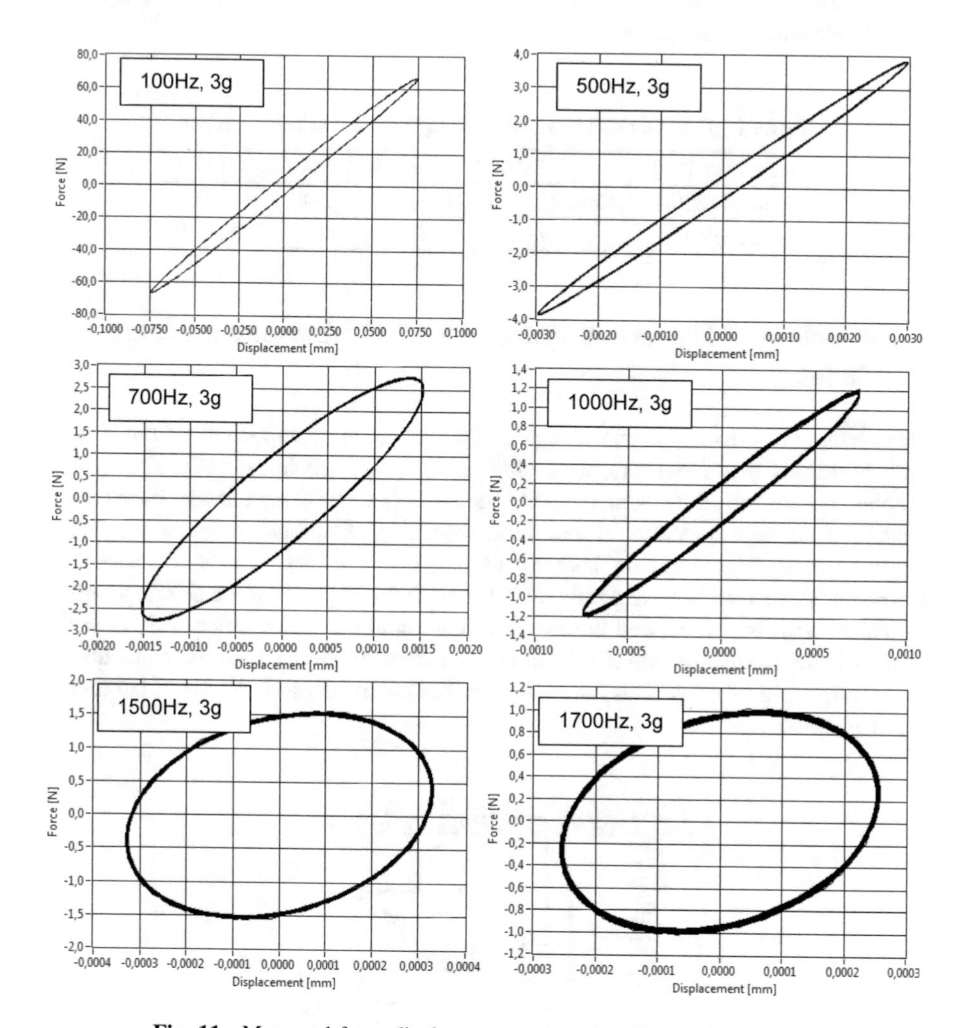

Fig. 11. Measured force-displacement curves for selected frequencies

3.2 Measured Force-Displacement Curves

For experimental measurement was set following conditions:

- Excitation sinusoidal signal,
- Excitation frequency range 100–1700 H,

- Excitation amplitude of acceleration 3g.

Force-displacement curves for selected frequencies are presented in the following figure (see Fig. 11). Curves are created from logged data.

Dynamic stiffness and loss angle were also obtained from logged data. Table 1 shows dynamic stiffness and loss angle for the selected frequency. Values of dynamic stiffness are increased and around 1500 Hz is the highest dynamic stiffness. It is caused by the resonance of the test specimen.

Table 1. Dynamic stiffness and loss angle for the selected frequency

Frequency [Hz]	100	500	700	1000	1500	1700
Dynamic stiffness [N/mm]	899	1298	1820	1598	4658	3884
Loss angle [°]	−176	−175	155	−169	105	75

4 Modeling of Hysteresis Properties of Rubber-Metal Motor Mount

Hysteresis is a property of wide range of systems as the components of the systems with hysteresis may have a response differently to applied forces [17]. This type of problem can be solved by using the FEM model [18], or by using a model based on the mathematical representation of nonlinear behavior. A problem where the output response to a given input is known but parameters of the process equations are unknown is called an inverse problem [19]. This is the opposite of the classic problems in which input is defined and the response is monitored. The problem solved in this chapter is the estimation and identification of parameters that define the mathematical prescription of a function. This function describes the hysteresis behavior of the rubber-metal mounts.

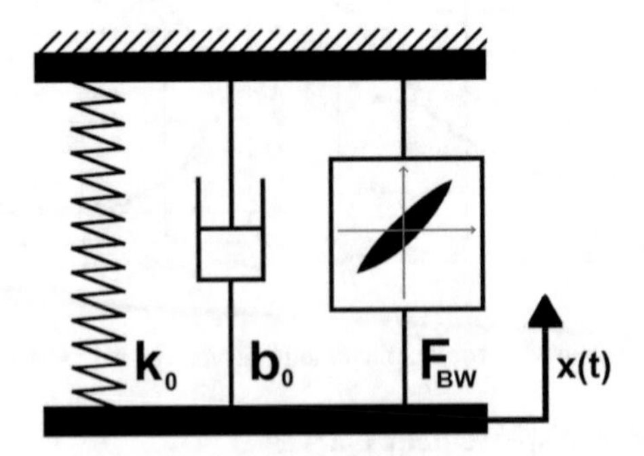

Fig. 12. Model of hysteresis

Bouc-Wen model (see Fig. 12) was proposed initially by Bouc early in 1967 and generalized by Wen in 1976. Its typical equations are expressed as follows:

$$\dot{z}(t) = A\dot{x}(t) - \gamma\dot{x}(t)|z(t)|^n - \beta|\dot{x}(t)|z(t)|z(t)|^{n-1} \tag{3}$$

$$F^{el}(t) = ak_i x(t) \tag{4}$$

$$F^h(t) = (1-a)k_i z(t) \tag{5}$$

$$F_{BW} = F^{el}(t) + F^h(t) \tag{6}$$

$$F_R = k_0 x(t) + b_0 \dot{x}(t) + F_{BW} \tag{7}$$

The overall response is representing by force F_R, it is the equation of motion of a single-degree-of-freedom. That force includes the force of the damping, stiffness and hysteresis component. The hysteresis component of the response is representing by force F_{BW}, this force contains an elastic $F^{el}(t)$ and a hysteretic part $F^h(t)$. Where the ratio of post-yield to pre-yield (elastic k_i) stiffness, the value of a is $0<a<1$. Hysteresis in the hysteretic part represents dimensionless hysteretic parameter $z(t)$ consist of some parameters, that describe the shape of hysteresis. Four hysteresis parameters A (controlling hysteresis amplitude), β, γ and n, and their interactions determine the basic hysteresis shape [20]. Relationships between β and γ determine whether the model is hardening or softening behavior, it is according to Table 2.

Table 2. Relationships between β and γ for $n = 1$, [20]

Interaction between β and γ parameters	Influence on hysteresis
• $\left.\begin{matrix}\beta+\gamma>0\\ \gamma-\beta<0\end{matrix}\right\rangle$	Weak softening
• $\left.\begin{matrix}\beta+\gamma>0\\ \gamma-\beta=0\end{matrix}\right\rangle$	Weak softening on loading mostly linear unloading
• $\left.\begin{matrix}\beta+\gamma>\gamma-\beta\\ \gamma-\beta>0\end{matrix}\right\rangle$	Strong softening on loading and unloading narrow loop
• $\left.\begin{matrix}\beta+\gamma=0\\ \gamma-\beta<0\end{matrix}\right\rangle$	Weak hardening
• $\left.\begin{matrix}\beta+\gamma<0\\ \beta+\gamma>\gamma-\beta\end{matrix}\right\rangle$	Strong hardening

To correctly define the model of hysteresis of rubber mount, it is important to identify the model parameters (see Fig. 13).

Parameter estimation represents an optimization problem, where the objective function is measured data. To parameter estimation was used Matlab Parameter Estimation Toolbox. As the optimization method was chosen Nonlinear least squares with Trust-Region-Reflective algorithm, where parameter and function tolerance were defined as value 0.001. The parallel pool during estimation was also used to reduce the computing time. Results of simulation model are compared to measured force-displacement curves (see Fig. 14.)

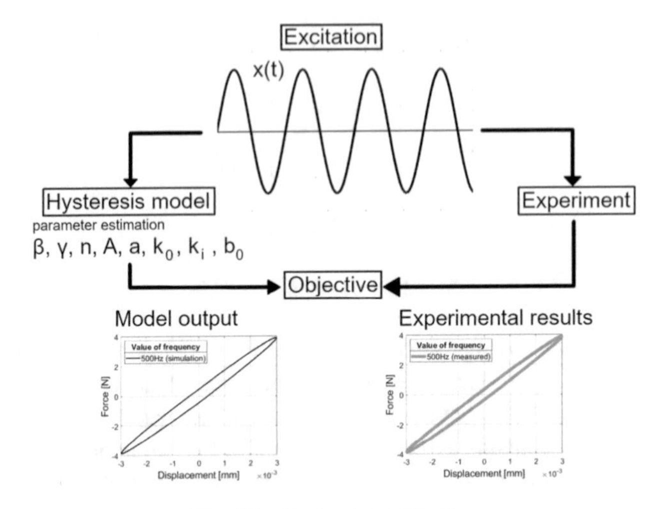

Fig. 13. System identification

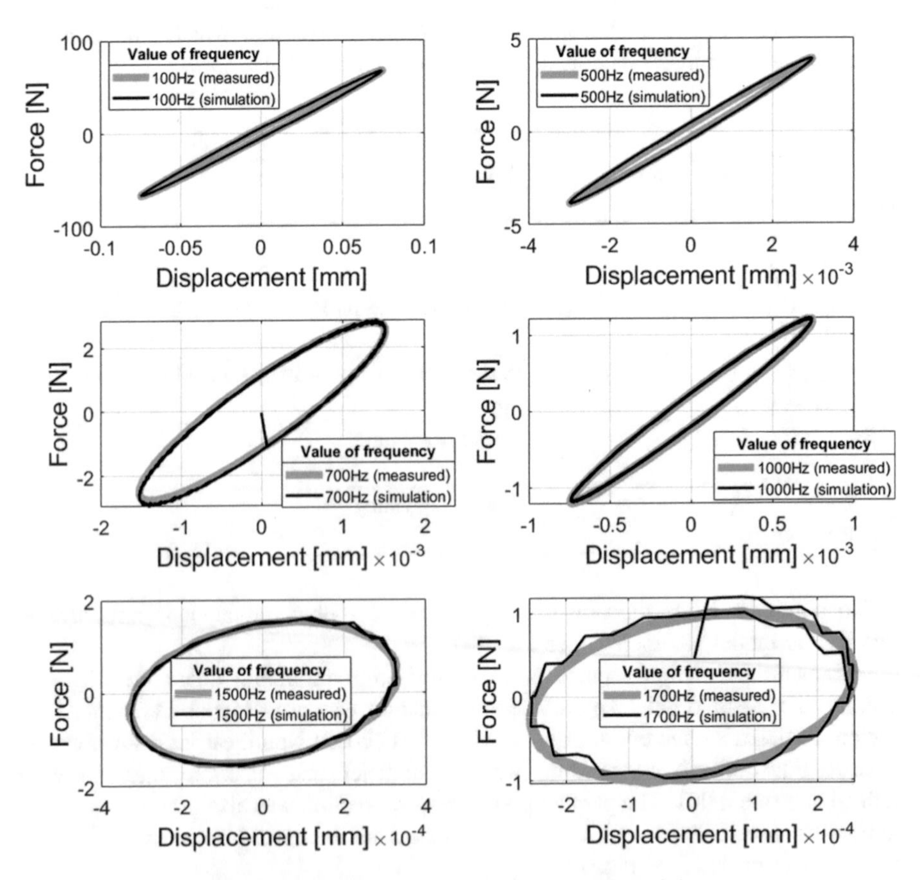

Fig. 14. Comparison of simulation to measured data

4.1 Evaluation of Computed Hysteresis Properties

The error of the model is determined by comparing the power response of the experimental measurement and force response of the model according to Eq. (8),

$$\varepsilon = \frac{\sum_{i=1}^{k} (F_{R_i} - F_{EXP_i})}{k} \tag{8}$$

where k is the number of rows in vector, F_{EXP_i} is the i-th value of the response force from experiment and F_{R_i} is the i-th value of the response force from the simulation model. Calculating the percentage deviation of the conformity σ of the model experiment is performed according to Eq. (9).

$$\sigma = \sqrt{\frac{1}{k}\sum_{i=1}^{k} \varepsilon^2} \tag{9}$$

Table 3. Results of errors of modeled and measured forces

	100 Hz	500 Hz	700 Hz	1000 Hz	1500 Hz	1700 Hz
ε	0.0877	0.0751	0.0466	0.0213	0.0057	0.0903
σ	0.0049	0.0087	0.0024	0.0025	0.0066	0.0104

In Table 3 are presented the results of the errors of modeled and measured forces. From the results of the model, evaluation is shown that the model of hysteresis is approximated characteristics obtained from measure nearly ideally. The deviation is caused by the estimation of model parameters. Estimation of parameters is specific and required to create their own method of optimization.

5 Conclusion

One of the aims of this paper is to show differences in the area of electric powertrain NVH from internal combustion engine powertrain. Also, the differences effort on mount system of the powertrain. Rubber-metal mounts are appropriate to reduce structure-borne vibration from the electric powertrain. However, rubber-metal mounts for electric powertrain must have to fulfill different requests, like a wide operating frequency range (up to 2000 Hz), lower excitation amplitudes, different force loads, different torques, etc. This results in some problems such as dynamic hardening, resonances of rubber at high frequency. Therefore, it is necessary to determine the dynamic properties of the designed mount.

Experimental determination of dynamic properties is time and cost consuming which leads to the use of simulations of powertrain mounts. Another of the described research aims is to perform one of the simulation options. Bouc-Wen model was used for simulation of the dynamic properties of the rubber-metal motor mount. This model is often used and brings good accuracy. This was confirmed by comparison experiments. Identification of the model parameters is needed for each frequency which is a disadvantage of this model.

Other simulation options will be studied in next research phases in order to find an optimal simulation model for the determination of dynamic properties of rubber-metal powertrain mount.

Acknowledgment. The authors would like to express their gratitude to the Slovak Research and Development Agency and to the Ministry of Education, Science and Technological Development of Republic of Serbia for the support of the bilateral project No. SK-SRB 18-0045.

References

1. Popovic, P., Ivanovic, G., Mitrovic, R., Subic, A.: Design for reliability of a vehicle transmission system. Proc. Inst. Mech. Eng., Part D J. Automob. Eng. **226**(2), 194–209 (2012)
2. Miskovic, Z., Mitrovic, R., Maksimovic, V., Milivojevic, A.: Analysis and prediction of vibrations of ball bearings contaminated by open pit coal mine debris particles. Technical Gazzete (Tehnicki Vjesnik) **24**(6), 1941–1950 (2017)
3. Tasic, M., Mitrovic, R., Popovic, P., Tasic, M.: Influence of running conditions on resonant oscillations in fresh-air ventilator blades used in thermal power plants. Therm. Sci. **13**(1), 139–146 (2009)
4. Miskovic, Z., Mitrovic, R., Stamenic, Z.: Analysis of grease contamination influence on the internal radial clearance of ball bearings by thermographic inspection. Therm. Sci. **20**(1), 255–265 (2016)
5. Heißing, B., Ersoy, M.: Chassis Handbook: Fundamentals, Driving Dynamics, Components, Mechatronics, Perspectives, 1st edn. Vieweg+Teubner Verlag, Wiesbaden (2011)
6. Eisele, G., Genender, P., Wolff, K., Schürmann, G.: Electric vehicle sound design – just wishful thinking? In: Pischinger, S. (ed.) Proceedings of AAC 2010, Aachen Acoustics Colloquium, 23–24 November 2010, pp. 47–59. FEV Motorentechnik, Aachen (2010)
7. Kumar, G.: NVH challenges in hybrid and electric vehicles. Acoust. Bull. **40**(6), 44–48 (2015)
8. Danko, J., Magdolen, L., Masaryk, M., Bugar, M., Madaras, J.: Energy management system algorithms for the electric vehicle applications. In: Mechatronics 2013 - 10th International Conference, pp. 25-31. Springer, Cham (2014)
9. Sound and Vibration Homepage. http://www.sandv.com/downloads/1104goet.pdf. Accessed 07 June 2019
10. Crocker, M.J.: Handbook of Noise and Vibration Control, 1st edn. Wiley, New York (2007)
11. Test based Optimization of NVH for hybrid and electric vehicles. https://community.plm.automation.siemens.com/siemensplm/attachments/siemensplm/Simcenter_event_tkb/109/6/26_Vortrag.pdf. Accessed 22 May 2019
12. Avoid NVH issues in electric vehicles by using the source-transfer-receiver methodology. https://www.plm.automation.siemens.com/global/cz/topic/nhv-issues-in-electric-vehicles/58756. Accessed 22 May 2019
13. Zeng, X., Liette, J., Noll, S., Singh, R.: Analysis of motor vibration isolation system with focus on mount resonances for application to electric vehicles. SAE Int. J. Alt. Power. **4**(2), 370–377 (2015)

14. Shangguan, W.B.: Engine mounts and powertrain mounting systems: a review. Int. J. Veh. Des. **49**(4), 237–258 (2009)
15. Mitrovic, N., Petrovic, A., Milosevic, M., Momcilovic, N., Miskovic, Z., Maneski, T., Popovic, P.: Experimental and numerical study of globe valve housing. Chem. Ind. **71**(3), 251–257 (2017)
16. Danko, J., Milesich, T., Bucha, J.: Nonlinear model of the passenger car seat suspension system. J. Mech. Eng. **67**(1), 23–28 (2017)
17. Zhua, X., Lua, X.: Parametric identification of Bouc-Wen model and its application in mild steel damper modeling. In: Heung Fai, L. (ed.) Proceedings of the Twelfth East Asia-Pacific Conference on Structural Engineering and Construction, vol. 14, pp. 318–324. Elsevier Procedia (2011)
18. Miskovic, Z., Mitrovic, R., Tasic, M., Tasic, M., Danko, J.: Determination of the wing conveyor idlers' axial loads using the finite element method. In: Mitrovic, N., Milosevic, M., Mladenovic, G. (eds.) Experimental and Numerical Investigations in Materials Science and Engineering, CNNTech 2018. Lecture Notes in Networks and Systems, vol. 54, pp. 174–192. Springer, Cham (2019)
19. Sage, A.P.: System identification history, methodology, future prospects. In: System Identification of Vibrating Structures. American Society of Mechanical Engineers, New York (1972)
20. Sengupta, P., Li, B.: Modified Bouc-Wen model for hysteresis behavior of RC beam-column joints with limited transverse reinforcement. Eng. Struct. **46**, 392–406 (2013)

The Effect of Inert Gas in the Mixture with Natural Gas on the Parameters of the Combustion Engine

Andrej Chríbik[1(✉)], Marián Polóni[1], Matej Minárik[1], Radivoje Mitrovic[2], and Zarko Miskovic[2]

[1] Faculty of Mechanical Engineering, Institute of Transport Technology and Engineering Design, Slovak University of Technology in Bratislava, 812 31 Bratislava, Slovakia
andrej.chribik@stuba.sk

[2] Faculty of Mechanical Engineering, University of Belgrade, Kraljice Marije 16, 11120 Belgrade 35, Serbia

Abstract. The article discusses the influence of inert gases (carbon dioxide and nitrogen) in a mixture with natural gas on power and economic parameters of the internal combustion engine LGW 702 designed for micro-cogeneration units. The experimental measurements were made under various engine operating modes and under various compositions of fuel mixtures. The aim of the experiment was to analyze and assess the full impact of the inert components in the gaseous fuel, especially on the particular integral parameters, as well as on the internal combustion engine parameters relating to the course of burning the mixture. Experimental results indicate a decrease in performance parameters and an increase in specific fuel consumption with an increase in the proportion of internal gases in the mixture. Increasing proportion of inert gases leads to decreasing maximum pressure in the cylinder (a decrease approximately by 50% with the mixture CO2NG50 or by 30% with the mixture N2NG50, compared to natural gas) and the position of maximum pressure value is shifted further into the area of expansion stroke.

Keywords: Internal combustion engine · Synthesis gas · Natural gas · Combustion · Pressure

1 Introduction

The use of synthesis gas (syngas), being produced from renewable energy sources, as an alternative fuel for cogeneration units allows the reduction of the burden on the environment by greenhouse gases. Depending on the mode of production of syngas, a certain amount and composition of the gas can be produced [1]. The syngases are gases with basic components, such as hydrogen, methane, carbon monoxide, carbon dioxide, higher hydrocarbons and nitrogen. The blow by exhaust gases from the combustion chamber to crankcase creates an oil-water emulsion which causes the engine oil degradation as well as premature wear of the engine main sliding bearings [2].

N. Mitrovic et al. (Eds.): CNNTech 2019, LNNS 90, pp. 410–426, 2020.
https://doi.org/10.1007/978-3-030-30853-7_24

The article gives an account on the influence of inert gases (nitrogen N2 and carbon dioxide CO2) present in the gaseous alternative fuels on the engine parameters. This issue has been addressed in several papers, for example [4–8]. The high temperature of the engine oil causes baking of the oil creating deposits in the lubrication canals. Such buildups can be dangerous to camshaft rolling bearing when released without proper dispersion of the clump particles. The failure of rolling bearings and supported shafts, as well as their relations, are also addressed by the authors in the numerous papers, such as [9–13].

The aim of the experiment was to analyze and assess the full impact of the inert components in the gaseous fuel, especially on the particular integral parameters, as well as on the internal combustion engine parameters relating to the course of burning the mixture. Because of certain simplifications, as well as lower costs for realization of the experiments, the effect of inert gases was assessed in the fuel mixture formed by natural gas combined with the particular inert gas.

2 Experimental

Table 1 presents the essential physical and chemical properties of the mixture of inert gas with the natural gas (96% CH_4). Generally speaking, with the increasing proportion of inert gas mixed with natural gas comes a decrease in the mass lower heating value. This decrease of the heating value is also reflected in a drop of the engine performance parameters. Comparing the inert gases, the decrease in the mass lower heating value is more significant in the mixture CO2NG than in the mixtures N2NG.

Table 1. Physical and chemical properties of mixture of natural gas (NG) with inert gas (IG).

Parameter	Unit	Natural gas	CO2NG50	N2NG50
NG/IG	[% vol.]	100/0	50/50	50/50
Molar Mass	[g.mol^{-1}]	16.75	30.38	22.38
Density	[kg.m^{-3}]	0.696	1.263	0.930
Lower heating value of fuel	[kJ.kg^{-1}]	48 692	13 421	18 218
Air to fuel ratio	[kg.kg^{-1}]	16.95	4.62	6.25
Mass heating value of mixture	[kJ.kg^{-1}]	2 712	2 386	2 514
Volume heating value of mixture	[kJ.m^{-3}]	3 128	2 890	2 900

All experimental measurements (power, economic and internal engine parameters) were carried out on the spark-ignition engine Lombardini LGW 702.

It is a twin-cylinder water-cooled four-stroke engine with a displacement of 686 cm^3 and a compression ratio of 12.5:1. The mixture is prepared with the help of a mixer with diffuser. The richness of the mixture is regulated by means of a control unit with a feedback regulation [3]. The basic scheme of the measuring stand is shown in Fig. 1.

The course of the pressure in the combustion chamber of the engine was recorded by piezoelectric pressure sensor integrated into the spark plug from company Kistler.

The correction of dynamic pressure course was performed by scanning the intake manifold pressure with a piezo resistive pressure sensor.

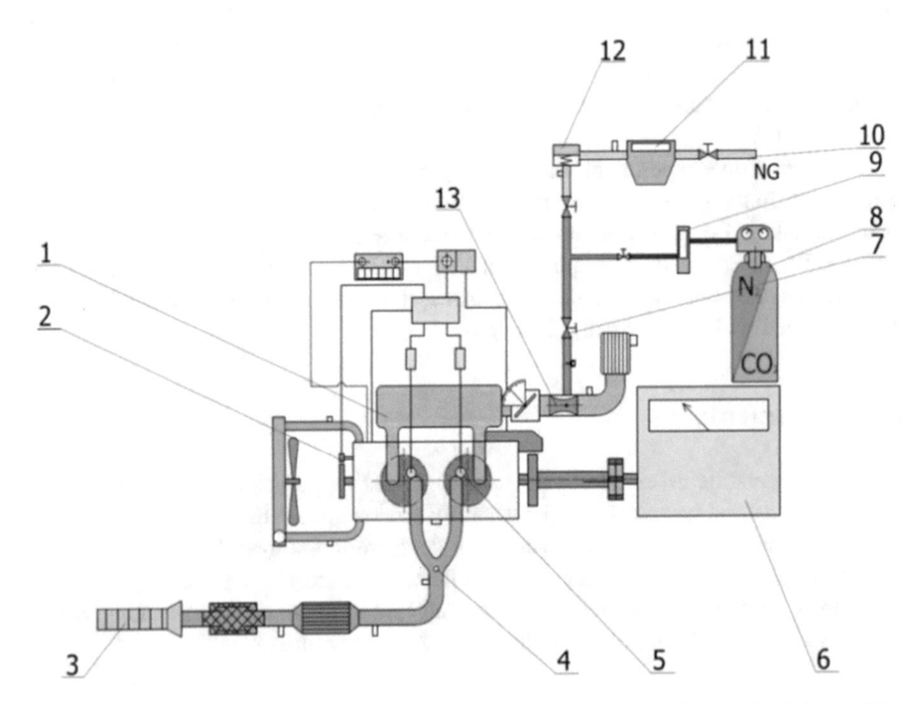

Fig. 1. The basic scheme of the combustion engine LGW 702 (1 - intake manifold, 2 - position sensor of the crankshaft, 3 - exhaust system, 4 - exhaust temperature sensor, 5 - spark plug with integrated pressure sensor, 6 - dynamometer, 7 - mixture richness regulation, 8 - pressure bottle of inert gas, 9 - mass flowmeter of inert gas, 10 - supply of natural gas, 11 - volumetric flowmeter of natural gas, 12 - zero pressure regulator, 13 - mixer with diffuser)

3 Experimental Results

The experimental measurements were made with various fuel compositions (volume ratio of the inert gas to natural gas) at various engine revolutions and at full load.

Figure 2 shows the course of the torque by composition of the mixture at various engine revolutions. Generally speaking, the addition of nitrogen to the natural gas decreases the torque more slowly than when carbon dioxide is added to the natural gas.

A comparison of the mixtures CO2NG50 and N2NG50 to natural gas itself at the revolutions of 1 500 min^{-1} has shown that the torque decreases by about 15.6 N.m with mixture CO2NG50, and by about 4.8 N.m with mixture N2NG50, compared to natural gas (44.7 N.m), at unchanged operating conditions.

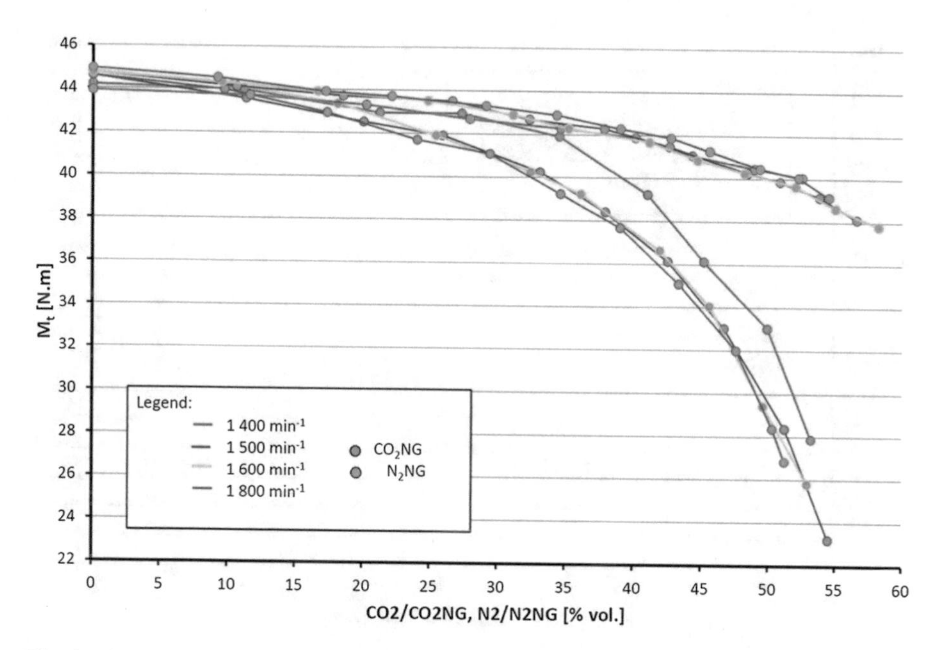

Fig. 2. Course of the reduced torque M_t depending on the various proportions of inert gas mixed with natural gas for a variety of internal combustion engine revolutions (Conditions: full load, stoichiometric ratio, angle of ignition advance φ_{ign} = 25°CA BTDC)

Figure 3 shows the relation between the specific fuel consumption and the composition of mixture for various engine revolutions. The lowest specific fuel consumption is with the combustion of natural gas (258 $g.kW^{-1}.h^{-1}$) at the revolutions of 1500 min^{-1}.

The combustion of the mixture CO2NG50 under the same operating conditions requires the specific fuel consumption 1349 $g.kW^{-1}.h^{-1}$, and the mixture N2NG50 requires 708 $g.kW^{-1}.h^{-1}$.

Increasing proportion of inert gas also increases the hourly fuel consumption. As an example, under the combustion of natural gas, this value is 1.8 $kg.h^{-1}$. However, under the combustion of the mixture CO2NG50 this value will raises by 233%, to around 6 $kg.h^{-1}$. The combustion of the mixture N2NG50 leads to the increase of hourly fuel consumption by 145%, to 4.4 $kg.h^{-1}$ compared to the combustion of natural gas.

The effective efficiency of the internal combustion engine at the revolutions 1500 min^{-1}, burning natural gas, was at 28.6% and the efficiency of CO2NG50 was 20.5%, and the efficiency of burning the mixture N2NG50 was 26.5%.

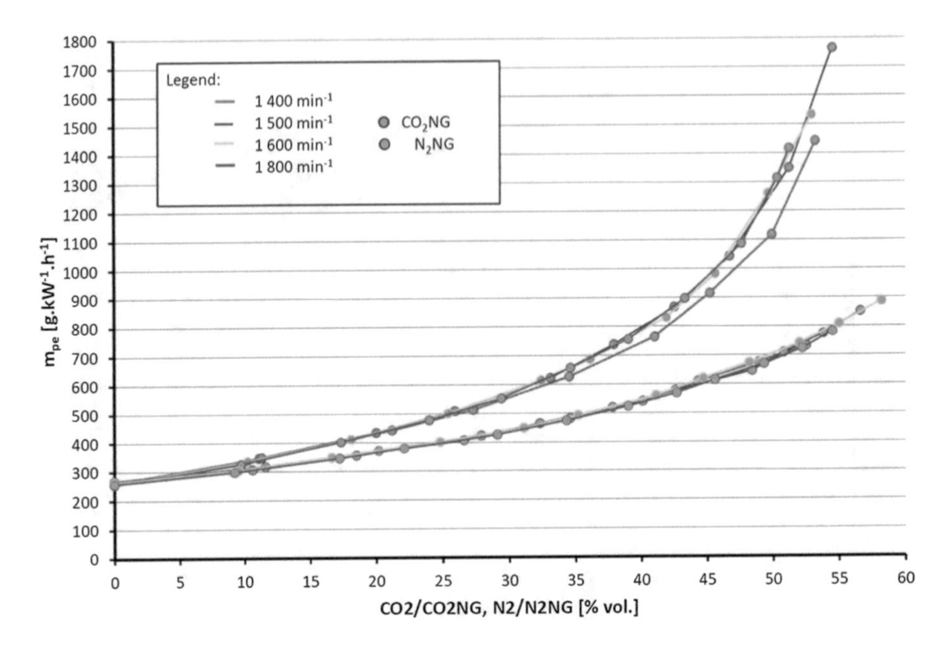

Fig. 3. The course of the reduced specific fuel consumption m_{pe} as dependent on various proportions of inert gas in mixture of natural gas for various engine revolutions (Conditions: full load, stoichiometric ratio, angle of ignition advance φ_{ign} = 25°CA BTDC)

Increasing proportion of inert gas in natural gas increases the time delay of ignition, as well as the total time of fuel mass fraction burned. Combustion itself takes place deep in the expansion stroke of the piston. This extension of the burning time also results in an increase of the exhaust gas temperature at the outlet of the cylinder of the engine (position of the temperature sensor No. 4 at Fig. 1).

Figure 4 shows the course of the temperature of exhaust gases as dependent on fuel composition. Generally, with increasing engine revolutions the temperature of the exhaust gases increases. For comparative revolutions of 1500 min^{-1}, with natural gas the exhaust temperature is at 592 °C. Increasing proportion of carbon dioxide increases the exhaust gas temperature. With the mixture CO2NG50 the rise of temperature has been 14% to approximately 672 °C, under same operating conditions. Adding nitrogen to natural gas led to a temperature reduction of the exhaust gas. For the mixture N2NG50 the temperature was 578 °C at 1500 min^{-1}, which is a temperature decrease by 2.5%, compared to the combustion temperature of natural gas.

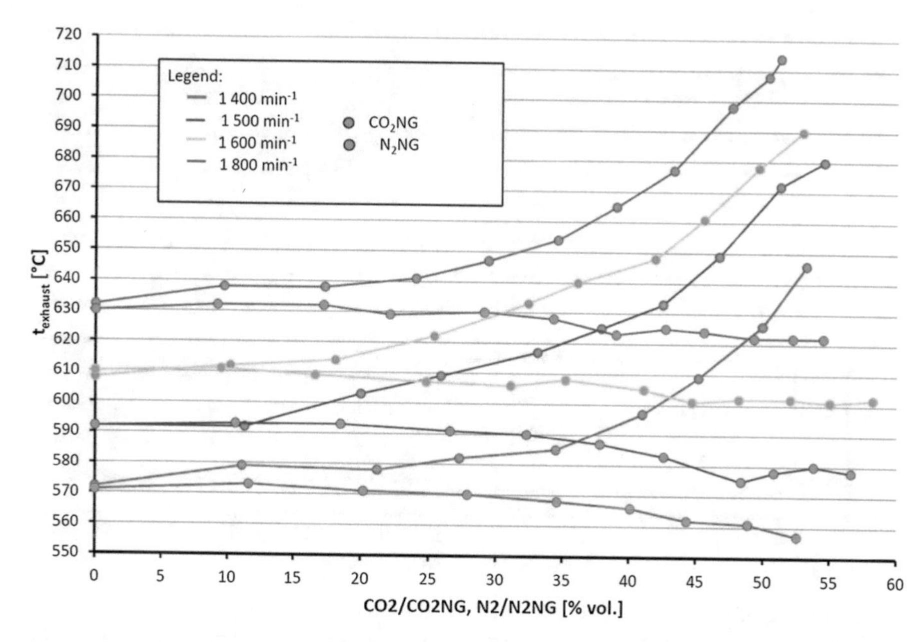

Fig. 4. Course of the exhaust gas temperature as dependent on various portions of inert gas mixed with natural gas, for various engine revolutions (Conditions: full load, stoichiometric ratio, angle of ignition advance φ_{ign} = 25°CA BTDC)

Figure 5 shows the course of engine effective efficiency for different engine fuels. The highest efficiency is achieved with natural gas (about 29%). With gradually rising proportion of nitrogen in the fuel, the overall efficiency is virtually unchanged. Conversely, with increasing carbon dioxide, the overall engine efficiency decreases to 17.2% in the case of 55% volumetric proportion of carbon dioxide in the fuel mixture.

Along with measurement of the integral parameters, analysis of the pressure in the engine combustion chamber and of the impact of the inert gas on the parameters associated with the actual course of combustion mixture was carried out.

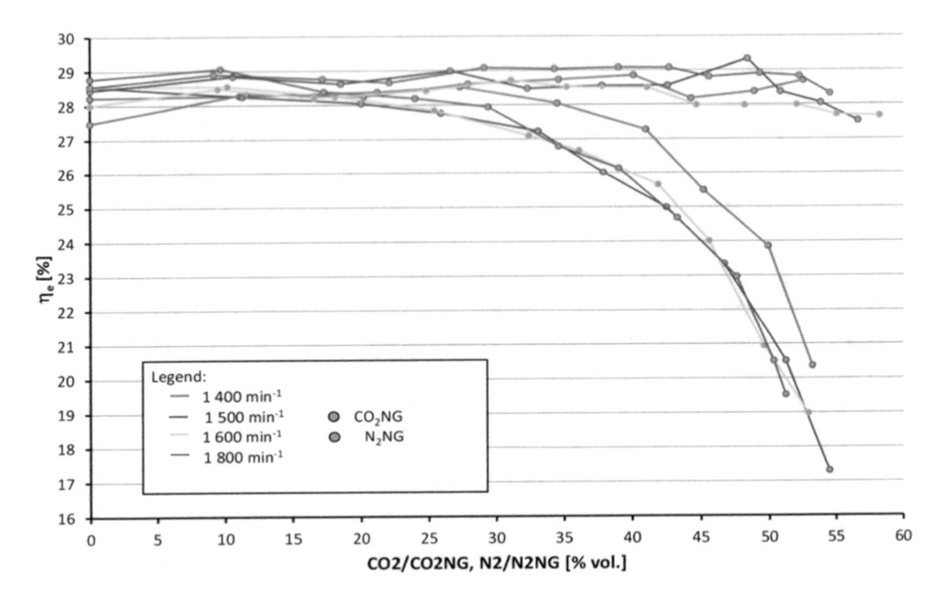

Fig. 5. Course of the effective efficiency η_e depending on the various proportions of inert gas mixed with natural gas, for a variety of internal combustion engine revolutions (Conditions: full load, stoichiometric ratio, angle of ignition advance φ_{ign} = 25°CA BTDC)

3.1 Pressure Course Analysis

Important data in terms of a comprehensive analysis of the effect of the fuel mixture on the combustion engine parameters are the obtained pressure curves from the combustion chamber of the engine and their analysis. For each measurement, analysis of 195 consecutive cycles of pressure development in the combustion chamber was performed. The signals were taken from the measured pressure, from the actual position of the crankshaft, from the position of top-dead center (TDC) and from the spark breakdown and were processed by a program created in Matlab. The pressure correction was made on the basis of the recorded dynamic pressure and the reference pressure in the intake manifold, so that the absolute pressures in the cylinder could be precisely determined.

The process of combustion in the combustion chamber of the engine was evaluated on the basis of single-zone-zero-dimensional thermodynamic model. The model was developed for a closed system, i.e. the intake and exhaust valves were closed, on the principle of energy conservation. The course of the release of heat from the fuel mixture was determined on the basis of analyzed course of pressure in the combustion chamber by Rassweiler-Withrow method. The method is based on the fact that the pressure increase in the combustion chamber comprises both the increase in combustion pressure and the increase in pressure, caused by changed volume as the piston moves in the cylinder (Eq. (5)). It is based on the general law of energy conservation for the open system:

$$dU = dQ - dW + \sum_i h_i.dm_i \quad (1)$$

Where:

dU - change of internal energy of matter in the system
dQ - heat delivered to the system
dW - the work produced by the system
hi.dmi - *i*-th component of enthalpy of mass flow across system boundaries

The change of heat in the system consists of the heat released from the chemical energy (dQCH) during the combustion process, and from the heat taken from the system (dQht) by the effect of heat transfer to the walls of the combustion space. The piston work (dWp), which is performed by the gases on the piston, is considered positive. Then:

$$dQ_{CH} = dU + dW_p - \sum_i h_i.dm_i + dQ_{ht} \quad (2)$$

For ideal gas, the energy change is a function of the mean gas temperature and is determined by the following form:

$$dU = m\, c_v\, dT + u\, dm \quad (3)$$

where:

m - the total filling mass
c_v - specific heat capacity at constant volume

Using the previous Eq. (3) and the state equation of gas, and after modification of the Eq. (2), the following form is obtained:

$$dQ_{CH} = \frac{1}{\kappa - 1} Vdp + \frac{\kappa}{\kappa - 1} pdV + \left(u - \frac{rT}{\kappa - 1}\right) dm - \sum_i h_i dm_i + dQ_{ht} \quad (4)$$

During combustion it is assumed that the last three members of previous Eq. (4) are zero-valued. The resulting relation between cylinder pressure components is as follows:

$$dp = \frac{\kappa - 1}{V} dQ_{CH} - \frac{\kappa p}{V} dV = dp_c + dp_p \quad (5)$$

where:

dp_c - pressure change to benefit combustion
dp_p - pressure change to benefit volume change

As a follow-up, the fuel burn-off pattern can be derived from the following relation:

$$x_\alpha = \frac{\sum_{i=0}^{\alpha} \Delta p_{ci}}{\sum_{i=0}^{EOC} \Delta p_{ci}} \qquad [-] \tag{6}$$

Where EOC is the moment of end of fuel burning.

The start and the end of combustion were determined by change in entropy. This method is described in more detail in the literature [14]. The basis lies in the change of entropy at the beginning and at the end of burning in the closed cycle of the combustion engine.

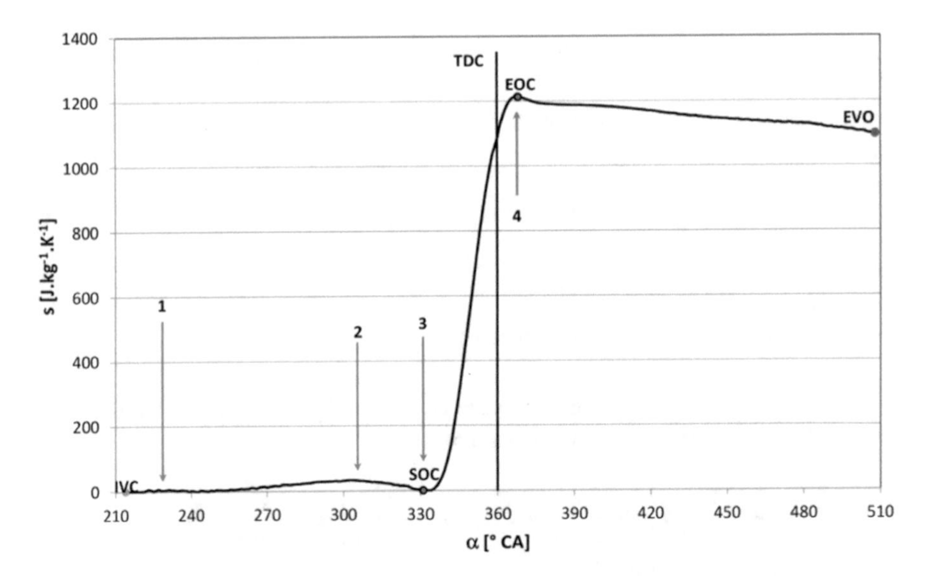

Fig. 6. Dependence of entropy change *(s)* on crankshaft rotation angle *(α)* (fuel: natural gas, full load, stoichiometric ratio, revolutions n = 1500 min^{-1}, SOC - start of combustion, EOC - end of combustion, IVC - intake valve closing, EVO - exhaust valve opening, TDC - top dead center)

After closing the intake valve, the mixture is heated by both the compression and the heat transfer from the combustion chamber walls to the gaseous medium. In case the temperature of the mixture is lower than the temperature of the cylinder walls, the heat transfer is orientated into the mixture and entropy increases (between points 1 and 2, Fig. 6). Conversely, the change in entropy decreases when the temperature of the gases exceeds the wall temperature (from point 2 to point 3). At the beginning of the combustion, when the heat release rate is faster than the heat transfer to the walls of the cylinder, entropy begins to increase again (between points 3 and 4).

In the last stage of burning, when the heat transfer exceeds the rate of heat generation by combustion, i.e. the combustion ends. At this point, entropy begins to decline. The local maximum value represents the end of combustion (point 4). In general, the combustion starts at the minimum entropy value and ends at the maximum entropy value.

The pressure measured continuously allowed for evaluation of uniformity of combustion engine running, which is characterized by the so-called coefficient of variation (COV). The coefficient of variation is calculated as the ratio of standard deviation to the arithmetic average of the parameter examined, thus:

$$COV = \frac{\sqrt{\frac{1}{n-1}\sum_{i=1}^{n} x_i - \overline{x}}}{\overline{x}}.100 \qquad [\%] \tag{7}$$

where n is the total number of consecutive cycles, xi is the i-th value of the tested parameter in a given cycle and $\bar{x}$ is the average value of the tested parameter from a given number of cycles. The cycle variability has a significant impact on the overall life of the internal combustion engine, as well as on its performance parameters and related fuel consumption.

Figure 7 shows the pressure course for various gas mixtures in the combustion chamber. The resulting pressure for mixtures of natural gas with inert gas have been initially measured at optimum start of ignition for natural gas (SOI - 25°CA BTDC), and then the pressure courses for CO2NG50 and N2NG50 have been measured at optimum spark advance angles. The maximum pressure for natural gas is 6.029 MPa at 13.5°CA after TDC. During combustion of the CO2NG50 mixture, the maximum pressure at spark advance 25°CA BTDC is reduced to 2.980 MPa at the angle 9.6°CA after TDC.

The optimum ignition angle for the particular mixture is 42°CA BTDC. At this angle of SOI, the maximum pressure 5.833 MPa at 10.2°CA BTDC rose up. If combusting the N2NG50 mixture, the maximum pressure at SOI 25°CA BTDC is reduced to 4.467 MPa at 14.2°CA after TDC, compared to natural gas. The optimum SOI angle for the given mixture is 32°CA BTDC. At this angle, the maximum pressure 5.871 MPa at 11.1°CA after TDC rises up. This optimization of SOI underlines the importance of changing SOI angles in case of changing fuel.

The following figure (Fig. 8) shows the courses of the burned fuel mass proportion (MFB), depending on the angle of the crankshaft rotation for the mixtures of inert gases with natural gas. The number of analyzed cycles was 195. From the burning of the fuel it can be seen that the increase of inert gases in natural gas leads to a longer ignition delay and the total extension of the burning period, which leads to a slower heat release, especially in the last phase of burning.

The ignition delay (from SOI to 5% of MFB) was the shortest in the case of natural gas operation (18.2°CA), while the highest ignition delay value was at CO2NG50 (32.9°CA), under optimized SOI angle, which means an increase of 80%. If the SOI angle for the CO2NG50 mixture remains at 25°CA BTDC, the ignition delay is reduced to 25.2°CA, but the main combustion period ($\alpha_{10-90MFB}$) is extended from 28° CA to 45.8°CA for the SOI angle 25°CA BTDC. The main combustion period for natural gas is 23°CA.

The time interval between the spark-over and the moment, when 10% of fuel has been burned, is the shortest in the case of natural gas (21.6°CA) operation. The longest burning in the above-mentioned interval is for CO2NG50, with optimum SOI angle (35.9°CA). The period from the moment of spark-over to the moment, when 50% of the fuel is burned, is the shortest when burning natural gas (33.0°CA). The longest run period burns 50% MFB in CO2NG50 (48.5°CA). For the N2NG50 mixture with optimum SOI angle, this value is 39.4°CA. The burning period, from the spark-over to the moment, when 90% of the fuel has been burned, is again the shortest with natural gas (44.6°CA) and vice versa the longest burning period is the one when burning CO2NG50 (75.5°CA), or CO2NG50, with the optimum SOI angle (63.8°CA). For the N2NG50 mixture, this value is 52.3°CA, or, in case of optimum SOI angle, this value is 57.7°CA.

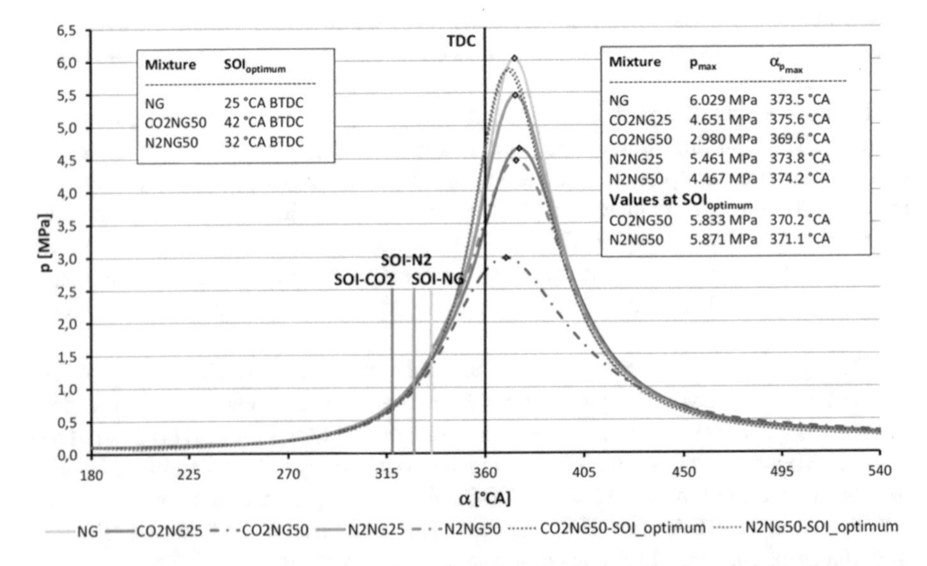

Fig. 7. Comparison of average pressure courses (p) in the combustion chamber for various mixtures of natural gas with inert gases in dependence of crankshaft angle (α) (conditions: full load, stoichiometric ratio, revolutions n = 1 500 min^{-1})

Figure 8 also shows the effect of composition of mixture on the coefficient of variation (COV). The statistical analysis of the COV for the particular periods, when 5%, 10%, 50% and 90% of the fuel was burned sequentially, shows that the mixture CO2NG50, or N2NG50, has had the greatest variance in combustion.

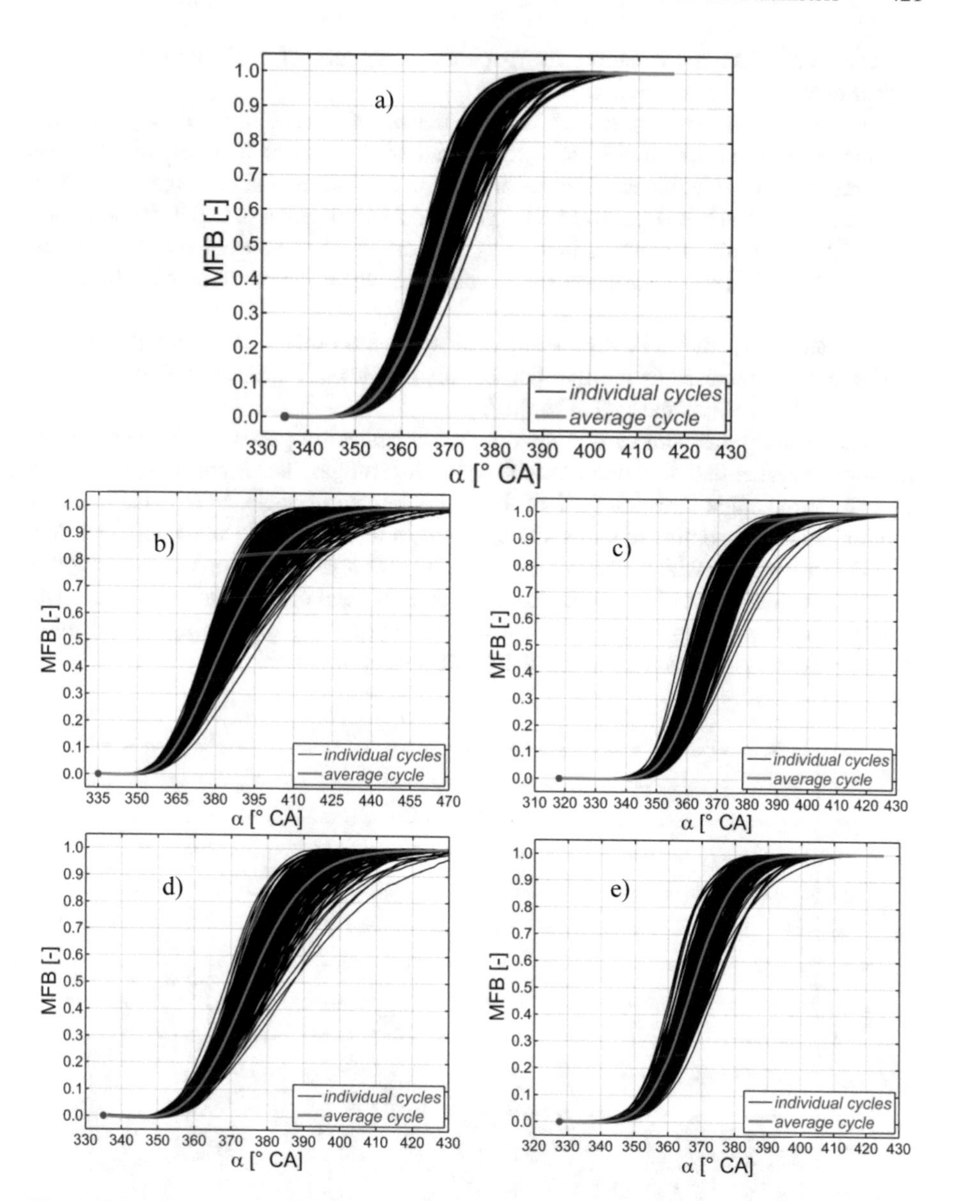

Fig. 8. Mass fraction burned (MFB) of the fuel depending on the crankshaft angle (α) for the fuel mixtures (a) natural gas – NG, (b) CO2NG50, (c) CO2NG50-SOI$_{optimum}$, (d) N2NG50, (e) N2NG50- SOI$_{optimum}$, (conditions: number of cycles 195, full load, stoichiometric mixture with air, revolutions 1 500 min^{-1})

Thus irregularities in courses of combustion occur at non-optimum SOI angle. The lowest coefficients of COV for particular amounts of fuel burned were found in operation of the internal combustion engine running on natural gas.

The coefficient of variation for particular amounts of fuel burned generally decreases with optimized SOI angle.

Figure 9 shows the values of the coefficient of variation (COV_{MFB}) for the crankshaft angle at which 5%, 10%, 50% and 90% mass of fuel for different gas mixtures are burned, at the same angle of ignition advance and at the angle of ignition advance, which has been optimized for a given fuel composition (CO2NG50-optimum, N2NG50-optimum). In general, increasing proportion of combusted fuel leads to an increase of coefficient of variation for the angle at which the given amount of fuel is burned.

For natural gas, the COV for the angle, at which 5% of the fuel is burned, is 0.32% and by gradual increase of the fuel burned this coefficient increases to 0.80% for the angle at which 90% of the fuel is burned.

When natural gas with carbon dioxide is combusted, in general, the coefficient of variation is higher than that one for combusting natural gas. The highest COV value is the one for the angle at which 90% of the fuel is burned (2.36%). In case the angle of ignition advance is optimized, this value decreases to 1.13%. The same applies to the N2NG50 mixture, in which case, after optimizing the angle of ignition advance; for this mixture, the coefficient of variation decreases at each point of burning the mixture. The exception is the start of combustion, where the optimizing of the angle of ignition advance has no significant effect on the coefficient of variation.

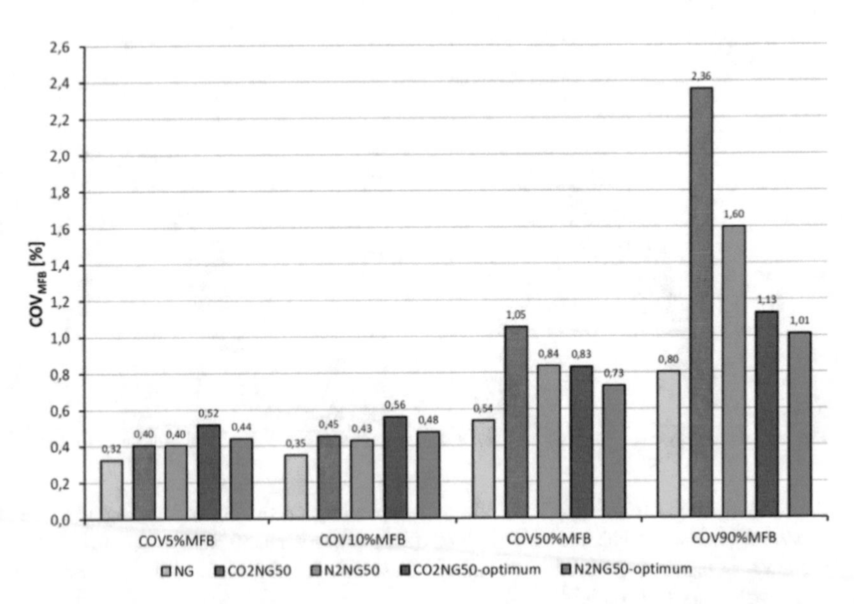

Fig. 9. Values of coefficient of variation (COV_{MFB}) for the crankshaft angle, at which 5%, 10%, 50% and 90% of the natural gas fuel and blends of natural gas with inert gases, before and after the angle of ignition advance is optimized

If we analyze the dependence of the course of maximum pressure in the combustion chamber from the crankshaft angle at which 50% of the fuel is burned relative to different angle of ignition advance (from 18°CA BTDC to 40°CA BTDC), we get a course as shown in Fig. 10. In general, the highest performance parameters are achieved in the range of 8 to 10°CA after TDC. The average maximum pressure value for the highest performance parameters is achieved between 5.5 MPa and 6.0 MPa.

The highest values of the maximum pressure in the cylinder are reached at a large angle of ignition advance, when the largest amount of fuel is burned before the top dead center. The maximum cylinder pressure (8.5 MPa) is achieved while burning natural gas, when the angle position at 50% MFB is about 20°CA BTDC. This value stabilizes and no longer changes with increasing angle of ignition advance. By gradual reduction of the angle of ignition advance, the angle values, at which 50% of the fuel is burned, also shift into the expansion stroke region, and, as a follow-up, the values of maximum cylinder pressure for all mixtures are also reduced.

The lowest value of the maximum pressure in the cylinder (3 MPa) is approximately 30°CA after the top dead center and the gradual increase does not change the maximum pressure. The graph also shows a variability of individual cycles. The lowest variability of the maximum pressure is achieved in combustion of natural gas and, on the contrary, the highest variability can be found in combustion of a mixture of natural gas with nitrogen. The variability of the cycle of maximum pressure for the mixture of natural gas with carbon monoxide has a low value and is similar to that one of natural gas.

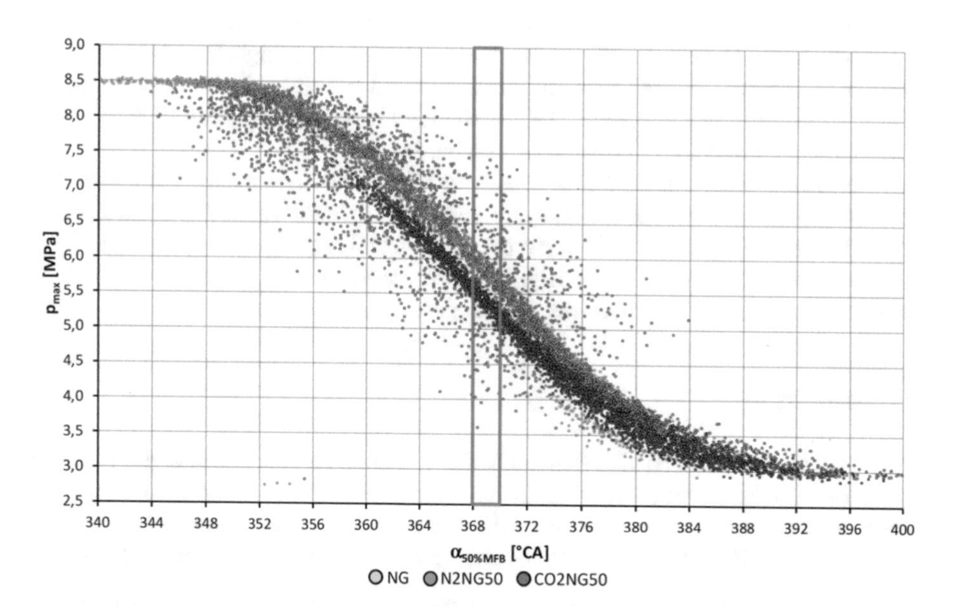

Fig. 10. Comparison of maximum pressure (p_{max}) in the combustion chamber for various mixtures of natural gas with inert gases in dependence on crankshaft angle, at which 50% of fuel ($\alpha_{50\%MFB}$) is burned, (conditions: full load, stoichiometric ratio, revolutions n = 1 500 min[1])

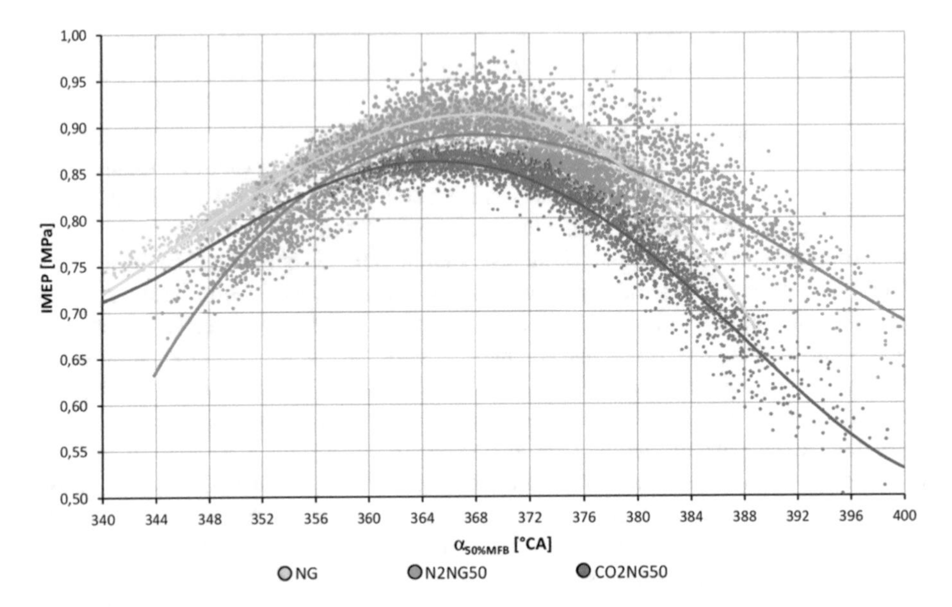

Fig. 11. Comparison of indicated mean effective pressure (*IMEP*) for various mixtures of natural gas with inert gases in dependence on crankshaft angle, at which 50% of fuel ($\alpha_{50\%MFB}$) is burned (conditions: full load, stoichiometric ratio, revolutions n = 1 500 min^{-1})

Figure 11 shows the progress of the indicated mean pressure as dependent on the crankshaft angle at which 50% of the fuel is burned. The highest indicated mean pressure is achieved in operation on natural gas and has the value of approximately 0.92 MPa at the angle 8°CA after TDC. A slightly lower value of IMEP (0.89 MPa) is achieved when the N2NG50 mixture is combusted.

The lowest IMEP (0.865 MPa) applies to CO2NG50. The graph also shows the variability of the mean effective pressure value; the greatest variability relates to the N2NG50 mixture, in which case some cycles achieve higher IMEP values than those achieved with natural gas. Conversely, the lowest value of the coefficient of variation IMEP is achieved while combusting natural gas.

4 Conclusion

The experiments on combustion of natural gas with different ratios of inert gas (CO2 or N2) have brought the following findings regarding the performance, consumption and parameters related to the combustion of the mixture in the cylinder (comparison of the mixture CO2NG50 or N2NG50 with natural the gas under revolutions of 1500 min^{-1}):

- Increase in the proportion of carbon dioxide or nitrogen decreases the engine performance parameters (comparison of the mixtures CO2NG0 and CO2NG50 demonstrates a decrease in torque of 35% and comparison of the mixtures N2NG0 and N2NG50 demonstrates a decrease in torque of 11%).

- Increasing proportion of CO_2 or N_2 in the fuel mixture leads to the increase in the specific fuel consumption (an increase of 420% with the mixture CO2NG50 or an increase of 175% with the mixture N2NG50, compared to natural gas).
- Increasing proportion of inert gases leads to decreasing maximum pressure in the cylinder (a decrease approximately by 50% with the mixture CO2NG50 or by 30% with the mixture N2NG50, compared to natural gas) and the position of maximum pressure value is shifted further into the area of expansion stroke.
- With increasing proportion of carbon dioxide or nitrogen mixed with natural gas a decrease of pressure rise rate by 62% occurs in case of CO_2 or 43% in case of N_2.
- Adding carbon dioxide or nitrogen to the mixture with natural gas extends both the total period of combustion and each of the stages of combustion, and at the same time the coefficient of variation (COV) increases for the position of the crankshaft when 5%, 10%, 50% and 90% of mass fraction (MFB) of the fuel is burned.

Acknowledgement. This work was supported by the Slovak Research and Development Agency under Contracts-No. APVV-17-0006, APVV-0015-12 and Contract-No. APVV-14-0399, and was also supported by the Slovak Scientific Grant Agency under the Contracts-No. VEGA 1/0301/17 and KEGA 026STU-4/2018.

References

1. Ahmed, I.I., Gupta, A.K.: Pyrolysis and gasification of food waste: syngas characteristics and char gasification kinetics. Appl. Energy **87**(1), 101–108 (2010)
2. Stanković, M., Marinković, A., Grbović, A., Mišković, Ž., Rosić, B., Mitrović, R.: Determination of Archard's wear coefficient and wear simulation of sliding bearings. Ind. Lubr. Tribol. **71**(1), 119–125 (2018)
3. Chríbik, A., Polóni, M., Lach, J., Ragan, B.: Utilization of Synthesis gases in combustion engine. In: Conference KOKA 2015, Kočovce, pp. 229–238 (2015). ISBN 978-80-227-4424-9
4. Zeng, W., Ma, H., Lian, Y., Hu, E.: Experimental and modelling study on effects of N2 and CO2 on ignition characteristics of methane/air mixture. J. Adv. Res. **6**(2), 189–201 (2014)
5. Al-Hamamre, Z., Diezinger, S., Talukdar, P., Von Issendorff, F., Trimis, D.: Combustion of low calorific gases from landfills and waste pyrolysis using porous medium burner technology. Process Saf. Environ. Prot. **84**(4), 297–308 (2006)
6. Chan, Y.L., Zhu, M.M., Zhang, Z.Z., Liu, P.F., Zhang, D.K.: The effect of CO2 dilution on the laminar burning velocity of premixed methane/air flames. Energy Procedia **75**, 3048–3053 (2015)
7. Lackner, M., Winter, F., Agarwal, A.K.: Handbook of Combustion Vol. 3: Gaseous and Liquid Fuels. WILEY-VCH Verlag GmbH & Co. KGaA (2010)
8. Puškár, M., Jahnátek, A., Kuric, I., Kádárová, J., Kopas, M., Šoltésová, M.: Complex analysis focused on influence of biodiesel and its mixture on regulated and unregulated emissions of motor vehicles with the aim to protect air quality and environment. Air Qual. Atmos. Health **12**, 1–10 (2019)

9. Miskovic, Z., Mitrovic, R., Tasic, T., Tasic, M., Danko J.: Determination of the wing conveyor idlers' axial loads using the finite element method. In: Mitrovic, N., Milosevic, M., Mladenovic, G. (eds.) Experimental and Numerical Investigations in Materials Science and Engineering. CNNTech 2018, CNNTech 2018. Lecture Notes in Networks and Systems, vol. 54, pp. 174–192. Springer (2019)
10. Ristivojevic, M., Mitrovic, R., Lazovic, T.: Investigation of causes of fan shaft failure. Eng. Fail. Anal. **17**(5), 1188–1194 (2010). Elsevier - Pergamon
11. Momcilovic, D., Odanovic, Z., Mitrovic, R., Atanasovska, I., Vuherer, T.: Failure analysis of hydraulic turbine shaft. Eng. Fail. Anal. **20**(1), 54–66 (2012). Elsevier-Pergamon
12. Grujičić, R., Tomović, R., Mitrović, R., Jovanović, J., Atanasovska, I.: The analysis of impact of intensity of contact load and angular shaft speed on the heat generation within radial ball bearing. Therm. Sci. **20**(5), 1765–1776 (2016)
13. Lazovic, T., Mitrovic, R., Ristivojevic, M.: Influence of internal radial clearance on the ball bearing service life. J. Balk. Tribol. Assoc. Sci. Bulg. Commun. **16**(1), 1–8 (2010)
14. Tazerout, M., Le Corre, O., Ramesh, A.: A new method to determine the start and end of combustion in an internal combustion engine using entropy changes. Int. J. Thermodyn. **3**(2), 49–55 (2000)

ISO 9001:2015 as a Framework for Creation of a Simulation Model for Business Processes

Petar Avdalovic[1], Jelena Sakovic Jovanovic[2](✉), Sanja Pekovic[3], Aleksandar Vujovic[2], and Zdravko Krivokapic[2]

[1] Mixed Holding Power Utility of the Republic of Srpska, Parent Joint - Stock Company, Stepe Stepanovića b.b., 89101 Trebinje, RS, Bosnia and Herzegovina
[2] Faculty of Mechanical Engineering Podgorica, University of Montenegro, 81000 Podgorica, Montenegro
sjelena@t-com.me
[3] Faculty of Tourism and Hospitality, University of Montenegro, 85330 Kotor, Montenegro

Abstract. Establishing quality management systems within an organization has a goal to achieve total performance and sustainable development. New release of ISO 9001:2015 standard has brought rudimentary changes, making opportunities for effective accomplishment of these goals. Numerous researches on this topic highlight the fact that the use of a technique for knowledge discovery has a big impact on the fulfilment of requests set by ISO 9001 standards. Based on the results of this research, the authors decided to investigate a possibility for creation of tools for knowledge discovery through the respect of principles, requests and recommendations of this standard. The development of such a tool (simulator) would have a goal to predict future states of business processes of an organization, which is the base for performing an efficient risk estimation and optimization of business processes. This paper describes an approach to the development of such a tool (simulator) based on key business processes in an organization and data on realization of these processes during the time, and placing them into a context by selected statistical methods.

Keywords: ISO 9001:2015 · Process approach · Simulation · Model

1 Introduction

A key novelty introduced by ISO 9001:2015 standard is related to the integration of risk estimation into a quality management system. A risk represents an uncertainty for the fulfilment of established goals, hence risk assessments are carried out with a goal to prevent unwanted conditions in a business system. Risk management demands predictions of conditions of business systems based on adequate methods and techniques. For the purposes of risk management concept, ISO 9001:2015 standard has highlighted the demands for the establishing of knowledge based organization.

This standard defines the specific knowledge for particular organization gained through experience, most often through solving problems in order to accomplish the

N. Mitrovic et al. (Eds.): CNNTech 2019, LNNS 90, pp. 427–441, 2020.
https://doi.org/10.1007/978-3-030-30853-7_25

organization's goals. Any knowledge management has a goal to extract useful knowledge from available data about a business system. Different methods and software systems such as: Data Warehouse, OLAP, Data Mining, Expert Systems and Process Mining are used for this purpose [1, 2].

In their previous researches, the authors of this paper had analysed a correlation of demands of ISO 9001:2015 standard and the practices in knowledge management [3], which resulted in a rank list yield in relation to the criteria based on this standard [4]. Knowledge management has developed and defined various methodological approaches to the realization of its goals. The practical realization of these methods and use of practical knowledge, by using adequate IT services, namely the group of user software, is defined under the term [5]. The results of this assessment have shown that the efficacy in decision making process is fairly increased by supporting itself on knowledge management practices. New knowledge can be generated from data stemming from tracing and measurements of business process performances according to the quality management system of the selected organization [4].

Standard ISO 9001:2015 strictly demands an analysis and estimation of collected data through the use of applicable statistical techniques. Having regard to the previously said, this paper is focused on the development of a practical simulation model for one organization through the satisfying requests from ISO 9001:2015 standard. It has been assumed that a quality management system in an organization can give enough data about the process's performances for the execution of a simulation. A simulation here means an applied methodology describing the behaviour of a system through the use of mathematical or graphical models [6]. Such a model would enable an adequate support in the decision making process necessary for the risk management and it would also be based on a knowledge management.

2 The Concept of Modelling and Simulation

Data mining is defined as a process of problem solving through analyses of data that are already present in databases [97 to the demands from ISO 9001:2015 standard for the development of a simulation model.

Reference literature no. [7–11] mostly present the use of data mining on data that are structured in the existing databases, not taking into account data in format of formalized entries demanded by QMS.

The demand no. 9.1. of ISO 9001 standard imperatively prescribes that a systematic tracing, measurement, analysis and estimation of business process performance should be performed within the QMS of one organization. This is specifically related to reference [12]:

- Collecting and analyses of adequate data;
- Estimating where an improvement and effectiveness could be accomplished;
- Analysis of adjustments to demands of clients;
- Analysis of characteristics of products and processes;
- Analysis of data from suppliers.

Given the purposes of fulfilment of this demand in the standard, organizations are obligated to collect and process big quantities of data, which, through certain transformations into structured data, then become usable for the execution of the data mining process.

2.1 The Basic Mathematical Model of a Process Organized System

Each business system can be described by a basic mathematical model in the following way:

$$y(t) = S(x(t)) \tag{1}$$

In this formulation, a non-empty group of inputs $x(t)$ corresponds to a non-empty group of outputs $y(t)$. The functions $x(t)$ and $y(t)$ have been defined in a time t for which the following relation applies $t_0 < t < t_1$. The inputs $x(t)$ (material, energy, information) can be corrected by the control of a state of outputs and management, or it is possible to work on this very system S, so as it could accomplish the function of its goal [6].

Business systems fulfil their obligations and realize goals through their business processes. The state of any system is determined by the group of values of individual elements of business processes in a certain moment. In other moment, some of these values will change and the system will transition into some other state. Therefore, it can be concluded that a process is a time-based range of changes in the system's states. The schematic presentation of one business process with its constituent elements is given in Fig. 1.

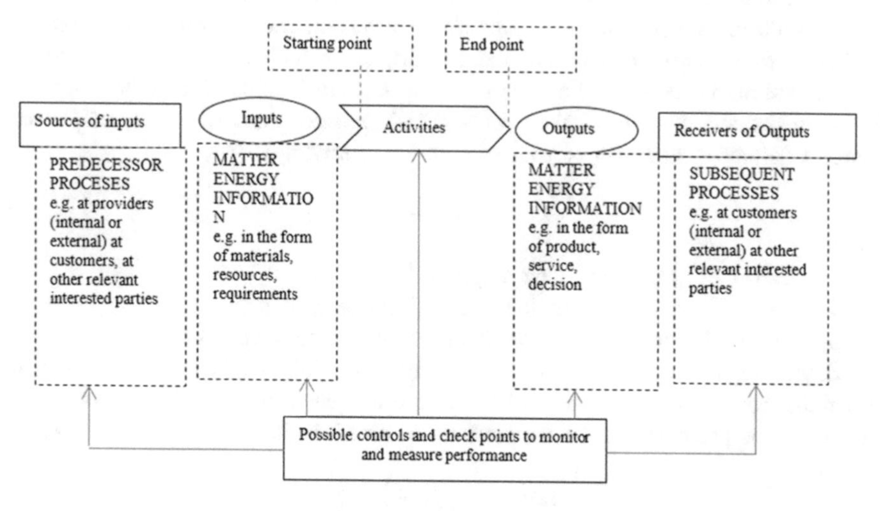

Fig. 1. The schematic presentation of elements of one process [13]

The previous figure illustrates a conceptual design of a business process. Business processes can be key and auxiliary business processes. A system realizes its goals through the key business processes [2]. Auxiliary processes have a purpose of giving support to key business processes. In further considerations, only key business processes will be observed.

Any process consists of one or more activities. Inputs into a process can be inputs into a system or outputs from another process that is situated inside or outside the system. Also, outputs from a process can be inputs into the next process or outputs from the process [14].

Requirement no. 8.4.2. of ISO 9001:2015 standard demands the following:

"An organization should:

(a) Ensure that the processes which are externally purchased stay within the management domain of its quality management system;
(b) Define the managements planned to be applied onto the external supplier and those planned to be applied on the resulting output elements."

According to this demand of the standard the boundaries of the observed system which, from the standpoint of QMS, can be narrower or wider than the boundaries of the business system that is considered, can also change. These boundaries mostly depend on the included external processes.

Each individual process can be presented by a certain function in the following form:

$$f : A \rightarrow B \quad (2)$$

A – Group of data about inputs into a process, domain

B – Group of data about outputs from the process, codomain.

A group of data presents all inputs about materials, energy, information, products and services which enter and exit the process, and which emerged during the considered period of execution of a process. These data are formulated as finite ranges. If it is assumed that a system consisted of *n* serially related processes, then each process inside a defined system can be presented in the following way:

$$f_i : A_i \rightarrow B_i \quad (3)$$

i – number of processes inside a system i = 1, 2, …, n,

A_i – group of data about the inputs in i-th process, domain

B_i – group of data about the outputs from i-th process, codomain

Having regard to the previously said, the basic formula of a business system having n serially connected processes which execute a transformation of an input into an output can be presented as the composition of functions having the following shape:

$$f_n(f_{n-1}(\ldots(f_1(x)))) = y \quad (4)$$

for each $x \in A_1$ and $y \in B_n$.

In the same way, one process can have inputs into several pre-processes, hence the basic formula of a business system made of n serially linked processes which perform transformation of inputs into outputs can be presented as the composition of functions having the following shape:

$$f_n(f_{n-1}(x_{n-1}),\ldots,f_1(x)) = y \tag{5}$$

for each $x \in A_i$ i $y \in B_n$.

Mathematical formulations (4) and (5) for connecting business processes present a base for developing a mathematical model of the considered organizations. The mathematical model of some organization directly depends on the established links between its processes. For the determination of functions the start is always from f_n function.

2.2 The Groups of Data

When variants of connecting processes are being defined, it is also necessary to define the groups of data which present information about inputs and outputs from the process.

A_i and B_i groups of input and output data can be presented in the form of matrices in the following way:

Input group of data (Input):

$$A_i = I_{j_i,k_i} \tag{6}$$

And output group of data (Output):

$$B_i = O_{j_i,l_i} \tag{7}$$

j – Number of measurements by each performance of elements of input and output from the process. The measurements are related to the changes of characteristic values of elements during the work of any process. $j = \{1, 2, \ldots, m\}$

k – Total number of characteristics of input elements of business process i taken into consideration. $k = \{1, 2, \ldots, maxk_i\}$

l – Total number of characteristics of output elements in business processes i taken into consideration. $l = \{1, 2, \ldots, maxl_i\}$

An available number of characteristics and measurements is taken into consideration for each process. The number of measurements for each characteristic within one process must be the same.

2.3 The Mathematical Model of a Process

Since some processes can have several input elements (material, energy, information), and each of these inputs describes several different data types (for example, for material it may be a type of material, hardness, elasticity, conductivity…) which could yield a huge groups of data. For the purposes of the rationalization of the model, it is necessary to remove all those data from further consideration, which are proven not to have a significant impact on the output from the process, as well as all those data about outputs which do not have significant impact on the next process. It can be done in two ways:

- The first way is to take into consideration all the input and output data and to create a correlation matrix in order to calculate a correlation coefficient for each possible pair of data of inputs and outputs in this matrix.
- The second way is that an analyst creates an impact model between all data inside the process and selects only those that have the most significant impact onto the output of the process. After that, a correlation matrix is generated again for the selected data about inputs and outputs. In this way, the analyst eliminates all the data for which he is sure that it does not have a more significant impact on output, and performs the calculation of a correlation coefficient for the remaining pairs of input and output data.

The development of each mathematical model starts with determining of the type of functional dependence. Functional dependence is determined on the basis of the scatter plot graph [15]. In the development of this model, the data of the selected organization were used, and based on them, the scattering diagrams were used. As it turned out that this is a liner function, the Pearson coefficient will be used to determine the strength and direction of the connections between the input and output variables for a particular process [16].

The same number of measurements for each data enables an uninterrupted normal distribution necessary for the calculated correlations.

The mathematical model of a simple process. An average value of elements in column k_i of matrix I is equal to:

$$\overline{x_{k_i}} = \frac{1}{m}\sum\nolimits_{j=1}^{m} x_{j_i k_i} \tag{8}$$

An average value of elements in column $\mathrm{k_i}$ of matrix O is equal to:

$$\overline{y_{l_i}} = \frac{1}{m}\sum\nolimits_{j=1}^{m} y_{j_i l_i} \tag{9}$$

A correlation coefficient is calculated for each pair of columns of matrices I and O $\{(x_{1_i}, y_{1_i}), \ldots, (x_{k_i}, y_{l_i})\}$, based on the following pattern:

$$r_{x_{k_i} y_{l_i}} = \frac{\sum_{j=1}^{n}\left(x_{j_i k_i} - \overline{x_{k_i}}\right)\left(y_{j_i l_i} - \overline{y_{l_i}}\right)}{\sqrt{\sum_{j=1}^{n}\left(x_{j_i k_i} - \overline{x_{k_i}}\right)^2}\sqrt{\sum_{j=1}^{n}\left(y_{j_i l_i} - \overline{y_{l_i}}\right)^2}} \tag{10}$$

A reduced correlation matrix is created in this way. Unlike a standard correlation matrix, where the biggest correlation coefficient is searched for an arranged pair of data where it is not known which ones are dependent and which ones are independent variables, in the case of this matrix, all the variables from an input were taken as independent variables, while all variables from the output of the process were taken as dependent variables. The matrix of correlation of inputs and outputs here quantifies impacts of individual input elements on output elements of the process.

Therefore, each member of a matrix represents the coefficient of correlation for an ordered pair of input variables and outputs from the process. The dimensions of the matrix type K are equal to the dimensions of the column of the input data matrix I, while the dimensions of the column of the matrix K are equal to the dimensions of the column of the output data matrix O.

$$K_i = \begin{bmatrix} r_{x_{1_i}y_{1_i}} & \cdots & r_{x_{maxk_i}y_{1_i}} \\ \vdots & \ddots & \vdots \\ r_{x_{1_i}y_{maxl_i}} & \cdots & r_{x_{maxk_i}y_{maxl_i}} \end{bmatrix} \tag{11}$$

After the calculation of the correlation coefficients, only a pair of input-output data whose correlation coefficient has the biggest absolute value within a matrix is selected:

$$r_{max} = max\left|r_{x_{k_i}y_{l_i}}\right| \tag{12}$$

For this established pair, an equation of simple linear regression [17, 18] which presents process i with one data of input and one data of output is created:

$$y_{j,l_i} = \beta_{0_i} + \beta_{k_i}x_{j,k_i} \tag{13}$$

Parameters $\beta_{0_i} i \beta_{k_i}$ are calculated by the following equations:

$$\beta_{k_i} = \frac{\sum_{j=1}^{m}\left(x_{j,k_i} - \overline{x_{k_i}}\right)\left(y_{j,l_i} - \overline{y_{l_i}}\right)}{\left(x_{j,k_i} - \overline{x_{k_i}}\right)^2} \tag{14}$$

$$\beta_{0_i} = \overline{y_{l_i}} - \beta_{k_i}\overline{x_{k_i}} \tag{15}$$

This approach enables the determination of the function for a simple process i where correlation coefficients between all data which describe inputs and outputs of the process were determined on the basis of the correlation matrix, and the biggest ones which indicate the fact that the results from the process will be best described through the impact presented by the data x_{k_i} and the data about output from the process y_{l_i} were determined. There is $x_{k_i} = y_{l_i - 1}$;

From the group of output variables of the next pre-processing, only one variable taken into consideration in the next step was determined in this way.

The Mathematical Model of a Complex Process. Unlike the simple process, the complex process can have several inputs. These processes can be boundary, namely to be situated at the input of the system, and central processes which are situated deeper in the system's structure. For the boundary processes, a question arises- which combination of data about inputs from the system's environment has the best correlation with the output from that process. For those complex processes situated deeper in the system's structure this question is not significant, because according to the process model it was already known which pre-processes had their impacts and through which data.

Therefore, in further considerations, the focus will be only on processes which simultaneously make the system's boundaries. The determination of the biggest correlation coefficient is also an initial step in this case, where it is necessary to determine a correlation coefficient for each input-output combination. An approach to solving of this problem was given through the researches dealing with the measurements of correlation of multi variables [19]. According to this variable and its definition *w* correlation coefficient of multi variables can be obtained by the following equation:

$$r^2_{a_1 a_2 \ldots a_m} = 1 - \sum\nolimits_{j_1 < j_2 < \ldots < j_m} \left(det\left[a'(m) \lfloor j_1, j_2, \ldots, j_m \right] \right)^2 \tag{16}$$

$a_1 a_2 \ldots a_m$ are present variables of inputs and outputs. Based on this definition, n presents a total number of variables and in the considered case, it amounts to:

$$n = maxk_i + 1 \tag{17}$$

In the previous equation, 1 presents one output variable that had already been determined through the analysis of the previous process. A number of correlation coefficient for planned number of input variables is calculated by the following equation:

$$C_k^g = \frac{k!}{(k-g)! \cdot k!} \tag{18}$$

The combinations without repetition of k elements of *g-th* class present the ways where g characteristics can be selected from k characteristics ($2 \leq g \leq k$), which in the same time has an impact on the process, where an order of the selected elements is not important, and where each element can also appear only once, namely without repetitions. Based on the above-mentioned, Eq. 16 yields the following value:

$$m = g + 1 \tag{19}$$

The correlation coefficients were calculated for each combination of input variables and one output variable. If all the input variables are taken into consideration, a multiple linear model [17, 19] for process i can be obtained.

$$y_{j,l_i} = \beta_{0_i} + \beta_{k_i} x_{j,k_i} + \ldots + \beta_{maxk_i} x_{j,maxk_i} \tag{20}$$

The basic mathematical model of one business system has been obtained by tracing of the process model and by formulating of linear functions for each process with the determination of variables giving the best result, as well as with their combination by formulas (4) and (5), completely based on satisfying demands set in ISO 9001 standard.

3 Case Study

The model has been developed within a company whose basic activities are trade and supply with electricity [20]. It has been noticed that in this company the data about a monthly realization of its production deflects from the planned values, which cause incorrect estimations of incomes from the electricity sale which were sometimes 30–40% bigger or smaller than the planned ones. It was the reason for comprehensive research in this field. Since the company had planned its incomes based on data provided by related organizations and used these data as relevant, the company often had problems with the estimation of incomes in certain months. Its management demanded from expert services to make more objective estimations, which was impossible without an adequate tool which would ensure avoidance of subjectivity in processing, and would raise the results to a satisfactory level. An initial idea was to utilize all the available data and to make a simulation tool based on them, that would be a significant support in the prediction of the system's state.

3.1 Model

By studying the opportunities for the development of the simulation model it was found out that the principle of data-based simulation was the most simple and efficient for realization. The basic problem was to provide a necessary data level because the existing SCAD had not preserved historic data for longer than 6 months, while the ERP system was based only on financial indicators, not taking into account power indicators. Based on the studying of the documentation of a quality system in the mentioned organization it was found out that from the entries of different procedures it was possible to obtain significant data which were necessary to be translated from one non-structured shape (reports, minutes, notes…) into a structured shape, as well as to develop a special data warehouse [21]. An approach to the development of the model on one segment of the defined system will be presented in the following chapters.

A map of business processes which would be extended to all key processes of related organization was the initial point. New boundaries of the system were defined in this way. In further considerations only key processes of one part of the company dealing with generation of electricity were taken into account [22], and these were the processes for which a non-concordance of data had been found out but it was later eliminated.

The development of a process model and impact model. Based on the procedures, a model of key macro processes and a model of mutual impacts of elements in the complete company were developed. Figure 2 gives a graphical presentation of macro processes in an accumulation hydroelectric power plant, impacts of external factors on these processes as well as their mutual impacts.

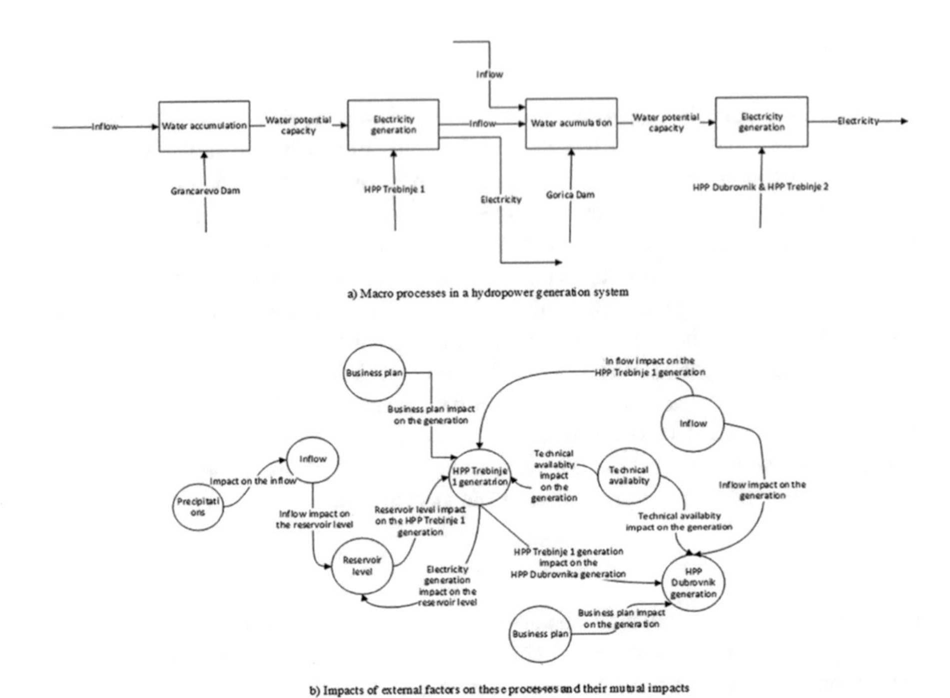

Fig. 2. The graphical presentation of macro processes in a hydroelectric power plant, impacts of external factors on these processes and their mutual impacts.

The basic characteristic of the accumulation hydroelectric power plant is its sensitivity of production to small water intake. However, there is a range of other factors that could not be quantified even though they lead to the deflection of the realized production from the planned one. Total of 6 macro processes with 31 available variables were defined. The have been related to monthly realizations for last 5 years.

The development of a mathematical model. The determination of a planned value at the exit of this subsystem, namely the production of the hydroelectric power plant, was the starting consideration for the purposes of the execution of the data analysis. Based on the measurements of a correlation coefficient it was concluded that, besides three generation facilities, the data about an equivalent operation in the grid most significantly influences the total output. The coefficient of correlation between the data about four independent variables: output of the first, second and third facility and their equivalent operation time in the grid and one dependent variable which gives the total output of the system is 0,99 which is a satisfactory level.

$$
\begin{aligned}
y_{1_6} = & -0{,}005332195 x_{1_6} + 0{,}04015396 x_{2_6} + 1{,}941599376 x_{3_6} + 0{,}2802167 x_{5_6} \\
& - 0{,}3018606
\end{aligned}
$$

The total number of input data for the sixth process was 16. The decision was made to use ones which best describe the output from the sixth process. The sixth process was defined by the total output in all three electricity production facilities and the equivalent operation time in the grid. After the definition of all the links between these processes it was possible to connect the inputs into the sixth process with other processes, and to obtain the following equation:

$$x1_6 = y1_2,$$

which represents the output in the first:

$$x2_6 = y1_3,$$

the second, and:

$$x3_6 = y1_4,$$

The third generating facility.

The output of the first facility was defined by the following mathematical expression:

$$\begin{aligned} y1_2 = & -342{,}925 + 2{,}263258x4_2 - 1{,}353016541x2_2 + 0{,}2552501x5_2 \\ & + 0{,}08241729x6_2 + 0{,}124713x7_2 \end{aligned}$$

$x4_2$ - Altitude at the beginning of a month
$x2_2$ - Altitude at the end of a month
$x5_2$ - Average intake into accumulation during a month
$x6_2$ - agreed production of the second facility
$x7_2$ - planned output of the power system
The calculated correlation coefficient for given equation is 0,72, while:

$$x6_2 = x2_6$$

For the second power facility, it was found out that there was no correlation between the data about the output and other data. The biggest correlation coefficient of 0,34 was determined for the data about the output in the first and the second power facility. An additional analysis found out that the services in the total considerations do not take the into account the data relevant for the output in the second facility, but they only join a half of that facility's output to its system. This would mean that the second generating facility is a part of production system but not a part of the business system- which is the focus of this analysis. This has led to the situation where the data about the production of this power facility must assume and take as an input parameter in further development of this model.

The data about the production of the third power facility had shown the biggest correlation dependence on the data about inter-intake, with a correlation degree of 0,57.

$$y_{1_4} = -0{,}5833 + 0{,}0592x_{1_4}$$

$$x_{1_4} = x_{5_2}$$

However, since it is a small power plant, its share in the total output is almost negligible- which can be proved by a multitude correlation coefficient that together with it amounts that are up to 0,9906, while without it amounts are 0,9899. In further considerations we assume the value of $x_{1_4} = 1$.

The mathematical model of the mentioned subsystem is obtained by composition of linear functions:

$$\begin{aligned} y_{1_6} = {} & 3{,}4682817464 - 0{,}012068133x_{4_2} + 0{,}007214548x_{2_2} - 0{,}0013610433x_{5_2} \\ & + 0{,}0397144949x_{6_2} - 0{,}000664994x_{7_2} + 0{,}2802167x_{5_6} \end{aligned}$$

In this way, the mathematical model was reduced to the following 6 independent variables:

x_{4_2} - Altitude of water accumulation at the beginning of a month
x_{5_2} - Average water intake into the accumulation in a month
x_{6_2} - agreed output of the second power facility
x_{7_2} - planned output of the system
x_{2_2} - Altitude of water accumulation at the end of a month
x_{5_6} - Equivalent operation in the grid

4 The Simulation and Verification of the Model

After the mathematical model had been created, its verification begun on the accomplishments in production for four months of 2018. The results are given in Table 1.

Table 1. The verification of the model and comparison of its results of prediction and realization

Period	x_{7_2}	x_{4_2}	x_{2_2}	x_{5_2}	x_{5_6}	x_{6_2}	y_{1_6}	Production
September	83,02	372,95	368,27	0,07	146,31	26,68	43,63	44,62
October	86,92	368,27	364,75	15,61	234,09	52,27	69,25	71,4
November	91,85	364,75	374,64	132,16	290,54	67,10	85,61	88,62
December	94,55	374,64	377,18	87,75	326,85	64,17	95,62	99,69

The verification has shown that this model yields satisfactory results. The predicted results were significantly closer to the realization ones than the planned value, which was taken into consideration of total input power. The comparison of results is given in Table 2.

Table 2. The comparison of the results of simulation and plan in comparison to the accomplished production

Period	Planned production	Predicted production	Accomplished production	Difference between accomplished and planned production	Difference between accomplished and predicted production
September	83,02	43,63	44,62	−38,40	0,99
October	86,92	69,25	71,40	−15,52	2,15
November	91,85	85,61	88,62	−3,23	3,01
December	94,55	95,62	99,69	5,14	4,07

The results presented in Table 2 best illustrate the significance of the creation of such simulation model for the prediction of states in a company's business process. Namely, a significant overlap of real data and the results obtained by the use of the mathematical simulation model was noticed. On the other hand, the results obtained by using traditional planning methods show significant deflections in comparison to real data, hence they cannot be a real base for risk estimations, decision making based on facts and general management in an organization as demanded by ISO 9001:2015 standard.

5 Conclusion

The previous researches in the area of developing tools for support to decision making process based on the use of business process simulations were carried out for individual but not simultaneously for all the business aspects. This paper presents an approach to the creation of a simulation model for a company which would reflect causal connections in this company, taking into account external processes influencing the company, with the goal to predict future states of a company and efficient management. The historical data about the realization of business processes collected through the quality system of the company present the source of data for the development of this model by which the organization's knowledge is put into operation in the very organization in concordance with the demands from ISO 9001:2015 standards.

In this regard, there is a goal to develop a unique simulation model which would integrate all the aspects of operation of a complex business system (technical and technological, economic, ecological, regulatory), which would make a base both for the prediction of states in this system and for making of objective and unbiased decisions. Based on the example of one electric power company, this paper has shown a significant overlap of the results obtained by the use of such simulation mathematical model and the real results on the example of one macro process. This has shown that the simulation mathematical model can be a reliable tool for a management of a company in a way demanded by ISO 9001:2015 standard, by managing the organization's knowledge and risk estimation. Also, in the spirit of a quality system, the simulation model can be continually improved through additional demands for data collection, as well as it can be extended onto the entire system.

Further research will be focused on an adaptable modelling through an automatic selection of the best multivariate models within one process.

References

1. Rygielski, C., Jyun-Cheng, W., Yen, D.C.: Data mining techniques for customer relationship management. Technol. Soc. **24**(4), 483–502 (2002)
2. Van der Alst, W.M.: Process Mining. Springer, Heidelberg (2001)
3. Avdalović, P., Jovanović, J., Vujović, A., Krivokapić Z.: Menadžment znanjem i standard ISO 9001, Kvalitet i izvrsnost 1–2, Beograd (2017)
4. Jovanović, J., Avdalović, P., Krivokapić, Z., Vujović, A.: AHP method application in the correlation analysis of knowledge management and requirements of ISO 9001:2015. In: Proceeding of 34th International Scientific Conference on Economic and Social Development, Moscow, pp. 147–156 (2018). ISSN 1849-7535
5. Ribière, V.M., Khorramshahgol, R.: Integrating Total Quality Management and Knowledge Management. J. Manag. Syst. **XVI**(1), 39–54 (2004)
6. Barjis, J.: Enterprise and Organizational Modeling and Simulation. Springer, Heidelberg (2010)
7. Witten, I.H., Frank, E., Hall, M.A.: Data Mining - Practical Machine Learning Tools and Techniques. Elsevier, Amsterdam (2011)
8. Ahlemeyer-Stubbe, A., Coleman, S.: A Practical Guide to Data Mining for Business and Industry. Wiley, New York (2014)
9. Sokolowski, J.A., Banks, C.M.: Principles of Modeling and Simulation - A Multidisciplinary Approach. Wiley, Hoboken (2009)
10. Kantardzic, M.: Data Mining - Concepts, Models, Methods and Algorithms. Wiley, New York (2011)
11. Pyzdek, T.: Using Data Mining and Knowledge Discovery With SPC. http://qualityamerica.com/LSS-Knowledge-Center/statisticalprocesscontrol/using_data_mining_and_knowledge_discovery_with_spc.php. Accessed 15 May 2019
12. Hoyle, D.: ISO 9000 - Quality systems Handbook. Elsevier, Oxford (2009)
13. ISO 9001:2015: Sistemi menadžmenta kvalitetom – Zahtjevi, ISO organization (2015)
14. Aguilar-Savien, R.S.: Business process modelling: reviewand framework. Int. J. Prod. Econ. **90**, 129–149 (2004)
15. Morbitzer, C., Strachan, P., Simpson, C.: Application of data mining techniques for building simulation performance prediction analysis. In: Proceeding of Eight International IBPSA Conference, Eindhoven (2003)
16. Jianji, W., Nanning, Z.: Measures of correlation for multiple variables. https://arxiv.org/abs/1401.4827. Accessed 10 May 2019
17. Hastie, T., Tibshirani, R., Friedman, J.: The Elements of Statistical Learning - Data Mining, Inference, and Prediction. Springer, New York (2013)
18. Evans, M.K.: Practical Business Forecasting. Blackwell, Oxford (2003)
19. Jianji, W., Nanning, Z.: Measures of Correlation for Multiple Variables. https://arxiv.org/abs/1401.4827.last. Accessed 10 May 2019
20. Elektroprivreda Republike Srpske. www.ers.ba. Accessed 15 May 2019

21. Reddy, S.G., Srinivatsu, R., Porna, C.R.M., Reddy, R.S.: Data warehousing, data Mining, OLAP and OLTP technologies are essential elements to support decision making process in industries. Int. J. Comput. Sci. Eng. **2**(9), 2865–2873 (2010)
22. Hidroelektrane na Trebišnjici. https://henatrebisnjici.com/sistem-hidroelektrana. Accessed 15 May 2019

Process Mapping of Glass Envelope Design of Geometrically Complex Form

Tatjana Kosic[1(✉)], Igor Svetel[1], and Mauro Overend[2]

[1] Innovation Center of Faculty of Mechanical Engineering, University of Belgrade, 11000 Belgrade, Serbia
tkosic@mas.bg.ac.rs

[2] Department of Engineering, Cambridge University, Cambridge CB2 1PZ, UK

Abstract. The design and construction of glass facades is a very complex sector in the construction industry, whose multidisciplinary character is reflected in the application of new technologies of glass and glass facade production, in solutions for multitude requirements of the envelope performances sometimes very complex geometry, as well as in wide range of various glass materials. The aim of the presented research is to create links and interactive relationship between the design process on the conceptual level, materialization and realization of geometrically complex form of glass envelope, which integrate different issues such as design, manufacture, performance and economy. Based on the survey about how architects, engineers and contractors deal with different design aspects of curved glass structures with a special focus on design and construction process, the investigation includes an overview of the findings of the interviews as a basis for the development of support tools – process map, which could help efficient storage, access and transfer of project information. The current paper presents creation and verification of a model of process mapping which will be supportive in further research of the application of both flat and curved glass in all types of geometrically complex glass envelopes. The process map is especially useful for defining deadlines, understanding the role of different actors and their responsibilities and the sequence of activities, as well as identifying different requirements during the design and construction process.

Keywords: Glass · Curved glass · Process map · Survey · Geometrically complex form of building envelope

1 Introduction

The design and construction of glass facades and envelopes, as a relatively new sector in the construction industry, is a very complex sector whose multidisciplinary character is reflected in the application of new technologies of glass and glass facades production, in solutions for multitude requirements of the envelope performances sometimes very complex geometry, as well as in a wide range of various glass materials. The aim of the presented research is to create links and interactive relationship between the design process on the conceptual level, materialization and realization of geometrically complex form of glass envelope, which integrate different issues such as design, manufacture, performance and economy. Based on the survey about how architects,

N. Mitrovic et al. (Eds.): CNNTech 2019, LNNS 90, pp. 442–459, 2020.
https://doi.org/10.1007/978-3-030-30853-7_26

engineers and contractors deal with different design aspects of curved glass structures with a special focus on design and construction process, the investigation includes an overview of the interview findings as a basis for the development of support tools, which could help efficient storage, access and transfer of project information. All this implies that the process of design and construction relevant to geometrically complex glass envelopes, including curved as one of the most complex should be analyzed and described through creation and verification of a model of the process map.

The most significant procedures, i.e. characteristic steps for the design and construction of geometrically complex forms of glass envelopes have been identified in the research, with the aim of creating a process map by the method of graphic presentation of specific processes in the business process model. Model - Building SMART's adaptation of the Business Process Map Notation (BPMN) [1] was created using the appropriate Visio Professional software and then its elements were verified through conducted interviews with experts in Europe and Serbia. In addition, by using the survey method, specific experiential data was obtained. The map itself will be supportive in further research of the application of both flat and curved glass in all types of geometrically complex glass envelopes.

The initiative of this study was given by the first author and has been supported by the Glass & Facade Technology Research Group [2] at the University of Cambridge, as well as the European research network on the application of structural glass (COST TU0905 Structural Glass - Novel Design Methods and Next Generation Products) [3].

2 Research Practice

2.1 Design and Construction Process of Geometrically Complex Form of Glass Envelope

Architectural structures of curvilinear forms set new requirements for application of curved and flat glass as an element of geometrically complex envelope of architectural structures. At first glance, the application of curved glass provides an exceptional freedom in the design of modern wavy shape, but set of constraints arises when it comes to the aspect of design, manufacture, use, performance and economy. Characteristics such as radius of curvature, minimum and maximum dimensions of the glass element, local regulations, available coatings, optical quality, and selection of the glass types significantly affect the final glass product. Concerning geometry of the glass envelope surface, there are no limitations today in the process of their modeling. In fact, architectural practice continuously has followed the development of the geometry, and many architectural trends were inspired by the latest developments in this field. The word 'free form' says it is possible to create new forms, if the architects and designers are familiar with the geometry of basic geometrical forms (Fig. 1), as well as with all elements of geometry.

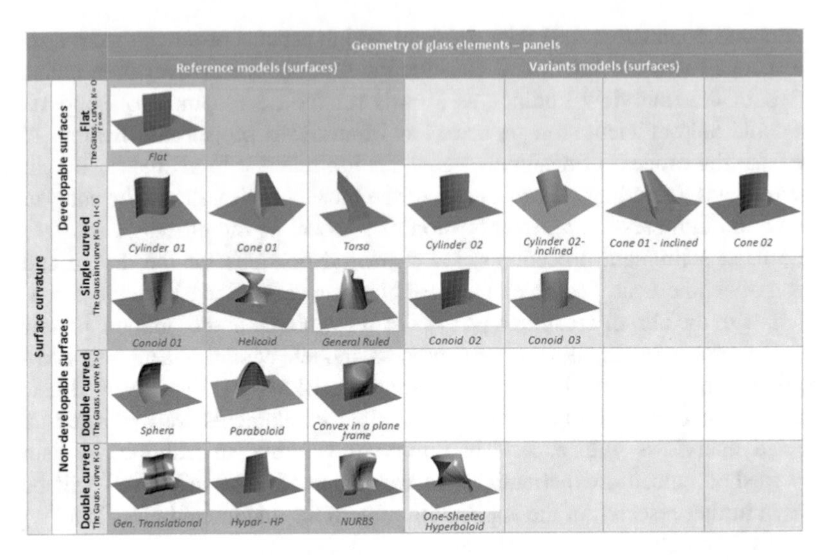

Fig. 1. Typology of the glass elements according to geometry and type of curvature [4].

However there are problems in the practical fabrication of the geometrically complex shapes, because unlike the abstract geometric forms, construction elements have physical characteristics that prevent the creation of any geometry [4]. This is especially emphasized in the case of glass that is brittle and easily breakable material and therefore unable to be produced in all shapes and sizes. In addition to the geometric aspects, the design and construction of curved surfaces involve many aspects typical for the material itself, which is particularly related to the thermal properties, production techniques, glass-shaping and finishing, as well as the effect that glass as material enters into completed building. More often, the relation between shape and fabrication poses new challenges and requires more sophistication from the underlying geometry [5]. As a consequence, numerous architects have returned to being highly engaged in the fabrication process to ensure the design intent is carried through into the making [6]. Therefore, design and selection of glass envelope technologies is frequently a compromise between the intent of the architect, fulfilment of performance requirements, simpler fabrication method, shipping limitations and the project budget.

2.2 Creation of Process Map

The use of glass in the free form glass envelopes involves many complex considerations in the early stages of the design process such as concept of the glass surface, geometry, aesthetics, structural and thermal characteristics, choice of the shape of glass elements and manufacturing technique, as well as compatibility of all treatments with bending process in the case of curved glass. Therefore, it is necessary to clarify their mutual influence and dependence in order to maximize the potentials and possibilities of the application of both curved and flat glass in the geometrically complex glass envelopes. From the above facts, the created map of design and construction process should be support for further research of glass application in materialization of the building envelope.

The main contribution of the process mapping implies defining deadlines, understanding the role of different actors, their responsibilities and the sequence of activities, as well as identifying different requirements during the design and construction process. This refers particularly to strategic, process, and IT requirements, i.e. forming databases for IT systems. As well, it could be used as a basis for further development of support tools, such as BIM - Building Information Modeling, as well as tools for optimization of glass surfaces. Finally, the map is useful for assessing whether new research results are compatible with current industry processes or there is need for improvement in order to take advantage of new research (innovation).

The aim of this research is to further develop and expand the existing map of facade design and construction [7], respectively, by mapping the design and construction process of the geometrically complex glass envelope, to show the main issues in the application of both flat and curved glass in the building envelope. In order to create a process map, it has been necessary to apply and combine existing knowledge from different research areas regarding glass facades, based on: created map of newly constructed facades, developments in the glass facade sector as a result of increasing application of the new technologies in industry of glass and glass facades, as well as "state of the art" in research of the application of glass and glass constructions. Also, the map is the basis for the development of support tools that help the efficient storage, access and transfer of information on the proposed curved glass envelope model. Further, using database and created tools within the map, it is possible at an early stage of the design to evaluate the proposed model of the curved glass envelope in relation to its design, structural and production constraints in order to select the final variant of geometrically complex glass envelope that is feasible and in accordance with the original architectural idea.

The first part of the research related to the mapping of design and construction process of geometrically complex glass envelopes (CGE) implies the design of the map. The map is created using the method of graphic presentation of specific processes in the business process model - BuildingSMART's adaptation of the Business Process Map Notation (BPMN), developed by the international organization BuildingSMART [8], which aims to improve the exchange of information between software applications used in the construction industry. This organization has developed the Industry Foundation Classes (IFC) as a neutral and open specification for Building Information Model. Creating a map has shown that some of the main issues of glass application in the building envelope are constraints that appear due to characteristics of the material itself as well as production process that limit the size (dimensions), the load capacity and possible curvature of the glass element achieved by appropriate shaping technique (cold or thermal bending process); all in order to obtain the desired shape.

The entire process is analyzed, defined and presented in the following phases:

Phase 1: Prepare preliminary CGE design proposal, including a sub-phase of outline cladding proposal,

Phase 2: Develop detailed CGE design proposal,

Phase 3: Prepare final CGE design proposal,
Phase 4: Prepare tender documentation and evaluate tender returns,
Phase 5: Coordinate design process with fabrication process,
Phase 6: Monitor construction work to completion.

Also, mapping of the process was carried out both on database and information about the method of design and construction of glass facades in Serbia and Europe, as well as according to the British RIBA 2013 Work Plan [9], which resulted in a comprehensive review and reconsideration of all phases and works in the process of building design and construction. The RIBA Plan of Work organizes the process of information, design, construction, maintenance, operational work and the use of a building in several key phases. It issues the tasks and results needed to be met at each stage, which may vary or overlapping to fulfill specific project requirements.

The final goal of the analyzed process is to give the comprehensive overview and to underline the most important steps characteristically for the design and construction of geometrically complex forms of glass envelopes. The process itself, which, besides combining knowledge from different areas, has included with particular attention the CAD-CAM fabrication of the curved glass envelope, as well as a different approach to the curved glass shaping technique. As well, the most important steps in the process were distinguished through the analysis of relevant case studies [4].

The second part of the research relating to the mapping of design and construction process was realized during interviews in which the map was verified and evaluated.

2.3 Survey Method

In the study, the survey that includes 12 interviews with professionals was carried out. The respondents were mainly engineers and architects with great experience in designing glass facades and structures, and some of them also had experience working on projects in which the curved glass was applied. Interviews were conducted in person and in own arrangement.

In the first part of interview, the respondents were asked to review and comment the process map, step by step, in order to confirm or suggest changes at different stages in the mapping process. Everyone freely talked about topics they considered most relevant in connection with the design and construction of curved glass envelopes. In the second part of interview, the respondents were asked to answer the questions within the survey.

Survey Questions. It was made a proper balance between the number and kind of questions (regarding the time required for answers) on one hand and the relevance and usefulness of the response on the other. This resulted in an eight question survey. The opening two questions inquiries about the general premises of the respondent with regard to different type of glass (in relation to their geometry) and curved glass structures. In the consecutive questions 3 through 5, the respondent was asked to list main considerations, challenges and opportunities for curved glass envelope engineering. Answers on questions 6 and 7 were supposed to show practical solutions of curved glass structures through the appropriate structural and glass fixing systems application. Finally, question eight gave comments on thermal performances of curved glass. Table 1 lists the survey questions.

Table 1. Survey questions.

No.	Question
1	What is your experience with the different type of glass (flat, single curved, double curved) and their characteristics (visual, physical, structural, manufacturing and economical)? Which do you find more challenging?
2	How has the design and construction process of curved glass structures in buildings evolved in recent years?
3	What are the main considerations when you design/specify a curved glass element?
4	How do these considerations differ from the design/specifications of a flat glass element?
5	What are the challenges and opportunities for curved glass envelope engineering?
6	Which structural systems are most appropriate/not appropriate (frames, simple trusses, mast trusses, strong-backs, glass fins, grid shells, tensegrity, cable trusses, cable nets, shell with cable) for curved glass panels and why?
7	Which glass fixing systems are most appropriate/not appropriate (framed systems - panel, veneer, unitized and frameless systems - point-fixed drilled and point-fixed clamped) for curved glass panels and why?
8	Do you consider the thermal performance when designing/specifying curve glass panels?

Distribution. The survey was distributed among architects, engineers and researchers known to have experience with the use of structural glass and glass facades, more specifically the members of several professional institutions, companies and glass manufacturers. They are: Foster + Partners, London; Glass Light and Special Structures, London; Eckersley O'Callaghan, London; Meinhardt Façade Technology, London; Malishev Wilson Engineers - Creative Engineering & Structural Glass Design, London; FH Johanneum University, Graz; Faculty of Architecture, University of Belgrade; Pavle Company, Belgrade and Concav Convex, Belgrade.

As survey is conducted during the period of one month, 12 responses were collected in that time period.

Respondents. The respondents are active in various professions within the field of glass structures. In this survey, distinction is made between (A) engineers at consulting offices, (B) engineers involved in facade construction, (C) researchers at university and (D) engineer at glass processing firm (Table 2). The survey was held at an international level; exactly the respondents are active in three different countries, as shown in Table 2. Although project and engineering skills, as well as experience would inevitably vary between respondents, all responses were valued equally.

Table 2. Respondent profession and respondent's country of practice.

Respondent profession	No. resp.	Country of practice	No. resp.
(A) Architects at consulting office	4	United Kingdom	8
(B) Engineers involved in facade construct	4	Austria	1
(C) Researchers at university	2	Serbia	3
(D) Engineers at glass processing firm	2		
Total	12	Total	12

2.4 Survey Results

Responses Analysis. Not all questions were answered by each respondent. Therefore, Table 3 lists the number of responses per question.

Table 3. Number of responses per question.

Question	1	2	3	4	5	6	7	8
No. of responses	9	9	9	8	7	8	8	10

A Brief Statement of the Main Points of Responses. The answers to the questions are summarized and presented by issues.

Summary of Responses to Question 1. The majority of respondents agreed that the application of curved glass presents a greater challenge. One respondent pointed out that the application of all types of glass represents an equal challenge. All respondents agreed that annealed glass presents more economical solution which does not have an optical distortion. However, this type of glass is sensitive to thermal shock and can lead to easy cracking. Besides, due to the way it breaks, creating large pieces of very sharp edges, this type of glass does not provide the necessary security. Heat strengthened glass tolerated higher loads, but it breaks similarly to annealed glass. Thermally toughened glass has a higher degree of resistance and it breaks as safety glass. The process of lamination of glass also provides a high degree of safety since the glass does not scatter during fracture. Some respondents identified the thermal process of curved glass shaping as a technique during which additional problems arise due to its exposure to higher temperatures leading to its expansion, and thus to the appearance of optical distortions and changes in dimensions and edges of glass. Only three respondents recognized that new technique of cold bending of glass does not lead to such problems, but in this way it is not possible to create all geometry of double curved glass. One respondent indicated that complex curved forms can be achieved also by application of flat glass, so that the design process represents compromise between the wishes of the architect and the possibilities provided by the application of a particular type of glass in relation to economic cost. Curved glass panels are difficult to produce to be identical and ideally processed, which is particularly expressed in relation to available different technologies of production. The common comment is that should take into account that the glass always break, but it should be considered how this will happen. Regarding structural characteristics, by application of curved glass is possible to achieve more resistant structures, as all respondents agreed. Also, in the design stage, different ways of bearing glass panels have to be considered because hole drilling in the glass for point supported (bolted) structure contributes to additional concentrated stress in glass creating opportunities for a break in the weakened areas. The general response is that there is still no systematized experience data for curved glass.

Summary of Responses to Question 2. In total, seven respondents particularly emphasized that development of technology have contributed to the larger dimensions of glass panels, smaller optical distortions and economically more favorable characteristics. Besides, the development of technology has led to the emergence of ultrathin and ultra-rigid Gorilla glass (0.7–1.5 mm) which shows excellent visual characteristics and can be further processed by cold or thermal bending for building application. Unfortunately, the limitation for its application in architecture is still small dimensions of the glass panel (max. 1.5–2.0 m). Two respondents commented that there are fewer questions concerning the possibility of fabrication of certain dimensions of curved glass panels, but rather questions of the functional and economic justification of their application. All respondents agreed that as the application of curved glass structures increasing, especially in the last ten years, the directions of development are the reduction of the economic cost, the rescission of optical distortion on the curved glass and the achievement of larger dimensions. The general response is that the best solutions are still achieved through the cooperation of designers with facade consultants, facade constructers and glass manufacturers.

Summary of Responses to Question 3. The answers were more diverging in nature than previous questions. According to respondents, the main considerations are: dimension of glass panel; structural system; stress value in the glass; techniques of shaping (bending); residual capacity after fracture; visual quality of glass; glass performances; solar factor; glass shading; glazing type (single, double and triple glazing – insulated glass unites panels); possibility of application of laminated glass, as well as determined color, ceramic frit, coatings and interlayers; visual distortion; rationalization of geometry; reduction of economic costs; the possibility of panel replacing after glass fracture without endangering the stability of the entire structure. The general response is that all aspects of the application of the curved glass are equally important, and the order of different aspect approach depends on engineer himself.

Summary of Responses to Question 4. The majority of the answers (7 of 8) claimed that the difference is primarily at an economic cost that is considerably higher for curved glass, while at the same time all technical characteristics and calculations are more complex. Some particularly complex characteristics of curved glass are emphasized by respondents as follows:

Using better structural characteristics of curved glass as a challenge.
Requirements for greater visual quality inevitably influence in increasing costs.
Greater sensitivity of curved glass as a material.
Greater stiffness of curved glass can contribute to reduction of glass thickness.
Drilling the holes in curved glass for bolted assembly.
Only a few types of low-emission coatings (mostly hard) are applicable on curved glass.

When applying curved insulated glass unites (IGU) panels, it is necessary to take into account the existence of internal pressure, that is, the existence of a difference in climatic conditions in which the panel is manufactured and assembled in the factory from the climatic conditions that prevail at the site or place of assembly, and that can lead to glass elements breakage.

Summary of Responses to Question 5. One respondent especially stated that the challenge could be, in addition to a little known technique of cold bending, a vision of use of traditional technique of glass blowing for shaping different forms from which the panels would be cut off. Nevertheless, the majority of the respondents (4 of 7) claimed that research on new materials is more challenging including new glass types (Dichroic glass) that improves design characteristics, as well as ultrathin Gorilla glass which possibility for free form application in architecture is in the research phase. Two respondents commented that improvements could be also achieved by the process of optimization of curved surface of glass structure which allows penalization by greater percentage of flat and single curved panels as opposed to a much smaller percentage of double curved panels or their avoidance. None of the respondents didn't take into consideration new developed 3D printing of glass which allows design and fabrication of geometrically complex and adaptable structures and which product - type of lens can reduce heat dispersion and thus reduce global warming on an urban scale [10]. The general response is that each curved glass application represents a new challenge, given that the architects' requests are still out of technical conditions of building construction. Manufacturers put an effort to bring the production technique closer to the architects' requirements, as well as to educate architects about technical production possibilities.

Summary of Responses to Question 6. According to the majority of interviewees (5 of 8), a significant feature of the application of curved glass is use of its geometric stiffness in order to achieve greater spans and to reduce the number of elements of supporting structure. However, according to one interviewee, complete glass structure which implies glass columns, beams and ribs or spatial structures of a large span cannot be applied in the case of large curvature of glass surface. As well, the smallest limitations for curved glass application is set up by frame systems, while for application of cable nets it is necessary to predict the possibility and to determine the maximum deflection. Glass is a material that does not tolerate large deformations, as four respondents emphasized. More precisely, it has its structural performance and maximally allowed stress values in relation to different static influences. As the engineer usually determines a primary structure by guiding other requirements, on the parts of the building where glazing is provided, the structure is subsequently verified and possibly corrected, while inversely it is much rarer. General response is that considering a number of different constructive systems, according to which the structural glass facades are classified [11], it could be concluded that there is no system that would be comprehensively the best for application on curved glass envelopes, yet all depends on the requirements and conditions of application.

Summary of Responses to Question 7. According to most interviewees, the framed system still stands out as the most appropriate considering curved and freely shaped glass which, opposite to tendency to return into original form, remains attached along the edge on substructure. On the other hand, architect's requirements for increasing transparency of the envelope surface remain a particular challenge for facade contractors. All respondents agreed that in case of curved glass panels that are point fixed on the substructure, it is important to take into account the order of the activity: 1. hole drilling; 2. glass bending using one of the shaping techniques (hot and cold bending); 3. curved glass tempering – pre-stressing. According to some respondents (3 of 8), one of the options is a linear or dotted glued glass on a substructure, especially placed on a frame in the factory and as such mounted on a construction site. General response is that connections and fixing methods for the application of curved glass must be analyzed and numerically verified, since the curved shape of glass panel contributes to the load distribution in different ways, acting as a glass shell.

Summary of Responses to Question 8. This question was answered by the most of respondents (10). Thermal performances must be analyzed in order to avoid thermal shock effect in the glass [12], where they are always considered, even in the case of curved glass. However, as four interviewees emphasized, it is possible to apply only a small number of low-emission coatings in thermal bending and therefore the possibility to achieve a good solar factor (*g-value*) is to be considered. Almost all the respondents agreed that thermal performance is taken into account without possibility of obtaining the correct results. The general response is that is necessary to explore the methods for a precise calculation of thermal characteristics of curved glass, respectively to create a mathematical model of heat transfer in a curved insulated glass panels that would give a link between geometry, type of insulated glass panels and thermal characteristics in order to evaluate the energy performances of buildings with geometrically complex glass envelope. It would be useful to evaluate the maximum amount of energy savings achievable by different type of glazing, especially adaptive glazing by modulating all its thermo-optical properties (T_{vis}, *g-value* and *U-value*) for specific boundary conditions [13], as in case of flat glass panels (e.g. low-emission glazing with argon filler vs. double glazing) when energy savings could be precisely defined [14].

3 Process Map of Design and Construction of Geometrically Complex Glass Envelopes

It should be emphasized that the process of verification was carried out among the participants from the A/E design team, consulting firms for facades, as well as contractors and glass manufacturers, which additionally contributed to the fact that all phases or levels of process mapping are sufficiently detailed and precise in the research, which result is the map. The same inaccuracies or additions identified in more than one interviews have been taken into account in the upgraded version of the map, in such a way that all modified or updated elements are marked.

Six phases and one sub-phase of the entire process of design and construction of geometrically complex glass envelopes (Fig. 2) are presented: 1. Prepare preliminary CGE design proposal (Fig. 3) including a sub-phase of outline cladding proposal (Fig. 4); 2. Develop detailed CGE design proposal (Fig. 5); 3. Prepare final CGE design proposal (Fig. 6); 4. Prepare tender documentation and evaluate tender returns (Fig. 7); 5. Coordinate design process with fabrication process (Fig. 8); 6. Monitor construction work to completion (Fig. 9).

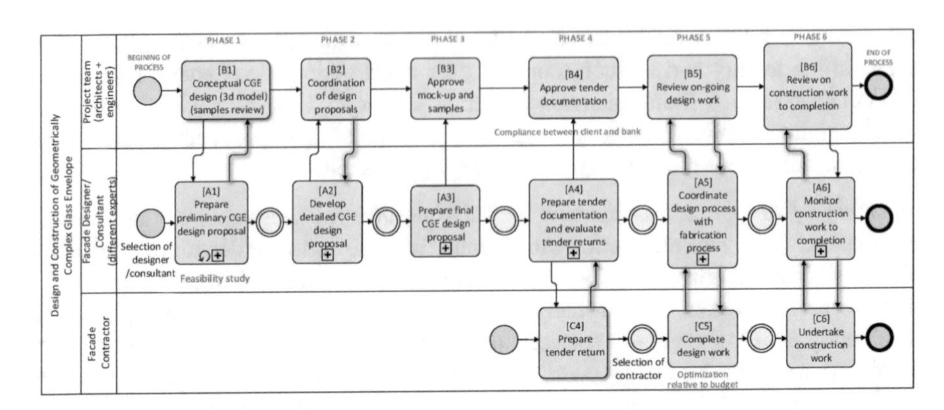

Fig. 2. Phases of design and construction process of geometrically complex glass envelopes.

The + mark presents the extension of certain activities within single diagram (phases) which is given by a separate diagram (sub-phases). Further, in digital form (Visio Professional Software), this mark gives a direct link to the diagram that presents the phase itself or the extension of the activity. Given that all participants in the process are divided into three basic groups:

Project team (architects + engineers),
Facade designer/consultant (different experts), and
Facade contractor;

so all the activities, carried out by a certain group of participants according to the phases, are indicated on the process map such as:

Project team - activities B1-B6
Facade designer/consultant - activities A1-A6, and
Facade contractor - activities C1-C6.

During the interview, different data/answers were obtained indicating that it is not possible to get the determined elements of the process, but the process is specifically adapted to each project. Bearing in mind the concept of glass surface, the geometry of the form, the constructive and the thermal characteristics, the form and the technique of shaping the glass elements, the compatibility of all treatments with the bending process, with the aim of creating a map of the process through diagrams, factors that define the application of glass in the materialization of geometrically complex envelope forms are identified.

In the phase of preparation of the initial design solution, after the design sketches it is necessary to carry out parametric modeling (analysis of geometry and quality control of contours and reflections), as well as the choice of panels, systems of supporting substructures and glass forming techniques before setting performance requirements of glass (design, constructive, thermal, acoustic and fire protection). After choosing the type of glazing and glass, as well as the attaching method, an important final activity is a preliminary analysis in terms of construction feasibility and budget that will enable the final draft of the project design proposal or the development of alternative solutions with a different choice of elements of the glass envelope. Also, it is important to note that this phase implies a possible initial thermal performance analysis, as well as a preliminary cost analysis.

At the stage of elaboration of a detailed design solution that involves the development of technical details and initial analysis of the constructive performances of the supporting structure and glass elements as well as thermal performance, an important characteristic of this phase is the determination of key criteria for permissible tolerances and displacements, followed by an assessment of the buildability of the projected system according to constraints. Also, it is important to note that this phase involves determining the final strategy for cleaning and maintaining the glass envelope.

In addition to conducting detailed analysis of the construction and thermal performance, as well as the elaboration of the final details of the design solution, the phase of final design solution development includes the preparation and execution of models and samples as conditions for the creation of criteria for project design and preparation of project specifications.

The phase of the tender documentation preparation shows that the composition of tender documents, which involves primarily the design of the conceptual solution of the geometric complex layer and the cost estimate, precedes the preparation of detailed drawings of the bearing structure and glazing elements and specifications of the glass cover performance as well as obtaining the contractor's comments on the design solution.

The phase of project process coordination (contractor activities) involves the design of individual products and their analysis, grouping and coding, followed by the initial ordering of components and the preparation of the construction work program. All the elements of this phase are controlled and adopted by a facade consultant.

The execution phase, specific in terms of curved glass elements, involves the preparation of workshop drawings that guide digital production (cutting) of glass and the technique of thermal design, and then finalization which is followed by assembly of components. In the case of a cold bending technique, the panel forming (TI panel) is carried out at the factory while the assembly implies direct bending on the construction site on the substructure. It is possible to bend the glass elements in the factory, but then it is necessary to attach them to the frame, since the glass tends to return to its original position.

A list of documents and reference materials required for the implementation of the activities (steps) or that represents their result (report) within a certain phase is quoted as the special contribution.

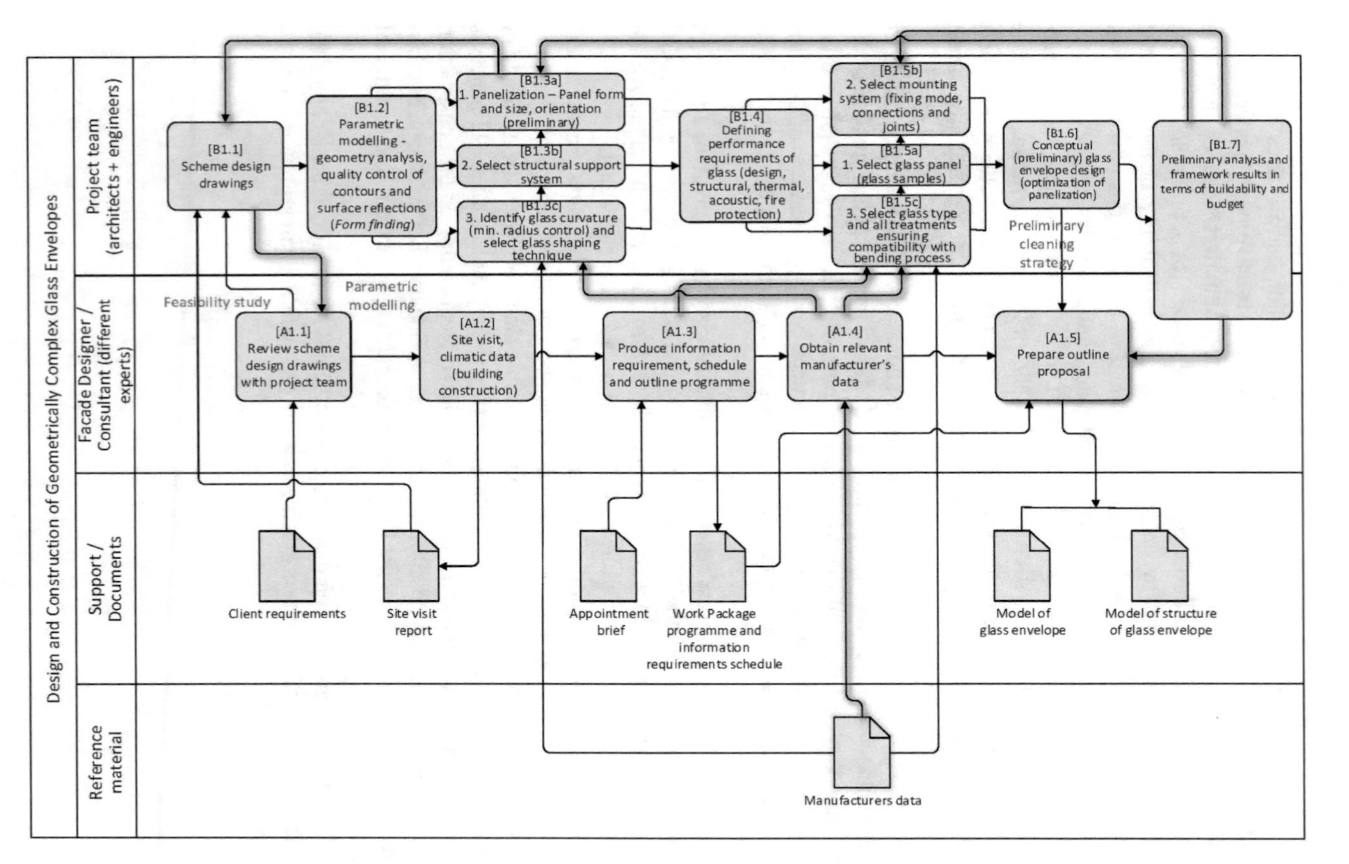

Fig. 3. Phase 1 - Prepare preliminary complex glass envelope design proposal

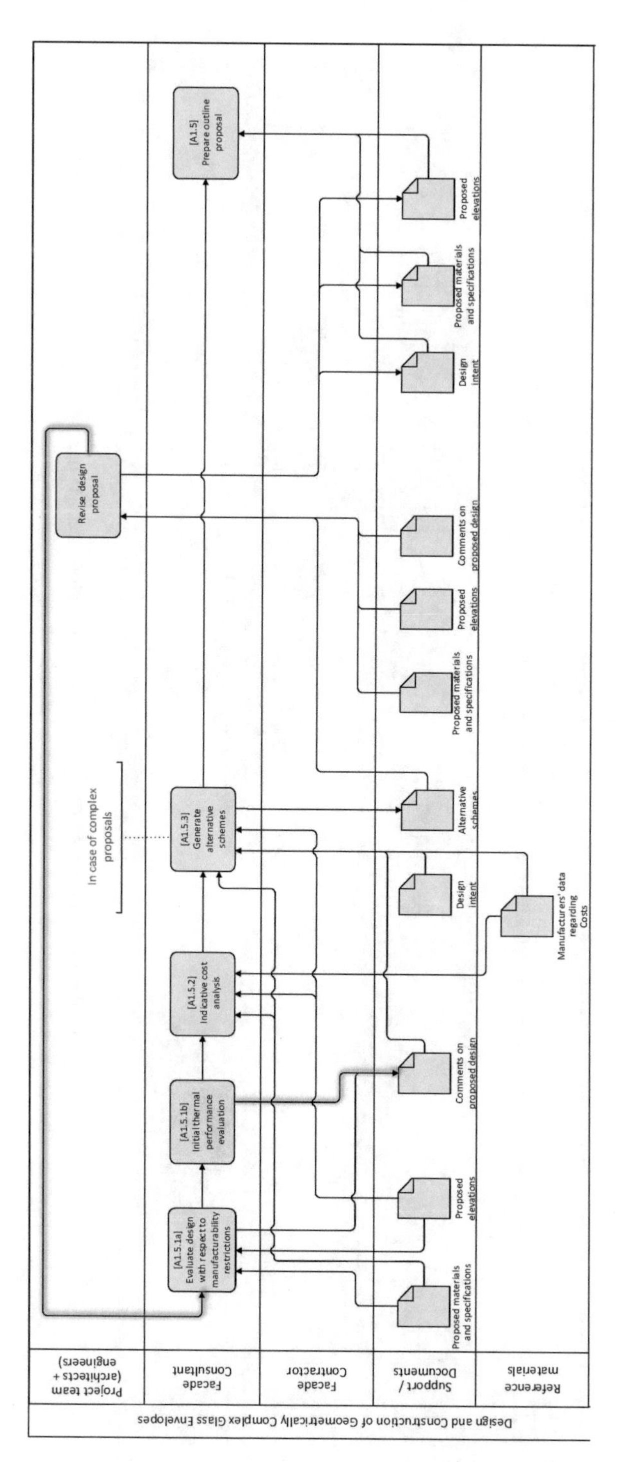

Fig. 4. Phase 1.5 - Prepare outline cladding proposal

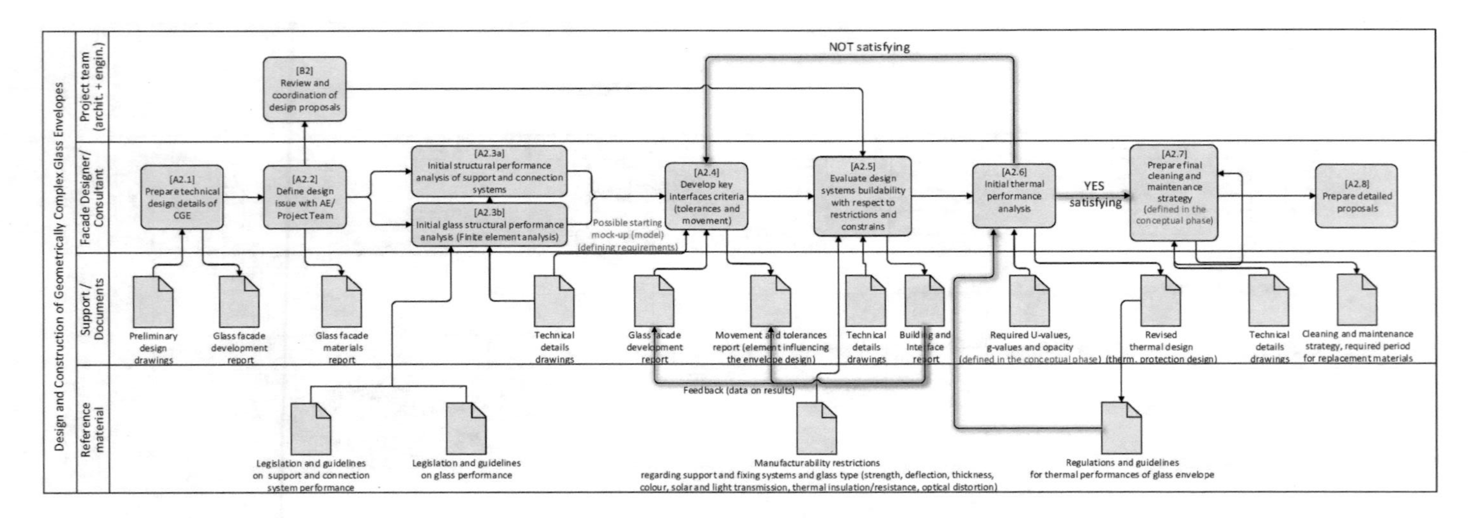

Fig. 5. Phase 2 - Prepare final complex glass envelope design proposal

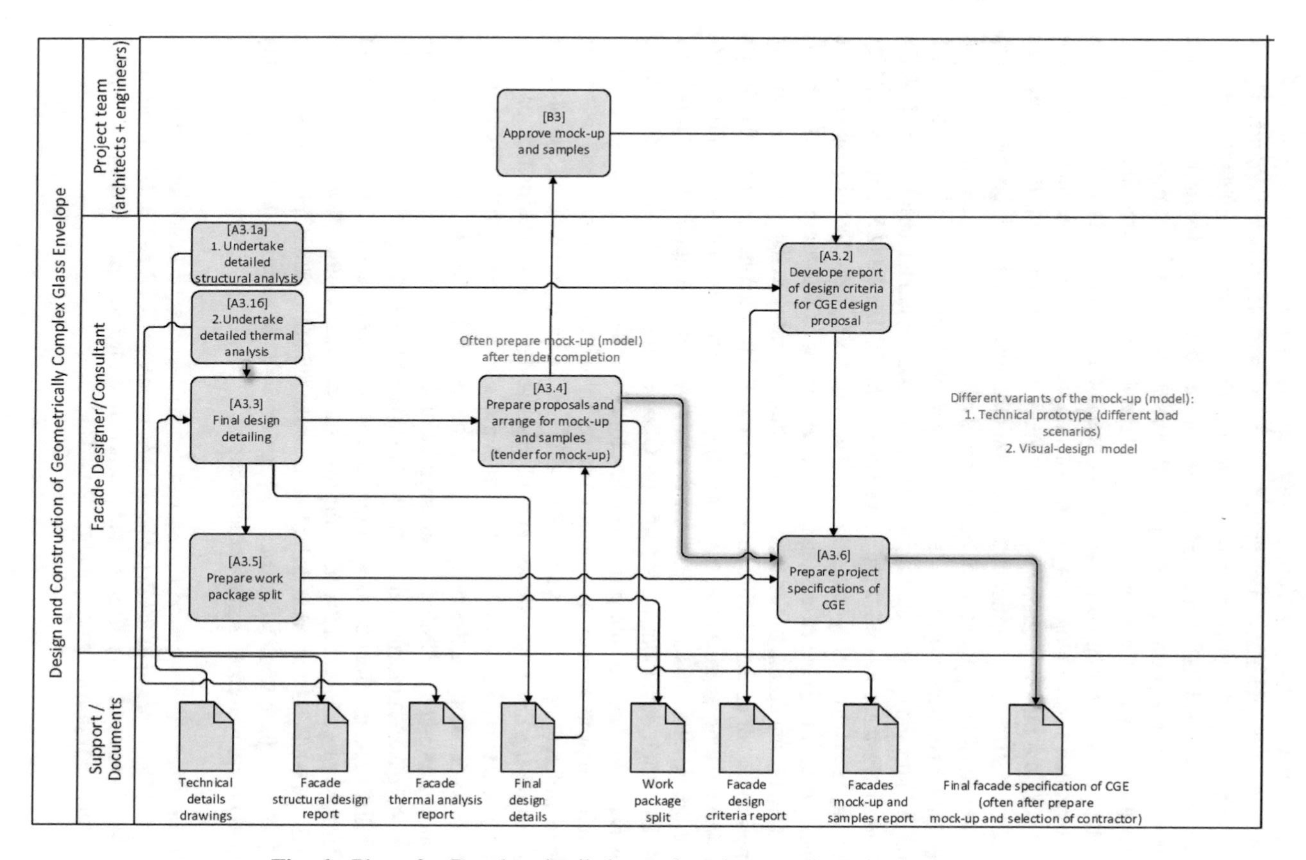

Fig. 6. Phase 3 - Develop detailed complex glass envelope design proposal

4 Conclusion

The analysis of collected data during survey clearly showed that there is a need for more research, comprehensive design aids and recommendation for engineering of geometrically complex glass envelopes. During the interviews, different data and answers were obtained indicating that it was not possible to get determined elements and their order in the process of design and construction, but the process especially adapts to each project. By examining the available databases, especially collected during interviews with engineers, architects, consultants and manufacturers of glass facades, the need for defining design and construction process of geometrically complex forms of the building envelopes was noticed. From that reason, the survey was enabled development and verification of the process map which will be supportive in further research of the application of both flat and curved glass in the materialization of building envelope. The map was created using a standardized way of graphic presentation of specific processes in the business process model (BuildingSMART's adaptation of Business Process Map Notation - BPMN). The BPMN notation enables software engineers to rapidly develop applications. This gives specific value to the developed process map because it enables development of the computer tools specifically crafted for the design of the curved glass facades, either as the part of the existing BIM applications or as the specific tools that use existing interoperability technologies to link with existing applications. Since existing software companies currently do not see curved glass design as an entry in the development of their applications, it is more practical that such development is based on Open source technologies, which would allow wider acceptance of the principles developed in this research [15].

As well, the process map is especially useful for defining deadlines, understanding the role of different actors and their responsibilities and the sequence of activities, as well as identifying different requirements during the design and construction process.

Finally it should be claimed that curved glass is a material that implies many complex considerations in the early stages of the design process involving glass surfaces concept, geometry, aesthetics, structural and thermal characteristics, selection of geometry of glass elements and glass-shaping (bending) techniques, as well as the compatibility of all the treatments with the bending process. The reasons of selection of particular type of glass geometry and construction methods should always be analyzed discussed and well understood in order to maximize the utilization of potentials of both flat and curved glass application and to provide safe, economical and aesthetically acceptable complex forms of glass constructions.

Acknowledgement. This work was supported by the Ministry of Education, Science and Technological Development of the Republic of Serbia under grant TR-36038. It is a part of the project 'Development of the method for the production of MEP design and construction documents compatible with BIM process and related standards'.

References

1. Voss, E., Jin, Q., Overend, M.: A BPMN-based process map for the design and construction of facades. J. Facade Des. Eng. **1**, 17–29 (2013)
2. Glass & Facade technology research group (gFT). https://www.gft.eng.cam.ac.uk/. Accessed 01 July 2019
3. TU0905 - Structural Glass - Novel design methods and next generation products. www.glassnetwork.org. Accessed 21 Oct 2016
4. Kosić, T.: Application of glass in materialization of geometrically complex forms of architectural building envelopes, Doctoral dissertation. Faculty of Architecture, University of Belgrade, Belgrade (2016)
5. Liu, Y., Pottman, H., Wallner, J., Yang, Y., Wang, W.: Geometric modeling with conical meshes and developable surfaces. ACM Trans. Graph. **25**(3), 681–689 (2006)
6. Dunn, N.: Digital Fabrication in Architecture. Laurence King Publishing Limited, London (2012)
7. Voss, E.: An Approach to Support the Development of Manufacturable Facade Designs, Doctoral dissertation. Department of Structural and Civil Engineering, University of Cambridge, Cambridge (2013)
8. BuildingSMART. https://www.buildingsmart.org. Accessed 21 Oct 2016
9. RIBA Plan of Work 2013. https://www.ribaplanofwork.com/. Accessed 01 July 2019
10. Klein, J., Stern, M., Franchin, G., Kayser, M., Inamura, C., Dave, S., Weaver, J.C., Houk, P., Colombo, P., Yang, M., Oxman, N.: Additive manufacturing of optically transparent glass. 3D Printing Addit. Manufact. **2**(3), 92–105 (2015)
11. Krstic-Furundzic, A., Kosic, T., Terzovic, J.: Architectural aspect of structural design of glass facades/glass skin applications. In: 3D International Proceedings on Architectural and Structural Application of Glass - Challenging Glass 3 Conference, pp. 891–900. IOS Press, Amsterdam (2012)
12. Wang, Q., Chen, H., Wang, Y., Sun, J.: Thermal shock effect on the glass thermal stress response and crack propagation. Procedia Eng. **62**, 717–724 (2013)
13. Favoino, F., Overend, M., Jin, Q.: The optimal thermo-optical properties and energy saving potential of adaptive glazing technologies. Appl. Energy **156**, 1–15 (2015)
14. Krstić-Furundžić, A., Kosić, T.: Asessment of energy and environmental performance of office building models: a case study. Energy Buildings Spec. Issue, Places Technol. **115**, 11–22 (2016)
15. Svetel, I., Đurović, A., Grabulov, V.: Prompt system redesign: shifting to Open source technology to satisfy user requirements. Comput. Sci. Inf. Syst. **7**(3), 441–457 (2010)

The Development of CAD/CAM System for Automatic Manufacturing Technology Design for Part with Free Form Surfaces

Goran Mladenovic[1(✉)], Marko Milovanovic[2], Ljubodrag Tanovic[1], Radovan Puzovic[1], Milos Pjevic[1], Mihajlo Popovic[1], and Slavenko Stojadinovic[1]

[1] Faculty of Mechanical Engineering, Department of Production Engineering, University of Belgrade, 11000 Belgrade, Serbia
gmladenovic@mas.bg.ac.rs

[2] Deutsches Elektronen-Synchrotron (DESY), Zeuthen, Germany

Abstract. Developed procedures for tool path generation and application representing a CAD/CAM system is presented in this paper. This application allows automatic manufacturing technology design for parts with free form surfaces for loaded CAD models of part and work piece in STL file format. Generated tool path will represent optimal tool path in accordance with multi criteria optimization methods. Developed application allows usage without any user's expertise in the field of CAD/CAM systems. The optimal tool path for manufacturing will be generated, which will be performed in the shortest time possible, having appropriate surface precision and quality. Developed procedures which are implemented in this system are the result of years of research in this field at the Department of Production Engineering, Faculty of Mechanical Engineering, University of Belgrade, Serbia. Also, development of cutting force model and experimental determination of cutting force coefficients for tool/work piece geometry and material combination is described. Research was conducted for combination of aluminium work piece and ball end mill of HSSE steel. The application is developed as a GUI interface, in MATLAB software. Mentioned application allows generation of NC codes for rough and finish machining for the tools stored in the application database. According to the generated NC codes, several parts were manufactured using the horizontal working centre in order to verify developed procedures for tool path generation and optimization. Based on the conducted experiment, it was concluded that the machining was performed in allowed tolerances and cutting conditions which confirm the usability of the developed software application.

Keywords: CAD/CAM systems · Free form surfaces · Tool path optimization

1 Introduction

Part with free form surfaces are widely used in all areas of mechanical engineering, but also in other areas in everyday living. It can be said that the usage of these parts is grooving in every day in all aspects of life. Beside this name it can be used sculptured

N. Mitrovic et al. (Eds.): CNNTech 2019, LNNS 90, pp. 460–476, 2020.
https://doi.org/10.1007/978-3-030-30853-7_27

surface which is also in usage. The first usage of parts with free form surfaces was in automobile industry more for aesthetic that functional reasons. Sometimes usage of part with free form surfaces is only for functional reasons such as turbine blade or in chip building for profiles which are result of engineering calculations. Usage of these part isn't only in macro space, it is also presented in micro space such as an example of various probes, a blood flow meter and similarly. Generally speaking, the usage of parts with free form surfaces is growing in exponential lever.

Machining of these part are usually done by milling on the CNC machine tools and according to this usage of CAD/CAM systems are inevitable for parametric description of surfaces. Surface roughness and quality is as function of generated tool path and according to this it is need for development of new CAD/CAM systems which will allow generation of optimal tool path. For this purpose, it is need for development of new methods for tool path generation, especially for tool path optimization with multi criteria method.

There are many methods for milling of parts with free form surfaces, but the most used is milling with ball end mill cutter. Firstly, it was done by 3 axis machining [1] which has some limitation for usage such as areas where cutting speed is equal to zero and according to that machining won't be done in defined limits and quality. Because of this lack, today it is widely usage of 5 axis machining [2] on horizontal and vertical working centers and milling machines.

The aim of this paper is to present developed procedures for tool path generation and optimization which are implemented in CAD/CAM system for automatic manufacturing technology design which development and usage is also described. For this purpose, it is descried experimental determination of cutting force coefficients for developed cutting force model.

2 Tool Path Generation and Optimization Methods

Free form surfaces are surfaces which are not analytically descriptive, they are usually descripted in parametric form. Definitions of free-form surfaces and objects are often intuitive rather than formal [3]. Free form surfaces are usually descried as Bezier, B spline or NURBS surfaces with coordinates of control points and weighed coefficients for every control point. Free form surfaces can't be completely descried by technical drawings [4].

As it was said, machining of free form surfaces are usually done by ball end mill cutter. For better surface quality, machining is usually conducted in several phases. These are several divisions of these machining phases, but in most cases, it is in use four phases. First phase is rough machining in which the most of material is machined by as much as bigger tool diameter. Scallop heights which are the rest after firs phase are machined in second phase which is semi finish machining. In that phase machined surface is an offset surface for the stock allow [5]. Final shape of surface which is defined by CAD model is generated in finish machining. Final stage is cleaning in which the rest of material is removed. In this phase as much as possible tool diameter is used. Second classification include only rough, finish and cleaning phases [6]. As a measurement for surface quality is used generated surface roughness (scallop height) and surface deviation. An ideal tool path is that which is generated uniform scallop distribution across the machining surface [7]. Smaller scallop heights than prescribed do not necessarily mean a better tool path, because the required increase in the number of tool passes increases machining time and, thereby, the cost of the part [8].

In all machining phases is need to avoid tool and workpiece collisions. Many algorithms are developed for this purpose [9–14]. Generally speaking, there are three classes of collisions in cases of 3 axis machining: local, and rear gouging and global collision [15]. Local gauging appears when tool radius is bigger than curvature radius in contact point (CC), Fig. 1a. Rear gauging is when the contact between tool and workpiece is outside the CC point, Fig. 1b. Global collision appears when the contact between tool and workpiece is somewhere outside the cutting zone, as it is for example tool holder, Fig. 1c. More about algorithms for avoidance of these collisions can be found in [16].

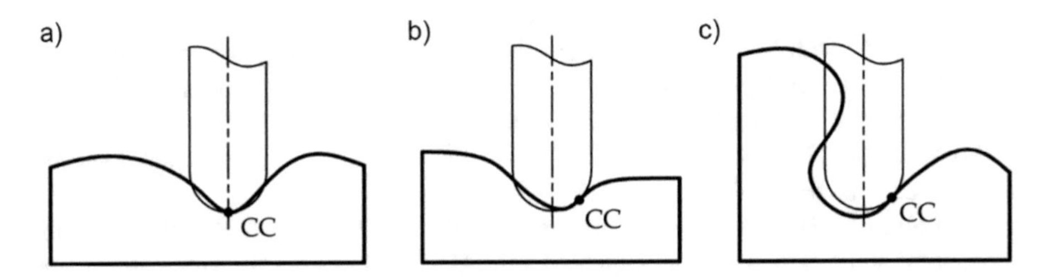

Fig. 1. Tool/work piece collisions in 3 axis machining [4]

Tool path generation is carried out through two segments: choosing of tool path topology and tool path parameters [17]. Tool path topology is determined by sets of cutter location points where tool moves from one to another and so far, it was developed several machining methods. Tool path topology directly effects to tool path length and according to that to total machining time [18]. Tool path parallel to the chosen direction, offset of contour of part or spiral are the mostly used. Special form of parallel topology is ZIG-ZAG strategy. Based od research [18, 19] tool paths parallel to the contour gives smaller machining time for the same machining parameters. Tool path parameters refer to side and forward step which directly effect on machining quality. Cutter location points are mostly connected by line segments, but there was attempts to connect with circular [10] and polynomial interpolation [20]. Currently, three methods are in use for free form surface machining: iso-parametric, iso-planar and iso-scallop [8].

Loney and Oyrow [21] first introduced to iso-parametric surface machining.

The main advantage of iso-parametric method is that cutter location points are obtained directly from the surface points by offsetting in direction of normal vector on the surface point for value of the ball end mill radius. The scallop heights aren't constant in this method so this is the mail disadvantage of this machining method because of differences between Cartesian and parametric space [8]. First solution for this disadvantage was reported by Elber and Cohen and that method was called adaptive iso-parametric [22].

Iso-planar method of machining means that the tool path is determined as an intersection of a freeform surface ($S_{u,v}$) with one of the coordinate planes of the Cartesian coordinate system [8]. As it is difficult to determine the points of intersection between the surface and the plane, the surface should be approximated by a set of planar triangular surfaces, where every triangle is defined by the coordinates of its three vertices [23]. This method is very robust and widely used in commercial CAM systems [24]. The side step (s_P) in this method equals the distance between the parallel planes and is determined from the conditions defined by the maximum allowable scallop height of the machined surface (R_{max}), ball end mill radius (R) and surface curvature (ρ) according to equation [23]:

$$s_P = \frac{|\rho|\sqrt{4R^2(\rho + R_{max})^2 - \left[(\rho + R)^2 - (\rho + R_{max})^2 - R^2\right]}}{(\rho + R)(\rho + R_{max})} \tag{1}$$

The size of the forward step (s_{CL}) is determined from the condition of the highest allowable deviation (Δh) and ball end mill radius (R) according to equation [23]:

$$s_{CL} = 2\sqrt{\Delta h(2R + \Delta h)} \tag{2}$$

Geometric representation of mentioned side and forward step with the used marks are presented on Fig. 2 [8].

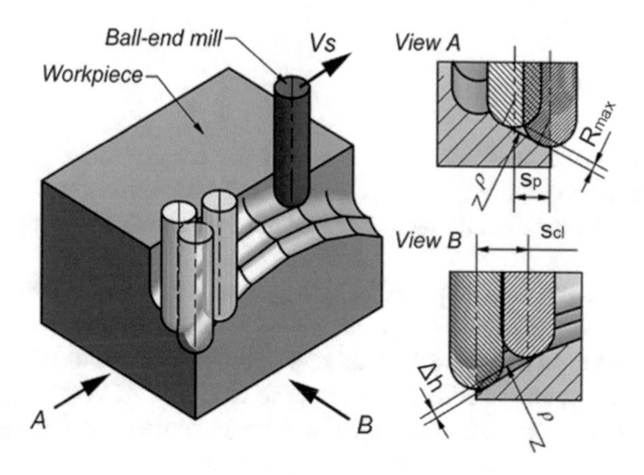

Fig. 2. Forward and side step [8]

In this method is also present disadvantage by non-uniform distribution of scallop height as at iso-parametric method [8].

Suresh and Lin [25, 26] first reported new machining method called iso-scallop which represent improved version of two above mentioned methods. The main advantage of this method is to get uniform scallop height across the whole surface. For defining next tool path, it should be known current tool path and the condition that the curve of scallop peaks is the same for both passes [8].

It was developed several optimization methods for tool path optimization so far. For this purpose, it was need for development of mathematical models for cutting simulation processes which is used in manufacturing technology design process via CAM systems. Some of commercial CAM software gives a possibility to choose machining parameters from its own database [27]. This reduce tool wearing and possibility of tool breakage.

Feedrate scheduling optimization method is the mostly in usage. It is mostly based on MRR (Material Removal Rate) [28–30]. This method can be conducted by two approach. Firs is based on geometrical analysis of volume for manufacturing. Second approach is based on cutting force model based on TWE (Tool Work piece Engagement) [31–35]. These models allow to predict value of cutting force before machining and according to that reduce or increase federate or spindle speed and such that write it into NC code.

Now days it is widely used multi criteria optimization method [36] which was firstly presented in 2011. The method involves the introduction of weight coefficients for each of the optimization parameters that can be: maximum allowed roughness, maximum allowed deviation, cutting forces, and the like [4]. Value of these weight coefficients can be from 0 to 1 or 0% to 100%. Value o means that parameter won't be taken in count and value 1 mean that parameter is the most important in tool path generation.

The primary objective of the multi-criteria optimization method is to find the tool paths which will respect the condition of minimizing the average value of the milling force, the surface roughness and the total machining time for the predetermined (by user) maximum values of the weight coefficients [8].

3 Concept of the System

Based on above mentioned and the analysis of the commercial CAD/CAM systems [8, 37–39], authors conducted many years research [40–45] in direction to develop a system which will provide automatic manufacturing technology design for parts with free form surfaces.

The mail goal was to develop CAD/CAM system which will provide automatically tool path generation for loaded CAD models of part and work piece. Generated tool path will be optimal according to multicriteria optimization method. Generated system allows usage without any user's expertise in the field of CAD/CAM systems.

Tool path generation using developed system is carried out through several procedures which will be described below.

3.1 Loading of Part and Work Piece CAD Models

At the start of manufacturing technology design it is need for loading CAD model of part with free form surfaces which is generated in some of commercial CAD software and saved as STL file format. After loading of part model, it will be converted in internal format where will be data with vertex coordinates and normal vectors of triangles. System also calculate volume of the part in order to conduct further calculations.

The next is to load CAD model of work piece. System provide only loading of CAD model in STL file format. After loading of work piece model system convert in to internal format using procedures for the generation of Z maps [46, 47]. In this stage system also calculate the volume of the work piece and after that calculate total volume for machining.

3.2 Tool Diameter Calculation

The next step is to determine maximal tool diameter for machining of loaded work piece. Machining with as bigger as possible tool diameter will provide minimal machining time and according to that minimal manufacturing costs. For that purpose, system has database with available tools for used machine tool [4, 42]. It is provided updating of database with tools for user at any time. Based on internal format of part model, system makes cross section of part with parallel planes with given resolution in two perpendicular directions. For each cross section plane, system calculate minimal curvature radius (ρ) on convex side of surface based on interpolation polynomials, Fig. 3.

Fig. 3. Cross section of surface and plane [4]

After calculations for all cross section planes it is necessary to determine minimal of all calculated curvature radiuses and according to that determine tool diameter which must be at least two times bigger than minimal curvature radius. If there isn't needed tool in the database, system uses tool with minimal diameter of all and gives warning that it will be manufactured approximate shape of the surface. Detailed procedure for tool diameter calculation is given in [4].

3.3 Determination of Tool Path Leading Planes

This system provides two direction for tool path. It is defined as planes parallel to the XZ or YZ planes. The goal is to define side step for machining after the input of required surface quality by software user. This planes will serve to generate cutter location points for NC code. The first plane is generated based the minimal coordinates vertex points. Every second plane is defined by side step which is calculated based the taper of the triangle and the horizontal plane. It can be two possibilities to calculate side step, according roughness formation model, Fig. 4.

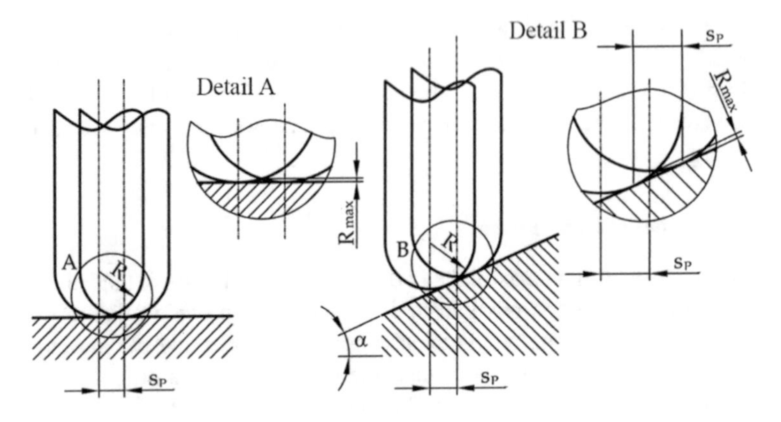

Fig. 4. Roughness formation model [4]

Based on procedures described in [4, 41] system calculate side step for every triangle on the cross section with the surface and running leading plane. Distance between previous and current plane will be minimal of all side steps calculated along the cross section triangles, Fig. 5.

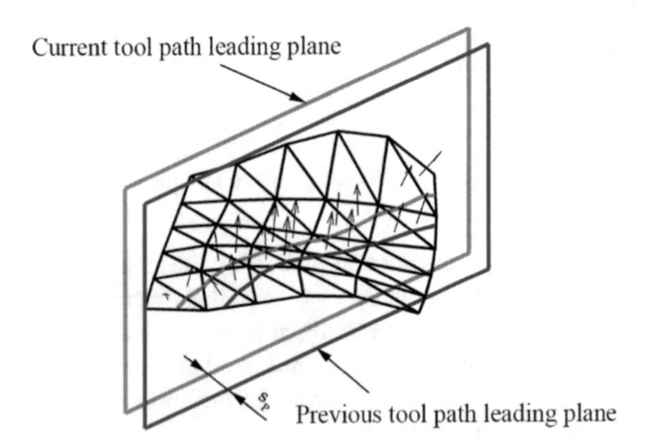

Fig. 5. Determination of next tool path leading plane [4]

This calculation ends with the final plane defined by maximal coordinates of vertex points. After finishing calculations in firs direction, system makes the same procedure for the second direction according to procedure described in [4].

3.4 Cut Width and Depth Determination

After the determination of tool path leading planes, it can be calculated depth and width of cut along the tool path. This should be done because the places with smaller depth of cut will demand bigger federate and the opposite. This procedure will be carried out only in cases when weight coefficient for cutting force is equal to 1, in other cases it won't be necessary because manufacturing will be with constant federate defined by user. All the data are stored in two- or three-dimensional matrix and will be further used in cutting simulation process to determine federate in all cutter location points.

3.5 Offset Surface Generation

Offset surface is used to determine cutter location point as cross section of offset surface and tool path leading planes. When it is known tool diameter it is easy to determine offset surface. In internal format of surface is written every vertex of triangles (V_i) and normal vectors (n_i) for all of the triangles (N). It is easy to move every vertex point in normal direction ($V_{ofst,i}$) by the value od tool radius (R) by the equation [4]:

$$V_{ofst,i} = V_i + R \cdot n_{sr,k}, \qquad i = 1, \ldots, N \tag{3}$$

But, when some vertex is the common for the several triangles (M) it should be done averaging of normal vector (n_{sr}) by the equation [48]:

$$\overrightarrow{n_{sr,k}} = \frac{\sum_{j=1}^{M} N_{k,j}}{\left|\sum_{j=1}^{M} N_{k,j}\right|}, 1 \le k \le M \tag{4}$$

Average normal vector for the example of fine triangles and generated offset surface are shown on Fig. 6.

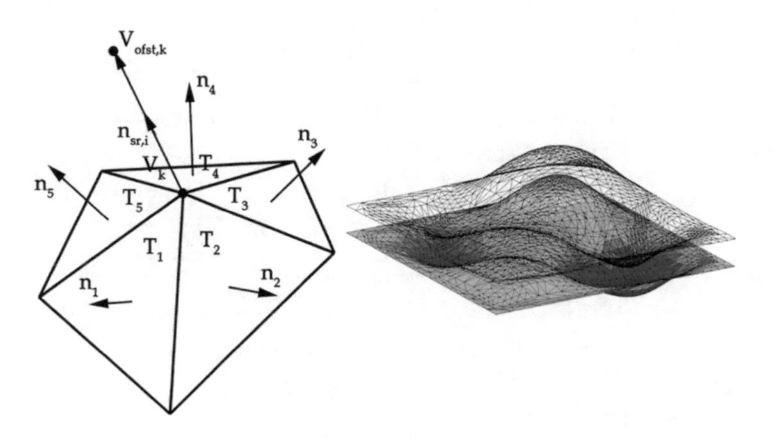

Fig. 6. Average normal vector and generated offset surface [4]

When vertex is only one for the triangle average normal vector is the normal vector from the STL file. System also provides taking into account stock allow (δ) for finish machining. In that case offset vertex should be moved by the value R + δ in normal direction.

3.6 Cutting Simulation Process

Cutting simulation process is given because milling with federate scheduling in order to get cutting condition with constant cutting force. This is useful because it can be defined maximal allowed cutting force per tool and according to that determine appropriate federate.

For the cutting simulation process is necessary to develop cutting force model. In this research the procedure for the discretization of the ball-end mill's edge was implemented by slicing the ball end mill into infinitesimal discs of thickness (dz) [49]. This model implies usage of cutting coefficients (K_{tc}, K_{rc}, K_{ac}, K_{te}, K_{re}, K_{ae}) for given work piece/tool material and geometry combination. In this research was used NSSE 8% Co ball end mill tool with 2 edges and tool diameters (D_{LG}) od 10, 12 and 14 mm. Work piece material was Aluminium (AlMg4.5Mn). Based on procedures described in [4, 8] it was experimentally determinated cutting coefficients and values for ball part of the mill are given on Fig. 7.

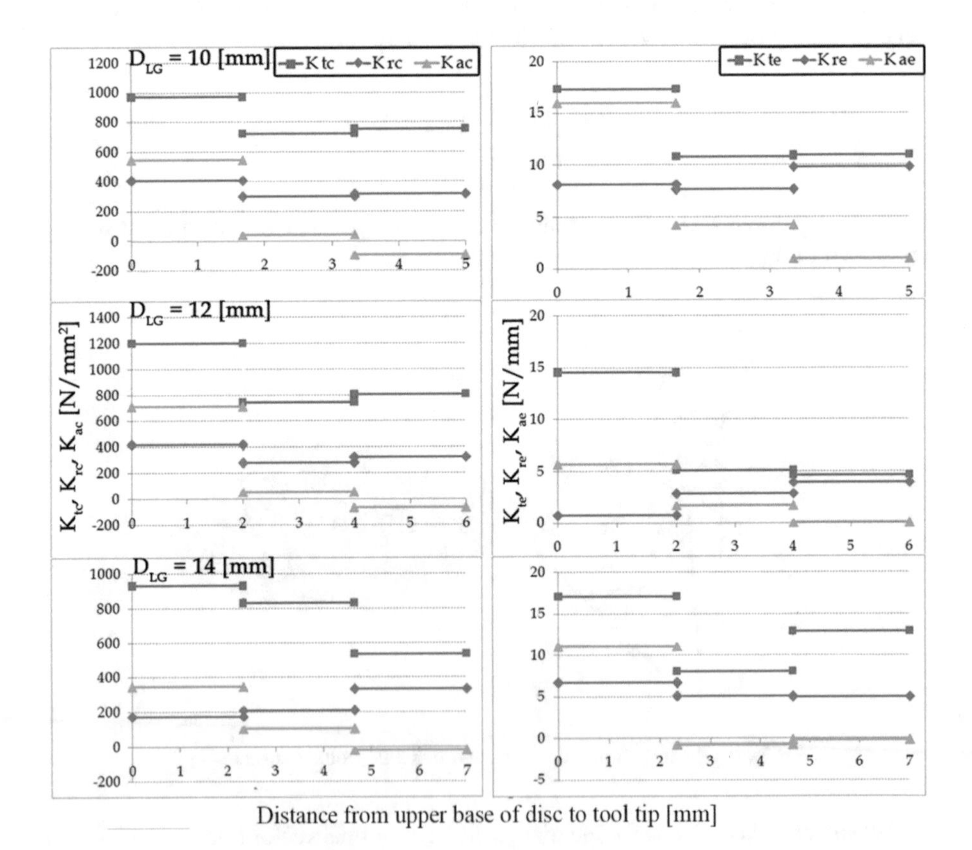

Fig. 7. Cutting coefficients values for ball part of mill [4]

Values for the cylindrical part of the mill are given in Table 1.

Table 1. Cutting coefficients values for cylindrical part of mill [4]

Tool diameter [mm]	K_{tc}	K_{rc}	K_{ac}	K_{te}	K_{re}	K_{ae}
	[N/mm^2]			[N/mm^2]		
10	734.4	350.1	−147.0	11.7	9.3	−0.2
12	931.2	454.1	−123.8	3.6	3.5	−0.7
14	669.9	275.9	−83.7	6.5	10	−0.6

Based on developed cutting force model it was possible to calculate federate values in every cutter location point according to previously generated matrix of width and depth of cut. This will allow milling with constant force along tool path.

3.7 Tool Path Generation

Based on all above mentioned, system can generate tool path. It is only remain to choose optimal tool path topology which will give manufacturing with minimal time. In system was implemented three strategies, and that: cutting in one direction, ZIG-ZAG strategy and spiral strategy, Fig. 8.

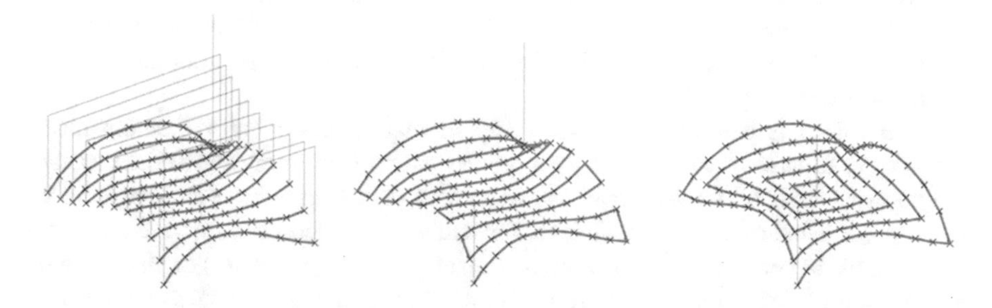

Fig. 8. Machining strategies implemented in system [4]

A report generation with machining time and length of generate tool path for all strategies are also supported by system. Manually definition of tool path topology is also enabled by software user.

4 Software Development

For software (CAD/CAM application) development was used MATLAB [50] software package. Application was developed as GUI interface and therefore no expert knowledge about CAD/CAM systems is necessary for its use, Fig. 9.

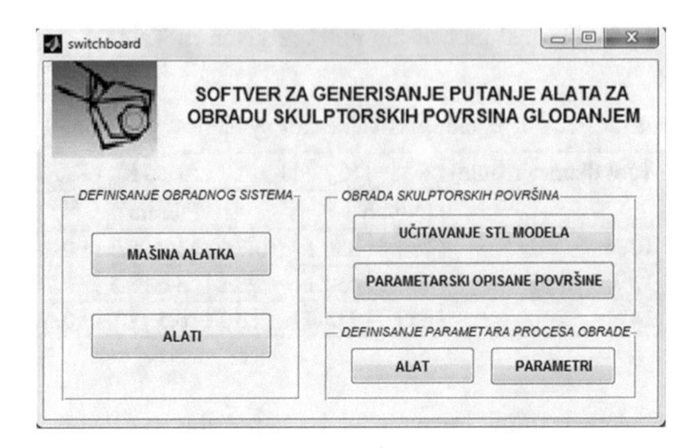

Fig. 9. Starting window of developed software [4]

In application was implemented all the procedures previously described. System was conceived so new windows and fields will open after the termination of current procedure according to order described in Sect. 3.

At the starting of application only visible field is for loading of part CAD model. After choosing the file, system generates internal format of file and executes procedures previously described. In this stage, system gives information of chosen tool for manufacturing and here user can accept that tool or manually define tool geometry. Besides that, system generates picture of loaded part and after that field for work piece loading is visible. After loading of work piece CAD model system generates internal format and make a picture in assembly with the part. It is also generated information about volume to remove by milling.

After this it is possible to automatically or manual designing of manufacturing technology. In case of automatically mode, system request from user desired surface roughness, deviation and feed rate. User can also define criteria for optimization by defining the value of weight coefficients for cutting force, side and forward step. It is only possible to define value 0 (without optimization of that parameter) or value 1 (with optimization of that parameter). Here is also need for defining of work piece material. After defining all these parameters, system makes analysis of technology for manufacturing of loaded part. For the chosen tool get maximal allowed cutting force for that tool and generate federate in every cutter location points. In this stage, user can accept that force value or define new for further calculations. Also, user must define clearance value for rapid movements of tool. Simultaneously it makes determination of optimal tool path for manufacturing if the manufacturing is possible in one pass. In this case system gives the report of chosen tool path topology and value of total machining time. In other case, system gives warning that manufacturing should be taken with rough machining first. Information about overruns of depth of cut is also provided and according to that user can in next iteration change maximal cutting force to avoid multi passes machining. Window after the completed manufacturing technology design using this system is presented on Fig. 10.

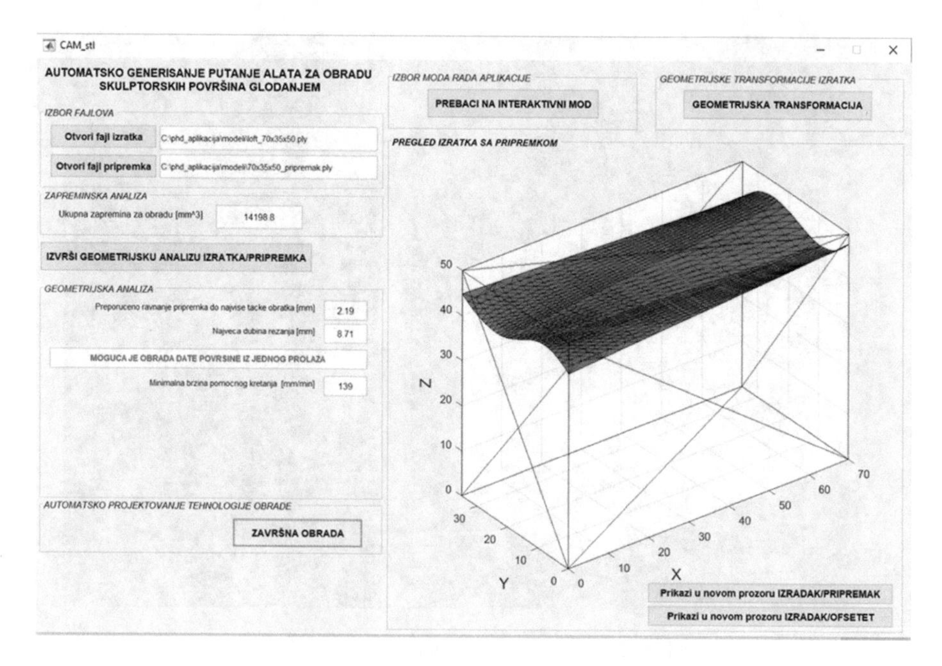

Fig. 10. Window of developed software

5 Experimental Verification

For the experimental verification of developed software is used horizontal working centre ILR HMC 500/40 at the Department for Production Engineering at the Faculty of Mechanical Engineering Belgrade, Serbia. Machining was performed according to NC codes generated by developed application. Several experiments were conducted. For the purpose of cutting force measurements was used Kistler 5007 dynamometer and system for data acquisition that consists of NI Compact DAQ USB cDAQ-9174 with a NI9215 module for analog input: ±10 V, 16-bit with 4 channels and 100 ks/s/ch. The measurement results were analysed in LabVIEW software [8]. Experimental setup for this experiment is presented on Fig. 11a. For this purpose, work piece was fastened via special produces plate to dynamometer. During the manufacturing it was concluded that the machining of these parts was performed with constant cutting force [45].

For experiment without cutting force measures was used machine clamp for positioning and tightening, Fig. 11b.

Precise dimensional inspection of manufactured parts was performed at Department of Physics - University of Liverpool, UK using OGP Smartscope CNC 624 multisensor metrology system [43]. The measurements of the manufactured parts were performed using two methods; contact (touch probe) and contactless (optical) whereby a point cloud of data was obtained. Using the MATLAB software package, a program code was written that generates a map of the deviation based on the difference between the loaded point cloud and the CAD model [45]. All the measurement was conducted

Fig. 11. Experimental setup for machining

according to procedures descripted in [51, 52]. In generation of maps of deviation was used procedure described in [53, 54]. It means that deviation bigger than zero refer to measurement point which is upper than the surface and deviation less than zero refer to measurement point which is below the surface defined by CAD model.

Generated map of deviation for the part manufactured by federate scheduling method is presented on Fig. 12.

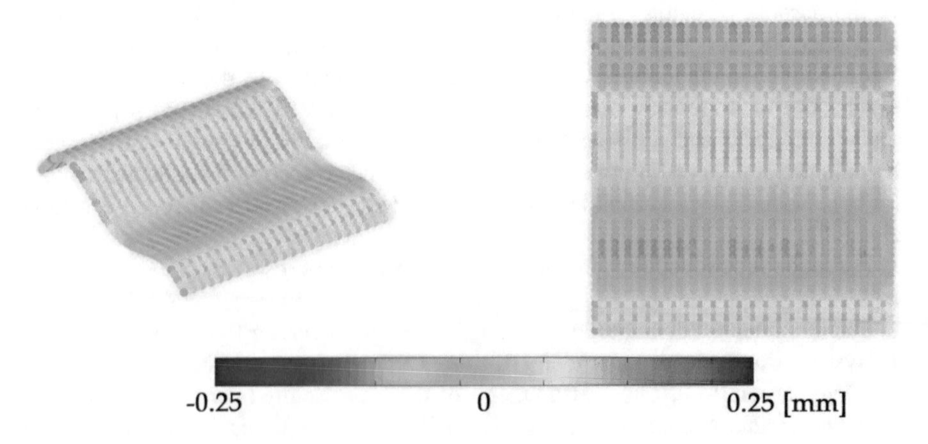

Fig. 12. Map of deviation for feedrate scheduling machining method [4]

6 Conclusion

In this paper was presented result of many years research in free form surfaces machining at the Department for Production Engineering, Faculty of Mechanical Engineering, Serbia. Developed procedures for tool path generation and optimization were presented, too. All the procedures were implemented in unique software solution which represents CAD/CAM system for automatic manufacturing technology design. This software allows to users who don't have expertise in this field, it is only need to know basic about machining.

Based on conducted experiment it was experimentally determinate cutting coefficients for given combination of work piece/tool material and geometry. Milling with constant cutting force was also confirmed based on the machining with the NC codes generated by the developed application. Several machining methods were conducted to verify developed and implemented procedures for tool path generation.

Using multisensory metrology system, it was determinate that the manufacturing was performed in defined tolerances and roughness defined by user.

Developed application is usable for the purpose for tool path generation when it is need for machining with constant cutting force according to multicriteria optimization methods. There were created condition for further research to develop procedures for rough machining in cases when it isn't possible machining with ball end mill in one pass.

Acknowledgement. The author wishes to thank the Ministry of Science and Technological Development of Serbia for providing financial support that made this work possible (Project TR 35022). We are also very thankful to people from University of Liverpool, Department of physics – Particle Physics Group and DESY for excellent collaboration and all the help in finalizing this project.

References

1. Chen, Z.C., Dong, Z., Vickers, G.W.: Automated surface subdivision and tool path generation for 3½½ - axis CNC machining of sculptured parts. Comput. Ind. **50**(3), 319–331 (2003)
2. Gray, P.J., Ismail, F., Bedi, S.: Arc-intersect method for 3½½-axis tool paths on a 5-axis machine. Int. J. Mach. Tools Manuf **47**(1), 182–190 (2007)
3. Campbell, R.J., Flynn, P.J.: A survey of free-form object representation and recognition techniques. Comput. Vis. Image Underst. **81**(2), 166–210 (2001)
4. Mladenovic, G.: Tool Path Optimization in Sculptured Surface Milling, Doctoral Dissertation, University of Belgrade - Faculty of Mechanical Engineering, Serbia (2015)
5. Lee, Y.S., Choi, B.K., Chang, T.C.: Cut distribution and cutter selection for sculptured surface cavity machining. Int. J. Prod. Res. **30**(6), 1447–1470 (1992)
6. Ren, Y., Yau, H.T., Lee, Y.S.: Clean-up tool path generation by contraction tool method for machining complex polyhedral models. Comput. Ind. **54**(1), 17–33 (2004)
7. Warkentin, A., Hoskins, P., Ismail, F., Bedi, S.: Computer aided 5-axis machining. In: Computer Aided Design, Engineering and Manufacturing: System Techniques and Applications [chapter 3]. CRC Press, Boca Raton (2001)

8. Mladenovic, G., Tanovic, L.J., Ehmann, K.F.: Tool path generation for milling of free form surfaces with feedrate scheduling. FME Trans. **43**(1), 9–15 (2015)
9. Rao, A., Sarma, R.: On local gouging in five-axis sculptured surface machining using flat-end tools. Comput. Aided Des. **32**(7), 409–420 (2000)
10. Pi, J., Red, E., Jensen, G.: Grind-free tool path generation for five-axis surface machining. Comput. Integr. Manuf. Syst. **11**(4), 337–350 (1998)
11. Wang, Y.J., Dong, Z., Vickers, G.W.: A 3D curvature gouge detection and elimination method for 5-axis CNC milling of curved surfaces. Int. J. Adv. Manuf. Technol. **33**(3–4), 368–378 (2007)
12. Lee, Y.S., Chang, T.C.: Automatic cutter selection for 5-axis sculptured surface machining. Int. J. Prod. Res. **34**(4), 977–998 (1996)
13. Morishige, K., Takeuchi, Y., Kase, K.: Collision-free tool path generation using 2-dimensional C-space for 5-axis control machining. Int. J. Adv. Manuf. Technol. **13**(6), 393–400 (1997)
14. Morishige, K., Takeuchi, Y., Kase, K.: Tool path generation using C-Space for 5-Axis control machining. J. Manuf. Sci. Eng. **121**(1), 144–149 (1999)
15. Lasemi, A., Xue, D., Gu, P.: Recent development in CNC machining of freeform surfaces: a state-of-the-art review. Comput. Aided Des. **42**(7), 641–654 (2010)
16. Wang, N., Tang, K.: Five-axis tool path generation for a flat-end tool based on iso-conic partitioning. Comput. Aided Des. **40**(12), 1067–1079 (2008)
17. Choi, B.K., Jerard, R.B.: Sculptured Surface Machining - Theory and Applications. Kluwer Academic Publishers, Dordrecht (1998)
18. Kim, B.H., Choi, B.K.: Machining efficiency comparison direction-parallel tool path with contour-parallel tool path. Comput. Aided Des. **34**(2), 89–95 (2002)
19. El-Midany, T.T., Elkeran, A., Tawfik, H.: Toolpath pattern comparison: contour-parallel with direction-parallel. In: Geometric Modeling and Imaging—New Trends, pp. 77–82. IEEE Conference Publications, London (2006)
20. Langeron, J.M., Duc, E., Lartigue, C., Bourdet, P.: A new format for 5-axis tool path computation, using Bspline curves. Comput. Aided Des. **36**(12), 1219–1229 (2004)
21. Loney, G.C., Ozsoy, T.M.: NC machining of free form surfaces. Comput. Aided Des. **19**(2), 85–90 (1987)
22. Elber, G., Cohen, E.: Toolpath generation for freeform surface models. Comput. Aided Des. **26**(6), 490–496 (1994)
23. Bojanić, P.: A tool path generation for three-axis ball-end milling of sculptured surfaces. In: Proceedings of the 33rd Conference on Production Engineering of Serbia, Belgrade, Serbia, pp. 115–118 (2009). (on Serbian)
24. Han, Z.L., Yang, D.C.H.: Iso-phote based tool-path generation for machining free-form surfaces. J. Manuf. Sci. Eng. **121**(4), 656–664 (1999)
25. Suresh, K., Yang, D.C.H.: Constant scallop-height machining of free-form surfaces. J. Eng. Ind. **116**(2), 253–259 (1994)
26. Lin, R.S., Koren, Y.: Efficient tool-path planning for machining free-form surfaces. J. Eng. Ind. **118**(1), 20–28 (1996)
27. Guzel, B.U., Lazoglu, I.: Increasing productivity in sculpture surface machining via off-line piecewise variable feedrate scheduling based on the force system model. Int. J. Mach. Tools Manuf **44**(1), 21–28 (2004)
28. Ip, R.W.L., Lau, H.C.W., Chan, F.T.S.: An economical sculptured surface machining approach using fuzzy models and ball-nosed cutters. J. Mater. Process. Technol. **138**(1–3), 579–585 (2003)

29. Li, Z.Z., Zheng, M., Zheng, L., Wu, Z.J., Liu, D.C.: A solid model-based milling process simulation and optimization system integrated with CAD/CAM. J. Mater. Process. Technol. **138**(1–3), 513–517 (2003)
30. Wang, K.K.: Solid modeling for optimizing metal removal of three-dimensional NC end milling. J. Manuf. Syst. **7**(1), 57–65 (1988)
31. Budak, E., Altintas, Y.: Peripheral milling conditions for improved dimensional accuracy. Int. J. Mach. Tools Manuf. **34**(7), 907–918 (1994)
32. Yazar, Z., Koch, K.F., Merrick, T., Altan, T.: Feed rate optimization based on cutting force calculations in 3-axis milling of dies and molds with sculptured surfaces. Int. J. Mach. Tools Manuf. **34**(3), 365–377 (1994)
33. DeVor, R.E., Kline, W.A., Zdeblick, W.J.: A mechanic model for the force system in end milling with application to machining airframe structures. Manuf. Eng. Trans., 297–303 (1980)
34. Kline, W.A., DeVor, R.E., Lindberg, J.R.: The prediction of cutting forces in end milling with application to cornering cuts. Int. J. Mach. Tool Des. Res. **22**(1), 7–22 (1982)
35. Fu, H.J., DeVor, R.E., Kapoor, S.G.: A mechanistic model for the prediction of the force system in face milling operations. J. Eng. Ind. **106**(1), 81–88 (1984)
36. Manav, C., Bank, H.S., Lazoglu, I.: Intelligent toolpath selection via multi-criteria optimization in complex sculptured surface milling. J. Intell. Manuf. **24**(2), 349–355 (2011)
37. Mladenović, G.: Analysis of machining strategy for the machining using commercial CAD/CAM software. In: Proceedings of the 37th JUPITER Conference, pp. 2.63–2.68. Faculty of Mechanical Engineering Belgrade, Belgrade (2011). (on Serbian)
38. Tanović, LJ., Puzović, R., Popović, M., Bojanić, P., Mladenović, G.: The application of the CAD/CAM/CAE program package in the design and manufacturing of casting pressure tools for parts made from polymers. In: Proceedings of the 37th JUPITER Conference, pp. 2.22–2.28. Faculty of Mechanical Engineering Belgrade, Belgrade (2011). (on Serbian)
39. Mladenović, G., Tanović, Lj., Pjević, M.: Free form machining – comparison of machining strategies. In: Proceedings of 39th JUPITER Conference, pp. 2.19–2.24. Faculty of Mechanical Engineering Belgrade, Belgrade (2014). (on Serbian)
40. Bojanić, P., Mladenović, G.: A iso-cusped tool path generation for 3-axes sculptured surface machining. In: Proceedings of the 36th JUPITER Conference, pp. 2.22–2.28. Faculty of Mechanical Engineering Belgrade, Belgrade (2010). (on Serbian)
41. Mladenović, G., Tanović, Lj., Pjević, M., Popović, M.: Sculptured surface miling - the development of CAD/CAM system. In: Proceedings of the 40th JUPITER Conference, pp. 2.27–2.32. Faculty of Mechanical Engineering Belgrade, Belgrade (2016). (on Serbian)
42. Mladenovic, G., Tanovic, Lj., Puzovic, R., Pjevic, M.: Software solution for automatic choise of cutting parameters in free form surfaces machining. In: Proceedings of the XIV International Conference Maintenance and Production Engineering, KODIP 2017, pp. 111–117. University of Montenegro, Faculty of Mechanical Engineering Podgorica, Budva, Montenegro (2017)
43. Mladenovic, G.M., Tanovic, L.M., Milovanovic, M.J., Jones, T.J.: CAD/CAM system for automatic manufacturing technology design of free form surface parts. In: The Book of Abstracts, 1st International Conference of Experimental and Numerical Investigations and New Technologies – CNN TECH 2017, p. 50. Innovation Center of Faculty of Mechanical Engineering, Zlatibor, Serbia (2017)
44. Mladenović, G., Tanović, Lj., Puzović, R., Pjević, M., Popović, M.: The development of software solution for automatic choise of machining parameters for free form surfaces parts. In: Proceedings of the 41th JUPITER Conference, pp. 2.19–2.24. Faculty of Mechanical Engineering Belgrade, Belgrade (2018). (on Serbian)

45. Mladenovic, G.M., Milovanovic, M.J., Tanovic, L.M., Jones, T.J., Pjevic, M.D.: Manufacturing and geometry measurement of parts with free form surfaces. In: The Book of Abstracts, 2nd International Conference of Experimental and Numerical Investigations and New Technologies – CNN TECH 2018, p. 47. Innovation Center of Faculty of Mechanical Engineering, Zlatibor, Serbia (2018)
46. Anderson, R.O.: Detecting and eliminating collisions in NC machining. Comput. Aided Des. **10**(4), 231–237 (1978)
47. Möller, T., Trumbore, B.: Fast, minimum storage ray-triangle intersection. J. Graph. Tools **2**(1), 21–28 (1997)
48. Malosio, M., Pedrocch, N., Molinari, T.L.: Algorithm to offset and smooth tessellated surfaces. Comput. Aided Des. Appl. **6**(3), 351–636 (2009)
49. Altintas, Y.: Manufacturing Automation – Metal Cutting Mechanics, Machine Tool Vibrations and CNC Design. Cambridge University Press, Cambridge (2000)
50. http://www.mathworks.com. Accessed 11 Apr 2015
51. Stojadinovic, S., Majstorovic, V., Durakbasa, N., Sibalija, T.: Towards an intelligent approach for CMM inspection planning of prismatic parts. Measurement **92**, 326–339 (2016)
52. Stojadinovic, S., Majstorovic, D.V.: An Intelligent Inspection Planning System for Prismatic Parts on CMMs. Springer, Cham (2019)
53. Savio, E., De Chifre, L., Schmitt, R.: Metrology of freeform shaped parts. CIRP Ann. Manuf. Technol. **56**(2), 810–835 (2007)
54. Poniatowska, M.: Free-form surface machining error compensation applying 3D CAD machining pattern model. Comput. Aided Des. **62**, 227–235 (2015)

Study of Mechanical and Physical Properties of Clothing in Maintenance

Mirjana Reljic[1,2](✉), Aleksandra Mitrovic[2,3], Stanisa Stojiljkovic[4], Milena Stojiljkovic[4], and Kocareva Marina[2]

[1] CIS Institute, Belgrade, Serbia
reljicmira@gmail.com
[2] The College of Textile - Design, Technology and Management, Belgrade, Serbia
[3] Faculty of Information Technology and Engineering, University Union "Nikola Tesla", 11000 Belgrade, Serbia
[4] University of Nis, Faculty of Technology, Leskovac, Serbia

Abstract. The fabrics have a great impact on human comfort. Climatic conditions and physical activity also are important factors. Clothes care by washing is very usual. In this paper, the water vapor permeability and thermal resistance of the clothes during exploitation were tested as the main parameters. The correlation of the physical, mechanical and thermophysiological properties was tested on one non-coated fabric and three different coated fabrics, as well as the changes of water vapor resistance (Ret) during the washing. Standard methods have been applied for testing physical and mechanical properties of fabrics. The hot plate measurements have been used to test the water vapor resistance (Ret) and thermal resistance (Rct). The obtained results indicate that certain properties of yarns for the basic fabric, as well as the characteristics of different polymeric materials (coatings and membranes) used in the production of fabrics and the constructional characteristics of fabrics significantly influence the water vapor resistance and its change during washing. The applied laminating and coating processes have led to the improvement of the physical and mechanical properties of the basic textile material. The influence of water vapor resistance for all tested samples at different temperature, relative humidity and air speed on Ret value of clothes was noticed. These results also determine the application field of tested textile material and the level of protection, i.e. the acceptability class for specified climatic conditions.

Keywords: Textile materials · Water-vapor resistance · Thermal resistance · Hot plate

1 Introduction

A human is considered a highly complex thermophysiological system. The anatomy and physiology of a human are adapted to living in a moderate and warm environment where a human body can perform the function of body temperature regulation. During exercise and exposure to heat, human body abundantly perspires in order to cool down, and body temperature is adjusted through changes in blood circulation. In a moderately

N. Mitrovic et al. (Eds.): CNNTech 2018, LNNS 90, pp. 477–497, 2020.
https://doi.org/10.1007/978-3-030-30853-7_28

cold environment, blood flow to the extremities and skin decreases, and an insulating fat layer of a human body serves to maintain body temperature.

The clothes are worn for different reasons: functional aspects (protection from different sorts of perils – protection from bad weather, thermal insulation, protection from wind and rain, protection from injuries etc.), comfort, cultural aspects, observing cultural norms, aesthetics etc. Apart from the fact that it helps to maintain body temperature, the clothing helps keep a person wearing it dry and comfortable by absorbing sweat and other body fluids as well as by providing protection against bad weather. When the function of clothes is not just to protect against heat or cold, a conflict may occur between a protective function of clothes and a thermal function of a body. This conflict affects discomfort that can lead to physical strain, and in extreme cases it can result in injuries from high or low temperatures or a disease [1].

Peculiar polymer properties make them useful in a wide variety of applications. Polymers are available in all fields of engineering from avionics through biomedical applications, cosmetics, textile, etc. [2–8]. Most of the material's properties are determined by a combination of fabric and polymer components together.

Influences for coated or laminated protective sportswear such as: dimensional stability, resistance to delamination waterproofness, breathability, spray rating, general durability to flexing/cleaning/ageing, abrasion resistance, easy care [9] are also affected by physico-mechanical properties of materials, namely, tensile strength and tear resistance.

Contemporary projecting methods consider construction and structural fabric parameters. However it is possible to get results that deviate from real values due to incorrect adjustment and control of tension forces of warp ends during weaving process [10].

1.1 Thermophysiological Properties of Textile Fabrics with Polymeric Materials

Temperature difference between the skin and external temperature is the cause of heat loss. Sweat evaporates from the skin into the environment through clothes. Since human skin is always "wet" to a certain degree, there is always a flow of vapor from a wet tissue beneath the skin towards drier air above (diffusion through the skin), so it is very important for the clothes to have an evaporation property. Several vapor permeability indices related to the clothes are defined [11].

The comfort of clothes in the process of wearing is one of the most important properties, especially of those clothes that are in a direct contact with the skin. In fact, there is a difference between a comfort at contact and a physiological comfort. The comfort at contact is mostly defined by the surface properties of the textile materials from which a clothing item is made, whereas the physiological comfort is mostly defined by the capacity of heat and moisture transmission through the textile surface.

Textile materials form a specific microclimate on the surface of a human body and provide thermal comfort in the process of wearing. All properties of the clothes that have an impact on a microclimate can be designated as comfort factors, and these include: thermal insulation of the used textile materials, moisture permeability of the used textile materials, thermal insulation of mid-layers inside and beneath the clothes, moisture permeability of mid-layers inside and beneath the clothes, airing of clothes by means of convection, airing of clothes by means of forced convection during body movements, reflexion from the clothes, that is, absorption of radiation by the clothes.

Since the properties of clothes primarily depend on the properties of textile materials that are made of, it is particularly important to have an understanding of the influence of the structure and properties of textile materials on the comfort parameters of clothes, in order to design and produce textile materials of desired properties for a specific purpose.

The role of manikins in the clothes development is becoming more and more a key part in the clothes design and assessment. Although they are still specialized and expensive, many countries have centers for performing tests on manikins. The heat transfer coefficients have been determined for many types of clothes by testing on thermal manikins, from which databases have also been created [7, 12, 13]. Nowadays, thermal manikins are widely and routinely used for the development, testing and certification of clothing items and clothing assemblies [14]. These manikins provide a realistic simulation of an entire body and heat transfer, enable 3D heat transfer measurement, integrate heat losses in a realistic way and provide a quick and reliable method which is repeatable and objective for measuring the heat transfer coefficient of clothes. Holmér also refers to the international and European standards where testing methods on thermal manikins are described [14].

Three-layer clothing system has to be waterproof and water vapor permeable for complete protection from rain or water. Under dry conditions, microporous fabrics have a higher water vapor transfer rate than hydrophilic fabrics due to laminated or coated solid film onto a fabric that incorporates an active component. This active component comprises up to 40% by weight of the film [15]. The type of the seams with type of thread used in the garment, and the method of sealing the constructed seams are necessary factors for accomplishing the maximum seams waterproofness. Combinations of numerous parameters that have influence on the seam quality usually determine severally technical-technological parameters of sewing process. These parameters should improve productivity and seam quality [16]. In an interesting study, Kim [17] has concluded that water vapor transfer through waterproof breathable fabrics is not affected only by the environmental but also by the property of the membrane due to waterproof breathable laminated or coated fabrics with different types of waterproof breathable membrane. For the production of clothing items made of layered/laminated materials, it is very important how seams are connected.

Porosity of fabrics and the porosity of all investigated comfort properties of woven fabrics are affecting air permeability. They are determined by the raw material composition, type of weave and fabrics surface condition [5]. Lifespan of clothes is a very important characteristic. To extend it as much as possible, the behavior of clothes during exploitation needs to be marked as excellent. A piece of clothing marked as excellent means that thermal comfort i.e. diffusive and thermal property during use will not significantly change [18]. It is very important to foresee the maintenance of functional properties of textile and other materials that clothing item is made of, since after that period certain properties, such as thermal resistance, are being lost due to maintenance/washing. Water vapor permeability is consequential parameter of thermal comfort of clothing. Changes in the water vapor resistance in the washing process have not been studied so far.

This paper investigated the change of water vapor resistance (Ret) during washing of four fabrics used for the production of clothing-jacket. The purpose of this research was to determine functional dependence of Ret values on the number of washing cycles. This functional dependence could give prediction of the behavior of a clothing assembly during its usage (determination of its life span for a particular purpose). The main goal of this study was to determine thermophysiological comfort of four tested textile materials made from PES fibers.

2 Materials and Method

Four fabrics were used. Samples of PES coated with different materials are marked in Table 1.

Table 1. Marking of fabric samples made from PES fibers

Marking sample	Product name
Sample No 1.	Basic material 100% PES (without membrane and coating)
Sample No 2.	100% PES with PTFE membrane
Sample No 3.	100% PES with PU membrane
Sample No 4.	100% PES three-layer fabric with PU coating and fleece knitwear (knitwear 100% PES one-side raising)

PES - polyester
PTFE - polytetrafluoroethylene
PU - polyurethane

Parameters of PES fibers and properties of fabrics prior to first washing cycle were tested. Also, change of mass per unit area of fabrics at washing (up to 10, 30 and 50 washing cycles) and change of the properties of water vapor permeability and thermal resistance of fabrics expressed through Ret values during maintenance, i.e. washing of fabrics (up to 10, 30 and 50 washing cycles) were examined.

The experiment was carried out on the hotplate (ELEKTRO UMI, Serbia) and in the washing machine (Elektrolux, WASCATOR FOM 71 CLS). Washing procedure of fabric 4N-Normal was applied according to ISO 6330 standard [19]. The experimental part was carried out in CIS Institute (Belgrade, Serbia).

Standardized methods were used in this research for characterizing and examining the quality of fibers and fabrics. Methods of biophysical analysis of clothing on a hot plate were used. Hot plate method is described in detail in paper [20]. Changes of the mass per unit area of the fabrics during 10, 30 and 50 washing cycles were measured according to ISO 3801 [21].

Thermal and moisture management show that performances of examined knitwear largely depend on the properties of raw materials they are made of, the characteristics of knitwear and structures, especially the thermal properties and the air and water vapor transmission [22]. Comfort rating according to Ret is described in detail in paper [20].

3 Results and Discussion

The research was conducted as follows: defining of weaves of the base PES fabric, physico-mechanical and chemical properties of samples made of 100% PES and laminated with different membranes and a coating – samples 1 to 4 (Table 1), physical-mechanical properties of samples 1 to 4 at 10, 30 and 50 washing cycles (Table 2), thermophysiological properties of samples made of 100% PES and laminated with different membranes and coatings (Table 3), changes in thermophysiological properties of samples 1 to 4 at 10, 30 and 50 washing cycles (Table 4).

Examining the influence of qualitative properties of fabrics on thermophysiological properties of clothes is very complex. During the experiments, a strong link of all fabric parameters to the thermophysiological properties of clothes was noticed. The experiments were classified in two groups, namely: physical-mechanical and thermophysiological.

Table 3 shows the results of testing Ret values (m^2Pa/W) (T = 35 °C; v = 1.0 m/s; RH = 40%, ISO 11092) [30] at 50 washing cycles for samples 1 to 4, and provides overview of the samples approximation (Fig. 1).

Table 3 shows the results of testing Rct values (m^2K/W) (T = 20 °C; v = 1.0 m/s; RH = 65%, ISO 11092) [30] at 50 washing cycles for samples 1 to 4, and provides overview of the samples approximation (Fig. 2).

Figures 3, 4 and 5 present physical-mechanical properties of the fabric (samples 1 to 4). Sample 1 is a base material (100% PES), which is laminated with a PTFE membrane in sample 2, with a PU membrane in sample 3, and coated with PU coating and laminated with fleece (knitwear) in sample 4. Therefore, the increase in mass per unit area is expected (Fig. 3), as well as the increase in tensile strength in warp and weft direction (Fig. 4) relative to the base material sample 1.

The difference in the mass per unit area is presented in Table 4. The difference in the mass per unit area of sample 2 and sample 1 is 31.2 g/m^2 and it represents the mass per unit area of the PTFE membrane and adhesives. Mass per unit area of sample 3 is by 61.7 g/m^2 larger than the mass per unit area of sample 1, which means that the PTFE membrane is lighter than the polyurethane membrane applied in sample 3. Sample 4 is a three-layered fabric and compared to the base one, its mass per unit area is larger by 98.3 g/m^2. The difference in mass per unit area of samples 1 and 4 is accounted for by the mass of coating and knitwear in sample 4.

After monitoring the tensile strength values in warp and weft direction, it has been established that the values go up from sample 1 to sample 4, which is expected considering the stronger structure of samples 2–4, and it is particularly pronounced in sample 4 which contains the third layer - knitwear (Fig. 4).

Table 2. An overview of the physico-mechanical and chemical characteristics of samples made from 100% PES and laminated with different membranes and coating

		Sample No 1	Sample No 2	Sample No 3	Sample No 4
Composition polyestar ISO 1833-24 [23] (%)		100% PES	100% PES with PTFE membrane	100% PES with PU memebrane	100% PES three-layer fabric with PU coating and fleece knitwear
Thickness (mm) ISO 5084 [24]		0.18	0.21	0.21	0.38
Number of threads per unit length (cm^{-1}) – warp – weft ISO 7211-2 [25]		69.0 47.6	70.0 48.3	68.0 46.0	68.0 47.3
Linear density of yarn removed from fabric – warp – weft SRPS ISO 2060:2012 [26] (tex)		8.5 × 1 8.5 × 1	8.7 × 1 9.2 × 1	8.7 × 1 9.3 × 1	8.6 × 1 8.9 × 1
Mass per unit area (g/m^2) ISO 3801 [21]	– Before washing	107.3	138.5	169.0	205.6
	– After 10 washing	107.0	137.2	168.9	205.1
	– After 30 washing	106.3	136.4	167.7	204.1
	– After 50 washing	105.9	136.0	166,8	203.1
Tensile strength (N) – warp – weft SRPS EN ISO 1421 [27]	– Before washing	1017 692	1043 713	1110 834	1181 847
	– After 10 washing	981 675	997 692	1097 823	1146 840
	– After 30 washing	952 664	948 683	1091 795	1036 823
	– After 50 washing	931 606	902 595.2	1059 725	1023 811
Tear resistance (N) – direction of warp – direction of weft SRPS EN ISO 4674-1, method B [28]	– Before washing	31.0 28.0	34.1 32.6	39.2 36.1	33.2 28.7
	– After 10 washing	28.9 25.7	33.8 32.4	32.5 31.8	31.2 27.0
	– After 30 washing	27.6 25.4	32.9 31.7	30.0 30.7	28.4 22.1
	– After 50 washing	27.5 24.7	31.8 30.5	27.5 26.8	26.2 16.2

Table 3. Changes in thermophysiological characteristics of sample 1–4 at 10, 30 and 50 washing cycles

		Sample No 1	Sample No 2	Sample No 3	Sample No 4
Permeability of fabrics to air (mm/s) ISO 9237 [29]	– Before washing	46.25	0.50	0.35	1.25
	– After 10 washing	33.10	0.55	0.35	1.30
	– After 30 washing	27.80	0.60	0.50	1.40
	– After 50 washing	28.1	0.65	0.55	1.45
Ret (m^2 Pa/W) ISO11092 [30]	– Before washing	4.44	4.86	19.55	27.28
	– After 10 washing	5.16	5.64	20.44	28.06
	– After 30 washing	6.68	8.04	21.70	30.79
	– After 50 washing	7.73	10.02	25.48	33.51
Rct (m^2 K/W) ISO11092 [30]	– Before washing	0.011	0.027	0.011	0.030
	– After 10 washing	0.009	0.027	0.010	0.029
	– After 30 washing	0.008	0.025	0.008	0.027
	– After 50 washing	0.007	0.023	0.008	0.026
Resistance to surface wetting (spray test) (rating) ISO 4920 [31]	– Before washing	5	5	5	5
	– After 10 washing	5	5	5	5
	– After 30 washing	4	5	3	4
	– After 50 washing	2	3	2	3
Water repellency of fabrics by the Bundesmann rain-shower test (%) ISO 9865 [32]	– Before washing	16.05	5.34	7.31	6.79
	– After 10 washing	22.7	10.11	18.71	13.63
	– After 30 washing	25.6	14.15	22.75	17.62
	– After 50 washing	28.5	18.25	25.61	19.63
Resistance to water penetration – Hydrostatic pressure test (cm H_2O) ISO 811 [33]	– Before washing	36.3	2830.0	1862.8	2353.7
	– After 10 washing	31.4	2796.2	1581.2	1837.4
	– After 30 washing	19.4	2721.6	1469.8	1525.9
		19.3	2511.2	1401.1	1498.3

(*continued*)

Table 3. (*continued*)

		Sample No 1	Sample No 2	Sample No 3	Sample No 4
	– After 50 washing				
Water vapor transmission of materials (g/m^2/ 24 h) ASTM E-96/E96 M-16 [34]	– Before washing	7688.6	4727.3	5427.2	4791.1
	– After 10 washing	6874.5	3799.1	5096.2	3738.8
	– After 30 washing	3636.4	3.272.7	4809.0	3333.3
	– After 50 washing	2532.7	3060.6	4522.0	2666.7
Absolutely dry mass per unit area (g/m^2)		105.5	137.4	166.7	198.3
Mass per unit area of the sample measured after settling 1 h in the water vapor flow resistance measuring apparatus (g/m^2)		105.6	140.2	172.8	208.1
Fi (%)		0.09	2.04	3.66	4.94
i_{mt} (Rct/Ret)		0.0025	0.0056	0.0006	0.0011

Fi - the quantity of water vapor absorbed by the sample of the material during one hour exposure on the skin model (standard wet flat plate - transfer of the humidity index), (%)
i_{mt} - water vapor permeability index, (dimensionless size)

Table 4. Monitoring changes of mass per unit area, tensile strength and tear resistance after washing

Sample mark	Reduction of mass per unit area		Reduction of tensile strength		Reduction of tear resistance	
	g/m^2	%	In the direction of warp %	In the direction of weft %	In the direction of warp %	In the direction of weft %
1	1.4	1.3	8.40	12.4	11.2	11.7
2	2.5	1.8	13.5	16.5	6.70	6.40
3	2.2	1.3	4.60	13.1	29.8	25.7
4	2.5	1.2	13.4	4.20	21.1	43.5

Figure 5 shows that in the process of lamination, i.e. layering, the tear resistance increase in warp and weft direction, except in case of the warp tear resistance in sample 4. The base fabric sample 1 has the smallest mass per unit area, and is the weakest. Other three fabrics are reinforced with membranes, i.e. coatings and adhesives. Since the membrane thickness could not be measured, higher tensile strength values in warp and weft direction in sample 3 compared to the base fabric sample 1 are attributed to the mass and strength of PU membrane, as well as to the applied adhesives. An insignificant increase in the tear resistance in weft direction in sample 4 compared to the base material (sample 1) is explained by the structure of the applied layer, which is not as strong as the membrane in sample 3.

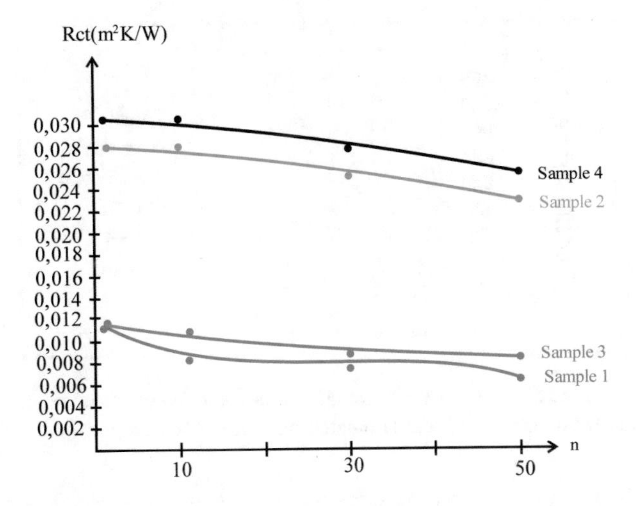

Fig. 1. Change in the Rct value after washing for four samples

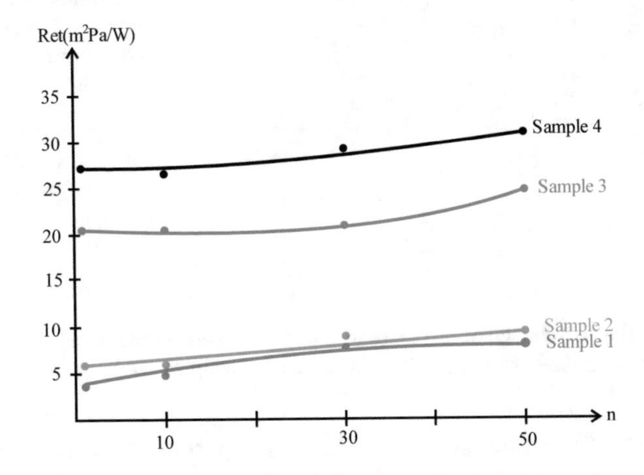

Fig. 2. Change in the Ret value after washing for four samples

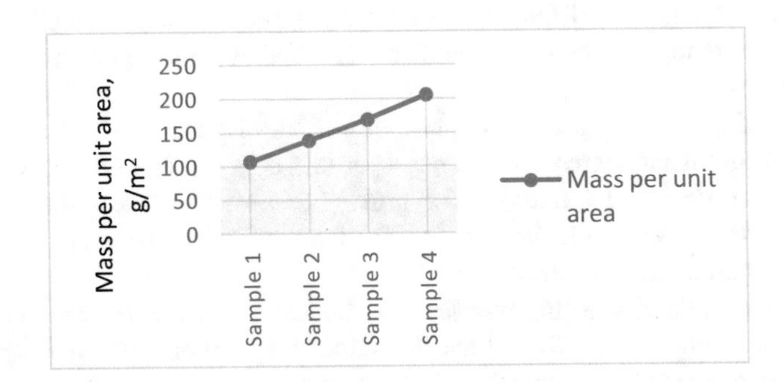

Fig. 3. Monitoring the changes mass per unit area of fabric made from 100% PES and covered with different materials (samples 1–4)

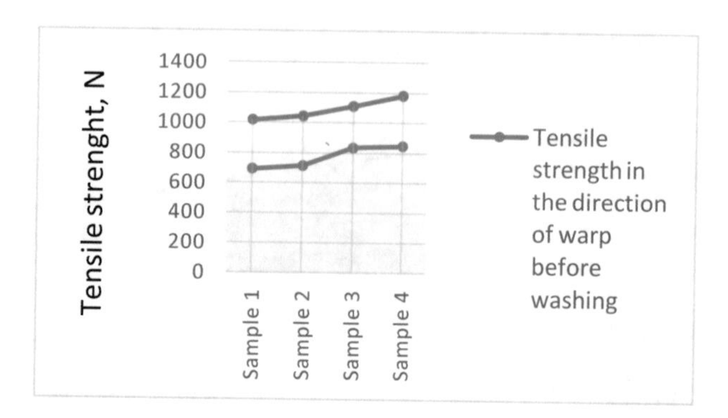

Fig. 4. Monitoring the change of tensile strength in the direction of warp and in the direction of weft fabric made from 100% PES and laminated with different materials (samples 1–4)

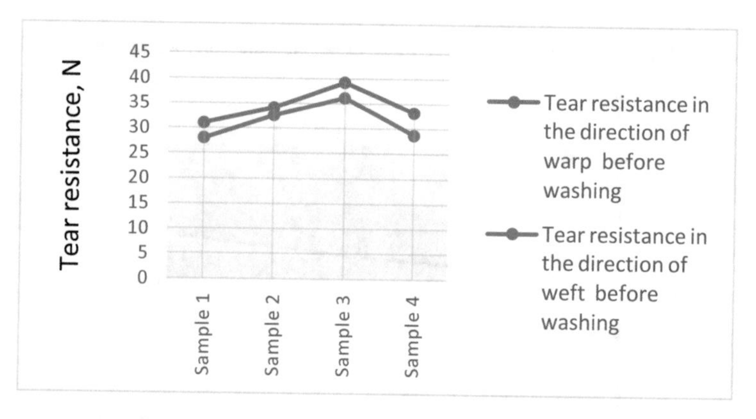

Fig. 5. Monitoring the change of tear resistance in the direction of warp and weft of fabric made from 100% PES and laminated with different materials (samples 1–4)

3.1 Tracking of the Changes in the Fabric Properties of Samples 1–4 During a Different Number of Washings

For a simple overview of the chart, two *y* axes are created in Fig. 6, where one *y* axis presented on the left side of the chart, shows mass per unit area before washing, and other *y* axis presented on the right side of the chart, shows mass per unit area after 50 washing cycles.

By observing the changes in the fabric properties of samples (1–4) at 50 washing cycles, an insignificant decrease in the mass per unit area can be noticed, which values range from 1.4 g/m^2 (1.3% mass) to 2.5 g/m^2 (1.2% mass). This is the result of the removal of short fibers with the solution for the final processing in the process of washing and percolation (centrifuging) (Fig. 6).

In addition, a decline in the physical-mechanical properties is observed, namely: tensile strength (Figs. 7 and 8) and tear resistance (Figs. 9 and 10) in warp and weft direction, as presented in Table 2, which shows a decrease in the values of the

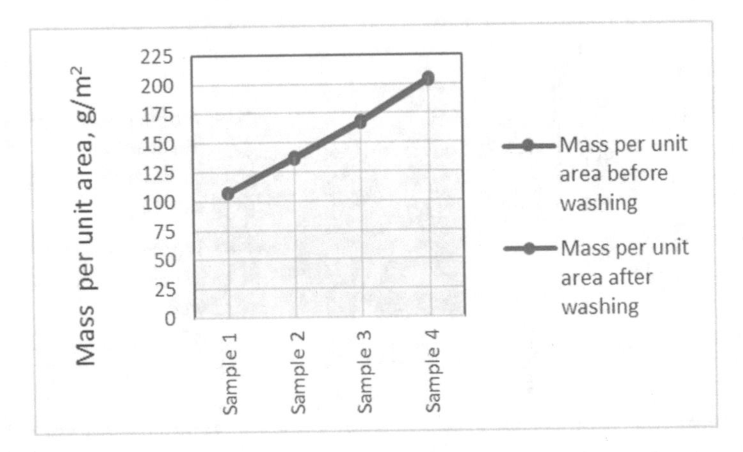

Fig. 6. Monitoring the change of mass per unit area before and after lot of number of washings for fabric samples 1–4

measured tensile strength and tear resistance expressed in %. The decrease in tensile strength and tear resistance occurred partially due to the decrease in mass per unit area, effect of centrifugal force, detergent and friction during washing.

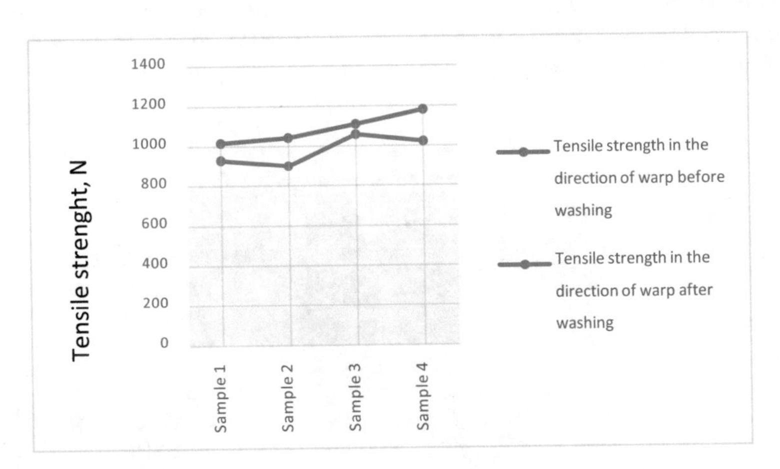

Fig. 7. Monitoring the change of the tensile strength in the direction of warp before and after lot of number of washings for fabric samples 1–4

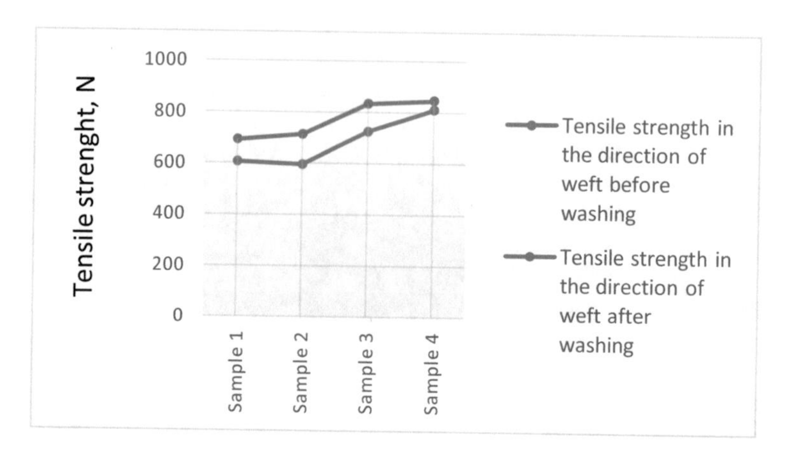

Fig. 8. Monitoring the change of the tensile strength in the direction of weft before and after lot of number of washings for fabric samples 1–4

The decrease in the values of tear resistance in warp direction is expressed in %, and ranges from 4.6% to 13.5%, whereas in weft direction it ranges from 4.2% to 16.5%. The decrease in the values of the fabric tensile strength is expressed in %, and in weft direction it ranges from 6.4% to 43.5%, whereas in warp direction it ranges from 6.7% to 29.8%. The reason for the decrease in tensile strength (in warp and weft direction), and tear resistance values (in warp and weft direction) can be explained by the effect of centrifugal force, detergent and friction during washing.

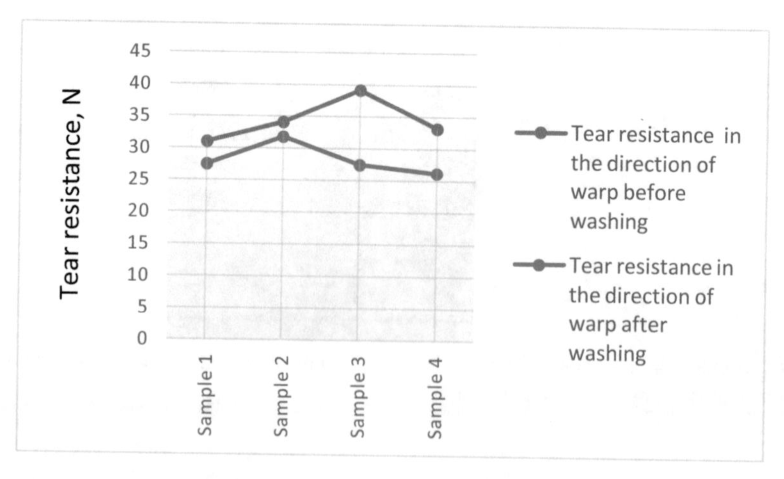

Fig. 9. Monitoring the tear resistance in the direction of weft before and after lot of number of washings for fabric samples 1–4

Table 4 shows that after 50 washing cycles, tensile strength in sample 3 (in warp direction) decreased the least, which can be explained by the structure of PU membrane, solid strength and compactness with the base fabric. Also, tensile strength of the fabric in weft direction in sample 4 experienced the smallest decrease expressed in %. Therefore, in sample 4, sample 1 also contains the third layer i.e. knitwear (Fig. 8). After 50 washing cycles, tear resistance, observed through the decrease in values expressed in % are the lowest in sample 2, which indicates that PTFE membrane was well connected to the base fabric, and it made a very good result. The biggest change expressed in % occurred in sample 4, where the decrease of tear resistance in weft direction amounted to as much as 43.5%, which can be explained by the structure of coating (it is thinner and not so firmly connected to the base material) and little strength of knitwear as the third layer (Fig. 9). It should be noted that in order to speed up the testing process at 50 washing cycles, each fabric sample was divided into 10 samples, so small differences occurred in the results of the properties measurements after each washing cycle.

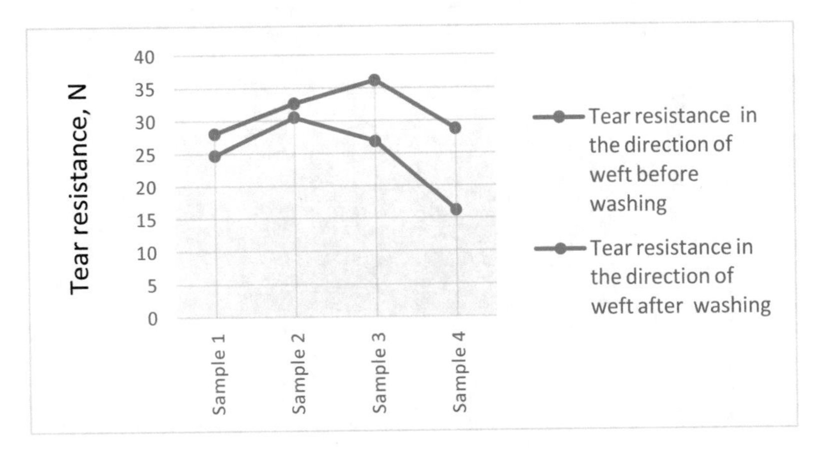

Fig. 10. Monitoring the tear resistance in the direction of warp before and after lot of number of washings for fabric samples 1–4

Table 4 provides a comparative overview of the decrease of mass per unit area and tensile strength in warp and weft direction as well as of tear resistance in warp and weft direction, after 50 washing cycles in samples 1–4, expressed in %.

Changes in thermophysiological properties of samples made of 100% PES and layered with different materials (samples 1–4) are presented in Figs. 11, 12, 13, 14 and 15.

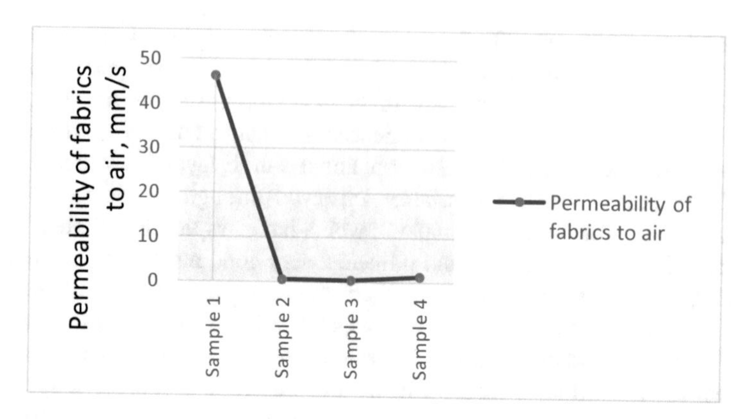

Fig. 11. Permeability of fabrics to air on samples 1–4 made from 100% PES and laminated with different materials

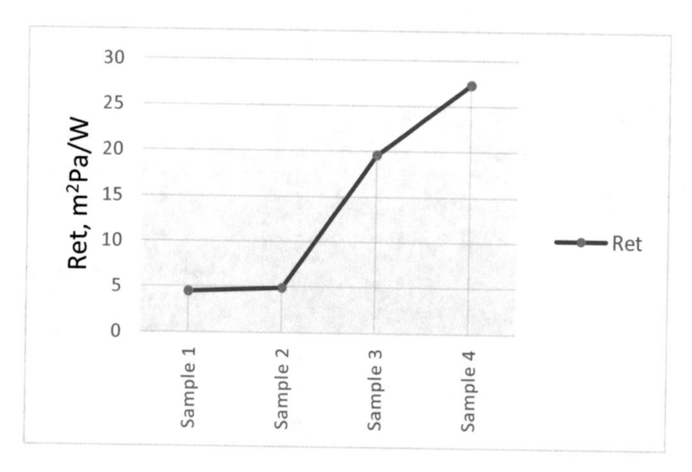

Fig. 12. Ret on samples 1–4 made from 100% PES and laminated with different materials

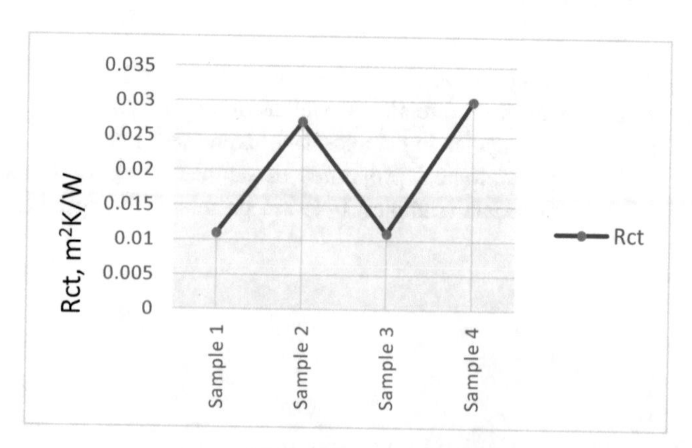

Fig. 13. Rct on samples 1–4 made from 100% PES and laminated with different materials

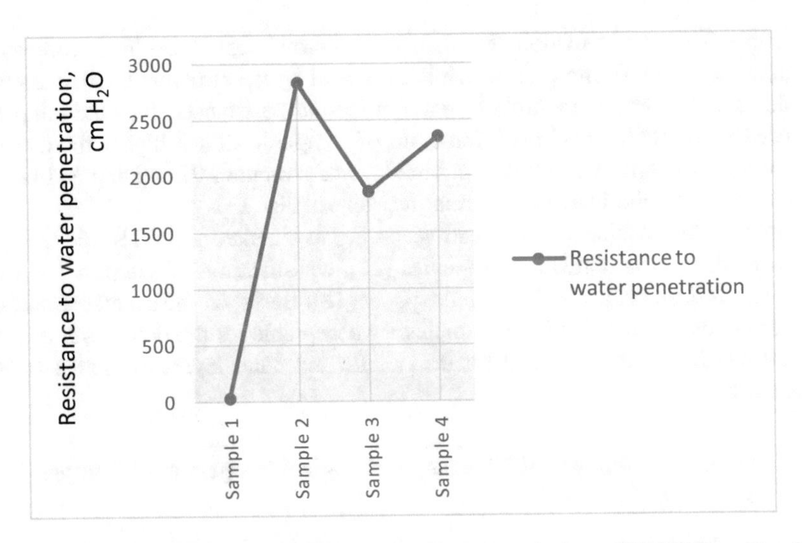

Fig. 14. Resistance to water penetration on samples 1–4 made from 100% PES and laminated with different materials

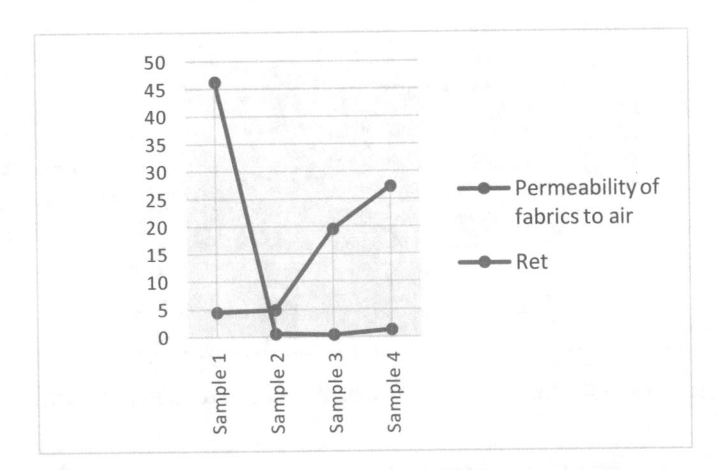

Fig. 15. Permeability of fabrics to air and Ret on samples 1–4 made from 100% PES and laminated with different materials

Figure 11 shows that the fabric marked as sample 1 has much higher air permeability quality than samples 2, 3 and 4 (Table 3). Therefore, sample 1 is the nonlaminated fabric i.e. it is used as the base fabric in the lamination process. According to Fig. 12, water vapor resistance is lowest in sample 1 since it is the base and non-laminated fabric. Sample 2, laminated with PTFE membrane, has very similar Ret results to sample 1. The reason is the structure of PTFE membrane which allows water vapor to penetrate through the material. Ret value is highest in sample 4 because it is a three-layered sample. The third layer is made of knitwear which gets shaggy and swollen in the final processing, reducing the water vapor permeability.

Sample 3 does not exhibit the change in thermal resistance (Rct) compared to nonlaminated fabric (sample 1), which is explained by the structure of PU membrane. Samples 2 and 4 show a multiple increase in thermal resistance, since the sample 2 is laminated with PTFE membrane. Structure of sample 2 allows higher thermal resistance, whereas sample 4 consists of a three-layered structure, PU coating and knitwear which also affects the increased thermal resistance (Fig. 13).

Based on the obtained results, and according to standard CEN/TR 16422- Classification of the thermoregulation properties [21], classification of samples 1–4 can be performed. It is described in detail in the paper [18]. Based on the stated classification, it can be concluded that the fabric samples are acceptable for the outer layer of clothes for cold weather conditions, and for the middle and outer layers of clothes in warm environments.

3.2 Tracking of Changes of Thermophysiological Properties of Samples 1–4 During Washing

Figures 16, 17 and 18 show the changes in results of Ret, Rct and water resistance values for samples 1–4 during an increased number of washings (Table 3).

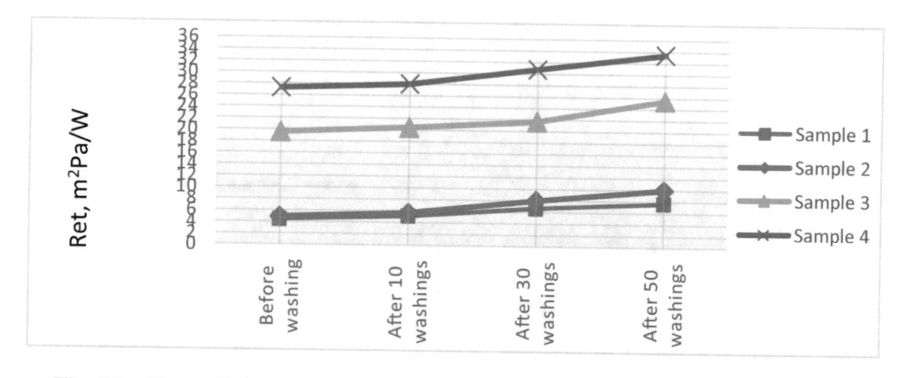

Fig. 16. Change Ret values while increasing the number of washings (samples 1–4)

For a simple overview of the chart, two *y* axes were created in Fig. 18, where *y* axis presented on the left side of the chart, shows the obtained water resistance values during an increased number of washings for samples 2, 3, and 4 whilst *y* axis presented on the right side of the chart, shows the change in water resistance values during an increased number of washings for sample 1.

Figure 16 suggests that in all samples, with an increase in the number of washings, values of resistance to water vapor penetration through the material (Ret) also increase. Sample 4 exhibits the highest Ret value, whereas sample 1 exhibits the lowest Ret value. Since sample 4 is a three-layered fabric, the base layer and knitwear were subject to shrinking, leading to a slightly decreased diffusion of water vapor through the sample, i.e. to an increased resistance to water vapor permeability. Samples 2 and 3 exhibit slightly lower Ret values. These samples are two-layered and laminated fabrics

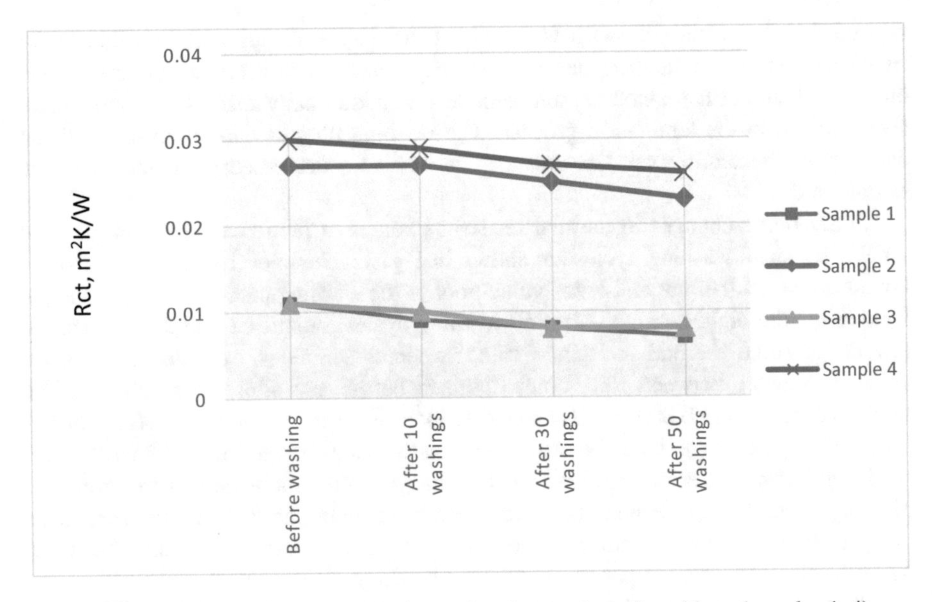

Fig. 17. Change Rct values while increasing the number of washings (samples 1–4)

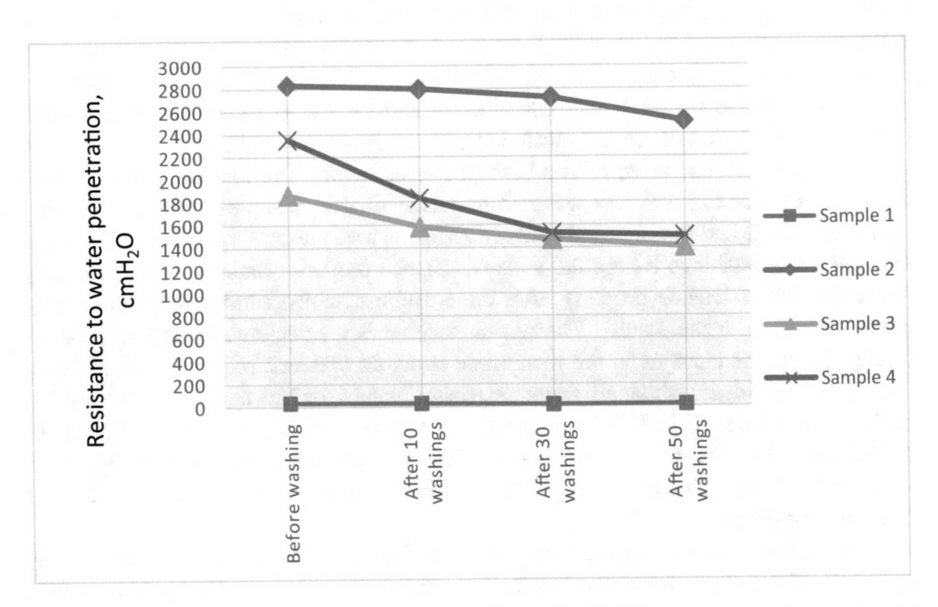

Fig. 18. Change value of resistance to water penetration while increasing the number of washings (samples 1–4)

which is the reason of the changes that occur in the washing process. Namely, during washing, membranes suffer certain changes in the structure, when their water vapor permeability property slightly decreased and Ret values increased. Also, it can be noticed that the results of samples 1 and 2 (Table 3) were quite similar, due to the smallest difference in the mass per unit area. PU membrane in itself has larger mass per unit area than PTFE membrane, and sample 3 is additionally made heavier by adhesives applied in the lamination process. The washing process caused an insignificant decrease of the fabric mass per unit area as well as a decreased permeability of the sample to air.

Comparative changes in thermal resistance values in fabric samples 1–4 made of 100% PES at 50 washing cycles are shown in Fig. 17. Samples 1 and 3 exhibited the lowest measured thermal resistance value prior to the washing process, whereas sample 4 exhibited the highest value. After 50 washing cycles, samples 1 and 3 exhibited the lowest measured thermal resistance value, whereas sample 4 exhibited the highest value. Difference between Rct values obtained before and after the washings in all samples ranges from 0.003 m^2K/W to 0.005 m^2K/W. If the difference in Rct values is observed as percentage, it can be concluded that the results range from 13.3 to 36.3% is not insignificant in a sense that it affects the changes in thermophysiological properties of samples. Chemical agents, added for enhancing hydrophobicity of the fabric, are being washed out in the washing process. For that reason, percent of water absorption according to Bundesmann gets higher during washing (Table 3).

Changes in water resistance values of samples 1–4 made of 100% PES at 50 washing cycles are presented in Fig. 18. Decrease in water resistance values is caused by chemical agents being washed out from the samples in the process of washing, thus water resistance is lower (Table 3). Before and after the washing process, sample 2 had the highest water resistance values whilst sample 1 had the lowest values. Considering the fact that sample 1 is the base nonlaminated fabric, it was expected. The good results in terms of water resistance are attributed to the membrane itself, that is, the coating used for lamination/layering of the base polyester fabric. The structure of the membrane and the coating prevents water from permeating the samples. After 50 washing cycles, sample 2 had the highest results in terms of water resistance, which means that even after the washings it kept the water resistance property. Sample 3 had a slightly reduced water resistance property after the washings, whilst sample 4 exhibited significant changes in the results. The reason for that is a three-layered structure of the fabric, that is, the changes in the membrane structure and knitwear as the third layer. The difference in the obtained water resistance values before and after 50 washing cycles ranges from 17 to 932.1 cm H_2O, so the difference is significant from a theoretical view, but considering the high values obtained after 50 washing cycles in samples 2, 3 and 4, it can be concluded that the samples possess a very good water resistance property.

For a simple overview of the chart, two *y* axes were created in Fig. 19 *y* axis on the right side of the chart shows the obtained air permeability values during an increased number of washings of samples 2, 3 and 4, and other *y* axis on the left side of the chart, shows the obtained air permeability values during an increased number of washings for sample 1. The comparison of air permeability values in Fig. 19 indicated that only sample 1 possesses a good air permeability property which slightly decreased during the washing process (Table 3).

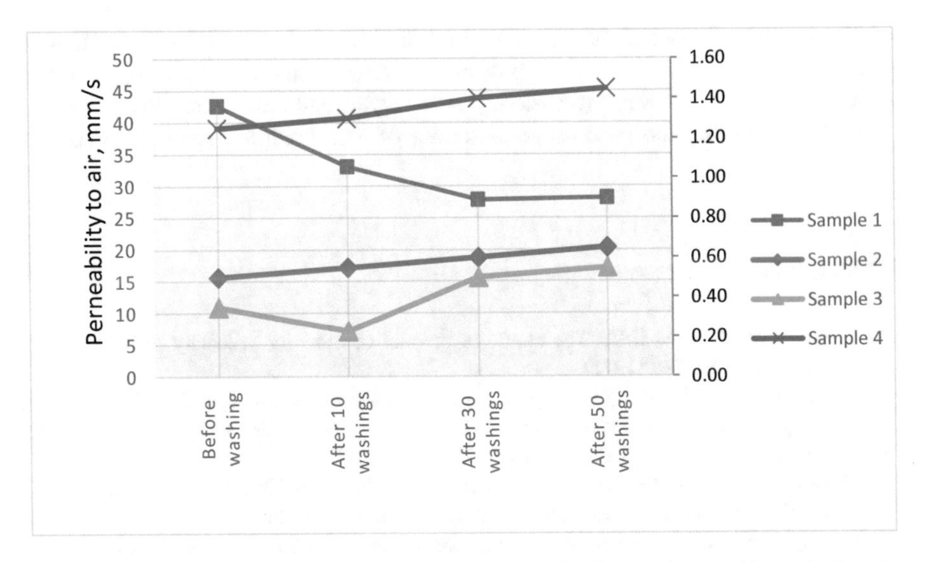

Fig. 19. Monitoring the change of value permeability fabrics to air (mm/s) increasing the number of washing (samples 1–4)

The sample 1 is not laminated which enables good air permeability and changed its structure at 50 washing cycles. Other samples 2, 3 and 4 exhibit very low air permeability values as they contain a membrane or a coating. During washing, the same samples exhibit increased air permeability because adhesives do not allow major fiber curl on the samples.

4 Conclusion

Based on the analysis, comparison and statistics of the obtained results relating to the thermophysiological property of textile materials, the following conclusions can be made:

- The conducted experiments indicated that certain properties of fabrics significantly affect the change in the resistance of the fabrics to heat and water vapor penetration. Important fabrics parameters are mass per unit area, thickness, density of strings etc. The fabrics structure is particularly important because of a great influence on other physico-mechanical and thermophysiological properties of all fabrics.
- The changes in fabrics properties during 50 washing cycles indicated an insignificant decrease of the mass per unit area and a decline of physical-mechanical properties.
- The applied processes of lamination and layering have enhanced the physical-mechanical properties of the base fabric.
- The comparison of the water resistance property of laminated and layered fabrics to the base fabric showed that a particular fabric can be specifically applied for the outer layer of a clothing assembly.

- The conducted research and presented results have expanded the existing knowledge based on which it will be possible to predict comfort parameters of similar textile products for warm (middle and outer layer) and cold (outer layer) environments, and to estimate the usage properties of a clothing assembly for a specific purpose.

References

1. Rohles, F.H., Nevins, R.G.: The nature of thermal comfort for sedentary man. ASHRAE Trans. **77**(1), 239–246 (1971)
2. Kalagasidis Krušić, M., Milosavljević, N., Debeljković, A.(Mitrović A.), Üzüm, Ö.B., Karadağ, E.: Removal of Pb2+ ions from water by poly (acrlymide-co-sodium methacrylate) hydrogels. Water Air Soil Pollut. **223**, 4355–4368 (2012)
3. Milosavljević, N., Debeljković, A.(Mitrović A.), Kalagasidis Krušić, M., Milašinović, N., Üzüm, Ö.B., Karadağ, E.: Application of poly(acrlymide-co-sodium methacrylate) hydrogels In copper and cadmium removal from aqueous solution. Environ. Prog. Sustain. Energy **33**, 824–834, (2014)
4. Debeljković, A.D.(Mitrović A.), Matija L.R., Koruga Đ. Lj.: Characterization of nanophotonic soft contact lenses based on poly (2-hydroxyethyl methacrylate) and fullerene. Hemijska industrija **67**, 861–870 (2013)
5. Asanović, K.A., Cerović, D.D., Mihajlović, T.V., Kostić, M.M., Reljić, M.M.: Quality of clothing fabrics in terms of their comfort properties. Indian J. Fibre Text. Res. **40**(4), 363–372 (2015)
6. Zhu, L.M., Yu, D.G.: Drug delivery systems using biotextiles. In: Biotextiles as Medical Implants (2013)
7. Ghori, S.W., Siakeng, R., Rasheed, M., Saba, N., Jawaid, M.: The role of advanced polymer materials in aerospace. In: Sustainable Composites for Aerospace Applications, pp. 19–35 (2018)
8. Yao, S., Yuan, J.: Thermally conductive dielectric polymer materials for energy storage. In: Dielectric Polymer Materials for High-Density Energy Storage, pp. 323–349 (2018)
9. Shishoo, R.: Textiles in Sports. Woodhead Publishing Limited, Cambridge (2005)
10. Stepanovic, J., Ćirkovic, N., Radivojevic, D., Reljic, M.: Defining the warp length required for weaving process. Industria Textila **63**(5), 227–231 (2012)
11. Oohori, T., Berglund, L.G. Gagge, A.P., Simple relationships among current vapour permeability indices of clothing with a trapped-air layer. In: Environmental Ergonomics (1988)
12. ISO 9920 Ergonomics of the thermal environment – Estimation of the thermal insulation and evaporative resistance of a clothing ensemble (1995)
13. Olesen, B.W., Dukes-Dubos, F.N.: International standards for assessing the effect of clothing on heat tolerance and comfort. In: Performance of Protective Clothing, pp. 17–30 (1988)
14. Holmér, I., Nilsson, H., Havenith, G., Parsons, K.C.: Clothing convective heat exchange: Proposal for improved representation in standards and models. Ann. Occup. Hyg. **43**(5), 329–337 (1999)
15. Lomax, G.R.: Breathable, waterproof fabrics explained. Textiles **4**, 12–16 (1991)
16. Popov, D.B., Ćirković, N., Stepanović, J., Reljić, M.: The analysis of the parameters that influence the seam strength. Industria Textil **3**, 131–136 (2012)

17. Kim, J.O.: Dynamic moisture vapor transfer through textiles, Part III: effect of film characteristics on microclimate moisture and temperature changes. Text. Res. J. **69**(3), 193–202 (1999)
18. Ruckman, J.E., Murray, R., Choi, H.S.: Engineering of clothing systems for improved thermophysiological comfort. Int. J. Cloth. Sci. Technol. **11**, 37–52 (1999)
19. ISO 6330 - Textiles - Domestic washing and drying procedures for textile testing (2012)
20. Reljic, M., Stojiljkovic, S., Stepanovic, J., Lazic, B., Stojiljkovic, M.: Study of water vapor resistance of Co/PES fabrics properties during maintenance. In: Mitrovic, N., Milosevic, M., Mladenovic, G. (eds.) CNNTech -2018. LNNS, vol. 54, pp. 72–83. Springer, Cham (2019). https://doi.org/10.1007/978-3-319-99620-2_6
21. ISO 3801 - Textiles – Woven fabrics – Determination of mass per unit length and mass per unit area (1977)
22. Reljić, M., Lazić, B., Stepanović, J., Djordjić, D.: Thermophysiological properties of knitted fabrics for sports underwear. Struct. Integr. Life **17**(2), 145–151 (2017)
23. ISO 1833-24 Textiles – Quantitative chemical analysis – Part 24: Mixtures of polyester and certain other fibres (method using phenol and tetrachloroethane (2010)
24. ISO 5084 Textiles – Determination of thickness of textiles and textile products (1996)
25. ISO 7211-2 Textiles – Woven fabrics – Construction – Methods of analysis – Part 2: Determination of number of threads per unit length (1984)
26. ISO 2060 Textiles - Yarn from packages - Determination of linear density (mass per unit length) by the skein method (1994)
27. EN ISO 1421 Rubber- or plastics-coated fabrics - Determination of tensile strength and elongation at break (2016)
28. EN ISO 4674-1 Rubber- or plastics-coated fabrics - Determination of tear resistance - Part 1: Constant rate of tear methods (2016)
29. ISO 9237 Textiles – Determination of the permeability of fabrics to air (1995)
30. ISO 11092 Textiles – Physiological effects – Measurement of thermal and water-vapor resistance under steady-state conditions (sweating guarded-hotplate test) (2014)
31. ISO 4920 Textile fabrics – Determination of resistance to surface wetting (spray test) (2014)
32. ISO 9865 Textiles – Determination of water repellency of fabrics by the Bundesmann rain-shower test (1991)
33. ISO 811 Textile fabrics – Determination of resistance to water penetration – Hydrostatic pressure test (2018)
34. ASTM E-96/E96 M-16 Standard Test Methods for Water Vapor Transmission of Materials (2016)

Author Index

N. Mitrovic et al. (Eds.): CNNTech 2019, LNNS 90, pp. 499–500, 2020.
https://doi.org/10.1007/978-3-030-30853-7